Signaling and Communication in Plants

Series Editors

František Baluška
Department of Plant Cell Biology, IZMB, University of Bonn, Kirschallee 1, D-53115 Bonn, Germany

Jorge Vivanco
Center for Rhizosphere Biology, Colorado State University, 217 Shepardson Building, Fort Collins, CO 80523-1173, USA

For further volumes:
http://www.springer.com/series/8094

Preface

Terrestrial plants are unable to relocate when faced with biological, physical, or chemical stress and have, therefore, developed efficient signaling mechanisms to respond to and ensure survival under adverse circumstances. The auxotrophic lifestyle obviously allows plants to fuel more primary active ATP-dependent tranporters (= pumps) than animals. Moreover, unlike bacteria that rely entirely on their cell walls, higher plants needed more secondary transport systems to control cell homeostasis and osmotic pressure. Finally, plant-specific organelles and vacuoles provide storage pools for ions and catabolites that are filled by cotransporters. This correlates with an increased and more divergent number of transporter genes, mainly pumps and secondary active transporters, in higher plants compared with genomes from bacteria or animals.

Although plant sensing and responding mechanisms might be considered as simple and even slow compared with nervous systems in animals, the basic mechanisms of signal transduction and sensing are very similar. At the same time, work in the recent years has identified novel, plant-specific signaling molecules and mechanisms. This volume focuses on the role of transporters and pumps in regulation of movement, long-range transport, and compartmentalization of water, solutes, nutrients, and classical signaling molecules. Function, regulation, and membrane-transporter interaction of prominent transporters and pumps and their individual roles in plant signaling controlling plant physiology and development are discussed.

Zurich, May 2010 — Markus Geisler
Granada, May 2010 — Kees Venema

Contents

Part I
Membranes and Water Transport

Plant Aquaporins: Roles in Water Homeostasis, Nutrition, and Signaling Processes

Gerd Patrick Bienert and François Chaumont

Abstract Exchange across biological membranes is controlled by the composition of the lipid bilayer, diffusion-facilitating channels, and active transport proteins. In 1992, a protein facilitating the passive diffusion of water across membranes was discovered in humans and named aquaporin-1. Since then, an increasing number of proteins belonging to the same superfamily of membrane intrinsic proteins have been identified and characterized as ubiquitous indispensable players in transmembrane water fluxes and water homeostasis. Compared to all other kingdoms of life, plants possess a high number of isoforms, clustered into seven subfamilies. A fascinating diversity of small, water-soluble, and uncharged compounds, ranging from gases to metalloids, has been identified as substrates for plant aquaporins. This chapter summarizes a variety of features and transport properties of these membrane pores illustrating their physiologically crucial contribution to water homeostasis, nutrition, and signaling processes.

1 Introduction

Plant cells and their internal organelles are separated from their environment by biological membranes, generating independent, but highly controllable, functional units and microenvironments. However, the functionality of these units is strongly dependent on the controlled exchange of information and substances across these barriers. Biological membranes consist of a lipid bilayer with a highly hydrophobic interior, which restricts the diffusion of charged and polar molecules. Because of the nature of the membrane, it can be stated that the bigger and more polar a compound, the worse it will permeate through the lipid bilayer. Thus, only small

G.P. Bienert and F. Chaumont (✉)
Institute of Life Sciences, Université catholique de Louvain, Croix du Sud, 4-15, 1348 Louvain-la-Neuve, Belgium
e-mail: gerd.bienert@uclouvain.be; francois.chaumont@uclouvain.be

M. Geisler and K. Venema (eds.), *Transporters and Pumps in Plant Signaling*, Signaling and Communication in Plants 7,
DOI 10.1007/978-3-642-14369-4_1,

and nonpolar compounds can easily permeate unaided through biological membranes. Regardless of the chemical nature of the compound, some are transported across the membrane against the electrochemical gradient resulting in their accumulation above the equilibrium concentration. The transport of compounds, especially that of larger, polar, and charged molecules, has therefore to be facilitated and controlled by membrane proteins (see other chapters in this book).

For a long time, it was thought that small polar substances were transported across biological membranes solely by passive diffusion and that there were no protein-facilitated pathways. This was thought to be the case for the uncharged water molecule, which is the most abundant substance in cells and organisms and is indispensable for life. However, it soon became obvious that the nonfacilitated passive diffusion of water across membranes was not in agreement with the observed high membrane permeability of various cells (Agre et al. 1993). Additionally, the activation energy of water movement across some cell membranes was found to be much lower than the values obtained for pure artificial membranes ($E_a > 10$ kcal/mol). Similar discrepancies between in vivo and calculated or in vitro measured permeability parameters were also observed for other small uncharged molecules (Dordas et al. 2000). These observations could only be explained by protein-mediated facilitated diffusion of these solutes. In 1956, Stein and Danielli hypothesized that the high membrane water permeability of red blood cells must be due to hydrophilic protein pores (Stein and Danielli 1956). Although their protein nature remained undiscovered for a long time, the existence of such water channels was demonstrated in 1992 by Preston in Agre's laboratory, who functionally characterized the first aquaporin (AQP) (CHIP28, renamed AQP1) from human red blood cells (Preston et al. 1992), a major breakthrough in the field. It was soon observed that AQP1 had sequence similarity to other known, but functionally uncharacterized, proteins from mammals and plants. Intensive research concentrated on the identification of an increasing number of related AQP proteins in different organisms and their characterization with respect to cell water homeostasis. The crucial role of AQPs, at this time confined to water transport processes, became clear, and the award of the Nobel Prize in Chemistry to Peter Agre in 2003 acknowledged their discovery and their relevance for life.

In the following years, a variety of important metabolic small uncharged solutes were recognized as substrates for AQPs, and several of these channels were proven to play important roles in the uptake, translocation, sequestration, or extrusion of these solutes. AQPs have been identified in vertebrates, insects, plants, fungi, protozoa, bacteria, and even viruses, thus covering all kingdoms of life (Zardoya 2005; Gazzarrini et al. 2006). Based on sequence homologies and in agreement with their major transport activities, phylogenetic analyses gave rise to two clear clades: the "water-permeable" AQPs, forming the AQP cluster, and the "glycerol-permeable" aquaglyceroporins, forming the glycerol facilitator-like protein (GLP) cluster. While the aquaglyceroporins from prokaryotes and eukaryotes most likely comprise a monophyletic clade, the AQPs represent a heterogeneous group of all the nonaquaglyceroporin isoforms (Danielson and Johanson 2010) Together, these two clades constitute the superfamily of major intrinsic proteins (MIPs), which form these fascinating hydrophilic pathways responsible for the passage of small uncharged molecules through biological membranes. In all kingdoms of life, MIPs play key roles in highly diverse and important physiological processes. In microbes, the functions of MIPs range from osmoadaptation or turgor regulation to enhanced cellular tolerance to rapid freezing and drug resistance (Petterson et al. 2005). Studies on mammalian AQP knock-out mice have pointed to the important functions of MIP-facilitated water and glycerol transport in transepithelial fluid transport, the influx and efflux of water in the brain, cell migration, sensory signaling, the glycerol content of diverse tissues, skin hydration, cell proliferation, carcinogenesis, and fat metabolism (reviewed in Verkman 2009).

2 Plant Aquaporins

One remarkable difference between MIPs from plants and other organisms is the much larger isoform diversity in plants. While 13 isoforms have been identified in mammals, many more have been found in the genomes of higher plants, with 35 in *Arabidopsis* (Johanson et al. 2001), 33 in rice (Sakurai et al. 2005), 37 in tomato (Sade et al. 2009), 45 in poplar (Gupta and Sankararamakrishnan 2009), and at least 36 in maize (Chaumont et al. 2001). Even the evolutionarily early land plant *Physcomitrella patens* has 23 different isoforms (Danielson and Johanson 2008).

Plant MIPs can be subdivided into seven evolutionarily distinct subfamilies. While the plasma membrane intrinsic proteins (PIPs), tonoplast intrinsic proteins (TIPs), small basic intrinsic proteins (SIPs), nodulin26-like intrinsic proteins (NIPs), hybrid intrinsic proteins (HIPs), and X intrinsic proteins (XIPs) all belong to the AQP cluster of MIPs, the glycerol facilitator (GlpF)-like intrinsic proteins (GIPs) represent the only known plant protein belonging to the GLP cluster. To date, EST sequence data have revealed that all land plants, ranging from nonvascular plants to angiosperms, contain MIPs belonging to the PIP, TIP, NIP, and SIP subfamilies. During the evolution of higher plants, the XIP subfamily was lost in monocots and some dicot species. HIPs and GIPs have not yet been found in any vascular plant (Fig. 1).

2.1 *Plasma Membrane Intrinsic Proteins*

PIPs constitute the most homogeneous group amongst the AQP subfamilies. The 15 isoforms found in poplar and the 13 found in each of *Arabidopsis*, rice, and maize can be further subdivided into two groups, named PIP1 and PIP2. Although *Zea mays* PIP1s (ZmPIP1s) have a longer N terminus, a shorter C terminus, and a smaller extracellular loop A than ZmPIP2s. The overall sequence identity is higher than 50% (Chaumont et al. 2001). PIPs are mainly localized to the plasma membrane (Table 1). PIP2 proteins are highly efficient water channels when

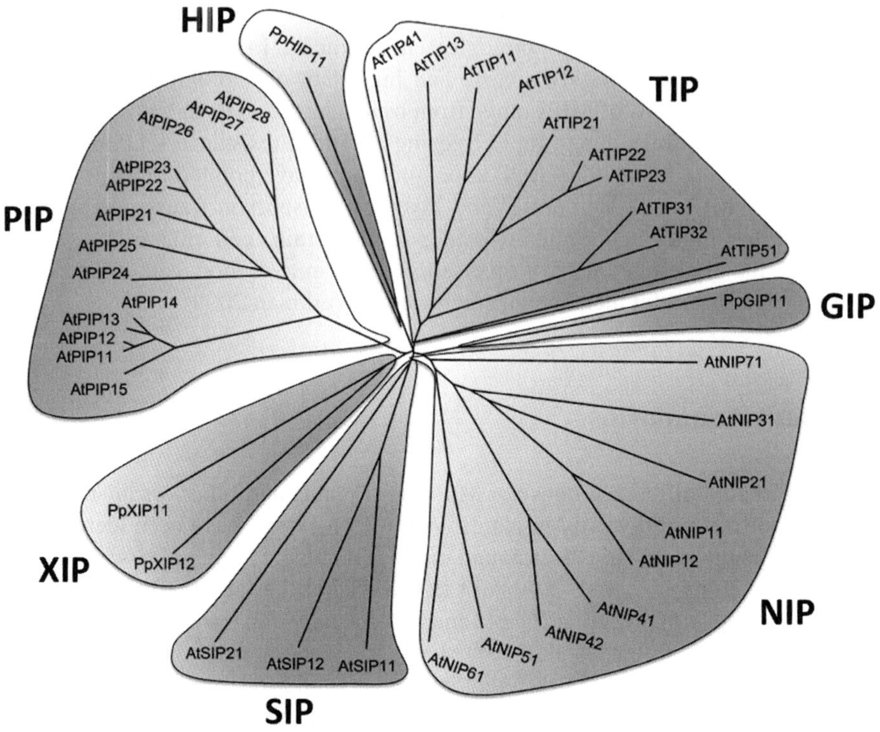

Fig. 1 Phylogenetic classification of plant aquaporins. PIP, TIP, NIP, and SIP isoforms have been identified in all plant species so far. Typical aquaporin sequences from *Arabidopsis* (At; *A. thaliana*) were used to generate the phylogenetic tree (Johanson et al. 2001). Additionally, aquaporin sequences of XIP, HIP, and GIP isoforms, which are absent in diverse plant species, for example, *Arabidopsis*, were used from the moss *Physcomitrella patents* (Pp) (Danielson and Johanson 2008)

expressed in heterologous systems, while PIP1s are often inactive or possess a much lower activity (reviewed in Chaumont et al. 2005; Kaldenhoff and Fischer 2006). A few isoforms have been shown to additionally facilitate the transport of urea, hydrogen peroxide (H_2O_2), carbon dioxide (CO_2), or boric acid (Table 2) (Uehlein et al. 2003; Bienert et al. 2008b; Dynowski et al. 2008a, b; Uehlein et al. 2008; Fitzpatrick and Reid 2009). The expression of the different PIP isoforms is quite high, and PIPs are expressed ubiquitously throughout the plant. Many factors influence their expression (Table 3).

2.2 *Tonoplast Intrinsic Proteins*

The TIP subfamily derives its name from their main localization in the tonoplast. TIPs can account for up to 40% of the total tonoplast proteins (Higuchi et al. 1998). Different TIPs seem to be specific for different vacuoles, such as storage or lytic

Table 1 Subcellular localization of some plant MIP isoforms

Subcellular localization	MIP isoform	References
Plasma membrane	PIPs in general	
	AtNIP5;1	Takano et al. (2006)
	OsNIP2;1	Ma et al. (2006)
	AtNIP2;1	Choi and Roberts (2007)
	TIPs	Whiteman et al. (2008), Marmagne et al. (2004)
Tonoplast	TIPs in general	
	PIPs	Barkla et al. (1999), Shimaoka et al. (2004)
Endoplasmic reticulum	AtSIPs	Ishikawa et al. (2005)
	AtNIP2;1	Mitzutani et al. (2006)
	ZmPIP1s	Zelazny et al. (2007)
Mitochondrion	rAQP8[a]	Soria et al. (2010), Calamita et al. (2005)
	rAQP9[a]	Amiry-Moghaddam et al. (2005)
Chloroplast	NtAQP1	Uehlein et al. (2008)
Small vesicle-like structures	TIPs and PIPs	Siefritz et al. (2001), Vera-Estrella et al. (2004), Boursiac et al. (2005)
Peribacteroid membrane	GmPIP1;1	Fortin et al. (1987), Guenther et al. (2003),
	MtNIP1	Catalano et al. (2004)
	LjTIP1;1	Wienkoop and Saalbach (2003)

[a]Mammalian isoforms

vacuoles (Paris et al. 1996; Jauh et al. 1999; Park et al. 2004). However, this specificity might be restricted to certain plants or developmental stages (Olbrich et al. 2007). Mass spectrometry studies have shown that TIPs are also present in the plasma membrane fraction (Table 1) (Jauh et al. 1999; Santoni et al. 2003; Whiteman et al. 2008). In *Arabidopsis*, maize, rice, and poplar, 10, 13, 11, and 11 isoforms have been found, respectively, and they can be clustered into five phylogenetic groups, TIP1–TIP5 (Johanson et al. 2001). This phylogenetic division is applicable to all vascular plants. TIPs possess a very high water channel activity in all plant species (reviewed in Kaldenhoff and Fischer 2006). Additional substrate specificities for urea, ammonia, and H_2O_2 have been observed when certain isoforms were heterologously expressed (Table 2) (Liu et al. 2003; Jahn et al. 2004; Loqué et al. 2005; Bienert et al. 2007; Dynowski et al. 2008a, b).

2.3 Small Basic Intrinsic Proteins

The SIP subfamily constitutes a very small, but the most divergent, AQP subfamily, with 3–6 isoforms in *Arabidopsis*, maize, rice, or poplar. SIPs show very low sequence identity with the other subfamilies and form two groups, which have been localized to the endoplasmic reticulum. While AtSIP1 isoforms have been shown to channel water across vesicle membranes (Ishikawa et al. 2005), no substrate for SIP2 isoforms has yet been identified and their physiological role remains obscure.

Table 2 Substrates of plant MIP isoforms

Transported solute	MIP isoform[a]	References
Water	Most PIP2s possess a high water permeability	
	SoPIP2;1	Törnroth-Horsefield et al. (2006)
	ZmPIP2;1, ZmPIP2;5, ZmPIP2;4	Fetter et al. (2004)
	In general, TIPs possess a high water permeability	
	ZmTIP1;1	Chaumont et al. (1998)
	AtTIP1;1	Maurel et al. (1993)
	NIPs possess a rather weak water permeability	
	OsNIP2;1	Mitani et al. (2008)
	GmNIP1;1	Wallace and Roberts (2005)
	AtSIP1;1, AtSIP1;2	Ishikawa et al. (2005)
Glycerol	NtAQP1	Biela et al. (1999)
	SsAQP1	Moshelion et al. (2002)
	NtTIPa	Gerbeau et al. (1999)
	OsTIP1;2, OsTIP3;2, OsTIP4;1	Li et al. (2008)
	PpGIP1;1	Gustavsson et al. (2005)
	AtNIP1;1, AtNIP1;2	Weig and Jakob (2000)
	GmNIP1;1	Dean et al. (1999), Rivers et al. (1997)
Urea	NtTIPa	Gerbeau et al. (1999)
	AtTIP1;1, AtTIP1;2, AtTIP2;1, AtTIP4;1	Liu et al. (2003)
	CpNIP1	Klebl et al. (2003)
Lactic acid	AtNIP2;1	Choi and Roberts (2007)
Antimonite	AtNIP5;1, AtNIP6;1, AtNIP7;1, LjNIP5;1, LjNIP6;1, OsNIP2;1	Bienert et al. (2008b)
	AtNIP1;1	Kamiya and Fujiwara (2009)
Arsenite	OsNIP1;1, OsNIP2;1, OsNIP2;2	Ma et al. (2008a)
	AtNIP5;1, AtNIP6;1, AtNIP7;1, LjNIP5;1, LjNIP6;1, OsNIP2;1	Bienert et al. (2008b)
	AtNIP1;1	Kamiya et al. (2009)
Methylated arsenic species	OsNIP2;1	Li et al. (2009)
Silicic acid	OsNIP2;1	Ma et al. (2006)
	OsNIP2;2	Yamaji et al. (2008)
	ZmNIP2;1, ZmNIP2;2	Mitani et al. (2009)
	HvNIP2;1	Chiba et al. (2009)
Boric acid	AtNIP5;1	Takano et al. (2006)
	AtNIP6;1	Tanaka et al. (2008)
Hydrogen peroxide	AtTIP1;1, AtTIP1;2	Bienert et al. (2007)
	AtPIP2;1, AtPIP2;4	Dynowski et al. (2008a, b)
Carbon dioxide	NtAQP1	Uehlein et al. (2003, 2008)
	HvPIP2;1	Hanba et al. (2004)
Ammonia	TaTIP2;1, TaTIP2;2	Jahn et al. (2004)
	AtTIP2;1, AtTIP2;3	Loque et al. (2005)
Nitric oxide	hAQP1[b]	Herrera et al. (2006)

[a]This list is not exhaustive
[b]Mammalian isoform (*h* human)

Table 3 Biotic and abiotic factors influencing the expression of some plant MIPs

Factors influencing expression	MIP isoform	References
Water supply and osmotic conditions	AtMIP	Boursiac et al. (2005), Alexandersson et al. (2005)
	OeTIP1;1, OePIP1;1, OePIP2;1	Secchi et al. (2007)
	NtPIP1;1, NtPIP2;1	Mahdieh et al. (2008)
Hormones		
Gibberilic acid	PsTIP1	Ozga et al. (2002), Kolla et al. (2004)
Abscisic acid	OsPIPs	Lian et al. (2006)
	OsTIPs	Li et al. (2008)
	AtPIP1	Kaldenhoff et al. (1993)
Auxin	PgTIP	Lin et al. (2007
Ethylene	RhPIP2;1	Ma et al. (2008b)
	VvPIPs	Chervin et al. (2008)
Light	AtPIP1	Kaldenhoff et al. (1993)
	DcPIPs, DcTIPs	Sato-Nara et al. (2004)
Circadian rhythm	ZmPIP2s	Lopez et al. (2003)
	NtPIPs	Siefritz et al. (2004)
Temperature	ZmPIP1s, ZmPIP2s	Aroca et al. (2006)
	RcPIP2;1, RcPIP2;2	Peng et al. (2007)
	OsPIPs	Yu et al. (2006)
Nutrients	AtNIP5;1	Takano et al. (2006)
	OsNIP2;1	Ma et al. (2006)
Tissue-specific and developmental expression	ZmPIPs	Hachez et al. (2006a, b, 2008), Heinen et al. (2009), Chaumont et al. (2001)
	OsPIPs	Sakurai et al. (2008)
	AtMIPs	Alexandersson et al. (2005)
	PtNIP1	Ciavatta et al. (2001)
	OsNIP2;1	Ma et al. (2006)
Symbiotic and pathogen interactions	MtNIP1	Uehlein et al. (2007)
	GmNIP;1	Fortin et al. (1987)
	LjTIP1	Wienkoop and Saalbach (2003)
	GhMIPs	Dowd et al. (2004)
	PtPIPs	Marjanovic et al. (2005)

2.4 *Nodulin26-Like Intrinsic Proteins*

One of the first identified and biochemically investigated plant MIPs was the soybean GmNIP1;1 (Fortin et al. 1987); however, it was not recognized as a water channel. The NIP subfamily encompasses highly divergent isoforms. This diversity can be seen both at the amino acid sequence level and in terms of their substrate specificities. NIPs display the greatest diversity in terms of the amino acid residues comprising the selective filters (see below) (Bansal and Sankararamakrishnan 2007). This sequence diversity represents the basis for their involvement in the transport of diverse solutes, such as water, ammonia, urea, glycerol, organic acids, and metalloids such as boric acid, antimonite, arsenite, and silicic acid

(Table 2). Interestingly, *Arabidopsis* and rice lines in which the expression of *NIP* genes has been knocked out or downregulated show clear phenotypes demonstrating physiologically crucial roles of the encoded proteins in the uptake and translocation of the nutritionally important metalloids boron and silicon and in the extrusion of the highly toxic metalloids arsenite and antimonite (Takano et al. 2006; Ma et al. 2006; Yamaji et al. 2008; Isayenkov and Maathuis 2008). NIPs are found in the endoplasmic reticulum or the plasma membrane (Table 1). NIP expression is not as ubiquitous as that of PIPs or TIPs, but restricted to defined cell types or tissues, consistent with their specific roles (Takano et al. 2006; Ma et al. 2006; Ciavatta et al. 2001).

2.5 X Intrinsic Proteins

As indicated by the subfamily name, XIPs are an as yet uncharacterized monophyletic group of MIPs from plants and fungi (Gupta and Sankararamakrishnan 2009; Danielson and Johanson 2008). Hydropathy prediction programs suggest that XIPs are localized to the plasma membrane of plant cells (Danielson and Johanson 2008). The selectivity filters of XIPs from moss and fungi clearly differ chemically from those in dicot plants, suggesting that the channeled solute and therefore the physiological function will be very different. Furthermore, expression analysis indicates that poplar XIPs do not show any tissue- or cell-specific difference in transcript abundance (Gupta and Sankararamakrishnan 2009; Danielson and Johanson 2008). Studies on solute transport, expression, posttranslational modifications, and other properties will have to be performed on XIPs to shed light on this novel MIP subfamily.

2.6 Hybrid Intrinsic Proteins

Isoforms of this subfamily have been identified in the moss *P. patens* and in the spikemoss *Selaginella moellendorffii*, but not in vascular plants (Danielson and Johanson 2008). HIPs seem to have been lost during the evolution of vascular plants. As indicated by the name, HIPs share similarities with both TIPs and PIPs. These proteins have not yet been characterized.

2.7 GlpF-Like Intrinsic Proteins

GIPs are closely related to aquaglyceroporins from bacteria (Gustavsson et al. 2005; Johanson and Danielson 2008). Two GIP sequences are known from *P. patens* and another closely related moss (Gustavsson et al. 2005). One possible explanation for the occurrence of this GLP homolog in moss species is horizontal gene transfer from bacteria. Consistent with the phylogenetic relationship with aquaglyceroporins, PpGIP1;1 heterologously expressed in Xenopus oocytes was demonstrated to conduct glycerol, whereas only very low water permeability was detected (Gustavsson et al. 2005).

3 Structural Features of Major Intrinsic Proteins

3.1 Aquaporin Structure

Structural information about MIPs was provided by the resolved atomic structures of mammalian AQP1 (Murata et al. 2000; Sui et al. 2001), AQP0 (Harries et al. 2004), AQP2 (Schenck et al. 2005), AQP4 (Hiroaki et al. 2006), AQP5 (Horsefield et al. 2008), and AQP9 (Viadiu et al. 2007), bacterial GlpF (Fu et al. 2000; Tajkhorshid et al. 2002), AqpZ (Savage et al. 2003), and AqpM (Lee et al. 2005), yeast PpAqy1 (Fischer et al. 2009), and protozoan PfAQP (Newby et al. 2008). To date, spinach SoPIP2;1 is the only plant AQP for which the structure has been resolved at atomic resolution (Törnroth-Horsefield et al. 2006). Despite the sometimes large sequence difference between the various crystallized MIPs, the overall structure is highly conserved (reviewed in Gonen and Walz 2006; Fu and Lu 2007). A closer examination of the sequences of MIPs shows a tandem repeat of three transmembrane helices and one half membrane spanning helix containing the conserved MIP Asn-Pro-Ala (NPA) signature. The two parts of the protein are orientated in opposite directions in the plasma membrane and form a six transmembrane (TM1–TM6) and two half membrane spanning helices. TM1–TM6 are connected by five loops (loop A–loop E), the cytoplasmic loop B and the extracellular loop E, forming the two short hydrophobic half-transmembrane helices. Both termini are located in the cytoplasm. The center of this protein forms a hydrophilic pore through the lipid bilayer, which is narrow enough to exclude the passage of hydrated ions. The passage of nonhydrated ions is energetically unfavorable, as the amino acids lining the pore do not allow hydrogen binding throughout the channel. However, the carbonyl backbone of the amino acids provides hydrogen binding partners for water or chemically similar solutes. Loops B and E, containing the NPA motifs, meet at the center of the pore and constitute a size exclusion zone, with a diameter of only 3 Å. The two asparagine residues constitute hydrogen bond donors for the hydrogen acceptors of the transported molecules and their orientation is structurally fixed in the pore. This arrangement forces the channeled compound into an orientation, which does not allow hydrogen bond formation between neighboring substrate molecules at the center of the channel. Substrates are therefore channeled in single file and possess a polarity and an electrostatic resistance, which prevents a possible proton wire and consequently proton transport through the pore. The available MIP structures have helped in understanding the mechanism of transport and explaining some of the experimentally observed posttranslational modifications of MIPs from plants and mammals, such as phosphorylation and protonation (Table 4) (Törnroth-Horsefield et al. 2006).

3.2 Aromatic/Arginine (ar/R) Constriction Region

A second and narrower constriction region for uncharged molecules, the so-called aromatic/arginine (ar/R) constriction region, is formed by four amino acids that contribute to a size exclusion barrier and the hydrogen bond environment necessary for effective transport of a substrate. One of the amino acids forming this tetrad is in transmembrane helix 2, another in transmembrane helix 5, and the other two in loop E. In plants, the ar/R selectivity filter is formed by a much larger number of amino acid combinations than in mammals or microbes (Wallace and Roberts 2004; Bansal and Sankararamakrishnan 2007). Mutational studies in homologous and heterologous expression systems have identified chemical prerequisites for the passage of water and other compounds or the exclusion of protons (Wallace and Roberts 2004; Beitz et al. 2006; Chen et al. 2006). In general, it can be concluded that typical water-conducting MIPs possess rather large and hydrophilic pore residues, which constrict the pore size to, for example, 1.86 Å in AQP1. The ar/R selectivity filter of aquaglyceroporins and NIPs is constituted of rather small and hydrophobic residues, resulting in a larger channel diameter of 3.14 Å in *E. coli* GlpF (EcGlpF) or even larger in NIPs.

3.3 Oligomer Formation

The structure described above is that of a MIP monomer. In membranes, MIPs assemble as tetramers with four functional pores. In mammals, only homo-tetramers are known, with the exception of hetero-tetramer formation by translational variants using different translation start points (Tajima et al. 2010). Human AQP1 is one of the best-studied water channels in terms of oligomerization. Three amino acids have been shown to be essential for its tetramerization (Buck et al. 2007). In EcGlpF, only one amino acid residue, present in the N terminus, has been demonstrated to be critical for the proper oligomerization and in vivo stability of the protein (Cymer and Schneider 2010).

Interestingly, data from plant PIPs indicate that several isoforms are able to form hetero-oligomers. Maize PIP1 isoforms are inactive when heterologously expressed in Xenopus oocytes, whereas PIP2s exhibit high water channel activity (Chaumont et al. 2000; Fetter et al. 2004). After coexpression, ZmPIP1 and ZmPIP2 induce an increase in the water permeability coefficient (P_f) of the membrane in a synergistic manner compared to ZmPIP2 expression alone. When transiently and singly expressed in maize protoplasts, ZmPIP1s and ZmPIP2s also differ in their subcellular localization (Zelazny et al. 2007). ZmPIP1s are retained in the secretory

Table 4 Posttranslational modifications of some plant MIP isoforms

Posttranslational modification	MIP isoform	Function	References
Glycosylation	McTIP1;2	Trafficking	Vera-Estrella et al. (2004)
	AQP2[a]	Trafficking	Hendriks et al. (2004)
Methylation	Several AtPIPs	Unknown	Santoni et al. (2006)
N-Acetylation	PvTIP3;1	Unknown	Daniels and Yeager (2005)
Phosphorylation	SoPIP2;1	Activity	Johansson et al. (1998), Törnroth-Horsefield et al. (2006)
	GmNIP1;1	Activity	Guenther et al. (2003)
	AQP2[a]	Trafficking	Moeller et al. (2009)
	ZmPIPs	Activity	Van Wilder et al. (2008)
	+ indirect physiological data		
Protonation	AtPIPs	Activity	Tournaire-Roux et al. (2003)
	AQP6[a]	Activity	Yasui et al. (1999)
	AQP3[a]	Activity	Zeuthen and Klaerke (1999)
	+ indirect physiological data		
Ions			
Zinc	bAQP0[a]	Activity	Nemeth-Cahalan et al. (2007)
Lead	AQP4[a]	Activity	Gunnarson et al. (2005)
Copper	hAQP3[a]	Activity	Zelenina et al. (2003)
Nickel	bAQP0[a]	Activity	Nemeth-Cahalan et al. (2007)
	hAQP3[a]	Activity	Zelenina et al. (2003)
Mercury	AtTIP1;1/AtTIP2;1	Activity	Daniels et al. (1994)
	hAQP1[a]	Activity	Preston et al. (1993)
			Savage and Stroud (2007)
	+ indirect physiological data		
Silver/Gold	ND	Activity	Niemietz and Tyerman (2002)
Calcium	SoPIP2;1	Gating	Törnroth-Horsefield et al. (2006)
	PIPs in general?		
	+ indirect physiological data		
Redox modifications	+ indirect physiological data	Activity	Henzler et al. (2004), Ye and Steudle (2006), Ampilogova et al. (2006)
Heteromer formation	PIP1 and PIP2 isoforms of several species	Trafficking	Fetter et al. (2004), Temmei et al. (2005), Zelazny et al. (2007), Vandeleur et al. (2009)
Cleavage of the N or C terminus	AQP0[a]	Switch from a transport protein to a cell adhesion protein	Gonen et al. (2004), Scheuring et al. (2007)

ND not determined
[a]Mammalian isoforms (*h* human, *b* bovine)

pathway and, more specifically, in the endoplasmic reticulum, whereas ZmPIP2s are targeted to the plasma membrane. Upon coexpression, ZmPIP1s are localized in the plasma membrane as a result of their physical interaction with ZmPIP2s, as shown by Förster resonance energy transfer analysis. Using oocytes and mammalian COS cells, similar results have been obtained for *Mimosa pudica* MpPIP1 and MpPIP2 and *Nicotiana tabacum* NtPIP1;1 and NtPIP2;1 (Temmei et al. 2005; Vandeleur et al. 2009). This hetero-oligomer formation of different PIP isoforms has so far been observed in plant AQPs, but not mammalian or microbial AQPs. Together, all these data suggest that the trafficking of plant AQPs plays a key role in the currently poorly understood physiological regulation of AQP functions. The structural basis of the physical interaction between PIPs has yet to be discovered.

The crystallization of members of the TIP, SIP, NIP, GIP, HIP, and XIP subfamilies will probably reveal unexpected and exciting details of plant AQP function, such as features responsible for gating and substrate selectivity, possible motifs for posttranslational modifications, and structural cues for the formation of oligomers.

4 Why Do Plants Contain So Many MIP Isoforms?

The previous section provided a broad picture of the multiplicity of known MIPs, their structural features, and cellular localization. In addition, the level of expression of plant MIPs is highly regulated during development and by environmental cues (Table 3), and the proteins are subjected to different posttranslational modifications regulating their trafficking and activity (Table 4) (reviewed in Chaumont et al. 2005; Maurel et al. 2008; Heinen et al. 2009). However, the physiological roles of MIPs in plants and how the discovered regulatory mechanisms are interacting are still largely unknown. Much of the research effort has been devoted to the characterization of MIPs with respect to cell water homeostasis. The roles of several isoforms in transmembrane water fluxes at the cellular level, long-distance translocation, and root or leaf hydraulic conductivity have been clearly demonstrated (reviewed in Maurel et al. 2001, 2008; Kjellbom et al. 1999; Hachez et al. 2006a, b; Heinen et al. 2009). However, recent findings about the ability of plant MIPs to channel a wide variety of small solutes suggest that these proteins are involved not only in cellular water relations, but also in detoxification processes, plant nutrition, and signal transduction.

This involvement of MIPs in many physiological key transport processes beside water transport is apparent in the moss *P. patens*. The direct environment of this moss, damp soil, is always supplied with water, so every cell is able to acquire nutrients directly from the surrounding water phase and no uptake organs have been formed during evolution. As mentioned previously, this plant expresses 23 different MIPs clustered in the seven identified plant MIP subfamilies, PIPs, TIPs, NIPs, SIPs, GIPs, HIPs, and XIPs (Danielson and Johanson 2008). This observation raises the question of why an ancient organism in which every single cell is directly connected to the surrounding water phase possesses 23 different water and glycerol channels. One obvious answer, apart from the very finely tuned control of the transport of a single substrate, is that MIP isoforms serve to channel different plant metabolites.

In the following sections, we will describe the known substrates of MIPs that have been identified either in transport assays *in planta* or in heterologous expression studies. These substrates belong to different important substance classes, including the vitally essential water molecule, diverse intermediate catabolic products, crucial nutrients, and signal messenger molecules. Current views on proven and potential functions of diverse MIPs in physiological transport processes are outlined.

4.1 MIP Function Related to Water Transport

The majority of plant MIPs are permeable to water. AQPs are high capacity channels with the ability to channel up to 10^9 water molecules per channel per second (Fujiyoshi et al. 2002). This allows a plant to tightly control water homeostasis by lowering by several orders of magnitude the activation energy needed to promote water transport across membranes. AQPs render liposome membranes and plant membranes 10–100 times more permeable to water than membranes lacking such channels (Dean et al. 1999; Sutka et al. 2005). AQPs have been shown to be responsible for the high P_f value differences observed in several cell types or in cells at different developmental stages of the same plant species (Chaumont et al. 2005). Independent of the level of the P_f value itself, the P_f is reduced after addition of the AQP blocker $HgCl_2$. P_f values correlate with the amount of AQPs in heterologous expression systems or in transgenic plants (reviewed in Hachez et al. 2006a, b). Together, these data show that AQPs regulate plant plasma membrane permeability. In general, vacuolar membranes also have very high P_f values (>200–500 μm/s). In plants, PIPs and TIPs represent the isoforms with the highest water permeability and are probably "the" key players in water homeostasis. The essential requirement for AQPs in water homeostasis is unquestioned. However, specific *PIP* and *TIP* knock-out *Arabidopsis* lines show no obvious developmental phenotypes (Javot et al. 2003; Schüssler et al. 2008; Beebo et al. 2009). This can be explained by the large number of isoforms that can probably compensate for the loss of a single isoform. The involvement of certain isoforms in water uptake from the soil, transcellular water transport, root-to-shoot water transport, cellular water homeostasis via cytoplasmic volume buffering, maintenance of cellular turgor, osmotic-driven growth control, and osmotic-driven organ movement under diverse environmental conditions was deduced from AQP expression patterns and characterization of plants overexpressing or silenced for specific isoforms (reviewed in Hachez et al. 2006a, b; Maurel et al. 2008; Heinen et al. 2009).

It has also been hypothesized that MIPs can act as sensors for gradients of osmotic or turgor pressure and that they transmit such information, in association with control systems, to signaling pathways. This function is secondarily related to the single-file water transport capacity itself and the flexible structure of MIP tetramers. This hypothesis may be termed the "Hill hypothesis," as it was formulated by Hill et al. (2004), who proposed that the ability to sense osmotic or turgor pressure gradients is inherent in the structure of each monomer in the tetramer.

As described above, each monomer forms a narrow hydrophilic channel, which separates the inner and outer atrium of the protein. A pressure or an osmotic gradient across the membrane will induce a pressure gradient between the two atria that results in water flux through the pore. Exclusion of solutes from the atrium will create a negative physical pressure within the atrium, producing an asymmetric deformation of the monomer or even the tetramer. Such changes in water or solute permeability due to changes in the osmotic or hydraulic pressure have been observed in different experimental systems and plant cell types (Wan et al. 2004; Ye et al. 2004; Kim and Steudle 2007). These studies showed that MIPs can sense an osmotic or pressure gradient across the membrane, probably by an asymmetric protein conformational change, and that they are able to act as sensors of such signals. Hill et al. (2004) suggested that MIPs are also involved in the generation of the response by directly interacting with downstream signaling components through a conformational change after signal perception. However, no study has yet identified such a sensor for osmotic or pressure gradients in plants, even though a general need for such a sensor has been formulated many times and downstream signaling pathways have been demonstrated in various organisms. Support for this MIP-mediated gradient sensing has come from two elegant studies on the osmotic effects on vacuolar ion release in guard cells of *Commelina communis* plants by MacRobbie (2006a, b), who demonstrated that the tonoplast of guard cells can sense an osmotic gradient and respond to water influx into the vacuole via increased vacuolar ion efflux. This suggests that MIPs, especially TIPs, which are expressed in the tonoplast of guard cells (Sarda et al. 1997), represent the most likely candidates for the described osmotic sensor. It has been speculated that the conformational change of TIPs caused by the pressure gradient across the channel induces ion efflux (MacRobbie 2006b). Two possibilities were proposed to explain the ion efflux: either TIPs become permeable to ions as a result of their deformation or TIPs are associated in the tonoplast with ion channels that are activated in response to the conformational change. In general, MIPs are not ion-permeable, although NOD26, AQP1, and AQP6 have been reported to transport ions (Anthony et al. 2000; Holm et al. 2005; Yasui et al. 1999). AQP6 shows ion permeability in intracellular vesicles after Hg treatment (Yasui et al. 1999). This is particularly interesting, as Hg treatment also induces ion permeability of the tonoplast in the guard cell (MacRobbie 2006b). These results can be explained through a conformation change induced by either a pressure gradient or Hg, which renders the TIP ion-permeable or induces a signal pathway resulting in ion flux. Hill et al. (2004) and MacRobbie (2006b) suggested that a TIP is involved in the sensing of osmotic signals; however, the downstream signaling pathway is not known.

4.2 MIP Function Related to Nitric Oxide Transport

It was recently discovered that MIPs facilitate the transmembrane transport of nitric oxide (NO), a hydrophobic gas that plays an important signaling role in a variety of physiological process in all organisms. Using an NO-specific fluorescent dye, Herrera et al. (2006) directly correlated the rate of NO influx into mammalian endothelial cells and liposomes with the abundance of AQP1. In plants, NO triggers signaling pathways involved in programmed cell death, pathogen defense, flowering, stomatal closure, and gravitropism (Neill et al. 2008). As earlier studies had reported that mammalian membranes represent no significant barrier for its diffusion, NO was thought to move from its site of production to its site of action by freely diffusing through the lipid bilayer of cell membranes without a need for a specific transporter (Liu et al. 2002). A study using various simulation techniques to examine the NO permeability of brain AQP4 suggested that the central pore formed by the association of the four monomers in a tetramer presents a lower barrier to NO gas permeation than pure lipid bilayers (Wang and Tajkhorshid 2010). The authors suggested that the central pore of AQP4 tetramer may act as a reservoir for NO messenger molecules, but doubted that AQP4 plays a physiological role in transmembrane NO transport, except in membranes with a low intrinsic gas permeability or when AQPs occupy a large portion of the membrane surface. The questions of whether plant MIPs are involved in the controlled distribution of the signaling molecule NO and whether plant membranes present a higher resistance to NO diffusion than mammalian membranes remain to be answered. Plant PIPs resemble the most to AQP1-like proteins from their structure and transport specificity. It will be interesting to investigate the permeability and the possible role of PIPs in NO signaling in MIP-deficient mutant plants and in transport assays. Future studies will show whether MIPs represent a key signal pore for NO, or whether this capacity represents a peculiar, but irrelevant, side effect.

4.3 MIP Function Related to Ammonia Transport

Approximately 1% of all man-made energy is used to produce ammonia gas, the direct or indirect precursor of most nitrogen-containing compounds. At physiological pH, the uncharged ammonia molecule is in thermodynamic equilibrium with its corresponding acid, the ammonium ion (NH_4^+). All assimilated nitrogen in plant cells is transformed into NH_4^+, the precursor for amino acids and all other N-containing molecules. NH_4^+ is transported across plant membranes via the ammonium protein transporter family. NH_3 has been suggested to cross membranes by free diffusion. However, the protein-mediated transport of NH_3 has some advantages compared to that of NH_4^+. Potassium ions, which interfere with the binding sites of the chemically similar NH_4^+ ion in transporter proteins, do not compete with NH_3 for transport. Additionally, transport can occur without the use of energy, as it is electrically neutral.

A yeast strain lacking its three intrinsic ammonium transporters and thus unable to grow on medium with low nitrogen levels was transformed with a cDNA library from wheat with the aim of identifying new transporters for NH_4^+ or NH_3 (Jahn et al. 2004). In addition to some already known high affinity ammonium transporters, low affinity transporters belonging to the TIP subfamily were identified in this screen. In a similar approach, TIPs from *Arabidopsis* were shown to be permeable to NH_3 (Loqué et al. 2005). NH_3 has comparable dimensions to water and its dipole moment is very close to that of water, suggesting that both may use the same channel (Jahn et al. 2004). A variety of MIPs were then shown to facilitate the transport of NH_3 in various heterologous test systems, and structural prerequisites for the ar/R selectivity filter were identified (Beitz et al. 2006; Dynowski et al. 2008a, b). In plants, nitrogen is one of the most limiting factors for growth. NH_3 gas is probably responsible for nitrogen loss from plants (Husted and Schjoerring 1996), so mechanisms preventing this are highly beneficial. Vacuoles could represent a major storage organelle where TIPs facilitate the uptake of NH_3, which is consequently trapped by being converted to the NH_4^+ anion in the acidic vacuolar environment. At low cytoplasmic NH_4^+/NH_3 concentrations, the vacuole might provide a nitrogen source for metabolism in the form of NH_4^+/NH_3. However, direct physiological data from TIP knockout mutants revealing such involvement of TIP-mediated NH_3 membrane transport are lacking (Schüssler et al. 2008). One reason for this lack of phenotype could be due to compensation mechanisms of closely related isoforms. The ability of several TIPs to channel NH_3 seems to be conserved across plant species and probably plays an important role in nitrogen transport.

4.4 MIP Function Related to Urea Transport

Urea is a secondary nitrogen metabolite in plants and has the highest nitrogen content of commonly used solid nitrogenous fertilizers. This lowest transportation cost per unit of nitrogen nutrient, together with its high solubility in water and readily assimilation by plants, makes urea a useful agronomical fertilizer. In *Arabidopsis*, a single gene codes for an active urea transport system, the proton–urea cotransporter AtDUR3 (Kojima et al. 2006). When urea is the sole nitrogen source in the medium, the *AtDUR3* knockout line becomes chlorotic compared to the wild type, in which *AtDUR3* transcript levels increase markedly under nitrogen limiting conditions (Kojima et al. 2007). How urea is relocalized and translocated across membranes for use as a nitrogen source is mostly unknown. As urea is uncharged, it diffuses slowly across membranes, and low affinity channels could be important for facilitated relocation between different compartments. Members of the NIP, PIP, and TIP subfamilies have been shown to facilitate the transport of urea across membranes (Gerbeau et al. 1999; Klebl et al. 2003; Liu et al. 2003; Wallace and Roberts 2005). Vacuoles might be used for short-term storage of urea or to avoid toxic concentrations in the cytoplasm. Phenotypes associated with TIP-mediated urea transport across the tonoplast of *Arabidopsis* have not been reported.

4.5 MIP Function Related to Carbon Dioxide Transport

Life on earth became possible through the evolution of photosynthesis and plant carbon fixation. Rubisco, a chloroplast bifunctional enzyme that catalyzes the transfer of CO_2 to ribulose 1,5 biphosphate in the Calvin cycle, requires a certain CO_2 threshold concentration to be efficient, and the CO_2 concentration in the vicinity of the enzyme is therefore the limiting factor for photosynthesis. It is assumed that the average concentration of CO_2 in chloroplasts is only half of the atmospheric CO_2 concentration of 0.038%. The diffusion of CO_2 from the atmosphere into the substomatal cavity via the stomata and to sites of carboxylation via the mesophyll is the main factor controlling CO_2 availability for Rubisco. Mesophyll conductance has a greater impact than stomatal conductance on the final concentration in the chloroplast and is regulated by the diffusion barriers for CO_2 formed by the intercellular space, apoplastic liquid phase, plasma membrane, cytosol, chloroplast membranes, and stroma. Plants are able to facilitate the diffusion of CO_2 into the chloroplast. However, the mesophyll conductance can be rapidly altered by several environmental factors, such as light, temperature, water, and CO_2 (Flexas et al. 2008). Another way to influence mesophyll conductance would be a change in the CO_2 permeability of the plasma membrane and chloroplast membranes. Recently, much attention has been paid to the involvement of MIPs in CO_2 conduction across cellular membranes. The first evidence for this came from a study by Terashima and Ono (2002), who showed that the AQP inhibitor $HgCl_2$ reduces the mesophyll conductance for CO_2. Thereafter, it was shown in the heterologous oocyte system that NtAQP1, a PIP1 isoform from tobacco, is able to significantly facilitate the transmembrane diffusion of CO_2 (Uehlein et al. 2003). Tobacco plants overexpressing NtAQP1 show increased net photosynthesis and leaf growth, whereas NtAQP1-silenced plants show decreased net photosynthesis and mesophyll conductance (Uehlein et al. 2003, 2008). However, the change in conductivity may not be explained solely by the modulated CO_2 membrane permeability due to AQPs, as a change in AQP quantity and localization could also influence water homeostasis by changing stomatal conductance. Uehlein et al. (2008) showed that, in tobacco, NtAQP1 is localized to the inner chloroplast envelope in addition to the plasma membrane. Water conductivity tests performed on isolated chloroplast or plasma membrane vesicles showed that the plasma membrane water permeability of NtAQP1-silenced plants is reduced to approximately 50% of that in wild type tobacco plants, while chloroplast water permeability is not changed. In contrast, the chloroplast envelope CO_2 permeability of NtAQP1-silenced plants is markedly decreased compared to that in the wild type, while the plasma membrane CO_2 permeability is unaffected. Additionally, this study indicated that the plasma membrane is about five times more permeable to CO_2 than the chloroplast envelope. It was concluded that the chloroplast envelope represents a higher diffusion barrier than previously estimated and that NtAQP1 markedly affects chloroplast CO_2 permeability in tobacco (Uehlein et al. 2008). An increased CO_2 conductance in plants induced by PIP overexpression has also been reported in rice (Hanba et al. 2004). Physiologically relevant MIP-mediated CO_2 transport across mammalian membranes is more controversial. While there is no debate about the fact that CO_2 can pass through mammalian AQPs, as shown in several transport and simulation studies, the question is whether AQPs can increase the diffusion of CO_2 across lipid membranes which, in the so far investigated mammalian systems, have an already very high intrinsic CO_2 permeability (Fang et al. 2002). In addition, the measured resistance to transmembrane CO_2 permeation in mammalian cell systems has been reported to be mainly caused by an unstirred layer next to the membrane and not by the expressed protein (e.g., AQPs) levels therein or the membrane composition itself (Missner et al. 2008). Physiological and molecular data for PIP-mediated CO_2 transport across the plasma membrane and chloroplast envelope point more toward a potential role of these channels in CO_2 transport in plants. Nevertheless, it remains to be determined whether the composition of the plant membrane causes a higher intrinsic CO_2 resistance than that seen with mammalian membranes.

4.6 MIP Function Related to Hydrogen Peroxide Transport

On the one hand, various studies have demonstrated the essential function of H_2O_2 as a signaling molecule controlling an amazingly wide spectrum of processes in plants (Neill et al. 2002). On the other hand, H_2O_2 is a reactive oxygen species (ROS), an oxidant that can react with various cellular targets and cause cell damage or even cell death. Cellular levels of H_2O_2 have therefore to be tightly regulated. The average half-life of H_2O_2 is long enough for it to act as a transportable signal, but short enough for its concentration to fluctuate and represent a potent on–off modulator. Its concentration in plant tissues is in the micromolar to low millimolar range, and its stability, in combination with its chemical reactivity, makes H_2O_2 a good multi-facetted messenger molecule. From the studies on yeast and bacteria, it is clear that the diffusion of H_2O_2 across membranes is limited, as they are quite impermeable to H_2O_2 (reviewed in Bienert et al. 2006). This permeability needs to be adjustable to allow signaling. It has been shown that yeast alters its permeability to H_2O_2 by a change in membrane composition (Branco et al. 2004; Matias et al. 2007). Like yeast or bacteria, mammals and plants have H_2O_2-impermeable membranes (Fritz et al. 2007). The transmembrane movement of H_2O_2 through AQPs was proposed in 2000 as a result of a biophysical study (Henzler and Steudle 2000). A broad survey aiming at the molecular identification of H_2O_2-permeable MIPs revealed that plant TIP1 and mammalian AQP8 are highly permeable to H_2O_2. Later, the ability to transport H_2O_2 was also demonstrated for a variety of other MIP isoforms (AtTIP2;3, AtNIP1;2, AtPIP2;1, and AtPIP2;4) from different plant subfamilies after expression in yeast (Dynowski 2008a, b). Molecular dynamic simulation analysis confirmed the H_2O_2 permeability of MIPs. H_2O_2 shares several chemical features with water (size, electrochemical properties, and ability to form hydrogen bonds), which makes them both candidates of MIP substrates (Bienert et al. 2006). Because of the different pore layouts of different MIPs, the level of H_2O_2 permeability varies considerably. While some isoforms, such as the TIP1s, are highly permeable to H_2O_2, TIP2s and PIPs are less permeable. A mutational approach revealed that amino acid mutations in AtPIP2;1 that abolish water permeability also abolish H_2O_2 permeability, suggesting that water and H_2O_2 share the same pathway through AQPs. Although the ability of specific MIPs to mediate H_2O_2 transport has been analyzed in detail, the physiological relevance of the MIP-mediated transport of this signaling molecule in plants and mammals is not yet resolved. Analysis of catalase activity and anthocyanin content in an *Arabidopsis* double mutant in which the H_2O_2-permeable TIP1;1 and TIP1;2 isoforms were both knocked out revealed a minor change in the steady-state level compared to the wild type under nonstressed conditions (Schüssler et al. 2008), suggesting a potential involvement of TIPs in redox metabolism. However, as the difference was minor, it is hard to make any conclusions from these observations. The presence of MIPs facilitating the diffusion of H_2O_2 across the tonoplast might seem surprising, given the presence of several effective cytoplasmic detoxification systems (reviewed in Mittler 2002 and Smirnoff 2005). However, vacuolar flavonoids have been shown to have highly antioxidative properties in vitro and in vivo and, in combination with vacuolar peroxidases, vacuoles potentially represent an effective ROS detoxification system (Smirnoff 2005). In this context, some very interesting observations were made in cytochemical studies. In pea plants, cadmium stress-induced H_2O_2 has been clearly detected in the tonoplast (Romero-Puertas et al. 2004). Similar "deposition" of H_2O_2 was seen in the tonoplast of leaf sheath cells of salt-stressed rice plants (Wi et al. 2006). These observations might represent snapshots of transmembrane H_2O_2 transport through TIP proteins.

In salt-stressed *Arabidopsis* plants, the formation of intracellular vesicles accumulating H_2O_2 can be observed in root cells (Leshem et al. 2006, 2007). These vesicles are subsequently targeted to the tonoplast, delivering the H_2O_2 into the vacuole. Interestingly, in the same conditions, *Arabidopsis* relocalized TIP1;1 proteins into intravacuolar vesicles (Boursiac et al. 2005). As TIP1;1 is permeable to H_2O_2 this internalization could prevent the efflux of H_2O_2 back into the cytoplasm. In agreement with this hypothesis, AtTIP2;1, which is less permeable to H_2O_2, was not internalized (Boursiac et al. 2005). However, this mechanistic interplay of H_2O_2-filled vesicles and AQP redistribution is far from being understood.

In terms of the involvement of MIPs in the regulation of plasma membrane permeability for H_2O_2, it will be interesting to study mutant plants in which H_2O_2-permeable PIPs are knocked out. Investigations on intercellular and membrane-crossing H_2O_2 signal transduction pathways and cell detoxification mechanisms should be prioritized.

4.7 MIP Function Related to Organic Acid Transport

Small organic acids represent one class of solutes that has been known for a long time to permeate through mammalian MIPs and speculated to be transported by plant AQPs (Tsukaguchi et al. 1998; Ouyang et al. 1991). Rat AQP9 is permeable to lactic acid in a pH-sensitive manner (Tsukaguchi et al. 1998). This lactic acid permeability at slightly acidic pH might be important during lactic acidosis in brain ischemia, in which AQP9 may contribute significantly to the clearance of excess lactate from the extracellular space under pathological conditions (Badaut et al. 2001). Recently, the aquaglyceroporin Fps1p from *Saccharomyces cerevisiae* was shown to mediate the transport of the undissociated form of acetic acid across yeast membranes (Mollapour and Piper 2007). Toxic external concentration levels of acetic acid trigger Hog1p-mediated phosphorylation of Fps1p, which results in the degradation of Fps1p in the vacuole, clearly indicating the involvement of this aquaglyceroporin in the regulation of acetic acid levels. Bacteria, fungi, and vertebrates contain isoforms of a monocarboxylate transport protein family, which is, interestingly, absent in plants (Halestrap and Price 1999). The expression of *AtNIP2;1* from *Arabidopsis* is upregulated 300-fold after anoxic stress, in which lactic acid fermentation precedes ethanolic fermentation. When heterologously expressed in oocytes, NIP2;1 was shown to be highly permeable to lactic acid, but only poorly permeable to water and glycerol (Choi and Roberts 2007). It is, therefore, tempting to speculate that AtNIP2;1 is important for the efflux of lactic acid under anoxic conditions. The first NIP described, namely GmNOD26 (GmNIP1;1), was suggested to be permeable to malate (Ouyang et al. 1991), as its phosphorylation status correlated closely with malate uptake across the peribacteroid membrane. Recently, a peptide transporter from *Alnus glutinosa* was identified as the first organic acid transporter in a peribacteroid membrane (Jeong et al. 2004). Nevertheless, an additional low affinity channel, such as a MIP, would result in a highly efficient uptake system. Further investigations on the ability of MIPs to channel organic acids are required. In particular, NIPs and the uncharacterized XIPs, which have rather wide and hydrophobic selectivity filters, represent potential candidates for such transporters.

4.8 MIP Function Related to Glycerol Transport

Glycerol is a triglyceride molecule that constitutes building blocks of biological membranes. In addition to its structural function, it is involved in energy metabolism as glycerol trisphosphate. Aquaglyceroporins from mammals and microbes have been demonstrated to play important roles in osmotic or freezing tolerance, regulation of the glycerol content of diverse tissues, and fat metabolism (Tamás et al. 1999; Rojek et al. 2008). Members from various plant MIP subfamilies, such as PpGIP1;1 from *Physcomitrella* (Gustavsson et al. 2005), and several TIP and PIP isoforms can facilitate the transport of glycerol across membranes (Gerbeau et al. 1999; Siefritz et al. 2001). Initially, NIP isoforms were characterized as predominantly glycerol facilitators and, for a long time, were seen as functional counterparts to aquaglyceroporins from microbes and mammals (Weig and Jakobs 2000). However, the lack of physiological evidence for NIP-mediated glycerol transport in plants and the conclusive evidence for NIP function in relation to metalloid transport make a main role of NIPs in glycerol transport unlikely. Glycerol is a very flexible molecule, with 576 known conformations (Law et al. 2005). It has been speculated that the ability of NIPs to channel glycerol is a simple consequence of the structural similarity of glycerol to metalloids, rather than to a physiological function (Porquet and Filella 2007; Bienert et al. 2008a).

4.9 MIP Function Related to Boric Acid Transport

Boron (B) has a special position among the essential elements. First, the difference between deficient and toxic concentrations in the soil solution is smaller than for any other nutrient element, and, second, it is the only nutrient that is assimilated as an uncharged molecule, boric acid. Its uptake and translocation have therefore to be highly regulated. Physiological experiments have demonstrated a channel-mediated boron transport system in plants. Nevertheless, the first experimental proof of a B transporter in plants revealed an active B efflux transporter (BOR1) with a high similarity to anion exchanger proteins (Takano et al. 2002). BOR1 homologs were also shown to be efflux transporters translocating B within the plant (Takano et al. 2008). Recently, the channel protein for boric acid uptake was identified in *Arabidopsis*. AtNIP5;1 proved to be essential for B uptake under B-limiting conditions (Takano et al 2006). In a microarray analysis, expression of *AtNIP5;1* was shown to be induced under low B conditions. The B transporting ability of AtNIP5;1 has been demonstrated in vivo and in heterologous expression systems. The cooperative functional interaction between AtNIP5;1, which takes up boric acid from the soil, and AtBOR1, which loads borate into the xylem, was shown to result in effective boron nutrition for the plant. In accordance with these results, only co-overexpression of AtBOR1 in an AtNIP5;1 activation tag line resulted in plants with a super-tolerance to a low B concentration (Kato et al. 2009). In contrast, overexpression of AtNIP5;1 under the control of the CaMV 35S promoter resulted in reduced overall growth. These results make clear that the overexpression of a gene without respecting the physiological context might lead to unexpected results. In this case, only the enhanced expression of AtNIP5;1 in cells in which it is endogenously expressed combined with modulation of AtBOR1 expression resulted in a positive outcome.

AtNIP6;1, the most similar gene to AtNIP5;1, is also permeable to boric acid, but possesses no water conductivity (Tanaka et al. 2008). Under conditions of B deprivation, AtNIP6;1 expression, mainly in nodal regions, is only increased in shoots, and *AtNIP6;1* knockouts only show reduced rosette leaf expansion and decreased B concentrations under low B conditions. This suggests that NIP6;1 plays a role in the xylem–phloem transfer of boric acid in the nodal regions. It is worth mentioning that AtNIP5;1 and AtNIP6;1 evolved differently in terms of their expression pattern and substrate specificity, apart from that for boric acid.

4.10 MIP Function Related to Silicic Acid Transport

Surprisingly, the first silicon (Si) transporter discovered in vascular plants was identified as a MIP, more precisely OsNIP2;1 from rice (Ma et al. 2006). Si, the second most abundant element in the Earth's crust, is essential for animals and diatoms and highly beneficial in a wide range of plant species. Si increases the tolerance of certain plants to an impressive variety of biotic and abiotic stresses and also increases their mechanical strength. In addition to its structural support function, Si has been suggested to interact with stress-related signaling systems to increase plant resistance to pathogens (Fautteux et al. 2005). It is taken up by the roots in the form of the bioavailable silicic acid [$Si(OH)_4$], a quite large uncharged molecule with restricted free diffusion across the lipid bilayer. After a long search, the genes encoding Si transporters were identified in a set of rice mutants defective in Si uptake. Ma and coworkers showed that a concerted cooperation of influx and efflux transporters is responsible for the uptake and translocation of Si in monocot species (Ma et al. 2006, 2007). While Si influx is mediated by MIPs, the efflux is mediated by proteins with similarities to anion efflux transporters. The roles of different MIP isoforms from rice, maize, and barley in Si uptake and distribution were unraveled, in part, using mutants (Ma et al. 2006). It will be interesting to determine which proteins are responsible for Si transport in dicot plants, as homologs of the Si transporting monocot NIPs have not yet been identified in dicots. The fact that the functions of MIPs, initially identified as essential water transporters, have been expanded to nutritionally important transport systems for large uncharged molecules, such as $Si(OH)_4$, points out their fundamental importance for life, especially in those plants that have evolved multiple isoforms.

4.11 MIP Function Related to Arsenite/Antimonite Transport

Aquaglyceroporins from diverse organisms have been shown to be involved in physiological important transmembrane glycerol transport. A comparison of glycerol, arsenite [$As(OH)_3$], and antimonite [$Sb(OH)_3$] revealed their high similarity in structural and electrochemical properties, which are also important for membrane transport (Porquet and Filella 2007). It was concluded that a MIP channel protein that is permeable for glycerol should also transport $As(OH)_3$ and $Sb(OH)_3$. The subsequent discovery that aquaglyceroporins play physiological roles in resistance to the highly toxic metalloid arsenic was therefore only a matter of time. The legume symbiont *Sinorhizobium meliloti* takes up arsenate via phosphate transporters and the arsenate is reduced to $As(OH)_3$ in the cytosol by the arsenate reductase ArsC. $As(OH)_3$ is transported by the bacterial MIP AqpS down the chemical concentration gradient out of the cell. AqpS, together with all other components involved in arsenic resistance, is encoded in the arsenic resistance operon (Yang et al. 2005). Several protozoan parasites decrease their expression of aquaglyceroporins and become resistant to arsenite- or antimonite-containing drugs, normally used as first line treatments (Mukhopadhyay and Beitz 2010). The increased resistance correlates with a reduced level of the metalloid in the cells. The parasite aquaglyceroporins were demonstrated to be permeable to these metalloids in transport assays.

The yeast aquaglyceroporin Fps1 allows the bidirectional flux of $As(OH)_3$ across the yeast membrane. It was shown that Fps1p mediates $As(OH)_3$ efflux in concert with another arsenic transporter, Acr3p, and plays a key role in the arsenic tolerance of the yeast. Yeast cells use Fps1-mediated $As(OH)_3$ efflux to ensure efficient detoxification of the cytoplasm when $As(OH)_3$ accumulates inside cells and prevent the detrimental influx of $As(OH)_3$ via posttranslational gating of Fps1, the closure of Fps1 being mediated by the Hog1-dependent kinase signaling cascade (Thorsen et al. 2006).

The variety of pore layouts in NIPs suggests that they may represent the until recently uncharacterized arsenite channels in plants. Four studies have independently demonstrated NIP-mediated $As(OH)_3$ transport in heterologous systems and directly in plants, using knockout mutants of rice and *Arabidopsis* (Bienert et al. 2008b; Isayenkov and Maathuis 2008; Kamiya et al. 2009; Ma et al. 2008a). Yeast growth and survival assays and direct uptake assays in both yeast and oocytes showed that NIP isoforms from *Arabidopsis*, rice, and *Lotus japonicus* from all three phylogenetic NIP groups are permeable to $As(OH)_3$. *AtNIP1;1* knockout mutants of *Arabidopsis* show increased resistance when grown on medium containing $As(OH)_3$ (Kamiya et al. 2009) and a field-grown *lsi1* mutant of rice had lower As concentrations in the straw than the wild type, but roughly the same levels in the grain and husk (Ma et al. 2008a). *AtNIP1;2, AtNIP5;1, AtNIP6;1,* and *AtNIP7;1* knockout lines do not show any increased tolerance to $As(OH)_3$, although they have been shown to be permeable to $As(OH)_3$ when expressed in heterologous systems (Bienert et al. 2008; Kamiya et al. 2009). These results revealed once again that a shared substrate specificity of proteins does not necessarily result in similar physiological functions. Although total arsenic uptake might be unchanged in a mutant knocked-out for one MIP channel compared to the wild type, translocation and subsequently compartmentalization could differ significantly. The characterization of more *NIP* genes in one plant species will provide a better understanding of the molecular mechanisms of $As(OH)_3$ fluxes in plants.

Given that, in contrast to the situation in prokaryotes and other eukaryotes, several eukaryotes, including plants, do not, as far as we are aware, encode specific active arsenite efflux transporters, but do contain arsenate reductases and metalloid-permeable MIPs and are mainly exposed to arsenate, rather than arsenite, it is tempting to hypothesize that MIPs constitute a major arsenic detoxification pathway. One physiological study in rice and tomato provided evidence for such a high capacity $As(OH)_3$ efflux system in roots (Xu et al. 2008). Recently, Zhao et al. (2010) showed that OsNIP2;1 accounts for 15–20% of the total $As(OH)_3$ efflux from rice roots, clearly indicating that plant AQPs are involved in arsenic detoxification processes. In addition to their involvement in the transport of arsenic, several NIP isoforms have been shown to be permeable to the toxic metalloid $Sb(OH)_3$ in a heterologous yeast system and *in planta* (Bienert et al. 2008; Kamiya and Fujiwara 2009). Nevertheless, proof is required that certain NIP isoforms play a role in plant arsenic and antimony transport metabolism and detoxification and that this ability is not an accidental side effect due to chemical and structural similarity to other metalloids.

5 Conclusion

Taking together all the data on the transport abilities of MIPs, it seems that, during evolution, there was a general trend to the production and the high conservation of two constriction regions (NPA boxes and ar/R selectivity filter), in which the pore-lining residues generate an environment with very strict selectivity against certain substrate groups, such as protons and other ions. In contrast, the concomitant selectivity for, or against, uncharged solutes was rather flexible and not that stringent. This selectivity against charged, but not uncharged, molecules, together with the multiplication of isoforms in plant species, opened up the broad substrate spectrum of today's MIPs. This, together with the array of physical and biochemical channel properties, was the decisive factor in their involvement in many physiologically important transport processes, including water homeostasis, plant nutrition, logistics and circulation of plant metabolites, and signaling processes. However, further investigation of the function and regulation of plant MIPs is still required. For instance, it is particularly important to evaluate the contribution of MIPs to signaling processes as sensors or channels transporting signaling molecules during specific physiological processes. The deciphering of the molecular and cellular mechanisms regulating these channels and in the proper transmission of signals is definitely required.

Acknowledgements This work was supported by grants from the Belgian National Fund for Scientific Research (FNRS), the Interuniversity Attraction Poles Program–Belgian Science Policy, and the "Communauté française de Belgique–Actions de Recherches Concertées." GPB was supported by an individual Marie Curie European fellowship.

References

Agre P, Preston GM, Smith BL, Jung JS, Raina S, Moon C, Guggino WB, Nielsen S (1993) Aquaporin CHIP: the archetypal molecular water channel. Am J Physiol 265:F463–F476

Alexandersson E, Fraysse L, Sjövall-Larsen S, Gustavsson S, Fellert M, Karlsson M, Johanson U, Kjellbom P (2005) Whole gene family expression and drought stress regulation of aquaporins. Plant Mol Biol 59:469–484

Amiry-Moghaddam M, Lindland H, Zelenin S, Roberg BA, Gundersen BB, Petersen P, Rinvik E, Torgner IA, Ottersen OP (2005) Brain mitochondria contain aquaporin water channels: evidence for the expression of a short AQP9 isoform in the inner mitochondrial membrane. FASEB J 19:1459–1467

Ampilogova YN, Zhestkova IM, Trofimova MS (2006) Redox modulation of osmotic water permeability in plasma membranes isolatedfrom roots and shoots of pea seedlings. Russ J Plant Physiol 53:622–628

Anthony TL, Brooks HL, Boassa D, Leonov S, Yanochko GM, Regan JW, Yool AJ (2000) Cloned human aquaporin-1 is a cyclic GMP-gated ion channel. Mol Pharmacol 257:576–588

Aroca R, Amodeo G, Fernadez-lliescas S, Herman EM, Chaumont F, Chrispeels MJ (2005) The role of aquaporins and membrane damage in chilling and hydrogen peroxide induced changes in the hydraulic conductance of maize roots. Plant Physiol 137–353

Aroca R, Ferrante A, Vernieri P, Chrispeels MJ (2006) Drought, abscisic acid and transpiration rate effects on the regulation of PIP aquaporin gene expression and abundance in *Phaseolus vulgaris* plants. Ann Bot (Lond) 98:1301–1310

Badaut J, Hirt L, Granziera C, Bogousslavsky J, Magistretti PJ, Regli L (2001) Astrocyte-specific expression of aquaporin-9 in mouse brain is increased after transient focal cerebral ischemia. J Cereb Blood Flow Metab 21:477–482

Bansal A, Sankararamakrishnan R (2007) Homology modeling of major intrinsic proteins in rice, maize and Arabidopsis: comparative analysis of transmembrane helix association and aromatic/arginine selectivity filters. BMC Struct Biol 7:27

Barkla BJ, Vera-Estrella R, Pantoja O, Kirch HH, Bohnert HJ (1999) Aquaporin localization – how valid are the TIP and PIP labels? Trends Plant Sci 4:86–88

Beebo A, Thomas D, Der C, Sanchez L, Leborgne-Castel N, Marty F, Schoefs B, Bouhidel K (2009) Life with and without AtTIP1;1, an arabidopsis aquaporin preferentially localized in the apposing tonoplasts of adjacent vacuoles. Plant Mol Biol 70:193–2009

Beitz E, Wu B, Holm LM, Schultz JE, Zeuthen T (2006) Point mutations in the aromatic/arginine region in aquaporin 1 allow passage of urea, glycerol, ammonia, and protons. Proc Natl Acad Sci USA 103:269–274

Biela A, Grote K, Otto B, Hoth S, Hedrich R, Kaldenhoff R (1999) The *Nicotiana tabacum* plasma membrane aquaporin NtAQP1 is mercury-insensitive and permeable for glycerol. Plant J 18:565–570

Bienert GP, Schjoerring JK, Jahn TP (2006) Membrane transport of hydrogen peroxide. Biochim Biophys Acta 1758:994–1003

Bienert GP, Møller AL, Kristiansen KA, Schulz A, Møller IM, Schjoerring JK, Jahn TP (2007) Specific aquaporins facilitate the diffusion of hydrogen peroxide across membranes. J Biol Chem 282:1183–1192

Bienert GP, Schüssler MD, Jahn TP (2008a) Metalloids: essential, beneficial or toxic? Major intrinsic proteins sort it out. Trends Biochem Sci 33:20–26

Bienert GP, Thorsen M, Schüssler MD, Nilsson HR, Wagner A, Tamás MJ, Jahn TP (2008b) A subgroup of plant aquaporins facilitate the bi-directional diffusion of $As(OH)_3$ and $Sb(OH)_3$ across membranes. BMC Biol 6:26

Boursiac Y, Chen S, Luu DT, Sorieul M, van den Dries N, Maurel C (2005) Early effects of salinity on water transport in arabidopsis roots. Molecular and cellular features of aquaporin expression. Plant Physiol 139:790–805

Branco MR, Marinho HS, Cyrne L, Antunes F (2004) Decrease of H_2O_2 plasma membrane permeability during adaptation to H_2O_2 in *Saccharomyces cerevisiae*. J Biol Chem 279:6501–6506

Buck TM, Wagner J, Grund S, Skach WR (2007) A novel tripartite motif involved in aquaporin topogenesis, monomer folding and tetramerization. Nat Struct Mol Biol 14:762–769

Calamita G, Ferri D, Gena P, Liquori GE, Cavalier A, Thomas D, Svelto M (2005) The inner mitochondrial membrane has aquaporin-8 water channels and is highly permeable to water. J Biol Chem 280:17149–17153

Catalano CM, Lane WS, Sherrier DJ (2004) Biochemical characterization of symbiosome membrane proteins from *Medicago truncatula* root nodules. Electrophoresis 25:519–531

Chaumont F, Barrieu F, Herman EM, Chrispeels MJ (1998) Characterization of a maize tonoplast aquaporin expressed in zones of cell division and elongation. Plant Physiol 117:1143–1152

Chaumont F, Barrieu F, Jung R, Chrispeels MJ (2000) Plasma membrane intrinsic proteins from maize cluster in two sequence subgroups with differential aquaporin activity. Plant Physiol 122:1025–1034

Chaumont F, Barrieu F, Wojcik E, Chrispeels MJ, Jung R (2001) Aquaporins constitute a large and highly divergent protein family in maize. Plant Physiol 125:1206–1215

Chaumont F, Moshelion M, Daniels MJ (2005) Regulation of plant aquaporin activity. Biol Cell 97:749–764

Chen H, Wu Y, Voth GA (2006) Origins of proton transport behavior from selectivity domain mutations of the aquaporin-1 channel. Biophys J 90:73–75

Chervin C, Tira-Umphon A, Terrier N, Zouine M, Severac D, Roustan JP (2008) Stimulation of the grape berry expansion by ethylene and effects on related gene transcripts, over the ripening phase. Physiol Plant 134:534–546

Chiba Y, Mitani N, Yamaji N, Ma JF (2009) HvLsi1 is a silicon influx transporter in barley. Plant J 57:810–818

Choi WG, Roberts DM (2007) Arabidopsis NIP2;1, a major intrinsic protein transporter of lactic acid induced by anoxic stress. J Biol Chem 282:24209–24218

Ciavatta VT, Morillon R, Pullman GS, Chrispeels MJ, Cairney J (2001) An aquaglyceroporin is abundantly expressed early in the development of the suspensor and the embryo proper of loblolly pine. Plant Physiol 127:1556–1567

Cymer F, Schneider D (2010) A single glutamate residue controls the oligomerization, function, and stability of the aquaglyceroporin GlpF. Biochemistry 49:279–286

Daniels MJ, Yeager M (2005) Phosphorylation of aquaporin PvTIP3;1 defined by mass spectrometry and molecular modeling. Biochemistry 44:14443–14454

Daniels MJ, Mirkov TE, Chrispeels MJ (1994) The plasma membrane of *Arabidopsis thaliana* contains a that is a homolog of the tonoplast water channel protein TIP. Plant Physiol 106:1325–1333

Danielson JA, Johanson U (2008) Unexpected complexity of the aquaporin gene family in the moss *Physcomitrella patens*. BMC Plant Biol 8:45

Danielson JA, Johanson U (2010) Phylogeny of major intrinsic proteins. In: Jahn TP, Bienert GP (eds) MIPs and their role in the exchange of metalloids. Landes Biosciences, Austin

Dean RM, Rivers RL, Zeidel ML, Roberts DM (1999) Purification and functional reconstitution of soybean nodulin 26. An aquaporin with water and glycerol transport properties. Biochemistry 38:347–353

Dordas C, Chrispeels MJ, Brown PH (2000) Permeability ad channel-mediated transport of boric acid across membrane vesicles isolated from squash roots. Plant Physiol 124:1349–1362

Dowd C, Wilson IW, McFadden H (2004) Gene expression profile changes in cotton root and hypocotyl tissues in response to infection with *Fusarium oxysporum f. sp. vasinfectum*. Mol Plant Microbe Interact 17:654–667

Dynowski M, Schaaf G, Loque D, Moran O, Ludewig U (2008a) Plant plasma membrane water channels conduct the signalling molecule H_2O_2. Biochem J 414:53–61

Dynowski M, Mayer M, Moran O, Ludewig U (2008b) Molecular determinants of ammonia and urea conductance in plant aquaporin homologs. FEBS Lett 582:2458–2462

Fang X, Yang B, Matthay MA, Verkman AS (2002) Evidence against aquaporin-1-dependent CO2 permeability in lung and kidney. J Physiol 542:63–69

Fauteux F, Rémus-Borel W, Menzies JG, Bélanger RR (2005) Silicon and plant disease resistance against pathogenic fungi. FEMS Microbiol Lett 249:1–6

Fetter K, Van Wilder V, Moshelion M, Chaumont F (2004) Interactions between plasma membrane aquaporins modulate their water channel activity. Plant Cell 16:215–228

Fischer G, Kosinska-Eriksson U, Aponte-Santamaría C, Palmgren M, Geijer C, Hedfalk K, Hohmann S, de Groot BL, Neutze R, Lindkvist-Petersson K (2009) Crystal structure of a yeast aquaporin at 1.15 angstrom reveals a novel gating mechanism. PLoS Biol 7:e1000130

Fitzpatrick KL, Reid RJ (2009) The involvement of aquaglyceroporins in transport of boron in barley roots. Plant Cell Environ 32:1357–1365

Flexas J, Ribas-Carbó M, Diaz-Espejo A, Galmés J, Medrano H (2008) Mesophyll conductance to CO2: current knowledge and future prospects. Plant Cell Environ 31:602–621

Fortin MG, Morrison NA, Verma DP (1987) Nodulin-26, a peribacteroid membrane nodulin is expressed independently of the development of the peribacteroid compartment. Nucleic Acids Res 15:813–824

Fritz R, Bol J, Hebling U, Angermüller S, Völkl A, Fahimi HD, Mueller S (2007) Compartment-dependent management of H(2)O(2) by peroxisomes. Free Radic Biol Med 42:1119–1129

Fu D, Lu M (2007) The structural basis of water permeation and proton exclusion in aquaporins. Mol Membr Biol 24:366–374

Fu D, Libson A, Miercke LJ, Weitzman C, Nollert P, Krucinski J, Stroud RM (2000) Structure of a glycerol-conducting channel and the basis for its selectivity. Science 290:481–486

Fujiyoshi Y, Mitsuoka K, de Groot BL, Philippsen A, Grubmüller H, Agre P, Engel A (2002) Structure and function of water channels. Curr Opin Struct Biol 12:509–515

Gazzarrini S, Kang M, Epimashko S, Van Etten JL, Dainty J, Thiel G, Moroni A (2006) Chlorella virus mt325 encodes water and potassium channels that interact synergistically. Proc Natl Acad Sci USA 103:5355–5360

Gerbeau P, Güçlü J, Ripoche P, Maurel C (1999) Aquaporin Nt-TIPa can account for the high permeability of tobacco cell vacuolar membrane to small neutral solutes. Plant J 18:577–587

Gonen T, Walz T (2006) The structure of aquaporins. Q Rev Biophys 39:361–396

Gonen T, Cheng Y, Kistler J, Walz T (2004) Aquaporin-0 membrane junctions form upon proteolytic cleavage. J Mol Biol 342:1337–1345

Guenther JF, Chanmanivone N, Galetovic MP, Wallace IS, Cobb JA, Roberts DM (2003) Phosphorylation of soybean nodulin 26 on serine 262 enhances water permeability and is regulated developmentally and by osmotic signals. Plant Cell 15:981–991

Gunnarson E, Axehult G, Baturina G, Zelenin S, Zelenina M, Aperia A (2005) Lead induces increased water permeability in astrocytes expressing aquaporin 4. Neuroscience 136:105–114

Gupta AB, Sankararamakrishnan R (2009) Genome-wide analysis of major intrinsic proteins in the tree plant *Populus trichocarpa*: characterization of XIP subfamily of aquaporins from evolutionary perspective. BMC Plant Biol 9:134

Gustavsson S, Lebrun AS, Nordén K, Chaumont F, Johanson U (2005) A novel plant major intrinsic protein in Physcomitrella patens most similar to bacterial glycerol channels. Plant Physiol 139:287–295

Hachez C, Moshelion M, Zelazny E, Cavez D, Chaumont F (2006a) Localization and quantification of plasma membrane aquaporin expression in maize primary root: a clue to understanding their role as cellular plumbers. Plant Mol Biol 62:305–323

Hachez C, Zelazny E, Chaumont F (2006b) Modulating the expression of aquaporin genes in planta: a key to understand their physiological functions? Biochim Biophys Acta 1758:1142–1156

Hachez C, Heinen RB, Draye X, Chaumont F (2008) The expression pattern of plasma membrane aquaporins in maize leaf highlights their role in hydraulic regulation. Plant Mol Biol 68:337–353

Halestrap AP, Price NT (1999) The proton-linked monocarboxylate transporter (MCT) family: structure, function and regulation. Biochem J 343:281–299

Hanba YT, Shibasaka M, Hayashi Y, Hayakawa T, Kasamo K, Terashima I, Katsuhara M (2004) Overexpression of the barley aquaporin HvPIP2;1 increases internal $CO_{(2)}$ conductance and $CO_{(2)}$ assimilation in the leaves of transgenic rice plants. Plant Cell Physiol 45:521–529

Harries WE, Akhavan D, Miercke LJ, Khademi S, Stroud RM (2004) The channel architecture of aquaporin 0 at a 2.2-A resolution. Proc Natl Acad Sci USA 101:14045–14050

Heinen RB, Ye Q, Chaumont F (2009) Role of aquaporins in leaf physiology. J Exp Bot 60:2971–2985

Hendriks G, Koudijs M, van Balkom BW, Oorschot V, Klumperman J, Deen PM, van der Sluijs P (2004) Glycosylation is important for cell surface expression of the water channel aquaporin-2 but is not essential for tetramerization in the endoplasmic reticulum. J Biol Chem 279:2975–2983

Henzler T, Steudle E (2000) Transport and metabolic degradation of hydrogen peroxide in Chara corallina: model calculations and measurements with the pressure probe suggest transport of H_2O_2 across water channels. J Exp Bot 51:2053–2066
Henzler T, Ye Q, Steudle E (2004) Oxidative gating of water channels (aquaporins) in Chara by hydroxyl radicals. Plant Cell Environ 27:1184–1195
Herrera M, Hong NJ, Garvin JL (2006) Aquaporin-1 transports NO across cell membranes. Hypertension 48:157–164
Higuchi T, Suga S, Tsuchiya T, Hisada H, Morishima S, Okada Y, Maeshima M (1998) Molecular cloning, water channel activity and tissue specific expression of two isoforms of radish vacuolar aquaporin. Plant Cell Physiol 39:905–913
Hill AE, Shachar-Hill B, Shachar-Hill Y (2004) What are aquaporins for? J Membr Biol 197:1–32
Hiroaki Y, Tani K, Kamegawa A, Gyobu N, Nishikawa K, Suzuki H, Walz T, Sasaki S, Mitsuoka K, Kimura K, Mizoguchi A, Fujiyoshi Y (2006) Implications of the aquaporin-4 structure on array formation and cell adhesion. J Mol Biol 2355:628–639
Holm LM, Jahn TP, Møller AL, Schjoerring JK, Ferri D, Klaerke DA, Zeuthen T (2005) NH_3 and NH_4^+ permeability in aquaporin-expressing Xenopus oocytes. Pflugers Arch 450:415–428
Horsefield R, Nordén K, Fellert M, Backmark A, Törnroth-Horsefield S, Terwisscha van Scheltinga AC, Kvassman J, Kjellbom P, Johanson U, Neutze R (2008) High-resolution x-ray structure of human aquaporin 5. Proc Natl Acad Sci USA 105:13327–13332
Husted S, Schjoerring JK (1996) Ammonia Flux between oilseed rape plants and the atmosphere in response to changes in leaf temperature, light intensity, and air humidity (interactions with leaf conductance and apoplastic NH_4^+ and H^+ concentrations). Plant Physiol 112:67–74
Isayenkov SV, Maathuis FJ (2008) The *Arabidopsis thaliana* aquaglyceroporin AtNIP7;1 is a pathway for arsenite uptake. FEBS Lett 582:1625–1628
Ishikawa F, Suga S, Uemura T, Sato MH, Maeshima M (2005) Novel type aquaporin SIPs are mainly localized to the ER membrane and show cell-specific expression in Arabidopsis thaliana. FEBS Lett 579:5814–5820
Jahn TP, Møller AL, Zeuthen T, Holm LM, Klaerke DA, Mohsin B, Kühlbrandt W, Schjoerring JK (2004) Aquaporin homologues in plants and mammals transport ammonia. FEBS Lett 574:31–36
Jauh GY, Phillips TE, Rogers JC (1999) Tonoplast intrinsic protein isoforms as markers for vacuolar functions. Plant Cell 11:1867–1882
Javot H, Lauvergrat V, Santoni V, Martin-Laurent F, Güçlü J, Vinh J, Heyes J, Franck KI, Schäffner AR, Bouchez D, Maurel C (2003) Role of a single aquaporin isoform in root water uptake. Plant Cell 15:509–522
Jeong J, Suh S, Guan C, Tsay YF, Moran N, Oh CJ, An CS, Demchenko KN, Pawlowski K, Lee Y (2004) A nodule-specific dicarboxylate transporter from alder is a member of the peptide transporter family. Plant Physiol 134:969–978
Johanson U, Karlsson M, Johansson I, Gustavsson S, Sjövall S, Fraysse L, Weig AR, Kjellbom P (2001) The complete set of genes encoding major intrinsic proteins in arabidopsis provides a framework for a new nomenclature for major intrinsic proteins in plants. Plant Physiol 126:1358–1369
Johansson I, Karlsson M, Shukla VK, Chrispeels MJ, Larsson C, Kjellbom P (1998) Water transport activity of the plasma membrane aquaporin PM28A is regulated by phosphorylation. Plant Cell 10:451–459
Kaldenhoff R, Fischer M (2006) Functional aquaporin diversity in plants. Biochim Biophys Acta 1758:1134–1141
Kaldenhoff R, Kölling A, Richter G (1993) A novel blue light- and abscisic acid-inducible gene of Arabidopsis thaliana encoding an intrinsic membrane protein. Plant Mol Biol 23:1187–1198
Kamiya T, Fujiwara T (2009) Arabidopsis NIP1;1 transports antimonite and determines antimonite sensitivity. Plant Cell Physiol 50:1977–1981
Kamiya T, Tanaka M, Mitani N, Ma JF, Maeshima M, Fujiwara T (2009) NIP1;1, an aquaporin homolog, determines the arsenite sensitivity of *Arabidopsis thaliana*. J Biol Chem 284:2114–2120

Kato Y, Miwa K, Takano J, Wada M, Fujiwara T (2009) Highly boron deficiency-tolerant plants generated by enhanced expression of NIP5;1, a boric acid channel. Plant Cell Physiol 50:58–66
Kim YX, Steudle E (2007) Light and turgor affect the water permeability (aquaporins) of parenchyma cells in the midrib of leaves of *Zea mays*. J Exp Bot 58:4119–4129
Kjellbom P, Larsson C, Johansson II, Karlsson M, Johanson U (1999) Aquaporins and water homeostasis in plants. Trends Plant Sci 4:308–314
Klebl F, Wolf M, Sauer N (2003) A defect in the yeast plasma membrane urea transporter Dur3p is complemented by CpNIP1, a Nod26-like protein from zucchini (*Cucurbita pepo L.*), and by *Arabidopsis thaliana* delta-TIP or gamma-TIP. FEBS Lett 547:69–74
Kojima S, Bohner A, von Wirén N (2006) Molecular mechanisms of urea transport in plants. J Membr Biol 212:83–91
Kojima S, Bohner A, Gassert B, Yuan L, von Wiren N (2007) AtDUR3 represents the major transport for high-affinity urea transport across the plasma membrane of nitrogen-deficient *Arabidopsis* roots. Plant J 52:30–40
Kolla VA, Suhita D, Raghavendra AS (2004) Marked changes in volume of mesophyll protoplasts of pea (Pisum sativum) on exposure to growth hormones. J Plant Physiol 161:557–562
Law JMS, Fejer SN, Setiadi DH, Chass GA, Viskolcz B (2005) Molecular orbital computations on lipids: an ab initio exploratory study on the conformations of glycerol and its fluorine congeners. J Mol Struct 722:79–96
Lee JK, Kozono D, Remis J, Kitagawa Y, Agre P, Stroud RM (2005) Structural basis for conductance by the archaeal aquaporin AqpM at 1.68 A. Proc Natl Acad Sci USA 102:18932–18937
Leshem Y, Melamed-Book N, Cagnac O, Ronen G, Nishri Y, Solomon M, Cohen G, Levine A (2006) Suppression of Arabidopsis vesicle-SNARE expression inhibited fusion of H_2O_2-containing vesicles with tonoplast and increased salt tolerance. Proc Natl Acad Sci USA 103:18008–18013
Leshem Y, Seri L, Levine A (2007) Induction of phosphatidylinositol 3-kinase-mediated endocytosis by salt stress leads to intracellular production of reactive oxygen species and salt tolerance. Plant J 51:185–197
Li GW, Peng YH, Yu X, Zhang MH, Cai WM, Sun WN, Su WA (2008) Transport functions and expression analysis of vacuolar membrane aquaporins in response to various stresses in rice. J Plant Physiol 165:1879–1888
Li RY, Ago Y, Liu WJ, Mitani N, Feldmann J, McGrath SP, Ma JF, Zhao FJ (2009) The rice aquaporin Lsi1 mediates uptake of methylated arsenic species. Plant Physiol 150:2071–2080
Lian HL, Yu X, Lane D, Sun WN, Tang ZC, Su WA (2006) Upland rice and lowland rice exhibited different PIP expression under water deficit and ABA treatment. Cell Res 16:651–660
Lin W, Peng Y, Li G, Arora R, Tang Z, Su W, Cai W (2007) Isolation and functional characterization of PgTIP1, a hormone-autotrophic cells-specific tonoplast aquaporin in ginseng. J Exp Bot 58:947–956
Liu X, Samouilov A, Lancaster JR Jr, Zweier JL (2002) Nitric oxide uptake by erythrocytes is primarily limited by extracellular diffusion not membrane resistance. J Biol Chem 277:26194–26199
Liu LH, Ludewig U, Gassert B, Frommer WB, von Wirén N (2003) Urea transport by nitrogen-regulated tonoplast intrinsic proteins in Arabidopsis. Plant Physiol 133:1220–1228
Lopez F, Bousser A, Sissoëff I, Gaspar M, Lachaise B, Hoarau J, Mahé A (2003) Diurnal regulation of water transport and aquaporin gene expression in maize roots: contribution of PIP2 proteins. Plant Cell Physiol 44:1384–1395
Loqué D, Ludewig U, Yuan L, von Wirén N (2005) Tonoplast intrinsic proteins AtTIP2;1 and AtTIP2;3 facilitate NH3 transport into the vacuole. Plant Physiol 137:671–680
Ma JF, Tamai K, Yamaji N, Mitani N, Konishi S, Katsuhara M, Ishiguro M, Murata Y, Yano M (2006) A silicon transporter in rice. Nature 440:688–691
Ma JF, Yamaji N, Mitani N, Tamai K, Konishi S, Fujiwara T, Katsuhara M, Yano M (2007) An efflux transporter of silicon in rice. Nature 448:209–212

Ma JF, Yamaji N, Mitani N, Xu XY, Su YH, McGrath SP, Zhao FJ (2008a) Transporters of arsenite in rice and their role in arsenic accumulation in rice grain. Proc Natl Acad Sci USA 105:9931–9935

Ma N, Xue J, Li Y, Liu X, Dai F, Jia W, Luo Y, Gao J (2008b) Rh-PIP2;1, a rose aquaporin gene, is involved in ethylene-regulated petal expansion. Plant Physiol 148:894–907

MacRobbie EA (2006a) Osmotic effects on vacuolar ion release in guard cells. Proc Natl Acad Sci USA 103:1135–1140

MacRobbie EA (2006b) Control of volume and turgor in stomatal guard cells. J Membr Biol 210:131–142

Mahdieh M, Mostajeran A, Horie T, Katsuhara M (2008) Drought stress alters water relations and expression of PIP-type aquaporin genes in *Nicotiana tabacum* plants. Plant Cell Physiol 49:801–813

Marjanovi Z, Uwe N, Hampp R (2005) Mycorrhiza formation enhances adaptive response of hybrid poplar to drought. Ann NY Acad Sci 1048:496–499

Marmagne A, Rouet MA, Ferro M, Rolland N, Alcon C, Joyard J, Garin J, Barbier-Brygoo H, Ephritikhine G (2004) Identification of new intrinsic proteins in arabidopsis plasma membrane proteome. Mol Cell Proteomics 3:675–691

Matias AC, Pedroso N, Teodoro N, Marinho HS, Antunes F, Nogueira JM, Herrero E, Cyrne L (2007) Down-regulation of fatty acid synthase increases the resistance of *Saccharomyces cerevisiae* cells to H_2O_2. Free Radic Biol Med 43:1458–1465

Maurel C, Reizer J, Schroeder JI, Chrispreels MJ (1993) The vacuolar membrane protein gamma-TIP creaters water specific chennels in Xenopus oocytes. EMBO J 12:2241–2247

Maurel C, Aquaporins CMJ (2001) A molecular entry into plant water relations. Plant Physiol 125:135–138

Maurel C, Verdoucq L, Luu DT, Santoni V (2008) Annu Rev Plant Biol 59:595–624

Missner A, Kügler P, Saparov SM, Sommer K, Mathai JC, Zeidel ML, Pohl P (2008) Carbon dioxide transport through membranes. J Biol Chem 283:25340–25347

Mitani N, Yamaji N, Ma JF (2008) Characterization of substrate specificity of a rice silicon transporter, Lsi1. Pflugers Arch 456:679–686

Mitani N, Yamaji N, Ma JF (2009) Identification of maize silicon influx transporters. Plant Cell Physiol 50:5–12

Mittler R (2002) Oxidative stress, antioxidants and stress tolerance. Trends Plant Sci 7:405–410

Mizutani M, Watanabe S, Nakagawa T, Maeshima M (2006) Aquaporin NIP2;1 is mainly localized to the ER membrane and shows root-specific accumulation in *Arabidopsis thaliana*. Plant Cell Physiol 47:1420–1426

Moeller HB, Knepper MA, Fenton RA (2009) Serine 269 phosphorylated aquaporin-2 is targeted to the apical membrane of collecting duct principal cells. Kidney 75:295–303

Mollapour M, Piper PW (2007) Hog1 mitogen-activated protein kinase phosphorylation targets the yeast Fps1 aquaglyceroporin for endocytosis, thereby rendering cells resistant to acetic acid. Mol Cell Biol 27:6446–6456

Moshelion M, Becker D, Biela A, Uehlein N, Hedrich R, Otto B, Levi H, Moran N, Kaldenhoff R (2002) Plasma membrane aquaporins in the motor cells of *Samanea saman*: diurnal and circadian regulation. Plant Cell 14:727–739

Mukhopadhyay R, Beitz E (2010) Metalloid transport by aquaglyceroporins: consequences in the treatment of human diseases. In: Jahn TP, Bienert GP (eds) MIPs and their role in the exchange of metalloids. Landes Biosciences, Austin

Murata K, Mitsuoka K, Hirai T, Walz T, Agre P, Heymann JB, Engel A, Fujiyoshi Y (2000) Structural determinants of water permeation through aquaporin-1. Nature 407:599–605

Neill S, Desikan R, Hancock J (2002) Hydrogen peroxide signalling. Curr Opin Plant Biol 5:388–395

Neill S, Barros R, Bright J, Desikan R, Hancock J, Harrison J, Morris P, Ribeiro D, Wilson I (2008) Nitric oxide, stomatal closure, and abiotic stress. J Exp Bot 59:165–176

Németh-Cahalan KL, Kalman K, Froger A, Hall JE (2007) Zinc modulation of water permeability reveals that aquaporin 0 functions as a cooperative tetramer. J Gen Physiol 130:457–464

Newby ZE, O'Connell J, Robles-Colmenares Y, Khademi S, Miercke LJ, Stroud RM (2008) Crystal structure of the aquaglyceroporin PfAQP from the malarial parasite *Plasmodium falciparum*. Nat Struct Mol Biol 15:619–625

Niemietz CM, Tyerman SD (2002) New potent inhibitors of aquaporins: silver and gold compounds inhibit aquaporins of plant and human origin. FEBS Lett 531:443–447

Olbrich A, Hillmer S, Hinz G, Oliviusson P, Robinson DG (2007) Newly formed vacuoles in root meristems of barley and pea seedlings have characteristics of both protein storage and lytic vacuoles. Plant Physiol 145:1383–1394

Ouyang LJ, Whelan J, Weaver CD, Roberts DM, Day DA (1991) Protein phosphorylation stimulates the rate of malate uptake across the peribacteroid membrane of soybean nodules. FEBS Lett 293:188–190

Ozga JA, van Huizen R, Reinecke DM (2002) Hormone and seed-specific regulation of pea fruit growth. Plant Physiol 128:1379–1389

Paris N, Stanley CM, Jones RL, Rogers JC (1996) Plant cells contain two functionally distinct vacuolar compartments. Cell 85:563–572

Park M, Kim SJ, Vitale A, Hwang I (2004) Identification of the protein storage vacuole and protein targeting to the vacuole in leaf cells of three plant species. Plant Physiol 134:625–639

Peng Y, Lin W, Cai W, Arora R (2007) Overexpression of a *Panax ginseng* tonoplast aquaporin alters salt tolerance, drought tolerance and cold acclimation ability in transgenic arabidopsis plants. Planta 226:729–740

Pettersson N, Filipsson C, Becit E, Brive L, Hohmann S (2005) Aquaporins in yeast and filamentous fungi. Biol Cell 97:487–500

Porquet A, Filella M (2007) Structural evidence of the similarity of $Sb(OH)_3$ and $As(OH)_3$ with glycerol: implications for their uptake. Chem Res Toxicol 20:1269–1276

Preston GM, Carroll TP, Guggino WB, Agre P (1992) Appearance of water channels in Xenopus oocytes expressing red cell CHIP28 protein. Science 256:385–387

Preston GM, Jung JS, Guggino WB, Agre P (1993) The mercury-sensitive residue at cysteine 189 in the CHIP28 water channel. J Biol Chem 268:17–20

Rivers RL, Dean RM, Chandy G, Hall JE, Roberts DM, Zeidel ML (1997) Functional analysis of nodulin 26, an aquaporin in soybean root nodule symbiosomes. J Biol Chem 272:16256–16261

Rojek A, Praetorius J, Frøkjaer J, Nielsen S, Fenton RA (2008) A current view of the mammalian aquaglyceroporins. Annu Rev Physiol 70:301–327

Romero-Puertas MC, Rodriguez-Serrano M, Corpas FJ, Gomez M, del Rio LA, Sandalio LM (2004) Cadmium induced subcellular accumulation of O_2 and H_2O_2 in pea leaves. Plant Cell Environ 27:1122–1134

Sade N, Vinocur BJ, Diber A, Shatil A, Ronen G, Nissan H, Wallach R, Karchi H, Moshelion M (2009) Improving plant stress tolerance and yieldproduction: is the tonoplast aquaporin SlTIP2;2 a key to isohydric to anisohydric conversion? New Phytol 18:651–661

Sakurai J, Ishikawa F, Yamaguchi T, Uemura M, Maeshima M (2005) Identification of 33 rice aquaporin genes and analysis of their expression and function. Plant Cell Physiol 46:1568–1577

Sakurai J, Ahamed A, Murai M, Maeshima M, Uemura M (2008) Tissue and cell-specific localization of rice aquaporins and their water transport activities. Plant Cell Physiol 49:30–39

Santoni V, Vinh J, Pflieger D, Sommerer N, Maurel C (2003) A proteomic study reveals novel insights into the diversity of aquaporin forms expressed in the plasma membrane of plant roots. Biochem J 373:289–926

Santoni V, Verdoucq L, Sommerer N, Vinh J, Pflieger D, Maurel C (2006) Methylation of aquaporins in plant plasma membrane. Biochem J 400:189–197

Sarda X, Tousch D, Ferrare K, Legrand E, Dupuis JM, Casse-Delbart F, Lamaze T (1997) Two TIP-like genes encoding aquaporins are expressed in sunflower guard cells. Plant J 12:1103–1111

Sato-Nara K, Demura T, Fukuda H (2004) Expression of photosynthesis-related genes and their regulation by light during somatic embryogenesis in *Daucus carota*. Planta 219:23–31

Savage DF, Stroud RM (2007) Structural basis of aquaporin inhibition by mercury. J Mol Biol 368:607–617

Savage DF, Egea PF, Robles-Colmenares Y, O'Connell JD, Stroud RM (2003) Architecture and selectivity in aquaporins: 2.5 a X-ray structure of aquaporin Z. PLoS Biol 1:E72

Schenk AD, Werten PJ, Scheuring S, de Groot BL, Müller SA, Stahlberg H, Philippsen A, Engel A (2005) The 4.5 A structure of human AQP2. J Mol Biol 350:278–289

Scheuring S, Buzhynskyy N, Jaroslawski S, Gonçalves RP, Hite RK, Walz T (2007) Structural models of the supramolecularorganization of AQP0 and connexons in junctional microdomains. J Struct Biol 160:385–394

Schüssler MD, Alexandersson E, Bienert GP, Kichey T, Laursen KH, Johanson U, Kjellbom P, Schjoerring JK, Jahn TP (2008) The effects of the loss of TIP1;1 and TIP1;2 aquaporins in *Arabidopsis thaliana*. Plant J 56:756–767

Secchi F, Lovisolo C, Uehlein N, Kaldenhoff R, Schubert A (2007) Isolation and functional characterization of three aquaporins from olive (*Olea europaea L.*). Planta 225:381–392

Shimaoka T, Ohnishi M, Sazuka T, Mitsuhashi N, Hara-Nishimura I, Shimazaki K, Maeshima M, Yokota A, Tomizawa K, Mimura T (2004) Isolation of intact vacuoles and proteomic analysis of tonoplast from suspension-cultured cells of *Arabidopsis thaliana*. Plant Cell Physiol 45:672–683

Siefritz F, Biela A, Eckert M, Otto B, Uehlein N, Kaldenhoff R (2001) The tobacco plasma membrane aquaporin NtAQP1. J Exp Bot 52:1953–1957

Siefritz F, Otto B, Bienert GP, van der Krol A, Kaldenhoff R (2004) The plasma membrane aquaporin NtAQP1 is a key component of the leaf unfolding mechanism in tobacco. Plant J 37:147–155

Smirnoff N (2005) Antioxidants and reactive oxygen species in plants. Blackwell, Oxford

Soria LR, Fanelli E, Altamura N, Svelto M, Marinelli RA, Calamita G (2010) Aquaporin-8-facilitated mitochondrial ammonia transport. Biochem Biophys Res Commun 393(2):217–221

Stein WD, Danielli JF (1956) Structure and function in red cell permeability. Discuss Faraday Soc 21:238–251

Sui H, Han BG, Lee JK, Walian P, Jap BK (2001) Structural basis of water-specific transport through the AQP1 water channel. Nature 414:872–878

Sutka M, Alleva K, Parisi M, Amodeo G (2005) Tonoplast vesicles of *Beta vulgaris* storage root show functional aquaporins regulated by protons. Biol Cell 97:837–846

Tajima M, Crane JM, Verkman AS (2010) Aquaporin-4 associations and array dynamics probed by photobleaching and single-molecule analysis of GFP-AQP4 chimeras. J Biol Chem 285:8163–8170

Tajkhorshid E, Nollert P, Jensen MØ, Miercke LJ, O'Connell J, Stroud RM, Schulten K (2002) Control of the selectivity of the aquaporin water channel family by global orientational tuning. Science 296:525–530

Takano J, Noguchi K, Yasumori M, Kobayashi M, Gajdos Z, Miwa K, Hayashi H, Yoneyama T, Fujiwara T (2002) Arabidopsis boron transporter for xylem loading. Nature 420:337–340

Takano J, Wada M, Ludewig U, Schaaf G, von Wirén N, Fujiwara T (2006) The Arabidopsis major intrinsic protein NIP5;1 is essential for efficient boron uptake and plant development under boron limitation. Plant Cell 18:1498–1509

Takano J, Miwa K, Fujiwara T (2008) Boron transport mechanisms: collaboration of channels and transporters. Trends Plant Sci 13:451–457

Tamás MJ, Luyten K, Sutherland FC, Hernandez A, Albertyn J, Valadi H, Li H, Prior BA, Kilian SG, Ramos J, Gustafsson L, Thevelein JM, Hohmann S (1999) Fps1p controls the accumulation and release of the compatible solute glycerol in yeast osmoregulation. Mol Microbiol 31:1087–1104

Tanaka M, Wallace IS, Takano J, Roberts DM, Fujiwara T (2008) NIP6;1 is a boric acid channel for preferential transport of boron to growing shoot tissues in Arabidopsis. Plant Cell 20:2860–2875

Temmei Y, Uchida S, Hoshino D, Kanzawa N, Kuwahara M, Sasaki S, Tsuchiya T (2005) Water channel activities of *Mimosa pudica* plasma membrane intrinsic proteins are regulated by direct interaction and phosphorylation. FEBS Lett 579:4417–4422

Terashima I, Ono K (2002) Effects of HgCl(2) on CO(2) dependence of leaf photosynthesis: evidence indicating involvement of aquaporins in CO(2) diffusion across the plasma membrane. Plant Cell Physiol 43:70–78

Thorsen M, Di Y, Tängemo C, Morillas M, Ahmadpour D, Van der Does C, Wagner A, Johansson E, Boman J, Posas F, Wysocki R, Tamás MJ (2006) The MAPK Hog1p modulates Fps1p-dependent arsenite uptake and tolerance in yeast. Mol Biol Cell 17:4400–4410

Törnroth-Horsefield S, Wang Y, Hedfalk K, Johanson U, Karlsson M, Tajkhorshid E, Neutze R, Kjellbom P (2006) Structural mechanism of plant aquaporin gating. Nature 439:688–694

Tournaire-Roux C, Sutka M, Javot H, Gout E, Gerbeau P, Luu DT, Bligny R, Maurel C (2003) Cytosolic pH regulates root water transport during anoxic stress through gating of aquaporins. Nature 425:393–397

Tsukaguchi H, Shayakul C, Berger UV, Mackenzie B, Devidas S, Guggino WB, van Hoek AN, Hediger MA (1998) Molecular characterization of a broad selectivity neutral solute channel. J Biol Chem 273:24737–24743

Uehlein N, Lovisolo C, Siefritz F, Kaldenhoff R (2003) The tobacco aquaporin NtAQP1 is a membrane CO_2 pore with physiological functions. Nature 425:734–737

Uehlein N, Fileschi K, Eckert M, Bienert GP, Bertl A, Kaldenhoff R (2007) Arbuscular mycorrhizal symbiosis and plant aquaporin expression. Phytochemistry 68:122–129

Uehlein N, Otto B, Hanson DT, Fischer M, McDowell N, Kaldenhoff R (2008) Function of Nicotiana tabacum aquaporins as chloroplast gas pores challenges the concept of membrane CO_2 permeability. Plant Cell 20:648–657

Van Wilder V, Miecielica U, Degand H, Derua R, Waelkens E, Chaumont F (2008) Maize plasma membrane aquaporins belonging to the PIP1 and PIP2 subgroups are *in vivo* phosphorylated. Plant Cell Physiol 49:1364–1377

Vandeleur RK, Mayo G, Shelden MC, Gilliham M, Kaiser BN, Tyerman SD (2009) The role of plasma membrane intrinsic protein aquaporins in water transport through roots: diurnal and drought stress responses reveal different strategies between isohydric and anisohydric cultivars of grapevine. Plant Physiol 149:445–460

Vera-Estrella R, Barkla BJ, Bohnert HJ, Pantoja O (2004) Novel regulation of aquaporins during osmotic stress. Plant Physiol 135:2318–2329

Verkman AS (2009) Knock-out models reveal new aquaporin functions. Handb Exp Pharmacol 190:359–381

Viadiu H, Gonen T, Walz T (2007) Projection map of aquaporin-9 at 7 A resolution. J Mol Biol 367:80–88

Wallace IS, Roberts DM (2004) Homology modeling of representative subfamilies of Arabidopsis major intrinsic proteins. Classification based on the aromatic/arginine selectivity filter. Plant Physiol 135:1059–1068

Wallace IS, Roberts DM (2005) Distinct transport selectivity of two structural subclasses of the nodulin-like intrinsic protein family of plant aquaglyceroporin channels. Biochemistry 44:16826–16834

Wan X, Steudle E, Hartung W (2004) Gating of water channels (aquaporins) in cortical cells of young corn roots by mechanical stimuli (pressure pulses): effects of ABA and of $HgCl_2$. J Exp Bot 55:411–422

Wang Y, Tajkhorshid E (2010) Nitric oxide conduction by the brain aquaporin AQP4. Proteins 78:661–670

Weig AR, Jakob C (2000) Functional identification of the glycerol permease activity of Arabidopsis thaliana NLM1 and NLM2 proteins by heterologous expression in *Saccharomyces cerevisiae*. FEBS Lett 481:293–298

Whiteman SA, Nühse TS, Ashford DA, Sanders D, Maathuis FJ (2008) A proteomic and phosphoproteomic analysis of *Oryza sativa* plasma membrane and vacuolar membrane. Plant J 56:146–156

Wi SG, Chung BY, Kim JH, Lee KS, Kim JS (2006) Deposition pattern of hydrogen peroxide in the leaf sheaths of rice under salt stress. Biol Plant 50:469–472

Wienkoop S, Saalbach G (2003) Proteome analysis. Novel proteins identified at the peribacteroid membrane from Lotus japonicus root nodules. Plant Physiol 131:1080–1090

Xu XY, McGrath SP, Meharg AA, Zhao FJ (2008) Growing rice aerobically markedly decreases arsenic accumulation. Environ Sci Technol 42:5574–5579

Yamaji N, Mitatni N, Ma JF (2008) A transporter regulating silicon distribution in rice shoots. Plant Cell 20:1381–1389

Yang HC, Cheng J, Finan TM, Rosen BP, Bhattacharjee H (2005) Novel pathway for arsenic detoxification in the legume symbiont *Sinorhizobium meliloti*. J Bacteriol 187:6991–6997

Yasui M, Hazama A, Kwon TH, Nielsen S, Guggino WB, Agre P (1999) Rapid gating and anion permeability of an intracellular aquaporin. Nature 402:184–187

Ye Q, Steudle E (2006) Oxidative gating of water channels (aquaporins) in corn roots. Plant Cell Environ 29:459–470

Ye Q, Wiera B, Steudle E (2004) A cohesion/tension mechanism explains the gating of water channels (aquaporins) in Chara internodes by high concentration. J Exp Bot 55:449–461

Yu X, Peng YH, Zhang MH, Shao YJ, Su WA, Tang ZC (2006) Water relations and an expression analysis of plasma membrane intrinsic proteins in sensitive and tolerant rice during chilling and recovery. Cell Res 16:599–608

Zardoya R (2005) Phylogeny and evolution of the major intrinsic protein family. Biol Cell 97:397–414

Zelazny E, Borst JW, Muylaert M, Batoko H, Hemminga MA, Chaumont F (2007) FRET imaging in living maize cells reveals that plasma membrane aquaporins interact to regulate their subcellular localization. Proc Natl Acad Sci USA 104:12359–12364

Zelenina M, Bondar AA, Zelenin S, Aperia A (2003) Nickel and extracellular acidification inhibit the water permeability of human aquaporin-3 in lung epithelial cells. J Biol Chem 278:30037–30043

Zeuthen T, Klaerke DA (1999) Transport of water and glycerol in aquaporin 3 is gated by H(+). J Biol Chem 274:21631–21636

Zhao FJ, Ago Y, Mitani N, Li RY, Su YH, Yamaji N, McGrath SP, Ma JF (2010) The role of the rice aquaporin Lsi1 in arsenite efflux from roots. New Phytol 186(2):392–399

Part II
Signaling Related to Ion Transport

Plant Proton Pumps: Regulatory Circuits Involving H^+-ATPase and H^+-PPase

A.T. Fuglsang, J. Paez-Valencia, and R.A. Gaxiola

Abstract Proton gradients are crucial for the transport of ions and solutes across the different membranes in plant cells. Several important developmental processes require a tightly controlled proton gradient across cellular membranes. This chapter focuses on two of the three primary proton transport proteins: the plasma membrane H^+-ATPase and the H^+-PPase.

This chapter is divided into two sections. The first section describes the state of plasma membrane H^+-ATPase research,with emphasis on the regulation by physiological stimuli, and proposes a novel mechanism of H^+-ATPase regulation. The second section focuses on the H^+-PPase and new evidence consistent with the involvement of H^+-PPases in plant growth and development. A hypothetical model is discussed.

1 P-Type H^+-ATPases

P-type H^+-ATPases are active transporters that utilize ATP as an energy source to transport H^+ across the plasma membrane. This, in turn, creates an electrochemical gradient that energizes channels and co-transporters (Duby and Boutry 2009; Gaxiola et al. 2007; Sondergaard et al. 2004). The plasma membrane H^+-ATPases belong to a large family of pumps, P-type ATPases, all of which are energized by ATP and form a phosphorylated aspartyl intermediate during the reaction cycle, therefore the name P-type. The P-type ATPase family is further divided

A.T. Fuglsang (✉)
Faculty of Life Sciences, University of Copenhagen, Thorvaldsensvej 40, 1871 Frederiksberg C, Denmark
e-mail: atf@life.ku.dk
J. Paez-Valencia and R.A. Gaxiola
ASU, School of Life Sciences, 874501,Tempe 85287, AZ, USA
e-mail: jpaezval@asu.edu, roberto.gaxiola@asu.edu

M. Geisler and K. Venema (eds.), *Transporters and Pumps in Plant Signaling*, Signaling and Communication in Plants 7,
DOI 10.1007/978-3-642-14369-4_2,

phylogenetically into subgroups dependent on their substrate specificity (Axelsen and Palmgren 1998). In this book you will also find information about some of the other P-type ATPase subgroups, namely calcium and lipid pumps.

The plasma membrane H^+-ATPase is a single subunit protein at approximately 950 amino acid residues. The protein contains ten trans-membrane helices and a large cytoplasmic domain. The cytoplasmic domain consists of four domains: the nucleotide binding domain (N-domain), the phosphorylation domain (P-domain), the actuator domain (A-domain) and the regulatory domain (R-domain). The N domain is a built-in protein kinase that phosphorylates the conserved aspartate residue on the P domain. The A domain is an intrinsic protein phosphatase that dephosphorylates the same aspartyl group. In 2007 a crystal structure at 3.6 Å of the *Arabidopsis thaliana* isoform 2 (*AHA2*) was published (Pedersen et al. 2007). The crystal structure includes the A, P, and N domains, but does not provide information on the structure of the R domain.

The R domain consists of the C-terminal of the protein and includes approximately 100 amino acids. Further, about 20 N-terminal residues might contribute to the function of the R-domain (Ekberg et al. 2010) (see below). The role of the regulatory domain was first recognized when it was observed that the removal of the C-terminal by trypsination results in an activated form of the H^+-ATPase (Palmgren et al. 1990). The regulatory domain forms an auto-inhibitory domain by binding to the large cytoplasmic domain and thereby inactivating the H^+-ATPase. By a systematic mutagenesis approach two regions have been pinpointed as important for the intramolecular interaction (Axelsen et al. 1999). These domains are called Region I and Region II (RI and RII, respectively). Mutations within these two regions cause a constitutive active pump, most likely because they affect the interaction with the intramolecular receptor for the C-terminal domain. In the inhibited state the protein is thought to exhibit a closed compact structure in contrast to the activated state where the C-terminal is released from the core part of the protein.

Recently it has been demonstrated that the N-terminal end is directly involved in controlling the pump activity state, and that N-terminal displacements are coupled to secondary modifications taking place at the C-terminal end (Ekberg et al. 2010). This suggests an intricate mechanism of *cis*-regulation with both termini of the protein communicating to obtain the necessary control of the enzyme activity state.

1.1 Arabidopsis *Encodes 11 Members of H^+-ATPases*

The *Arabidopsis* genome encodes 11 genes of plasma membrane H^+-ATPases named *AHA1–11* for Arabidopsis $H^{\pm}$-ATPase isoform number 1–11. The plasma membrane H^+-ATPase is essential for the plant cell and the large number of genes reflects the expression of different isoforms in different cell types and organs. *AHA1* and *AHA2* are the most abundant isoforms expressed all over the plant, with AHA1 mainly in the leaves and AHA2 mainly in the roots (data obtained from Genevestigator). Reverse genetics have only revealed limited information about the physiological role of H^+-ATPases most likely because the different isoforms can functionally substitute for each other. Very recently a study of the two major isoforms were published. *aha1* or *aha2* single knock out plants or plants with reduced *AHA1* and *AHA2* transcript does not possess any detectable phenotypes but the *aha1/aha2* double knock out is lethal (Haruta et al. 2010). Analysis of expression patterns based on available micro array data shows that most of the H^+-ATPase isoforms are expressed at a relative constant level and expression level does not change when a related isoform is deleted or reduced as found in the *aha1* and *aha2* plants (Haruta et al. 2010). Interestingly it was found that in the plants with reduced levels of either *AHA1* or *AHA2*, the remaining plasma membrane H^+-ATPase isoforms had a higher degree of phosphorylation of the pen-ultimate Thr residue. This indicates that most regulation of the enzyme activity occurs at the post-translational level.

1.2 Mechanism of Activation by 14-3-3 Proteins

The plasma membrane H^+ pump is subject to regulation by a number of proteins interacting directly with the pumps. The first proteins found to interact with the pump were 14-3-3 regulatory proteins. 14-3-3 proteins belong to a highly conserved protein family that typically bind to phosphorylated target proteins and regulate signaling in eukaryotic cells (Oecking and Jaspert 2009).

14-3-3 proteins bind to the C-terminal regulatory domain of the H^+ pump. Binding of 14-3-3 proteins to the H^+ pump is dependent on the phosphorylation of the penultimate Thr residue (Fuglsang et al. 1999, 2003; Olsson et al. 1998; Svennelid et al. 1999). The phosphorylation site within the very C-terminal end of the H^+-ATPase is an uncommon protein kinase recognition site H/S-Y-T-V. However, a number of similar 14-3-3 binding sites are now identified in other proteins and named mode III (Coblitz et al. 2005). In this study, they demonstrate that binding of 14-3-3 protein to the C-terminal end of several membrane proteins is required for their targeting to the plasma membrane (Coblitz et al. 2005; Shikano et al. 2005). However, earlier studies of AHA2 expressed in yeast did not show a role for 14-3-3 proteins in targeting this pump to the plasma membrane (Jahn et al. 2002), but a reinvestigation of this mechanism might reveal new information.

The fungal phytotoxin fusicoccin (FC) is a commonly used tool in the study of H^+-ATPase activity. FC stimulates H^+ pumping by locking the preformed complex of 14-3-3 proteins and H^+- ATPase in a nearly irreversible manner (Fuglsang et al. 2003). There are no reports if FC on its own can stimulate protein kinases. The penultimate Thr residue is phosphorylated in response to different physiological stimuli and this phosphorylation seems to be to major regulatory mechanism of the H^+ pump since other phosphorylations seem to regulate the 14-3-3 binding. Despite a huge effort in several laboratories the protein kinase responsible for phosphorylation of this particular Thr residue has not yet been identified.

1.3 Phosphoproteomic Studies of Plasma Membrane H^+-ATPases

Phosphoproteomic studies have revealed several phosphosites within the C terminus of several isoforms of the H^+-ATPases. Some are found responding to stimuli others are found in systematic analysis of plasma membrane fractions (Nuhse et al. 2003, 2007; Whiteman et al. 2008). Phosphorylated sites are indicated on the H^+ pump drawn in Fig. 1. The fact that the C-terminal is subjected to such a large number of diverse phosphorylation events suggests a complex mechanism of regulation involving a number of different protein kinases. Only a few of the sites have an assigned physiological role(s), one is the penultimate Thr-947 residue and the other is the Ser-931 residue, as discussed below. One method to link physiological stimuli with specific phosphorylation sites was made in a quantitative phosphoproteomic study (Niittyla et al. 2007). In this study *Arabidopsis* seedlings were grown in hydroponics and the composition of the media could thereby be tightly controlled. By growing the seedling in the dark, followed by the addition of sucrose, the response to sucrose starvation/addition could be monitored. Phospho-peptides were purified and characterized at different time points after sucrose addition and thereby changes in specific phospho-peptides were measured. One phospho-residue in the H^+-ATPase was found to change as response to the sucrose depletion/addition regime namely the Thr947 residue described in relation to 14-3-3 binding, again underlining the importance of this regulatory mechanism.

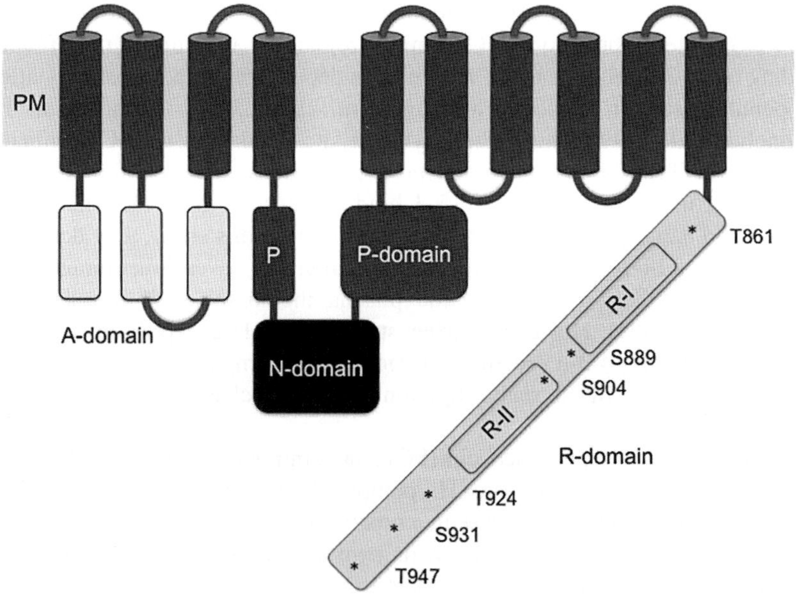

Fig. 1 Schematic drawing of the *A. thaliana* H^+-ATPase isoform 2 (AHA2). The nucleotide binding domain (*N-domain*), the phosphorylation domain (*P-domain*), the actuator domain (*A-domain*) and the regulatory domain (*R-domain*) is indicated with different coloring. In the R-domain the two regions involved in the auto inhibitory regulation is indicated (*R-I* and *R-II*). Phospho-sites identified *in planta* are marked in the R-domain. Numbering is according to amino acid residues in *AHA2*

1.4 *Controlling the Size of the Stomatal Pore*

1.4.1 Opening of Guard Cells

Pairs of guard cells form stomatal pores and regulate gas exchange between plant cells and the surrounding atmosphere. Light (primarily blue) stimulates stomata opening by activating the plasma membrane H$^+$-ATPase (Kinoshita et al. 2003; Kinoshita and Shimazaki 1999). Briefly, blue light induces rapid and highly sensitive stomata opening correlated with the phosphorylation of a plasma membrane H$^+$-ATPase pump and increased H$^+$ pumping, which results in the activation of voltage-gated K$^+$ channels by membrane hyperpolarization (reviewed by (Shimazaki et al. 2007)) along with the inhibition of S-type anion channels. H$^+$-ATPases are phosphorylated upon blue light treatment leading to the binding of regulatory 14-3-3 proteins to the C-terminal end of the H$^+$ pump (Fig. 2).

Receptors of blue light are phototropins (*PHOT1* and *PHOT2*). Phototropin contains in addition to the light sensing LOV domain(s) a serine/threonine protein kinase domain. In the presence of blue light, it is stimulated and autophosphorylated, resulting in the binding of 14-3-3 proteins. Thus, one consequence of phototropin autophosphorylation is 14-3-3 binding to the PHOT protein. One might speculate that the 14-3-3 proteins may be responsible for the transmission

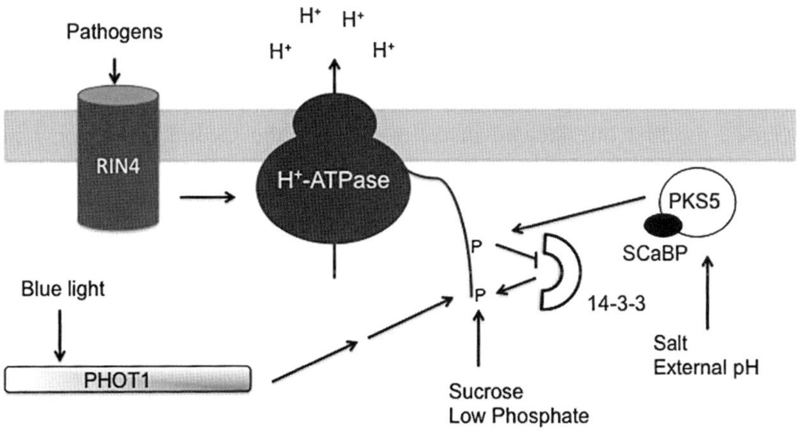

Fig. 2 A model cell showing the regulation of the plasma membrane H$^+$-ATPase by regulatory proteins. The H$^+$-ATPase is activated upon binding of 14-3-3 proteins to the phosphorylated penultimate Thr in the R-domain. This phosphorylation is responding to different stimuli. The *left* side of the figure illustrates processes taking place in the guard cells resulting in the opening of the stomatal pore. Here the blue light receptor, PHOT1, initiates an activation cascade activating the H$^+$-ATPase. Pathogens utilize the RIN4 protein to activate the H$^+$-ATPase through a direct interaction. The *right* side of the figure illustrates processes taking places in other parts of the plant (including roots). The protein kinase PKS5 is regulated by a combination of salt and high pH and targets a Ser residue upstream of the 14-3-3 binding site. Phosphorylation of this Ser residue prevents 14-3-3 binding

of the signal by facilitating a direct link between phototropins and guard cell H^+-ATPase (Kinoshita et al. 2003; Sullivan et al. 2009). An obvious thought was that the phototropins themselves phosphorylate the H^+-ATPases but several lines of research have demonstrated that other components are implicated in this activation. In *phot1/phot2* double mutants phosphorylation of the plasma membrane H^+-ATPase can be observed upon FC treatment indicating that other protein kinases are involved downstream of the phototropins (Ueno et al. 2005). Very recently the protein phosphatase PP2A was shown to deactivate Phot2 (Tseng and Briggs 2010). PP2A is also implicated in the regulation of H^+-ATPase (Fuglsang et al. 2006) and this suggests a coordinated regulation point in the blue light response.

In addition to activating the H^+ pumps the phototropins also mediate the inhibition of the plasma membrane anion channels (Marten et al. 2007) and as is the case for the activation of the H^+ pump, other players downstream have not yet been identified.

1.4.2 Closure of Guard Cells

Closure of stomata occurs as response to light to dark transition, high CO_2 levels and the hormone abscisic acid (ABA). Studies of mutants with ABA insensitive stomata have revealed that the plasma membrane H^+-ATPases in guard cells are important for the ABA induced closure. One of the mutant loci identified (*Open stomata 2*, *OST2*) were caused by a mutation leading to a constitutive activated form of the *A. thaliana* H^+-ATPase isoform 1 (*AHA1*) (Merlot et al. 2007). Two different *OST2* alleles were identified, one contained two missense mutations $L_{169}F$ and $G_{867}L$, the first one located in the A-domain, the second in Region I in the regulatory domain. The latter easily explains the constitutive activated form. The other allele contained a mutation $P_{68}S$ in the first transmembrane domain. By expression in yeast both alleles were shown to result in activated forms of AHA1.

The open stomata phenotype in *OST2* implies that the R- and S-channel currents are not sufficient to sustain plasma membrane depolarization to close stomata without curtailing the proton pump activity (Merlot et al. 2007). The mechanisms that would inactivate the pumps under normal circumstances are not known, but it has previously been suggested that the protein phosphatases type 2C are involved (Roelfsema et al. 1998).

1.4.3 Pathogens Modulate the H^+ Pumps to Invade Plants Through the Stomatal Pore

Stomata do not only open and close in order to exchange gases, they also form a gateway for pathogens to enter the interior of the leaf. When receptors at the cell surface recognize pathogens, one defense response is, therefore, to close the stomatal pore to prevent bacteria from entering the leaf interior (Melotto et al.

2006). Pathogenic bacteria have evolved strategies to suppress the closure of stomata. One example is the RIN4 protein known to negatively regulate plant response to pathogens (PAMP-triggered immunity, PTI) (Kim et al. 2005; Mackey et al. 2002). Recently it was demonstrated that plants use RIN 4 to regulate H^+-ATPase activity during immune responses, thereby controlling stomatal apertures during pathogen attack. RIN4 binds to the plasma membrane H^+-ATPase (AHA1/AHA2) resulting in activation of the pump (Liu et al. 2009a) (Fig. 2). In resistant plant genotypes the interaction between RIN4 and the H^+-ATPase is prevented, presumable by post-translational modification of the RIN4 protein. RIN4 homologs are found in many plant species and although this mechanism might very well be a general mechanism used by pathogens, it is still unknown if RIN4 plays a role in the opening and closure of stomata under noninfected conditions or if RIN4's role is solely regulated as response to pathogens.

1.5 PKS5: A Protein Kinase Preventing Binding of 14-3-3 Protein

The first protein kinase found as regulator of H^+-ATPase was PKS5. PKS5 belongs to a family of calcium regulated Serine/Threonine protein kinases (PKS/CIPK11) containing 25 members in *Arabidopsis* (Guo et al. 2001; Kolukisaoglu et al. 2004; Kudla et al. 2010). Another member of this protein kinase family SOS2/CIPK24 phosphorylates the Na^+/H^+ antiporter (SOS1) upon salt stress (Qiu et al. 2002), CIPK23 phosphorylates the K^+ channel AKT1 (Laloi et al. 2007; Li et al. 2006; Xu et al. 2006). This indicates a specialized role for this family of protein kinases towards regulation of ion-transporters by phosphorylation.

PKS5 phosphorylates Ser931 positioned between autoinhibitory Region II and the 14-3-3 protein binding site in the C-terminal end of AHA2. Phosphorylation of Ser931 prevents 14-3-3 binding even though the penultimata threonine residue (Thr947) is phosphorylated (Fig. 2). This finding added a second layer to the regulation of the H^+-ATPases in that inactivation of pump activity can occur both by dephosphorylation of Thr947 and by phosphorylation of Ser931. In both cases, 14-3-3 binding is prevented. The mechanism of regulating the binding of 14-3-3 protein was previously also identified in *Nicotiana tabacum* (Duby et al. 2009).

pks5 seedlings exhibit a pH tolerant phenotype tolerating pH in the media up to pH 8.5. The seedlings can adjust pH in the rhizosphere faster than wildtype plants by activating the H^+-ATPase. Growth at pH 8.5 is not physiologically relevant and probably a secondary effect of the *pks5* mutation, the real role of PKS5 is not yet fully understood.

1.5.1 ScaBP1: A Calcium Binding Protein Modulating PKS5 Action

As found for other members of the PKS/CIPK family also, PKS5 interacts with a Ca^{2+} binding protein from the CBL/ScaBP family. By the use of yeast two-hybrid assays, PKS5 was found to interact with the Ca^{2+} binding protein SCaBP1/CBL2 but not with any other member of the SCaBP/CBL family (Fuglsang et al. 2007). CBL2 has later been located to the tonoplast of the plant cell raising a question about the actual mechanism of interaction between PKS5 and CBL2. The function of the SCaBP/CBL proteins in relation to the PKS/CIPK kinases is still under debate. In some cases the SCaBP/CBL protein seems to regulate the activity of the PKS/CIPK kinase (Tominaga et al. 2010) in other cases the SCaBP/CBL protein possess a role in the recruitment of the protein kinase to the plasma membrane and thereby the phosphorylation target as demonstrated by the *sos1* recruitment system (SRS) for SOS1 (Quintero et al. 2002). Also it has been demonstrated that myristoylation targets CBL1 to the endoplasmic reticulum and that the following S-acylation is crucial for endoplasmic reticulum-to-plasma membrane trafficking (Batistic et al. 2008). When reconstituting the AHA2/PKS5/SCaBP1 signaling pathway in yeast is was found that SCaBP1 was required in order to observe a phenotype related to changed activity of the proton pump AHA2, on the other hand *in vitro* experiments demonstrated that recombinant PKS5 could phosphorylate recombinant AHA2 without the presence of SCaBP1(Fuglsang et al. 2007) (Fig. 2).

1.5.2 DnaJ: A Chaperone Like Protein Repressing PKS5 Activity

In order to understand the physiological role of PKS5 regulation a screen for interacting proteins was performed (Yang et al. 2010). Here a putative Co-chaperone DnaJ-like heat shock protein (AtJ3 homologue 3) was identified. This protein was shown to interact with PKS5 and repressing its protein kinase activity and thereby activating the H^+-ATPase. Environmental stresses often cause protein denaturation, therefore chaperones are key components helping to maintain proteins in their functional conformation during stress conditions. Knock out *atj3* seedlings did not show the same pH resistant phenotype as *pks5*-1 seedlings. Often alkaline conditions are associated with increased soil salinity and the effect of combined salt and high pH was therefore tested. Here atj3 seedlings demonstrated an increased sensitivity compared to wild type plants, at the same conditions *pks5*-1 plants were less sensitive than wild type. Further tests of *pks5/atj3* double mutants responded to salt at alkaline conditions in the same way as pks5 seedlings indicating that Atj3 functions upstream of PKS5. These data also suggest that PKS5 might be involved in regulation of H^+-ATPase activity in relation salt stress.

1.6 Nutrient Uptake and Responses to Changes in the Soil

A critical feature distinguishing plants from animals is that plants are sessile and thus have to cope with numerous environmental challenges. For example, plant roots are exposed to soil solutions that are constantly changing in pH as well as in the concentrations of mineral nutrients and toxic ions.

1.6.1 Response to Limited Phosphate

An example of regulation of the plasma membrane H^+-ATPase is found as response to limited amounts of phosphate. White lupin (*Lupinus albus* L.) can grow in soils with sparingly available phosphate (P) by producing specialized structures called cluster roots. To mobilize sparingly soluble P forms in soils, cluster roots release substantial amounts of carboxylates and concomitantly acidify the rhizosphere. It has been demonstrated that, citrate exudation increased transiently and reached a maximum after 5 h. This effect was accompanied by a strong acidification of the external medium and alkalinization of the cytosol. Fusicoccin stimulated citrate exudation, whereas vanadate, an inhibitor of the H^+-ATPase, reduced citrate exudation. The increase in proton secretion was due to both an increased transcription level of a H^+-ATPase gene as well as activating post-translational modifications of H^+-ATPase protein involving binding of activating 14-3-3 protein (Tomasi et al. 2009).

2 Plant H^+-PPases

Prototypical plant H^+-PPases (V-PPase EC 3.6.11) have an overall amino-acid sequence identity of 85% or greater and localize to the vacuolar, Golgi, and plasma membranes (Baltscheffsky et al. 1999; Cleland 1995; Drozdowicz et al. 2000; Jiang et al. 2001; Mitsuda et al. 2001a; Ratajczak et al. 1999). Plants have two phylogenetically distinct types of H^+-PPases: type I and type II. Type I H^+-PPases depend on cytosolic K^+ for their activity and are moderately sensitive to inhibition by Ca^{2+}, and type II H^+-PPases are K^+-insensitive but extremely Ca^{2+}-sensitive. Type I H^+-PPases have been shown to acidify the plant vacuole. The resulting H^+ and electrochemical gradient is instrumental for the storage of sucrose, organic acids, regulation of hydrostatic pressure through the storage of inorganic ions, and cytoplasmic detoxification (Maeshima 2001). H^+-PPases from various sources have been successfully purified and characterized as peptides ranging from 65 to 115 kDa (predicted) and 56–79 kDa (apparent) molecular weight. Variations in its predicted M*r* from the cDNA size and the apparent M*r* from PAGE are common to highly hydrophobic proteins and appear to be related to their extreme hydrofobicity and incomplete saturation by SDS (Maddy 1976). *A. thaliana* has one gene encoding for a type I H^+-PPase (AVP1) and another gene encoding for a type II H^+-PPase (AVP2) (Drozdowicz et al. 2000). Interestingly, *Arabidopsis* web sites report the existence of an *AVP1.2* gene product that results from an alternative splicing of the *AVP1* locus (http://www.arabidopsis.org/). However, there is currently no *in planta* evidence of its expression.

AVP1 is an extremely hydrophobic protein of 770 residues ($MW = 80{,}800$ Da) and its heterologous expression in yeast demonstrated that this polypeptide is sufficient for both H^+ pumping and PPi hydrolysis (Zhen et al. 1994). Although early models favored the presence of 13 transmembrane domains (TMD) for AVP1, later models suggest either 15 or 16 TMD (Maeshima 2000). Lines of evidence demonstrate that the catalytic pocket of H^+-PPases is facing the cytosolic side and is probably composed of 2–3 conserved segments. A fragment of DXXXXXXXKXE on cytosolic loop 3 (CL3) was suggested as the putative substrate-binding region (Nakanishi et al. 2001). NN′-dicyclohexylcarbomiide-binding residues (Glu-305 and Asp-283 on CL3 and Asp-504 on CL5) have been identified by the combination of site-directed mutagenesis and chemical modification as essential residues involved in enzymatic and proton translocating reaction of H^+-PPases (Kim et al. 1995; Yang et al. 1999; Zhen et al. 1997a). H^+-PPases require Mg^{2+} as a cofactor for the formation of the $MgPP_i$ complex and the resultant active conformation (Gordon-Weeks et al. 1996).

Today over 100 sequences from bacteria, archaea, and eukaryotes are available. Sequences alignments have revealed the existence of only two subfamilies of H^+-PPase described above. Type I family members are K^+-dependent and type II are K^+-independent enzymes (Belogurov and Lahti 2002; Drozdowicz et al. 2000). K^+-dependent H^+-PPases have been found in algae (Takeshige et al. 1988), protozoan, (Docampo et al. 2005) and higher plants (Sarafian et al. 1992). On the other hand, type II H^+-PPases exist in archaebacterium (Drozdowicz et al. 1999), photosynthetic bacterium (Au et al. 2006), fungus (Mimura et al. 2005), and *A. thaliana* (AVP2) (Drozdowicz et al. 2000). Intriguingly, members of the K^+-independent sub-family contain a Lys residue at the position equivalent to the residue 541 of AVP1. Furthermore, substitution of a neutral residue by Lys in the position of K^+-dependent H^+-PPase from *Carboxydothermus hydrogenoformans* confers K^+ independency (Belogurov and Lahti 2002). However, it has also been shown that other residues contribute to the K^+ binding site including G544 and various Cysteine (Cys) residues on the N terminus. In plants, the transmembranal domain 5 (TM5 residues 211–242) of the enzyme is highly conserved. A mutation in the motif GYG (residues from 229 to 231) ceased the cation effect on the H^+-PPase (Van et al. 2005). Alignment of amino-acid sequences demonstrated a relatively high degree of conservation of the C-terminal domains among H^+-PPases. Topological studies using yeast heterologous H^+-PPase expression suggest that both the C-termini and the N-termini face the lumen side and are opposite to the cytosolic catalytic domain that is cytosolic (Maeshima 2000).

Truncation of the C terminus induces dramatic decline in H^+-PPase enzymatic activity, H^+ translocation and coupling efficiency (Lin et al. 2005). In addition, deletion of the C terminus of the H^+-PPase increases its susceptibility to heat stress and apparent K^+ binding constant. Thus, it is likely that the C terminus plays and essential role in sustaining the physiological functions of H^+-PPase.

Unlike the vacuolar H^+/ion pumping ATPases that are large hetero-multimeric complexes, all the catalytic properties of H^+-PPases are imparted by a single polypeptide as demonstrated by the heterologous expression of H^+-PPases in yeast (Kim et al. 1994). However, H^+-PPases could work as homo-dimers or homo-multimeres as judged by native PAGE, cross-linking, and gel filtration data (Zhen et al. 1997b). Radiation inactivation analysis demonstrated that the proper dimeric structure of the H^+-PPase on tonoplast membranes is a prerequisite for both enzymatic activity and PP_i-supported H^+ translocation. One subunit of the dimeric complex is sufficient for PP_i hydrolysis but proton translocation requires the presence of both subunits (Tzeng et al. 1996). More recent atomic force microscopy (AFM) has been used to observe purified H^+-PPase reconstituted into planar lipid bilayer under physiological conditions. These results reveal a dimeric complex for the H^+-PPase where both the C termini of a dimeric subunit are on the same side of the membrane and are approximately 1.9–2 nm apart. In the proposed mechanistic model, the H^+ channel lies at the interface between the C termini of the H^+-PPase homodimer (Liu et al. 2009b). The high-resolution crystal structure of H^+-PPases is a pending assignment that will expedite the elucidation of the molecular mechanisms involved in the function and regulation of this primary H^+ pump.

2.1 *Vacuolar H^+-PPases in Fruits*

The vacuoles are organelles that fulfill highly specialized functions depending on tissue, cell type, and/or developmental stage. All vacuoles seem to contain vacuolar H^+-ATPases (V-ATPases) and H^+-PPases that differ in their function depending on the type of vacuole in which they reside (Martinoa et al. 2007). Generally, H^+-PPase activity is high in young tissues whereas V-ATPase activity is relatively constant during growth and maturation. In pear fruit the ratio of H^+-PPase to V-ATPase activity indicated that H^+-PPase is the major H^+-pump of young fruit vacuolar membranes. However, the contribution of the V-ATPase increases with time to become the major H^+-pump during the later stages of fruit development (Shiratake et al. 1997). Growing tissues and exponentially growing cells generate large amounts of pyrophosphate. It is tempting to speculate, that the H^+-PPase could be serving two purposes: the generation of the proton gradient required for vacuolar transport/expansion and the scavenging of PP_i to alleviate its well-documented inhibitory feedback effect. Generally the abundance and activity of the H^+-PPase is high in young tissues. However in some cases such as grape berries, the H^+-PPase is also the predominant vacuolar proton pump in mature cells. Grape berries are very acidic and it is intriguing that in tissues were vacuoles are highly acidic (pH $\geq$3) the H^+-PPase appears to be the predominant pump (Terrier et al. 1997). It has been suggested that the thermostability of the H^+-PPase could be the reason for its abundance in mature grape berries, since they are exposed to the sun and consequently reach high temperatures. In line of this hypothesis, the grape berry H^+-PPase is heat stable and exhibits a temperature optimum of -50°C (Martinoa et al. 2007).

In *Prunus persica* (peach), two different full-length clones of H^+-PPase have been isolated from fruit (*PPV1* and *PPV2*). The expression of *PPV1* is very low in contrast with the high expression of *PPV2* in the fruit. *PPV2* presents a clear biphasic pattern of expression during peach fruit development that correlates with the accumulation of citric or malic acid and maturation (Etienne et al. 2002). It has been suggested that *PPV2* is instrumental for both accumulation of organic acids and sugar storage. Other studies with grape berry revealed the existence of a H^+-PPase that is highly expressed during ripening and appeared to have a synergic behavior with the V-ATPase (Terrier et al. 2001).

2.2 Vacuolar H^+-PPase Is a Key Player for Plant Salt Tolerance

Vacuolar sodium sequestration is a conserved mechanism used by salt tolerant plant species. Overexpression of the type I H^+-PPase AVP1 in *Arabidopsis* resulted in plants with enhanced salt tolerance and drought resistance (Gaxiola et al. 2001). The salt tolerant phenotype of these plants was explained by an increased uptake of Na^+ into their vacuoles. The drought related phenotype was originally attributed to an enhanced vacuolar osmoregulatory capacity (Gaxiola et al. 2001). Since the arrival of this work other groups have subsequently demonstrated that overexpression of this and other plant genes encoding for a type I H^+ PPase can increase both salt- and drought-tolerance in heterologous systems including rice (Zhao et al. 2006), tobacco (Gao et al. 2006), cotton (Lv et al. 2008, 2009), alfalfa (Bao et al. 2008), maize (Li et al. 2008), and creeping bentgrass (Li et al. 2010). Interestingly, a study on the variation of salinity tolerance amongst *Arabidopsis* ecotypes reported a positive relationship between salt tolerance and the levels of AVP1 expression (Jha et al. 2010). Furthermore, an *Arabidopsis* mutant has been characterized in which energization of vacuolar transport solely relies on the activity of the H^+-PPase (Krebs et al. 2010). These lines remain salt tolerant and further confirm our supposition that AVP1 is important for salt tolerance (Gaxiola et al. 2001, 2007).

2.3 Vacuolar H^+-PPases in Maize Aleurone

Cereal endosperm is a model system for cell fate determination in plants. Cells at the outermost layer of the endosperm adopt an aleurone cell fate. An intriguing finding relates to the restricted expression of the maize *Vpp1* gene encoding for a H^+-PPase to the aleurone layer (Wisniewski and Rogowsky 2004). Its expression identifies it as an aleurone cell fate developmental marker, but its physiological role remains obscure. The aleurone layers are rich in lipids and accumulate hydrolytic enzymes in protein bodies during seed maturation. The aleurone cells are less vacuolated than the underlying starchy endosperm cells and they are the only cells in kernels that accumulate storage proteins in their vacuoles. Therefore, it has been suggested that *Vpp1* may play a role in the filling of these storage vacuoles also called aleurone bodies (Wisniewski and Rogowsky 2004). The protein storage vacuoles (PSV) contain three morphologically distinct regions: the matrix, the crystalloid and the globoid cavities (Weber and Newman 1980). Biochemical and structural characterization of protein storage vacuoles showed that the globoid cavity is defined as a unit membrane that is specifically marked by the presence of H^+-PPase and γ-TIP (Tonoplast Intrinsic Protein) (Jiang et al. 2001). The environment of the globoid cavity is optimal for the formation of phytic acid crystals. Since the globoid membrane contains a H^+-PPase and phytic acid molecules are known to carry pyrophosphate groups (Loewus and Murphy 2000), it has been suggested that the PSV could represent a plant functional equivalency to acidocalcisomes. Acidocalcisomes are the only organelles that have been conserved during evolution from prokaryotes to eukaryotes. Acidocalcisomes have been linked with several functions including the storage of cations and phosphorous, calcium homeostasis, and osmoregulation. Furthermore, its function is essential in the adaptation of parasites to environmental stress (Docampo et al. 2005). H^+-PPases are integral proteins of the acidocalcisomes of many different parasitic protozoa (*T. cruzi, T.brucei, Leishmania donovani, L. amazonensis, Phytomonas francai, Toxoplasma gondi, Plasmodium falciparum* and *Plamsodium berghei*) (Docampo et al. 2005). Considering aleurone cells are the only endosperm cells that maintain a metabolic activity at seed maturity and during its dispersal, it has been suggested that the H^+-PPase in the PSV could help seeds to face stress during germination.

2.4 Subcellular Localization of Plant H^+-PPases

In animal systems lacking H^+-PPases, the vacuolar H^+-pumping ATPases (V ATPases) acidify a wide array of intracellular compartments. In polarized cells such as osteoclast and renal tubular epithelial cells, V ATPases have been localized to the plasma membrane (PM) where they acidify discrete extra cellular compartments. V ATPases have been localized in virtually all the intracellular compartments except for the nucleus. As expected their function is very diverse (Zhou et al. 1999). The subcellular localization of the H^+-PPase in parasites is also versatile and not restricted to the acidocalcisomes. For example the H^+-PPase of *Trypanosome cruzi* has been shown to localize at the acidocalcisomes membrane, golgi apparatus (GA) and PM (Docampo et al. 2005). Furthermore, H^+-PPases can change intracellular localization during invasion of host cells (Drozdowicz et al. 2003). Plant type I H^+-PPases were first isolated from vacuoles and initially considered to be *bona fide* vacuolar markers (Maeshima 1991; Rea et al. 1992). However, density gradient centrifugation and phase partitioning of membrane fractions coupled with immunogold electron microscopy showed the presence of H^+-PPases in the plasma membrane of *Ricinus communis* seedlings and cauliflower inflorescence (Cleland 1995; Ratajczak et al. 1999). Furthermore, proteomic studies confirmed the plasma membrane localization of the *A. thaliana* H^+-PPase AVP1 (Alexandersson et al. 2004).

The *Arabidopsis* type II H^+-PPase encoded by the *AVP2/AVP1L* locus has been shown to localize exclusively to the GA (Mitsuda et al. 2001a). The expression of type II H^+-PPase has been documented in young seedlings, cotyledons, rosette-trichomes, sepals and stamen filaments. It has been suggested that the type II H^+-PPase may be required during cell expansion (Mitsuda et al. 2001a). It is likely that the type II H^+-PPase may aid the GA resident V-ATPase in the generation of the acidic environment under conditions where ATP availability is compromised. The existence of a GA H^+-PPase in roots of maize has been demonstrated by immunoelectron microscopy (Oberbeck et al. 1994). Recently, a cDNA encoding a putative AVP2-like maize H^+-PPase gene has been cloned using an elegant suppression subtractive hybridization (SSH) approach (Yue et al. 2008). This gene, named *ZmGPP,* is a good candidate for the GA resident H^+-PPase of *Zea mays*. *ZmGPP* is constitutively expressed in leaves, stems, roots, tassels and ears under normal growth conditions. Interestingly, the expression of *ZmGPP* is up-regulated in both shoots and roots of maize seedlings under dehydration, cold, and higher salt stress: suggesting *ZmGPP* could play an important role in abiotic stress tolerance in *Z. mays* (Yue et al. 2008).

2.5 Are There Other H^+-PPases in Plants?

For years, investigators have been interested in finding out whether endomembrane systems other than the vacuole and GA contain H^+-PPases. Vianello et al. (1991) reported a pyrophosphate dependent H^+ pumping activity from pea stem submitochondrial particles (Vianello et al. 1991). The mass of putative mitochondrial H^+-PPase was shown to be smaller than the vacuolar H^+PPase indicating the possible existence of a new type of H^+PPase. However, another study with radiation inactivation analysis showed that submitochondrial particles from etiolated mung bean seedlings contained a H^+-PPase with an estimated functional size of 170 kDa (Jiang et al. 2000). The discrepancy remains unsolved and to our knowledge no further information has been published. There is an isolated report about the existence of a H^+-PPase present on a endoplasmic reticulum-enriched vesicle fraction from etiolated mung bean seedlings. Antiserum prepared against the vacuolar H^+-PPase did not inhibit the activity of this novel proton pyrophosphatase which excludes a possible contamination of the membrane preparation with tonoplast vesicles (Kuo et al. 2005). Here again, no further information is available.

2.6 Transcriptional Regulation of H^+-PPases

Transcriptional regulatory networks that drive organ specific and cell-specific patterns of gene expression and mediate interactions with the environment represent a fundamental aspect of plant cell signaling. The transcriptional regulation of gene expression in eukaryotes is mediated by the recruitment of transcription factors (TFs) to *cis* regulatory elements. Transcription factors interact with DNA elements, other TFs, and the basal machinery to regulate the expression of target genes. TF binding sites (or *cis* elements motifs) are the functional DNA elements that influence temporal and spatial transcriptional activity. Multiple *cis*-elements comprise *cis* regulatory modules (CRMs). CRMs integrate signals from multiples TFs that result in a combinatorial control and highly specific pattern of gene expression. Therefore, identifying and understanding the function of *cis* elements and their combinatorial role in CRMs is essential for elucidating the mechanisms by which cells perceive and correctly respond to their environment (Priest et al. 2009). The expression levels of the H^+-PPase are precisely controlled at the transcriptional level in response to various environmental conditions or developmental stages (Maeshima 2000). It has been shown that *cis*-acting regions regulate the expression of AVP1 in pollen. AtCAMTA5 and AtCAMTA 1 (calmoduline-binding TFs) were shown to bind to the pollen-specific *cis*-acting region of *AVP1* promoter (Mitsuda et al. 2003). In the same work, the authors suggested that AVP1 expression in pollen might be regulated via Ca^{2+} signaling (Mitsuda et al. 2003). The *cis*-acting region of the *AVP1* gene was used to identify two novel proteins, AtVOZ1 and AtVOZ2 (*A. thaliana* Vascular plant One Zinc finger protein). The expression of At VOZ1 is restricted to the phloem, while AtVOZ2 expression has been detected in roots, stipules, stamen filaments, and anthers (Mitsuda et al. 2004).

2.6.1 Sugar Starvation

Transient expression assays using a GUS-reporter under the control of a 1,413 bp fragment of the *AVP1* promoter showed that its expression is regulated in response to several energy related stresses (Mitsuda et al. 2001b). The up-regulation of *AVP1* in response to a reduction in light intensity is reminiscent of the behavior of genes involved in sugar starvation. Of note, sugar responsible *cis* elements (i.e., AMY, BOX1, 2 CGACG boxes) are present in the regulatory region of the *AVP1* promoter (Mitsuda et al. 2001b). Up-regulation of H^+-PPase genes has been reported in sucrose-starved cells of *Oryza sativa* (Wang et al. 2007).

Early work with rice seedlings documented anoxia-triggered up-regulation of the H^+-PPase (Carystinos et al. 1995). Furthermore, a recent study showed that among the six rice H^+-PPase genes (*OVP1–6*) only *OVP3* was specifically up-regulated under anoxia (Liu et al. 2010). When the production of ATP drops sharply under anoxia due to decreased oxidative phosphorylation, flood-tolerant species such as rice adapt by switching from the anaerobic respiration to anaerobic fermentation. This results in a cytosolic acidification and inhibition of ATP-dependent proton pump activity (Gibbs and Greenway 2003). Stitt speculated that H^+-PPases could provide the driving force for vacuolar transport during oxygen deficit conditions that limit ATP supply for the function of the vacuolar H^+-ATPase (Stitt 1998).

2.6.2 P_i Starvation

The existence of a complicated transcriptional regulation system involved in plant responses to P_i starvation is well documented (Franco-Zorrilla et al. 2004). A rice TF (OsPTF1) involved in the response to phosphate starvation has been reported (Yi et al. 2005). OPTF1 is expressed in phloem cells of the primary root, leaves, and lateral roots. Overexpression of OsPTF enhances rice tolerance to P_i starvation. Interestingly, microarray data on this OsPTF transgenic rice plants showed a concomitant enhanced expression of H^+-PPases (Yi et al. 2005). These data are consistent with results that showed up-regulation of H^+-PPase activity in *Brassica napus* cell suspensions under phosphate starvation (Palma et al. 2000). In *A. thaliana* P_i starvation triggers increases in transcript and protein abundance of both AVP1 and the plasma membrane H^+-ATPase. Furthermore, the overexpression of AVP1 in *Arabidopsis*, tomato, and rice improves growth under P_i limitation (Yang et al. 2007).

2.7 *Puzzling Phenotypes Triggered by Altering the Expression of H^+-PPases in Plants*

Li et al. reported that the overexpression of the H^+-PPase AVP1 in *Arabidopsis* results in increased cell division at the onset of organ formation, root, and shoot hyperplasia as well as increases in auxin transport. Furthermore, *avp1-1* null mutants display severely disrupted root and shoot development and reduced auxin transport. Intriguingly, changes in the expression of AVP1 affect the abundance and activity of the PM H^+-ATPase that correlate with apoplastic pH alterations and rhizosphere acidification (Li et al. 2005; Yang et al. 2007). Rhizosphere acidification is a central mechanism for plant mineral nutrition. Accordingly, it has been shown that AVP1 transgenic *Arabidopsis*, tomato and rice plants outperform controls when grown under phosphate limitations and accumulate higher contents of potassium under all conditions tested (Yang et al. 2007). Of note, up-regulation of either the *A. thaliana* or *Thellungiella halophila* type I H^+-PPases triggers enhanced growth/biomass and photosynthetic capacity in a variety of agriculturally important crops (Bao et al. 2008; Gaxiola et al. 2001; Li et al. 2008; Lv et al. 2008, 2009; Park et al. 2005; Yang et al. 2007) grown under normal or stressful conditions such as nutrient limitations, water scarcity, and salinity. As described earlier, the salt tolerant phenotypes triggered by the overexpression of the H^+-PPase are consistent with its residence at the tonoplast. However, a vacuolar restricted H^+-PPase complicates the explanation of phenotypes such as an enhanced abundance and activity of the PM H^+-ATPase with the concomitant acidification of apoplast and rhizosphere or an enhanced biomass and photosynthetic capacity. As described earlier, H^+-PPase has been localized to the PM, and its function here warrants more attention.

2.8 Could the H^+-PPase Affect Sucrose Phloem Loading?

2.8.1 PP_i Concentrations Are Essential for Sucrose Phloem Loading

Sonnewald and coworkers suggested that the cytosolic concentration of PP_i in the phloem was essential for sucrose transport (Sonnewald 1992). Lerchl and collaborators further tested this hypothesis via the phloem-specific expression of a soluble pyrophosphatase from *E. coli* (*ppa1*) in tobacco plants (Lerchl et al. 1995). Characterization of these *ppa1* plants revealed that removal of cytosolic PP_i from phloem cells triggered the accumulation of sucrose in source leaves, chlorophyll loss, and reduced shoot and root growth. Interestingly, phloem-specific expression of a yeast invertase (suc2) circumvented the metabolic block of the *ppa1* plants restoring wild type phenotypes (Lerchl et al. 1995). These data are consistent with a model where sucrose phloem loading depends on the levels of cytosolic PP_i in companion cells (Fig. 3). Sucrose must be actively transported from mesophyll cells to companion cells via a sucrose/H^+ symporter that depends on the proton gradient generated by the plasma membrane H^+-ATPase (Srivastava et al. 2008). In order to have an adequate ATP supply for the maintenance of this transmembrane proton gradient, a percentage of the incoming sucrose must be cleaved into fructose and UDP-glucose by sucrose synthase (Lerchl et al. 1995) and subsequently oxidized through the cellular respiration pathway. In this pathway, both the PP_i: fructose 6-phosphate 1-phosphotransferase (PFP) and the UDP-glucose pyrophosphorylase (UGPase) work near equilibrium, so a decrease in the cytosolic concentration of PP_i should prevent the reactions leading to glycolysis and therefore compromise the energy production (Lerchl et al. 1995).

2.8.2 H^+-PPase and H^+-ATPase Localize in Close Proximity at the PM of Sieve Elements

A series of immuno-gold studies with phloem tissue of *R. communis* seedlings prompted the suggestion that the H^+-PPase could be involved in sucrose transport (Long et al. 1995; Robinson et al. 1996). Further work with double- labeling immunolocalization experiments indicated that the H^+-PPase and PM H^+-ATPase localize in close proximity at the PM of the sieve elements in *R. communis* (Langhans et al. 2001). These authors suggested that both H^+-pumps are required for sieve element membrane energization to maintain high sucrose, K^+, and amino acid concentrations. However, a theoretical paper by Julia Davies argued that the sieve tube H^+-PPase could not operate hydrolytically to pump H^+ into the apoplast based on the estimation of the free energy of the H^+-PPase pump action for *in vivo* conditions (apparent PP_i hydrolysis constant and a cytosolic PP_i concentration of 0.011 mM). Interestingly, Davies suggested that a reverse reaction (where the H^+-PPase uses the H^+-gradient at the plasma membrane to synthesize PP_i) was thermodynamically feasible (Davies et al. 1997). Of note, *in vivo* data obtained with

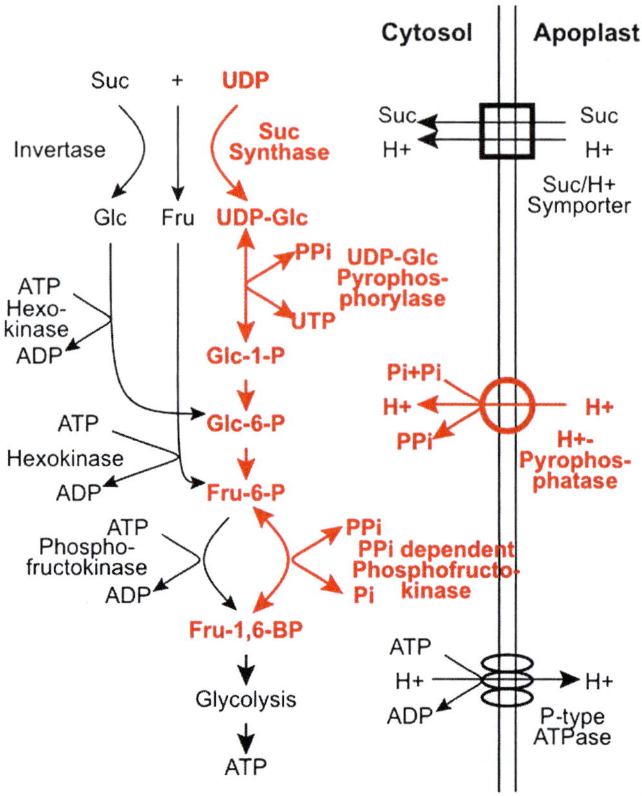

Fig. 3 Model for the function of a PM localized H^+-PPase in the sieve-element/companion cell. complex. In *red*, PP_i dependent ATP conserving pathway. This model is based on work reported elsewhere (Davies et al. 1997; Langhans et al. 2001; Lerchl et al. 1995; Stitt 1998)

the H^+-PPase from *Rhodospirillum rubrum* are consistent with the capacity of this enzyme to play two distinct roles depending on location; it can act as an intracellular proton pump in the acidocalcisomes or as a PP_i synthase in the chromatophore membranes during illumination (Seufferheld et al. 2004). Furthermore, Rocha-Facanha and Meis presented in vitro evidence with tonoplast fractions of maize coleoptiles and seeds consistent with the reverse function of the H^+-PPase (Rocha Facanha and de Meis 1998).

2.8.3 Hypothetical Model

It is tempting to speculate that the plant relative of the *R. rubrum* H^+-PPase can still "remember its prokaryote days" and work as a pump or as a PP_i synthase depending on location or milieu. A PM localized H^+-PPase in the sieve element/companion cell complex could use the proton motive force (pmf) to maintain cytosolic PP_i levels required for sucrose (Suc) respiration through an ATP-conserving pathway (shown in red in Fig. 3). Companion cells have high respiration rates to maintain the pmf and Suc loading. In the first ATP-conserving step, a portion of the loaded Suc is metabolized to fructose (Fru) and UDP-Glc (glucose) by Suc synthase (Susy), which is well established to have high levels of expression in companion cells (Nolte and Koch 1993; Yang and Russell 1990). Then, UDP-Glc pyrophosphorylase requires PP_i to metabolize UDP-Glc to Glc-1-P and UTP. In the second ATP-conserving step, a PPi-dependent phosphofructokinase uses PP_i and Fru-6-P to create Fru-1,6-BP and P_i. By utilizing this pathway, the companion cells reserve ATP for generating the proton motif force. Since the reactions catalyzed by both UDP-Glc pyrophosphorylase and PP_i-dependent phosphofructokinase are readily reversible, high concentrations of PP_i are needed to maintain the reaction moving toward glycolysis. The required PP_i could be produced as a byproduct of several reactions, whereas the PMF can only be generated by ATP. Therefore, the use of the PMF to regulate and maintain PP_i levels when necessary would help optimize efficient Suc respiration and leave more for transport.

Based on this model, we hypothesize that the upregulation of type I H^+-PPases enhances sucrose fluxes from source to sink tissues by improving phloem sucrose loading capacity. Sucrose produced by photosynthesis, is the cornerstone of higher plant metabolism in both source and sink organs. It is the main substrate for respiration, biosynthesis, and storage. Thus, an enhanced availability of sucrose in the phloem for transport could result in both larger and more energized root systems with an enhanced apoplast and rhizosphere acidification capacity. The latter will result in more efficient nutrient uptake capacity. It is likely that up-regulation of H^+-PPases in the phloem may improve sucrose transport to sink organs, and improve growth through several pathways related to higher availability of reduced carbon. In a sense, it will produce a domino effect for integral plant growth and development.

Acknowledgements We would like to apologize to authors whose work we did not discuss because of space constraints. RAG and JPV would like to thank J. Sanchez and B. Ayre for editing help. RAG and JPV were supported by Arizona State University start-up funds. A.T. Fuglsang would like to thank M. Palmgren for comments and suggestions to the manuscript.

References

Alexandersson E, Saalbach G, Larsson C, Kjellbom P (2004) Arabidopsis plasma membrane proteomics identifies components of transport, signal transduction and membrane trafficking. Plant Cell Physiol 45:1543–1556

Au KM, Barabote RD, Hu KY, Saier MHJ (2006) Evolutionary appearance of H^+-translocating pyrophosphatases. Microbiology 152(Pt 5):1243–1247

Axelsen KB, Palmgren MG (1998) Evolution of substrate specificities in the P-type ATPase superfamily. J Mol Evol 46(1):84–101

Axelsen KB, Venema K, Jahn T, Baunsgaard L, Palmgren MG (1999) Molecular dissection of the C-terminal regulatory domain of the plant plasma membrane H^+-ATPase AHA2: mapping of residues that when altered give rise to an activated enzyme. Biochemistry 38(22):7227–7234

Baltscheffsky M, Schultz A, Baltscheffsky H (1999) H^+-proton-pumping inorganic pyrophosphatase: a tightly membrane-bound family. FEBS Lett 452:121–127

Bao A-K, Wang S-M, Wu G-Q, Xi J-J, Zhang J-L, Wang C-M (2008) Overexpression of the Arabidopsis H+-PPase enhanced resistance to salt and drought stress in transgenic alfalfa (Medicago sativa L.). Plant Sci 176:232–240

Batistic O, Sorek N, Schultke S, Yalovsky S, Kudla J (2008) Dual fatty acyl modification determines the localization and plasma membrane targeting of CBL/CIPK Ca^{2+} signaling complexes in Arabidopsis. Plant Cell 20(5):1346–1362

Belogurov GA, Lahti R (2002) A lysine substitute for K^+. J Biol Chem 277(51):49651–49654

Carystinos GD, MacDonald HR, Monroy AF, Dhindsa RS, Poole RJ (1995) Vacuolar H^+-translocating pyrophosphatase is induced by anoxia or chilling in seedlings of rice. Plant Physiol 108:641–649

Cleland RE (1995) Auxin and cell elongation. In: Davies PJ (ed) Plant hormones. Kluwer, Dordrecht, pp 214–217

Coblitz B, Shikano S, Wu M, Gabelli SB, Cockrell LM, Spieker M, Hanyu Y, Fu H, Amzel LM, Li M (2005) C-terminal recognition by 14-3-3 proteins for surface expression of membrane receptors. J Biol Chem 280(43):36263–36272

Davies JM, Darley CP, Sanders D (1997) Energetics of the plasma membrane pyrophosphatase. Trends Plant Sci 2:9–10

Docampo R, de Souza W, Miranda K, Rohloff P, Moreno SNJ (2005) Acidocalcisomes – conserved from bacteria to man. Nat Rev Microbiol 3:251–261

Drozdowicz YM, Lu YP, Patel V, Fitz-Gibbon S, Miller JH, Rea PA (1999) A thermostable vacuolar-type membrane pyrophosphatase from the archaeon Pyrobaculum aerophilum: implications for the origins of pyrophosphate-energized pumps. FEBS Lett 460(3):505–512

Drozdowicz YM, Kissinger JC, Rea PA (2000) AVP2, a sequence-divergent, K^+-insensitive H^+-translocating inorganic pyrophosphatase from Arabidopsis. Plant Physiol 123:353–362

Drozdowicz YM, Shaw M, Nishi M, Striepen B, Liwinski HA, Roos DS, Rea PA (2003) Isolation and characterization of TgVP1, a type I vacuolar H^+-translocating pyrophosphatase from Toxoplasma gondii. The dynamics of its subcellular localization and the cellular effects of a diphosphonate inhibitor. J Biol Chem 278(2):1075–1085

Duby G, Boutry M (2009) The plant plasma membrane proton pump ATPase: a highly regulated P-type ATPase with multiple physiological roles. Pflügers Arch Eur J Physiol 457(3):645–655

Duby G, Poreba W, Piotrowiak D, Bobik K, Derua R, Waelkens E, Boutry M (2009) Activation of plant plasma membrane H^+-ATPase by 14-3-3 proteins is negatively controlled by two phosphorylation sites within the H^+-ATPase C-terminal region. J Biol Chem 284 (7):4213–4221

Ekberg K, Palmgren MG, Veierskov B, Buch-Pedersen MJ (2010) A novel mechanism of P-type ATPase autoinhibition involving both termini of the protein. J Biol Chem 285(10):7344–7350

Etienne C, Moing A, Dirlewanger E, Raymond P, Monet R, Rothan C (2002) Isolation and characterization of six peach cDNAs encoding key proteins in organic acid metabolism and solute accumulation: involvement in regulating peach fruit acidity. Physiol Plant 114 (2):259–270

Franco-Zorrilla JM, González E, Bustos R, Linhares F, Leyva A, Paz-Ares J (2004) The transcriptional control of plant responses to phosphate limitation. J Exp Bot 55(396):285–293

Fuglsang AT, Visconti S, Drumm K, Jahn T, Stensballe A, Mattei B, Jensen ON, Aducci P, Palmgren MG (1999) Binding of 14-3-3 protein to the plasma membrane H^+-ATPase AHA2 involves the three C-terminal residues Tyr946-Thr-Val and requires phosphorylation of Thr947. J Biol Chem 274(51):36774–36780

Fuglsang AT, Borch J, Bych K, Jahn TP, Roepstorff P, Palmgren MG (2003) The binding site for regulatory 14-3-3 protein in plant plasma membrane H^+-ATPase: involvement of a region promoting phosphorylation-independent interaction in addition to the phosphorylation-dependent C-terminal end. J Biol Chem 278(43):42266–42272

Fuglsang AT, Tulinius G, Cui N, Palmgren MG (2006) Protein phosphatase 2A scaffolding subunit A interacts with plasma membrane H^+-ATPase C-terminus in the same region as 14-3-3 protein. Physiol Plant 128(2):334–340

Fuglsang AT, Guo Y, Cuin TA, Qiu QS, Song CP, Kristiansen KA, Bych K, Schulz A, Shabala S, Schumaker KS, Palmgren MG, Zhu JK (2007) Arabidopsis protein kinase PKS5 inhibits the plasma membrane H^+-ATPase by preventing interaction with 14-3-3 protein. Plant Cell 19 (5):1617–1634

Gao F, Gao Q, Duan XG, Yue GD, Yang AF, Zhang JR (2006) Cloning of an H^+-PPase gene from Thellungiella halophila and its heterologous expression to improve tobacco salt tolerance. J Exp Bot 57(12):3259–3270

Gaxiola R, Li J, Undurraga S, Dang L, Allen G, Alper S, Fink G (2001) Drought- and salt-tolerant plants result from overexpression of the AVP1 H^+-pump. Proc Natl Acad Sci USA 98:11444–11449

Gaxiola RA, Palmgren MG, Schumacher K (2007) Plant proton pumps. FEBS Lett 581 (12):2204–2214

Gibbs J, Greenway H (2003) Mechanisms of anoxia tolerance in plants. I. Growth, survival and anaerobic catabolism. Funct Plant Biol 30:1–47

Gordon-Weeks SH, Steele RA, Leigh RA (1996) The role of magnesium, pyrophosphate, and their complex as substrate and activators of the vacuolar H^+-pumping inorganic pyrophosphatase. Plant Physiol 11:195–202

Guo Y, Halfter U, Ishitani M, Zhu J-K (2001) Molecular characterization of functional domains in the protein kinase SOS2 that is required for plant salt tolerance. Plant Cell 13(6):1383–1400, %R 10.1105/tpc.13.6.1383

Haruta M, Burch HL, Nelson RB, Barrett-Wilt G, Kline KG, Mohsin SB, Young JC, Otegui MS, Sussman MR (2010) Molecular characterization of mutant Arabidopsis plants with reduced plasma membrane proton pump activity. J Biol Chem 285(23):17918–17929

Jahn TP, Schulz A, Taipalensuu J, Palmgren MG (2002) Post-translational modification of plant plasma membrane H^+-ATPase as a requirement for functional complementation of a yeast transport mutant. J Biol Chem 277(8):6353–6358, %R 10.1074/jbc.M109637200

Jha D, Shirley N, Tester M, Roy SJ (2010) Variation in salinity tolerance and shoot sodium accumulation in Arabidopsis ecotypes ,linked to differences in the natural expression levels of transporters involved in sodium transport. Plant Cell Environ 33:793–804. doi:10.1111/j.1365-3040.2009.02105.x

Jiang SS, Yang SJ, Kuo SY, Pan RL (2000) Radiation inactivation analysis of H^+-pyrophosphatase from submitochondrial particles of etiolated mung beanseedlings. FEBS Lett 468 (2–3):211–214

Jiang L, Phillips TE, Hamm CA, Drozdowicz YM, Rea PA, Maeshima M, Rogers SW, Rogers JC (2001) The protein storage vacuole: a unique compound organelle. J Cell Biol 155 (6):991–1002

Kim EJ, Zhen RG, Rea PA (1994) Heterologous expression of plant vacuolar pyrophosphatase in yeast demonstrates sufficiency of the substrate-binding subunit for proton transport. Proc Natl Acad Sci USA 91(13):6128–6132

Kim EJ, Zhen RG, Rea PA (1995) Site-directed mutagenesis of vacuolar H^+-pyrophosphatase. Necessity of Cys634 for inhibition by maleimides but not catalysis. J Biol Chem 270 (6):2630–2635

Kim H-S, Grotz N, Parson B, Colangelo E, Atkinson AE, Hibbard S, Gehl M, Woelbel AM, Clark S, Maser P, Gong J, Gierth M, Lahner B, Mahmoudian M, Fana F, Nair M, Podell S, Tchieu J, Veretnik S, Schulz P (2005) The Pseudomonas syringae effector AvrRpt2 cleaves its C-terminally acylated target, RIN4, from Arabidopsis membranes to block RPM1 activation. PNAS 102(18):6496–6501

Kinoshita T, Shimazaki K (1999) Blue light activates the plasma membrane H^+-ATPase by phosphorylation of the C-terminus in stomatal guard cells. EMBO J 18(20):5548–5558

Kinoshita T, Emi T, Tominaga M, Sakamoto K, Shigenaga A, Doi M, Shimazaki K-I (2003) Blue-light- and phosphorylation-dependent binding of a 14-3-3 protein to phototropins in stomatal guard cells of broad bean. Plant Physiol 133(4):1453–1463

Kolukisaoglu U, Weinl S, Blazevic D, Batistic O, Kudla J (2004) Calcium sensors and their interacting protein kinases: genomics of the Arabidopsis and rice CBL-CIPK signaling networks. Plant Physiol 134(1):43–58

Krebs M, Beyhl D, Gorlich E, Al-Rasheid AS, Marten I, Stierhof YD, Hedrich R, Schumacher K (2010) Arabidopsis V-ATPase activity at the tonoplast is required for efficient nutrient storage but not for sodium accumulation. Proc Natl Acad Sci USA 107:3251–3256

Kudla J, Batistic O, Hashimoto K (2010) Calcium signals: the lead currency of plant information processing. Plant Cell 22(3):541–563

Kuo SY, Chien LF, Hsiao YY, Van Ru C, Yan KH, Liu PF, Mao SJ, Pan RL (2005) Proton pumping inorganic pyrophosphatase of endoplasmic reticulum-enriched vesicles from etiolated mung bean seedlings. J Plant Physiol 162(2):129–138

Laloi M, Perret AM, Chatre L, Melser S, Cantrel C, Vaultier MN, Zachowski A, Bathany K, Schmitter JM, Vallet M, Lessire R, Hartmann MA, Moreau P (2007) Insights into the role of specific lipids in the formation and delivery of lipid microdomains to the plasma membrane of plant cells. Plant Physiol 143(1):461–472

Langhans M, Ratajczak R, Lutzelschwab M, Michalke W, Wachter R, Fischer-Schliebs E, Ullrich CI (2001) Immunolocalization of plasma-membrane H^+-ATPase and tonoplast-type pyrophosphatase in the plasma membrane of the sieve element-companion cell complex in the stem of Ricinus communis L. Planta 213:11–19

Lerchl J, Giegenberger P, Stitt M, Sonnewald U (1995) Impaired photoassimilate partitioning caused by phloem-specific removal of pyrophosphate can be complemented by a phloem-specific cytosolic yeast-derived invertase in transgenic plants. Plant Cell 7:259–270

Li J, Yang H, Peer WA, Richter G, Blakeslee J, Bandyopadhyay A, Titapiwantakun B, Undurraga S, Mariya K, Richards EL, Krizek B, Murphy AS, Gilroy S, Gaxiola R (2005) Arabidopsis H^+-PPase AVP1 regulates auxin-mediated organ development. Science 310(5745):121–125

Li L, Kim B-G, Cheong YH, Pandey GK, Luan S (2006) A Ca^{2+} signaling pathway regulates a K^+ channel for low-K response in Arabidopsis. Proc Natl Acad Sci USA 103(33):12625–12630

Li B, Wei A, Song C, Li N, Zhang JR (2008) Heterologous expression of the *TsVP* gene improves the drought resistance of maize. Plant Biotechnol J 6:146–159

Li Z, Baldwin CM, Hu Q, Liu H, Luo H (2010) Heterologous expression of Arabidopsis H^+-pyrophosphatase enhances salt tolerance in transgenic creeping bentgrass (Agrostis stolonifera L.). Plant Cell Environ 33:272–289

Lin HH, Pan YJ, Hsu SH, Van RC, Hsiao YY, Chen JH, Pan RL (2005) Deletion mutation analysis on C-terminal domain of plant vacuolar H^+-pyrophosphatase. Arch Biochem Biophys 442 (2):206–213

Liu J, Elmore JM, Fuglsang AT, Palmgren MG, Staskawicz BJ, Coaker G (2009a) RIN4 functions with plasma membrane H^+-ATPases to regulate stomatal apertures during pathogen attack. PLoS Biol 7(6):e1000139

Liu TH, Hsu SH, Huang YT, Lin SM, Huang TW, Chuang TH, Fan SK, Fu CC, Tseng FG, Pan RL (2009b) The proximity between C-termini of dimeric vacuolar H^+-pyrophosphatase determined using atomic force microscopy and a gold nanoparticle technique. FEBS J 276 (16):4381–4394

Liu Q, Zhang Q, Burton RA, Shirley NJ, Atwell BJ (2010) Expression of vacuolar H^+-pyrophosphatase (OVP3) is under control of an anoxia-inducible promoter in rice. Plant Mol Biol 72(1–2):47–60

Loewus FA, Murphy PP (2000) Myo-inositol metabolism in plants. Plant Sci 150:1–19

Long AR, Williams LE, Nelson SJ, Hall JL (1995) Localization of membrane pyrophosphatase activity in *Ricinus communis* seedlings. J Plant Physiol 146:629–638

Lv S, Zhang K, Gao Q, Lian L, Song Y, Zhang JR (2008) Overexpression of an H^+-PPase Gene from Thellungiella halophila in cotton enhances salt tolernace and improves growth and photosynthetic performance. Plant Cell Physiol 49(8):1150–1164

Lv S-L, Lian L-J, Tao P-L, Li Z-X, Zhang K-W, Zhang J-R (2009) Overexpression of Thellungiella halophila H^+-PPase (TSVP) in cotton enhances drought stress resistance of plants. Planta 299(4):899–910

Mackey D, Holt B, Wiig A, Dangl J (2002) RIN4 interacts with Pseudomonas syringae type III effector molecules and is required for RPM1-mediated resistance in Arabidopsis. Cell 108 (6):743–754

Maddy AH (1976) A critical evaluation of the analysis of membrane protein by polycrylamide gel electrophoresis in the presence of SDS. J Theor Biol 62:315–326

Maeshima M (1991) H^+-translocating inorganic pyrophosphatase of plant vacuoles: inhibition by Ca^{2+}, stabilization by Mg^{2+} and immunological comparison with other inorganic pyrophosphatases. Eur J Biochem 196:11–17

Maeshima M (2000) Vacuolar H^+-pyrophosphatase. Biochim Biophys Acta 1465:37–51

Maeshima M (2001) Tonoplast transporters: organization and function. Annu Rev Plant Physiol 52:469–497

Marten H, Hedrich R, Roelfsema M (2007) Blue light inhibits guard cell plasma membrane anion channels in a phototropin-dependent manner. Plant J 50(1):29–39

Martinoa E, Maeshima M, Neuhaus HE (2007) Vacuolar transporters and their essential role in plant metabolism. J Exp Bot 58(1):83–102

Melotto M, Underwood W, Koczan J, Nomura K, He S (2006) Plant stomata function in innate immunity against bacterial invasion. Cell 126(5):969–980

Merlot S, Leonhardt N, Fenzi F, Valon C, Costa M, Piette L, Vavasseur A, Genty B, Boivin K, Muller A, Giraudat J, Leung J (2007) Constitutive activation of a plasma membrane H^+-ATPase prevents abscisic acid-mediated stomatal closure. EMBO J 26(13):3216–3226

Mimura H, Nakanishi Y, Maeshima M (2005) Disulfide-bond formation in the H^+-pyrophosphatase of Streptomyces coelicolor and its implications for redox control and enzyme structure. FEBS Lett 579:3625–3631

Mitsuda N, Enami K, Nakata M, Takeyasu K, Sato MH (2001a) Novel type Arabidopsis thaliana H^+-PPase is localized to the golgi apparatus. FEBS Lett 488:29–33

Mitsuda N, Takeyasu K, Sato MH (2001b) Pollen-specific regulation of vacuolar H^+-PPase expression by multiple cis-acting elements. Plant Mol Biol 46:185–192

Mitsuda N, Isono T, Sato MH (2003) Arabidopsis CAMTA family proteins enhance V-PPase expression in pollen. Plant Cell Physiol 44(10):975–981

Mitsuda N, Hisabori T, Takeyasu K, Sato MH (2004) VOZ; isolation and characterization of novel vascular plant transcription factors with a one-zink finger from Arabidopsis thaliana. Plant Cell Physiol 45(7):845–854

Nakanishi Y, Saijo T, Wada Y, Maeshima M (2001) Mutagenic analysis of functional residues in putative substrate-binding site and acidic domains of vacuolar H^+-pyrophosphatase. J Biol Chem 276(10):7654–7660

Niittyla T, Fuglsang AT, Palmgren MG, Frommer WB, Schulze WX (2007) Temporal analysis of sucrose-induced phosphorylation changes in plasma membrane proteins of Arabidopsis. Mol Cell Proteomics 6(10):1711–1726

Nolte KD, Koch KE (1993) Companion-cell specific localization of sucrose synthase in zones of phloem loading and unloading. Plant Physiol 101:899–905

Nuhse TS, Stensballe A, Jensen ON, Peck SC (2003) Large-scale analysis of in vivo phosphorylated membrane proteins by immobilized metal ion affinity chromatography and mass spectrometry. Mol Cell Proteomics 2(11):1234–1243

Nuhse TS, Bottrill AR, Jones AME, Peck SC (2007) Quantitative phosphoproteomic analysis of plasma membrane proteins reveals regulatory mechanisms of plant innate immune responses. Plant J 51(5):931–940

Oberbeck K, Drucker M, Robinson DG (1994) V-type ATPase and pyrophosphatase in endomembranes of maize roots. J Exp Bot 45:235–244

Oecking C, Jaspert N (2009) Plant 14-3-3 proteins catch up with their mammalian orthologs. Curr Opin Plant Biol 12(6):760–765

Olsson A, Svennelid F, Ek B, Sommarin M, Larsson C (1998) A phosphothreonine residue at the C-terminal end of the plasma membrane H^+-ATPase is protected by fusicoccin-induced 14-3-3 binding. Plant Physiol 118(2):551–555

Palma DA, Blumwald E, Plaxton WC (2000) Upregulation of vacuolar H+-translocating pyrophosphatase by phosphate starvation of Brassica napus (rapeseed) suspension cell cultures. FEBS Lett 486:155–158

Palmgren MG, Larsson C, Sommarin M (1990) Proteolytic activation of the plant plasma membrane H^+-ATPase by removal of a terminal segment. J Biol Chem 265(23):13423–13426

Park S, Cheng NH, Pittman JK, Yoo KS, Park J, Smith RH, Hirschi KD (2005) Increased calcium levels and prolonged shelf life in tomatoes expressing Arabidopsis H^+/Ca^{2+} transporters. Plant Physiol 139(3):1194–1206

Pedersen B, Buch-Pedersen M, Morth J, Palmgren M, Nissen P (2007) Crystal structure of the plasma membrane proton pump. Nature 450(7172):1111–1114

Priest HD, Filichkin SA, Mockler TC (2009) Cis-regulatory elements in plant signaling. Curr Opin Plant Biol 12:643–649

Qiu Q-S, Guo Y, Dietrich MA, Schumaker KS, Zhu J-K (2002) Regulation of SOS1, a plasma membrane Na+/H+ exchanger in Arabidopsis thaliana, by SOS2 and SOS3. Proc Natl Acad Sci USA 99(12):8436–8441, %R 10.1073/pnas.122224699

Quintero FJ, Masaru O, Shi H, Zhu J-K, Pardo JM (2002) Reconstitution in yeast of the Arabidopsis SOS signaling pathway for Na^+ homeostasis. Proc Natl Acad Sci USA 99 (13):9061–9066

Ratajczak R, Hinz G, Robinson DG (1999) Localization of pyrophosphatase in membranes of cauliflower inflorescence cells. Planta 208:205–211

Rea PA, Kim Y, Sarafian V, Poole RJ, Davies JM, Saunders D (1992) Vacuolar H^+-translocating pyrophosphatases: a new category of ion translocase. Trends Biochem Sci 17(9):348–353

Robinson DG, Haschke HP, Hinz G, Hoh G, Maeshima M, Marty F (1996) Immunological detection of tonoplast polypeptide in the plasma membrane of pea cotyledon. Planta 198:95–103

Rocha Facanha A, de Meis L (1998) Reversibility of H^+-ATPase and H^+-pyrophosphatase in tonoplast vesicles from maize coleoptiles and seeds. Plant Physiol 116:1487–1495

Roelfsema MRG, Staal M, Prins HBA (1998) Blue light induced apoplastic acidification of *Arabidopsis thaliana* guard cells: Inhibition by ABA is mediated through protein phosphatases. Physiol Plant 103(4):466–474

Sarafian V, Kim Y, Poole R, Rea P (1992) Molecular cloning and sequence of cDNA Encoding the pyrophosphate-energized vacuolar membrane proton pump of *Arabidopsis thaliana*.. Proc Natl Acad Sci USA 89:1775–1779

Seufferheld M, Lea CR, Viera M, Oldfield E, Docampo R (2004) The H+-pyrophosphatase of *Rhodospirillum rubrum* is predominantly located in polyphosphate-rich acidocalcisomes. J Biol Chem 279:51193–51202

Shikano S, Coblitz B, Sun H, Li M (2005) Genetic isolation of transport signals directing cell surface expression. Nat Cell Biol 7(10):985–992

Shimazaki K, Doi M, Assmann S, Kinoshita T (2007) Light regulation of stomatal movement. Annu Rev Plant Biol 58:219–247

Shiratake K, Kanayama Y, Maeshima M, Yamaki S (1997) Changes in H^+-pumps and a tonoplast intrinsic protein of vacuolar membranes during the development of pear fruit. Plant Cell Physiol 38(9):1039–1045

Sondergaard TE, Schulz A, Palmgreen MG (2004) Energization of transport processes in plants. roles of the plasma membrane H^+-.ATPase. Plant Physiol 136:2475–2482

Sonnewald U (1992) Expression of *E. coli* inorganic pyrophosphatase in transgenic plants alters photoassimilate partitioning. Plant J 2:571–581

Srivastava AC, Ganesan S, Ismail IO, Ayre BG (2008) Functional characterization of the Arabidopsis AtSUC2 sucrose/H+ symporter by tissue-specific complementation reveals an essential role in phloem loading but not in long-distance transport. Plant Physiol 148:200–211

Stitt M (1998) Pyrophosphate as an energy donor in the cytosol of plant cells: an enigmatic alternative to ATP. Bot Acta 111:167–175

Sullivan S, Thomson CE, Kaiserli E, Christie JM (2009) Interaction specificity of Arabidopsis 14-3-3 proteins with phototropin receptor kinases. FEBS Lett 583:2187–2193

Svennelid F, Olsson A, Piotrowski M, Rosenquist M, Ottman C, Larsson C, Oecking C, Sommarin M (1999) Phosphorylation of Thr-948 at the C terminus of the plasma membrane H^+-ATPase creates a binding site for the regulatory 14-3-3 protein. Plant Cell 11(12):2379–2391

Takeshige K, Tazawa M, Hager A (1988) Characterization of the H translocating adenosine triphosphatase and pyrophosphatase of vacuolar membranes isolated by means of a perfusion technique from Chara corallina. Plant Physiol 86(4):1168–1173

Terrier N, Deguillous C, Romieu C (1997) V-ATPase, inorganic pyrophosphatase and and anion transport on the tonoplast grape berries (Vitis viniferaL.). Plant Physiol Biochem 36:79–193

Terrier N, Sauvage FX, Ageorges A, Romieu C (2001) Changes in acidity and in proton transport at the tonoplast of grape berries during development. Planta 213(1):20–28

Tomasi N, Kretzschmar T, Espen L, Weisskopf L, Fuglsang AT, Palmgren MG, Neumann G, Varanini Z, Pinton R, Martinoia E, Cesco S (2009) Plasma membrane H^+-ATPase-dependent citrate exudation from cluster roots of phosphate-deficient white lupin, Plant Cell Environ 32 (5):465–475

Tominaga M, Harada A, Kinoshita T, Shimazaki K-I (2010) Biochemical characterization of calcineurin B-like-interacting protein kinase in vicia guard cells. Plant Cell Physiol 51(3): 408–421

Tseng T-S, Briggs WR (2010) The Arabidopsis rcn1-1 mutation impairs dephosphorylation of Phot2, resulting in enhanced blue light responses. Plant Cell 22(2):392–402

Tzeng CM, Yang CY, Yang SJ, Jiang SS, Kuo SY, Hung SH, Ma JT, Pan RL (1996) Subunit structure of vacuolar proton-pyrophosphatase as determined by radiation inactivation. Biochem J 316(1):143–147

Ueno K, Kinoshita T, Inoue S-I, Emi T, Shimazaki K-I (2005) Biochemical characterization of plasma membrane H^+-ATPase activation in guard cell protoplasts of Arabidopsis thaliana in response to blue light. Plant Cell Physiol 46(6):955–963, %R 10.1093/pcp/pci104

Van RC, Pan YJ, Hsu SH, Huang YT, Hsiao YY, Pan RL (2005) Role of transmembrane segement 5 of the plant vacuolar H^+-pyrophosphatase. Biochim Biophys Acta 1709:84–94

Vianello A, Zancani M, Biardot E, Petrussa E, Macri F (1991) Proton pumping inorganic pyrophosphatase of pea stem submitochondrial particles. Biochim Biophys Acta 1060:299–302

Wang HJ, Wan AR, Hsu CM, Lee KW, Yu SM, Jauh GY (2007) Transcriptomic adaptations in rice suspension cells under sucrose starvation. Plant Mol Biol 63(4):441–463

Weber E, Newman D (1980) Protein storage bodies organelles in plant seeds. Biochem Physiol Pflanz 175:279–306

Whiteman SA, Nuhse TS, Ashford DA, Sanders D, Maathuis FJM (2008) A proteomic and phosphoproteomic analysis of Oryza sativa plasma membrane and vacuolar membrane. Plant J 56(1):146–156

Wisniewski J-P, Rogowsky PM (2004) Vacuolar H+-translocating inorganic pyrophosphatase (Vpp 1) marks partial aleurone cell fate in cereal endosperm development. Plant Mol Biol 56:325–337

Xu J, Li H-D, Chen L-Q, Wang Y, Liu L-L, He L, Wu W-H (2006) A protein kinase, interacting with two calcineurin B-like proteins, regulates K^+ transporter AKT1 in Arabidopsis. Cell 125:1347–1360

Yang N-S, Russell D (1990) Maize sucrose synthase-1 promoter directs phloem cell-specific expression of Gus gene in transgenic tobacco plants. Proc Natl Acad Sci USA 87:4144–4148

Yang SJ, Jian SS, Kuo SY, Hung SH, Tam MF, Pam RL (1999) Localization of carboxylic residues possible involved in the inhibition of vacuolar H+-pyrophosphatase by N, N′dicyclohexylcarbodiimide. Biochem J 342:641–646

Yang H, Knapp J, Koirala P, Rajagopal D, Peer WA, Silbart L, Murphy A, Gaxiola R (2007) Enhanced phosphorus nutrition in monocots and dicots over-expressing a phosphorus-responsive type I H^+-pyrophosphatase. Plant Biotechnol J 5:735–745

Yang Y, Qin Y, Zhao F, Xie C, Liu D, Chen S, Fuglsang AT, Palmgren MG, Schumaker KS, Den XW, Guo Y (2010) AtJ3 regulates the plasma membrane H^+-ATPase in Arabidopsis through interaction with the PKS5 kinase. Plant Cell 22:1313–1332

Yi K, Wu Z, Zhou J, Du L, Guo L, Wu Y, Wu P (2005) OsPTF1, a novel transcription factor involved in tolerance to phosphate starvation in rice. Plant Physiol 138:2087–2096

Yue G, Sui Z, Gao Q, Zhang J (2008) Molecular cloning and characterization of a novel H^+-translocating pyrophosphatase gene in Zea mays. DNA Seq 19(2):79–86

Zhao F-Y, Zhang X-J, Li P-H, Zhao Y-X, Zhang H (2006) Co-expression of the Suaeda salsa SsNHX1 and Arabidopsis AVP1 confer greater salt tolerance to transgenic rice than the single SsNHX1. Mol Breed. doi:10.1007/s11032-006-9005-6

Zhen RG, Kim EJ, Rea PA (1994) Localization of cytosolically oriented maleimide-reactive domain of vacuolar H^+-pyrophosphatase. J Biol Chem 269(37):23342–23350

Zhen RG, Kim EJ, Rea PA (1997a) Acidic residues necessary for pyrophosphate-energized pumping and inhibition of the vacuolar H+-pyrophosphatase by N, N'-dicyclohexylcarbodiimide. J Biol Chem 272(35):22340–22348

Zhen RG, Kim EJ, Rea PA (1997b) The molecular and biochemical basis of pyrophosphate-energized proton translocation at the vacuolar membrane. Adv Bot Res 25:298–337

Zhou Z, Peng SB, Crider BP, Andersen P, Xie XS, Stone DK (1999) Recombinant SFD isoforms activate vacuolar proton pumps. J Biol Chem 274(22):15913–15919

Na^+ and K^+ Transporters in Plant Signaling

José M. Pardo and Francisco Rubio

Abstract Potassium is an essential macronutrient for plants that fulfills important functions related to enzyme activation, osmotic adjustment, regulation of membrane electric potential, growth and development. K^+ concentration in the soil solution may vary greatly and is always below the relatively constant concentration of 100 mM found in the cytoplasm of plant cells. To secure K^+ acquisition from the soil and its distribution within the plant, cell membranes are furnished with a suite of K^+ transport systems. Because Na^+ is the most abundant cation in saline soils and Na^+ is chemically very similar to K^+, cytosolic Na^+ toxicity if often concurrent with K^+ deficiency. In addition, the high external salt concentrations lower the water potential of the soil solution making water uptake by plant roots more difficult. To avert Na^+ toxicity, plants use an array Na^+ transport systems aimed at preventing the build up of high cytosolic Na^+ concentrations by restricting Na^+ uptake, enhancing Na^+ efflux, or mediating the sequestration of Na^+ into vacuoles. In addition, Na^+ transport systems for Na^+ movement within the plant facilitate its translocation to tissues and organs where toxicity is minimal. Here we present an updated overview of the main mechanisms involved in the regulation of K^+ and Na^+ transport systems that are key factors in securing K^+ nutrition and salt tolerance.

J.M. Pardo
Instituto de Recursos Naturales y Agrobiologia, Consejo Superior de Investigaciones Cientificas (IRNASE-CSIC), Reina Mercedes 10, 41012 Sevilla, Spain

F. Rubio (✉)
Centro de Edafologia y Biologia Aplicada del Segura, Consejo Superior de Investigaciones Cientificas (CEBAS-CSIC), Campus de Espinardo, 30100 Murcia, Spain
e-mail: frubio@cebas.csic.es

M. Geisler and K. Venema (eds.), *Transporters and Pumps in Plant Signaling*, Signaling and Communication in Plants 7,
DOI 10.1007/978-3-642-14369-4_3, © Springer-Verlag Berlin Heidelberg 2011

1 Introduction

Potassium is an essential macronutrient for plants that fulfills important functions related to enzyme activation, osmotic adjustment, regulation of membrane electric potential, growth and development. K^+ concentration in the soil solution may vary from 0.1 to 10 mM K^+, which is in contrast to the relatively constant concentration of 100 mM found in the cytoplasm of plant cells. To secure K^+ acquisition from the soil and its distribution within the plant, cell membranes are furnished with a suite of K^+ transport systems. An environmental stress condition that is widely affecting agricultural lands is salinity. Because Na^+ is the most abundant cation in saline soils and Na^+ is chemically very similar to K^+, K^+ is displaced by Na^+ under salinity conditions and thus cytosolic Na^+ toxicity together with K^+ deficiency occur. In addition, the high external NaCl concentrations of saline environments lower the water potential of external solutions making water uptake by plant roots more difficult. To avert Na^+ toxicity, plants are endowed with Na^+ transport systems aimed at preventing the build up of high cytosolic Na^+ concentrations by restricting Na^+ uptake, enhancing Na^+ efflux, or to mediate the accumulation of Na^+ in vacuoles. In addition, Na^+ transport systems for Na^+ movement within the plant facilitate its translocation to tissues and organs where toxicity is minimal. Therefore K^+ and Na^+ transport systems are key factors in securing K^+ nutrition and salt tolerance. Here we present an updated overview of the main mechanisms involved in the regulation of K^+ and Na^+ transport systems.

2 K^+ Transport

Potassium is the most abundant cation in plant cells and constitutes about 10% of plant dry weight (Wyn-Jones and Gorham 1986). K^+ is taken up from the soil solution by root epidermal and cortical cells. Once K^+ is inside the root symplast it may be stored in the vacuole, where it fulfills osmotic functions, or transported to the shoot via xylem. Shoot cells take up K^+ from apoplast and xylem and may also supply stored K^+ for redistribution via phloem. In this journey from the soil to the different plant organs, K^+ crosses different cell membranes through K^+-specific transport systems which regulate K^+ movements and are therefore key pieces of K^+ nutrition. Coordinated operation of the different transport systems within the plant to secure K^+ uptake from the soil and subsequent K^+ delivery to the different plant organs requires complex K^+ sensing and signaling mechanisms. Molecular identification of the K^+ transport systems begun in the early 90's and only recently the mechanism and components of their regulation have begun to be identified. This section will focus on the systems involved in K^+ uptake from the soil solution and the mechanisms involved in their regulation, although a brief description of systems involved in K^+ transport within the plant will be also presented.

2.1 K^+ Uptake from Diluted Solutions and Plant Signaling

Potassium concentration in the soil solution is not constant and varies spatial and temporally. This is in contrast with the concentration of K^+ in the cytoplasm of cells which should be maintained constant, around 50–100 mM to fulfill its functions. To secure K^+ uptake from a wide range of external K^+ concentrations, roots can put into operation transport systems with different affinities and concentrative capacities. In addition, these systems are highly specific for K^+ to avoid inhibition of K^+ uptake by the presence of other beneficial or toxic cations. Classically, high- and low- affinity transport systems have been described at the root (Epstein et al. 1963). High-affinity K^+ uptake was proposed to be mediated by transporters, probably by a K^+-H^+ symport mechanism, and low-affinity K^+ uptake by channels (Maathuis and Sanders 1993). High-affinity K^+ uptake is induced by low K^+ availability (Glass 1978), and repressed when K^+ is present at high concentrations (Glass 1976). These early observations in the 70's described for the first time the capacity of plants to sense and respond to the external supply of K^+ and suggested the existence of mechanisms for regulation of the systems involved in K^+ acquisition. Today, the regulatory elements have begun to be identified (Wang and Wu 2010).

The first K^+ transport systems from plants were identified in *Arabidopsis thaliana* by complementation of yeast mutants defective in endogenous K^+ uptake. This approach led to the isolation of cDNAs encoding the *Shaker*-like inward-rectifier K^+ channels AKT1 (Sentenac et al. 1992) and KAT1 (Anderson et al. 1992). Since AKT1 was expressed in roots (Lagarde et al. 1996) and low-affinity K^+ uptake was thought to be mediated by K^+ channels (Maathuis and Sanders 1993), AKT1 was ascribed to the low affinity K^+ uptake system. Subsequently, based on the homology between the *Escherichia coli* Kup K^+ transporter and the *Schwanniomyces occidentalis* high-affinity K^+ transporter SoHAK1, a cDNA from barley *HvHAK1* was isolated (Santa-María et al. 1997). *HvHAK1* was induced in barley roots by K^+ starvation and, when expressed in yeast, mediated high-affinity K^+ uptake of the same characteristics of that shown by barley roots. In *Neurospora crassa*, high-affinity K^+ uptake has been demonstrated to be mediated by a H^+-K^+ symport (Rodriguez-Navarro et al. 1986), encoded by the homolog gene *NcHAK1* (Haro et al. 1999). HAK1-type transporters showing the same characteristics to HvHAK1 have been isolated in many different plant species, including important crops and model species. All together, these results suggest that in plants high-affinity K^+ uptake is mediated by HAK1-type transporters. In *Arabidopsis*, tomato and *Thellungiella halophila* root high-affinity K^+ transport is mediated by orthologous proteins named HAK5 (Rodriguez-Navarro and Rubio 2006).

Studies with *Arabidopsis* T-DNA insertion lines in *AtHAK5* and *AtAKT1*, have allowed us to clearly establish the range of K^+ concentrations where these two systems operate (Hirsch et al. 1998; Spalding et al. 1999; Gierth et al. 2005; Qi et al. 2008; Rubio et al. 2008; Nieves-Cordones et al. 2009; Rubio et al. 2010). AtHAK5 is required for K^+ uptake at external concentrations below 10 μM. From 10 to 50 μM both systems contribute to K^+ uptake. These results were unexpected because they assigned to channels a role in K^+ uptake in the high-affinity range of concentrations. It should be noted, however, that some of these experiments were conducted in the presence of millimolar concentrations of NH_4^+ that likely inhibited HAK transporters and triggered the hyperpolarization of the plasma membrane, thereby magnifying the role of channels in K^+ uptake (Hirsch et al. 1998; Spalding et al. 1999). At higher external K^+ concentrations AtHAK5 function is less relevant and AtAKT1 seems to be the only system for K^+ uptake. When K^+ further increases, unknown systems come into operation and at 10 mM K^+, AtAKT1 function is not required for growth. Identification of these unknown systems is the next challenge, being candidate systems members of the cyclic nucleotide gated channels family (Kaplan et al. 2007), and of the cation proton exchanger family as CHX13 (Zhao et al. 2008) or CHX17 (Cellier et al. 2004), which will be further described below.

2.2 Transcriptional Regulation of High-Affinity HAK Transporters

When K^+ is withdrawn from the growing solution, *HAK1* genes are induced and K^+ re-supply rapidly down-regulates them. Studies with *Arabidopsis* plants show that *AtHAK5* mRNA may be detected as soon as 1 day (Ahn et al. 2004) or even 6 h (Shin and Schachtman 2004) after K^+ withdrawal. This rapid response may indicate that transcriptional regulation of *HAK1* genes responds to the presence or absence of K^+ in the external solution rather than reflecting the intracellular K^+ status. These short periods of K^+ starvation are sufficient to activate K^+ signaling pathways and to produce increases in *HAK1* mRNA levels. The use of sensitive techniques for mRNA quantification such as real-time PCR, allowed detecting mRNA levels produced at the beginning of the response to low K^+. However, these short K^+ starvation periods produce low levels of gene expression and therefore low rates of high-affinity K^+ transport (Martinez-Cordero et al. 2004; Shin and Schachtman 2004). Full gene induction and high rates of high-affinity K^+ uptake that supply significant K^+ for growth require longer K^+ starvation periods that reduce internal K^+ contents. Thus, starvation periods between 3 and 7 days that lead to about 50% decreases in internal root K^+ concentrations have been shown to be required for complete *HAK1* induction and development of high-affinity K^+ uptake in pepper (Martinez-Cordero et al. 2004, 2005), tomato (Nieves-Cordones et al. 2007) and *Arabidopsis* plants (Rubio et al. 2008).

Regulation of *HAK1* genes seems to be complex as it responds to many different environmental factors, some of them unrelated to K^+ availability. Root K^+ concentrations must drop below a threshold level to fully induce gene expression and high-affinity K^+ uptake (Martinez-Cordero et al. 2005). However growth in the presence of high NaCl concentrations led to root K^+ concentrations below that threshold and yet gene expression or high-affinity K^+ uptake was not induced (Nieves-Cordones et al. 2007, 2008). On the other hand, roots of plants grown with NH_4^+ in K^+-sufficient conditions show K^+ concentrations above that threshold, induction of HAK genes and high-affinity K^+ uptake. These NH_4^+ effects are observed in tomato plants (Nieves-Cordones et al. 2007, 2008) but not in *Arabidopsis* (Rubio et al. 2008). Hence,contrary to the initial hypothesis, no correlation between root K^+ concentration and gene expression or high-affinity K^+ uptake could be found. Interestingly, a negative correlation between root membrane potential and the tomato *LeHAK5* gene expression was discovered. In addition, rapid down-regulation of *LeHAK5* by depolarizing agents that do not change internal K^+ concentrations has been observed and therefore a role for the membrane potential in the regulation of the expression of these genes has been proposed (Nieves-Cordones et al. 2008) (Fig. 1).

Adding more complexity to the stimuli that induce *HAK5* genes, it has been observed that changes in plant mineral status affect the expression levels of these genes. A study in tomato plants showed that NO_3^- resupply to NO_3^--deprived plants strongly down-regulated *LeHAK5*, while a tomato homolog of the carrot K^+ channel *KDC1* was up-regulated (Wang et al. 2001). The same study showed that K^+ deficiency induced the *LeNRT1.2* and *LeNRT2.1* genes, encoding high-affinity NO_3^- transporters. By contrast, a transcriptomics study showed that K^+ starvation down-regulated the *Arabidopsis* genes *AtNRT2.1*, *AtNRT2.3* and *AtNRT2.6* encoding NO_3^- transporters, which were up-regulated after K^+ resupply (Armengaud et al. 2004). A more detailed study in *Arabidopsis* showed that K^+, NO_3^- and Pi

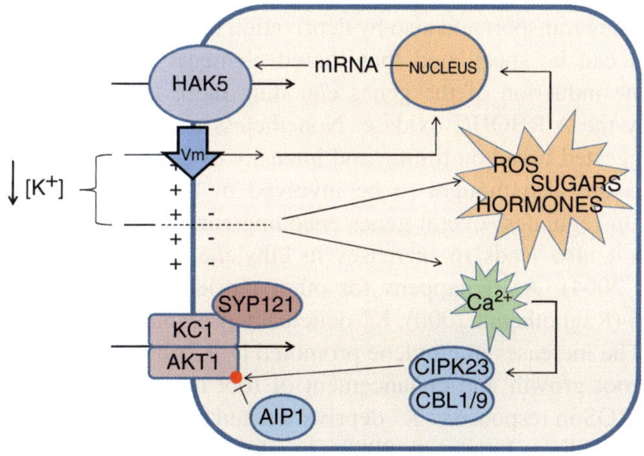

Fig. 1 Model of plant response to low external K^+. Low external K^+ induces transcriptional activation of the high-affinity K^+ transporter HAK5, and activation of the inward-rectifier K^+ channel AKT1. *HAK5* induction is mediated by signal cascades that include ROS, hormones as ethylene and probably jasmonic acid, and sugars as sucrose. In addition, the hyperpolarization of the plasma membrane produced by the low concentrations of K^+ in the external medium also promote transcriptional activation of *HAK5*. Activation of the AKT1 channel to provide K^+ from low concentrations involve interaction with the KC1 subunit and the SYP121 SANRE protein as well as phosphorylation by a Ca^{2+}-dependent CIPK/CBL complex. The AIP1 phosphatase negatively regulates the channel

starvation all induced *AtHAK5* as well as *AtNRT2.1*. Together these results indicate that the systems involved in the acquisition of mineral nutrients may respond to the general mineral status of the plant, probably by means of pathways that include shared and specific regulatory elements.

Finding general and specific sensors and components of these regulatory cascades is an appealing challenge in the field of mineral nutrition in plants. Several lines of evidence suggest that reactive oxygen species (ROS), hormones and sugars (Fig. 1) are involved in the regulation of mineral nutrient transporters, *AtHAK5* included. ROS production has been suggested as an early root response to K^+ deficiency, modulating gene expression and K^+ uptake (Shin and Schachtman 2004). K^+ deprivation leads to a rapid induction of the gene encoding the AtRBOHC NADPH oxidase, resulting in an increase in H_2O_2 that up-regulates, among other genes, *AtHAK5*. Exogenous application of H_2O_2 to K^+-sufficient plants of the AtRBOHC defective mutant *rhd2* did not induce *AtHAK5*, suggesting that both K^+ deprivation and H_2O_2 production were required to induce the gene. Recently, the AtRCI3 peroxidase has been described as another component of the low-K^+ signal transduction pathway, (Kim et al. 2010). AtRCI3 is up-regulated by K^+ deprivation, plants overexpressing AtRCI3 show more ROS production and *AtHAK5* expression and *atrci3* mutants show reduced ROS production and lower *AtHAK5* expression under K^+ deprivation than wild type plants. These results support the role of ROS in root response to K^+ deficiency. Interestingly, H_2O_2 is also increased in roots subjected to nitrogen and phosphorous deprivation (Shin et al. 2005). In addition, transporters for K^+, NO_3^- or Pi are strongly induced by the deprivation of the nutrient that they transport and also by deprivation of any of the two other nutrients. Therefore, it can be speculated that shared elements of the signaling pathway leading to the induction of the genes encoding these transporters include H_2O_2 and probably the AtRBOHC oxidase. Nonetheless a certain degree of specificity has been suggested based on timing and intensity of gene expression.

Hormones are also thought to be involved in K^+ deprivation signaling. K^+ deprivation up-regulates several genes encoding enzymes related to ethylene production and it also leads to increases in ethylene in starved plants (Shin and Schachtman 2004). As it happens for other nutrients as NO_3^- (Remans et al. 2006) and Pi (Raghothama 2000), K^+ deficiency promotes morphological changes in the root. The increases in ethylene promoted by K^+ deprivation led to inhibition of primary root growth and enhancement of root hair elongation. Ethylene acts upstream of ROS in response to K^+ deprivation and expression of *AtHAK5* depends on ethylene signaling (Jung et al. 2009). In addition, ethylene insensitive mutants are more sensitive to low K^+.

In the *Arabidopsis ethylene insensitive2-1* (*ein2-1*) mutant, the ethylene-mediated responses to low K^+ are not completely eliminated, suggesting that some K^+ deprivation–induced responses are either ethylene independent or AtEIN2 independent. In agreement with this idea, jasmonic acid (JA) and auxins have been also related to low K^+ signaling. A microarray study showed that the transcripts of many JA-related genes were induced in response to low K^+ stress (Armengaud et al. 2004). Transcript levels for the JA biosynthetic enzymes lipoxygenase, allene oxide synthase, and allene oxide cyclase were strongly increased during K^+ starvation and quickly decreased after K^+ resupply. A large number of well-known JA responsive genes showed the same expression profile, including genes involved in storage of amino acids (VSP), glucosinolate production (CYP79), polyamine biosynthesis (ADC2), and defense (PDF1.2). Auxin signaling may also play roles in low K^+ signaling because this hormone has been shown to be involved in lateral root development under K^+-deficient conditions (Shin et al. 2007). On the other hand, cytokinins have been suggested to play a role in the regulation of K^+ fluxes at the root (Shabala et al. 2009). Therefore, it is possible that low- K^+ signaling and probably mineral nutrient deficiency signaling in general, involve multiple hormone pathways that lead to morphological, physiological and molecular changes that provide the plants with the tools to adapt to nutrient deficiency.

Finally, sugars have also been shown to participate in the transcriptional regulation of the genes encoding transporters of several nutrients. The addition of sucrose to the growth medium induced the expression of genes encoding transporters of NO_3^- (*NRT* family, including *NRT1.1* and *NRT2.1*), NH_4^+ (*AMT* family), Pi (*PHT* family), SO_4^{2-} (*SULTR* family) and K^+ (*HAK5*) (Lejay et al. 2008). Regulation of these genes by sucrose involves several pathways that include carbon signal downstream HXK and the oxidative pentose phosphate pathway. The induction of *AtHAK5* by sucrose may be due to an osmotic effect, not related to the signaling role of sucrose. However, more complex interactions between light and carbon signaling seem to take place because *AtHAK5* induction by sucrose is counteracted by light. In conclusion, it is possible that integration of mineral nutrient uptake to match plant demand occurs through regulatory systems that respond to the general nutritional status of the plant, including photosynthetic activity of the shoot and sucrose as a signaling molecule.

Regulation of HAK transporters at the protein level has been scarcely studied. Negative regulation of the barley HvHAK1-mediated K^+ transport by the yeast PPZ1 and HAL4/5 phosphatases has been described (Fulgenzi et al. 2008). These results in yeast are difficult to explain because *Saccharomyces cerevisiae* lacks HAK transporters and therefore a signaling cascade to regulate their function. However, some *S. cerevisiae* phosphatases seem to use HvHAK1 as a substrate, indicating that these transporters may be regulated in planta by phosphorylation/dephosphorylation. On the other hand, mutational studies have described amino acid residues within HAK1 transporters which are important for their activity (Rubio et al. 2000; Senn et al. 2001; Garciadeblas et al. 2007; Mangano et al. 2008), indicating a possible regulation at the protein level.

All high-affinity K^+ transporters of the HAK family belong to phylogenetic group I (Rubio et al. 2000). However, there are other members of group I of unknown function and many other members distributed among other groups of this family. The functions of some of these other members have begun to be elucidated. AtKT2/AtKUP2, which belongs to group II, mediates low-affinity K^+ uptake in yeast (Quintero and Blatt 1997) and a semidominant mutation in AtKT2, *shy3-1*, produces short hypocotyls, small leaves and short flowering stem. These defects in *shy3-1* result from decreased cell expansion (Elumalai et al. 2002). A T-DNA insertion mutant in *AtKUP4,* which also belongs to group II, displayed strongly reduced root hair elongation and was called *tiny root hair* (*trh1*). Although Rb^+ (K^+) uptake was reduced in the *trh1* mutant, high K^+ did not restore the defective phenotype, indicating that K^+ availability was not the reason for the observed phenotype (Rigas et al. 2001). Interestingly AtKUP4 has been localized to the tonoplast (Jaquinod et al. 2007) and further research indicated that this transporter was involved in root-specific distribution of auxin (Vicente-Agullo et al. 2004). The mechanism by which a putative K^+ transporter as AtKUP4 interacts with root auxin transport remains to be clarified. Interestingly, the H^+-PPase AVP1 has been shown to control auxin transport and auxin-dependent development (Li et al. 2005a). In grape, two transporters of this family, VvKUP1 and VvKUP2, may be involved in K^+ accumulation during berry development (Davies et al. 2006) and another transporter from cotton, GhKT1, is up-regulated in rapidly expanding cotton fibers for turgor pressure build-up (Ruan et al. 2001). Other members of this family, as the rice OsHAK10, locate to the tonoplast and may be involved in K^+ transport from the vacuole to the cytosol under K^+ starvation conditions (Bañuelos et al. 2002). All these results highlight the important role of K^+ and K^+ transporters in turgor maintenance, cell expansion and growth. Group II K^+ transporters of this HAK family may play these functions operating at the tonoplast. Interaction of these transporters with hormone distribution may be an important point of growth regulation.

In addition to the HAK transporters, members of the cation proton antiporter-2 family (CPA2 or CHX family) have been putatively involved in K^+ acquisition. AtCHX17 is strongly induced by salt stress, K^+ starvation and ABA in roots and *chx17* knockout mutants accumulate less K^+ in roots in response to salt stress and K^+ starvation (Cellier et al. 2004). Another family member, AtCHX13, mediates relatively high-affinity K^+ uptake and *chx13* mutants were sensitive to K^+ deficiency conditions, whereas overexpression of *AtCHX13* reduced the sensitivity to K^+ deficiency (Zhao et al. 2008). CHXs proteins are thought to mediate K^+:H^+ antiport and the mechanism by which they contribute to K^+ uptake is unclear at present.

2.3 *Regulation of AKT1 Channels by Phosphorylation/ Dephosphorylation*

As mentioned above, transcriptional regulation by K^+ supply of the genes encoding AKT1 channels has been shown to depend on the species. In wheat *TaAKT1* is up-regulated by K^+ deprivation (Buschmann et al. 2000) whereas transcription of the *Arabidopsis AtAKT1* does not respond to K^+ supply (Lagarde et al. 1996) and the same is true for the pepper homolog *CaAKT1* (Martinez-Cordero et al. 2005). Although K^+ supply does not affect *AtAKT1* expression, it has been reported that salt stress strongly down-regulates *AtAKT1* (Kaddour et al. 2009), the *Mesembryanthemum crystallinum* homolog *MKT1* (Su et al. 2003), and the *Oryza sativa* homolog *OsAKT1* in the salt-sensitive IR29 variety, but not in the salt-tolerant Pokkali and BK varieties (Golldack et al. 2003). These results suggest a different response to salinity of the genes encoding AKT1 channels in plants with different salt tolerance. Down-regulation of *AKT1* genes by Na^+ could contribute to salt tolerance by preventing K^+ loss through AKT1 in plants with a salt-induced depolarized plasma membrane. This AKT1-mediated K^+ loss may be substantial when external K^+ concentration is low (Nieves-Cordones et al. 2009). However, the available information does not allow establishing a correlation between gene up- or down-regulation and plant salt tolerance.

Independently of the transcriptional regulation of the genes encoding AKT1, the important role of Ca^{2+}-dependent phosphorylation via a CIPK/CBL complex in the regulation of AKT1 activity has been recently described (Fig. 1). Many external signals including changes in K^+ status in the soil are sensed by plants through Ca^{2+} signaling mechanisms. Rapid changes in cytoplasmic Ca^{2+} are detected by Ca^{2+} sensors that transmit the signal downstream to trigger cellular responses (Luan et al. 2009). A screening for *Arabidopsis* mutants sensitive to low K^+ lead to the isolation of *lks1* which harbored a mutation in the gene encoding the CBL Interacting Protein Kinase 23 (CIPK23). *AtLKS1* (*AtCIPK23*) was strongly up-regulated in response to low K^+ and the *lks1* mutation reduced K^+ uptake and caused growth inhibition (Xu et al. 2006). AtCIPK23 was shown to interact with and phosphorylate AtAKT1 via interaction with the Ca^{2+} sensors AtCBL1 and AtCBL9, thereby activating K^+ currents through AtAKT1 and enhancing root K^+ uptake under low K^+ (Li et al. 2006; Xu et al. 2006). The interaction between AtCIPK23 and AtAKT1 was shown to involve the kinase domain of AtCIPK23 and the ankyrin repeat domain of AtAKT1. Further, it was shown that dephosphorylation of AtAKT1 by the PP2C phosphatase AtAIP inactivated AtAKT1 (Lee et al. 2007). All together these results provide evidence that Ca^{2+}-sensitive CBL and CIPK families together with 2C-type protein phosphatases form a protein phosphorylation/dephosphorylation network that regulates AtAKT1 channel for K^+ transport in plants. Interestingly, activation of AtAKT1 by the CBL/CIPK complex occurs under low K^+ supply supporting the idea that the AtAKT1 channel contributes to K^+ uptake in the high-affinity range of concentrations.

In a different screening for drought-tolerant phenotypes a loss-of-function allele of *AtCIPK23* was isolated. Studies of this *cipk23* mutant confirmed the role of CBL1/9-CIPK23 in root K^+ uptake via phosphorylation of AtAKT1. In addition, *cipk23* plants showed reduced transpirational water loss from leaves coinciding with an enhanced ABA sensitivity of guard cells during opening as well as closing reactions (Cheong et al. 2007). Whether AtAKT1 is a AtCIPK23 target for regulation of transpiration or additional ion transporters regulated by AtCIPK23 are involved in this process remains to be investigated in future studies. Other members of the CIPK family may contribute to regulation of K^+ nutrition. Studies on the transcriptional profiles of members of the CIPK family showed that *AtCIPK9* was induced by K^+ deficiency. The *cipk9* mutant was hypersensitive at low K^+ although

K^+ uptake and plant K^+ contents were not affected. Thus, a role for AtCIPK9 in the regulation of K^+ utilization or sensing process has been suggested (Pandey et al. 2007).

In addition to regulation of channel activity by phosphorylation/dephosphorylation, heteromultimerization may also contribute to modify channel properties and capacities for K^+ uptake (Lebaudy et al. 2008). The α-subunit AtKC1 has been shown to modulate AtAKT1 activity by displacing to more negative values its activation threshold, adjusting AtAKT1-mediated K^+ uptake (Reintanz et al. 2002; Duby et al. 2008). On the other hand, the AtSYP121 SNARE protein interacts with AtKC1, activating K^+ channel current and K^+ uptake. Mutation in any of the three genes *AtAKT1*, *AtKC1* or *AtSYP121* suppressed the inward-rectifying K^+ current in root protoplasts and K^+ acquisition and growth in plants (Honsbein et al. 2009) (Fig. 1).

Besides AKT1, other *Shaker*-like K^+ channels have been described that may be important for net uptake of K^+ at the root. The outward-rectifier AtGORK mediates K^+ efflux in guard cells and contributes to stomatal closure, but it is also expressed in root hairs and may mediate K^+ efflux from the root and be involved in K^+ sensing (Ivashikina et al. 2001; Ward et al. 2009).

3 Sodium Transport

Sodium transport processes have major roles in salinity tolerance, including organellar Na^+ sequestration, Na^+ extrusion by plasma membrane Na^+/H^+ exchangers, and exclusion of Na^+ from photosynthetic tissues and meristems. Multiple independent cation-uptake transporters have been suggested to mediate Na^+ uptake from the soil into roots, even though some of them, such as nonselective cation channels, appear to fulfill primary roles in nutrient uptake or signaling events. Sodium uptake, despite being regulated to some extent and minimized under salinity stress, is inevitable in the long term. A large fraction of Na^+ entering the root is effluxed back to the medium, and yet Na^+ will accumulate over time. Consequently, long-distance transport of Na^+ to redistribute this ion among tissues and organs with lesser sensitivity to Na^+ toxicity is paramount. Significant progress has been made in the identification of key elements in this whole-plant process and their contribution to salt tolerance. Ion exchangers mediating the extrusion of Na^+ from the cytosol have been identified and their regulation is being unraveled.

3.1 Sodium Influx at the Plasma Membrane

Plant cells are endowed with an array of ion transporters that regulate the flux of Na^+ in and out of the cell. The plasma membrane negative-inside electric potential characteristic of all living cells creates an electrochemical gradient for the downhill flux of Na^+ towards the cell cytoplasm. So far, the only known transporters that mediate substantial and Na^+-specific uptake belong to the HKT family. High-affinity Potassium Transport (HKT) proteins were first identified as suppressors of yeast mutants unable to grow in low K^+ media (Schachtman and Schroeder 1994), and hence their name, but subsequent research showed that HKT proteins are Na^+-specific transporters that mediate either Na^+ uniport or Na^+-K^+ symport (Rubio et al. 1995; Horie et al. 2001, 2009; Yao et al. 2010). Ion transport specificity partly depends on whether the HKT protein has a highly conserved serine (subfamily 1) or glycine (subfamily 2) residue in the first pore-loop of the protein and, likewise other transport proteins with multiple cation binding sites, on the extracellular Na^+/K^+ ratio (Horie et al. 2009; Yao et al. 2010). Generally, HKT members of subfamily 1 have a relatively higher Na^+-to-K^+ selectivity than subfamily 2 HKT proteins, but there are exceptions to this rule (reviewed in Horie et al. 2009). In *Arabidopsis*, the single copy gene *HKT1;1* encodes a Na^+-selective transporter (subfamily 1) that is preferentially expressed in the vasculature and controls Na^+ distribution between roots and shoots (Berthomieu et al. 2003; Sunarpi et al. 2005). Protein AtHKT1;1 is localized in the plasma membrane of xylem parenchyma cells where it is thought to unload Na^+ from xylem vessels, thereby minimizing the amount of Na^+ reaching the leaves and preventing the inhibition of salt-sensitive photosynthetic processes (Sunarpi et al. 2005; Munns and Tester 2008). Measurement of unidirectional Na^+ influx rates indicate that AtHKT1;1 contributes little to net Na^+ uptake by *Arabidopsis* root (Essah et al. 2003).

In contrast with dicot species that have few *HKT* genes in their genomes, multiple *HKT* genes are found in monocots and some of them do mediate Na^+ uptake from the soil into roots (Garciadeblas et al. 2003). The rice protein OsHKT1;5 (SKC1), and the *Triticum monococcum* proteins TmHKT1;4-A2 (HKT7-A2) and TmHKT1;5 (HKT8), all of them subfamily 1 orthologs of AtHKT1;1, have been shown to mediate the retrieval of Na^+ from the xylem sap and exclusion from leaves (Ren et al. 2005; Huang et al. 2006; Byrt et al. 2007). Although Na^+ exclusion is often regarded as a primary determinant of salinity tolerance in many species (Munns et al. 2006), only *HKT1;5* among these QTL-linked *HKT* genes and which corresponds to the *Kna1* locus of bread wheat *T. aestivum*, has been shown to impart salinity tolerance (Dvořak et al. 1994).

Members of subfamily 2 of HKT transporters mediate Na^+ uptake by roots of cereal plants. Down-regulation of wheat *TaHKT2;1* (previously *TaHKT1*) by anti-sense RNA produced a significant decrease in $^{22}Na^+$ short-term influx, reduced the net Na^+ content in roots, and enhanced growth under salinity of transgenic wheat plants when compared with control plants (Laurie et al. 2002). In rice, disruption of gene *OsHKT2;1* reduced growth compared with wild-type plants under low Na^+ and K^+ conditions. Approximately 80–90% of $^{22}Na^+$ influx into K^+-starved rice roots was mediated by OsHKT2;1 at low Na^+ concentrations (1–200 μM). *OsHKT2;1* was mainly expressed in the cortex and endodermis of roots, and the mRNA was up-regulated by K^+ starvation but rapidly repressed upon the addition of 30 mM NaCl (Horie et al. 2007). Hence, the relative contribution of OsHKT2;1 to Na^+ influx to net root Na^+ influx is minimized at an increase in the Na^+

concentration, presumably to prevent Na^+ toxicity. Furthermore, *oshkt2;1* mutant plants did not show any difference under salt stress compared to wild-type plants. These features may also be applicable to other subfamily 2 *HKT* genes in cereals whose transcript levels have been shown to increase by K^+-starvation but are down-regulated by salinity stress (reviewed in (Horie et al. 2009)). It is worth noting that at low K^+ availability, moderate levels of Na^+ actually promote plant growth, presumably by replacing K^+ for osmotic adjustment (Rodriguez-Navarro and Rubio 2006). In rice and other species of the *Poaceae* family, high-affinity Na^+ uptake is probably mediated by HKT transporters. In other plants, high-affinity Na^+ uptake is mediated by one or several transporters fundamentally different from HKT transporters and they remain to be identified (Haro et al. 2010). Thus, subfamily 2 HKTs, which are generally more permeable to K^+ compared with the subfamily 1 HKTs, are thought to mediate high-affinity Na^+-K^+ symport at low external K^+ concentrations, thereby providing "substitute nutritional" Na^+ under K^+ starvation but bearing little relevance to salinity tolerance (Horie et al. 2009).

Nothing is known about the molecular processes regulating the activity of HKT proteins. Plant HKTs are phylogenetically related to fungal TRK and to bacterial HtrB and TrkH K^+ transporters, although well conserved primary sequences are not found outside the selectivity filters in the P-loop domains that form the outer portion of the pore (Rodriguez-Navarro 2000). Fungal TRK proteins are thought to facilitate K^+ uptake by K^+-H^+ symport at low K^+ and low pH or as K^+ uniport at high pH, although they may permeate Na^+ depending on the conditions (Rodriguez-Navarro 2000). They also play a key role in the fine-tuning of the plasma membrane electric potential (Madrid et al. 1998). In *S. cerevisiae*, the activity of TRK1 and TRK2 proteins is regulated by two redundant protein kinases, HAL4 and HAL5 (Mulet et al. 1999). HAL4/HAL5 kinases are essential to sustain TRK-dependent K^+ transport, promoting the influx of K^+ and decreasing the membrane potential. However, these kinases are not required for TRK1 activity itself. Rather, they stabilize the transporter at the plasma membrane under low K^+ conditions, preventing its endocytosis and vacuolar degradation (Perez-Valle et al. 2007). The protein phosphatases PPZ1 and PPZ2 deactivate TRK1, and PPZ1 is in turn inhibited by its regulatory subunit HAL3 (de Nadal et al. 1998; Yenush et al. 2002). Another phosphatase, calcineurin, is required for the activation of TRK-dependent high-affinity K^+ transport under Na^+ stress (Mendoza et al. 1994). Other proteins involved in the regulation of K^+ transport include the protein kinase SKY1 (Forment et al. 2002) and the G protein of the Ras superfamily, ARL1 (Munson et al. 2004). However, the precise mechanism of regulation by these factors is uncertain. Also unknown are their functional homologues in plant cells which, by analogy, might be involved in the regulation of the HKT/TRK proteins. Also, the signals triggering regulatory events controlling the activity of the TRK system are poorly defined. The TRK1 protein is stabilized by high intracellular K^+ levels (Perez-Valle et al. 2007), but the signal promoting TRK-dependent K^+ transport is not the internal K^+ concentration itself (Ramos et al. 1990). Similar conclusions have been reached regarding the up-regulation of *HAK5* by K^+-starvation, which seems uncoupled to the K^+ status of the cell (see Sect. 2.2). Whether early regulatory events controlling

TRK proteins in fungi are also applicable to plant HKT proteins is uncertain since these transporters are primarily involved in Na^+ uptake, rather than of K^+.

Electrophysiological evidence suggest that nonselective cation channels (NSCCs) contribute to Na^+ uptake in plant cells because they show inward, noninactivating currents and low discrimination between a wide range of monovalent and divalent cations, channel properties likely associated with sustained cation uptake (reviewed in Demidchik and Maathuis 2007), but the molecular identification of the corresponding transport proteins is still pending. Among the different classes in which NSCC activities have been categorized based on electrophysiological features, the voltage-independent NSCCs (VI-NSCCs) are considered as prime candidates for Na^+ entry in roots, although their main role is thought to be Ca^{2+} uptake (Demidchik and Maathuis 2007). Na^+ currents through VI-NSCCs are directly blocked by external Ca^{2+} in a similar manner as Ca^{2+} inhibits Na^+ influx into root tissues (Essah et al. 2003; Wang et al. 2006). A direct comparison between *T. halophila* and *A. thaliana*, two closely related species that differ greatly in their salt tolerance, indicated that voltage-independent channels play a role in Na^+ uptake in both species (Volkov and Amtmann 2006). However, the permeability to Na^+ was lower in protoplast of root cells in *T. halophila* compared to *A. thaliana*, in agreement with the twofold lower rate of unidirectional influx of $^{22}Na^+$ in intact roots of the salt-tolerant species (Wang et al. 2006). Moreover, *T. halophila* cells succeeded in partly preventing a Na^+-induced depolarization of the plasma membrane, thereby averting excessive K^+ loss. Although these findings are suggestive, a direct link between gene products and in vivo observed channel currents is still largely absent to firmly establish the role of NSCCs in Na^+ uptake and salt tolerance.

Genomic sequencing has revealed the existence of two large families of putative NSCCs bearing sequence similarity to animal counterparts which had not been identified by previous analyses, namely the Cyclic Nucleotide-Gated Channels (CNGCs) and the Glutamate Receptor-like channels (GLRs) (Davenport 2002; Kaplan et al. 2007). Little functional data are available for GLRs, partly because ad hoc heterologous expression systems fail to yield functional channels (Davenport 2002), whereas understanding of CNGCs function is fragmented (Demidchik and Maathuis 2007). Plant CNGC proteins characterized so far in heterologous systems show permeability to K^+, Ca^{2+} and, with few exceptions, to Na^+ as well (Demidchik and Maathuis 2007). They show inward and noninactivating current, channel properties likely associated with sustained cation uptake in plants (Maathuis and Sanders 2001; Leng et al. 2002; Balague et al. 2003; Hua et al. 2003). The predicted structure of plant CNGCs resemble those of the *shaker* family of K^+ channels (six transmembranes and a P-domain between TM-5 and TM-6) but are defined by unique pore selectivity filters, unlike those found in any other family of channels, and by the presence at the C terminus of partly overlapping cyclic nucleotide- and calmodulin-binding domains (Hua et al. 2003; Demidchik and Maathuis 2007). This arrangement suggests opposing regulation by calmodulin and cyclic nucleotides that is reminiscent of the antagonistic effects of these signaling intermediaries on animal CNGC channel gating. In animal systems,

CNGCs are voltage-independent and activated by cyclic nucleotides (cAMP and cGMP). Likewise, activation of heterologously expressed plant CNGCs by cAMP and cGMP has been demonstrated (Leng et al. 2002; Talke et al. 2003). Physical interaction with calmodulin has also been shown for several plant CNGCs (Köhler et al. 1999) (Schuurink et al. 1998; Arazi et al. 1999) and evidences of the inhibitory effect of calmodulin binding has been obtained (Hua et al. 2003; Li et al. 2005b; Ali et al. 2006). However, CNGC-like activity in the plasma membrane of *Arabidopsis* root protoplasts was *deactivated* by cAMP and cGMP (Maathuis and Sanders 2001) and externally applied cGMP restricted Na^+ uptake and increased salt tolerance in *Arabidopsis* and pepper plants (Maathuis and Sanders 2001; Essah et al. 2003; Rubio et al. 2003). Furthermore, salt and osmotic stress cause rapid increases in cGMP levels in *Arabidopsis thaliana* (Donaldson et al. 2004). Together, these results indicate that, in planta, CNGC are inactivated to achieve salt tolerance. The bases for the discrepancies regarding the gating of CNGCs by cyclic nucleotides in plant cells and heterologous systems remain to be resolved. Perhaps, cyclic nucleotides indirectly affect the activity of NSCCs other than CNGCs in electrophysiological studies, resulting in the observed phenomena. In this case, CNGCs would not account for the NSCC activities that are measured in plant cell membranes.

Based on genetic evidence plant CNGCs are mainly involved in developmental processes and plant pathogen interactions, linking cytosolic cAMP elevation to Ca^{2+}-dependent signal transduction (Talke et al. 2003; Ma et al. 2009a), whereas reports relating CNGC activity to salt tolerance phenotypes are scant. The *Arabidopsis* CNGC1 and CNGC3 proteins transport Na^+ as well as K^+ in yeast, they are expressed in plant roots, and thus they may provide a physiologically significant Na^+ entry pathway from the soil solution (Hua et al. 2003; Ali et al. 2006; Gobert et al. 2006). Indeed, a *cngc1* knockout mutation in *Arabidopsis* led to improved salt tolerance, although no significant affectation of the shoot and root Na^+ content was observed (Maathuis 2006). Null mutants of *cngc3* had decreased seed germination in the presence of NaCl but not KCl or sorbitol, yet they showed slightly greater tolerance to NaCl and KCl at the seedling stage. Short-term uptake of Na^+ was greater in the *cngc3* mutant relative to the wild-type, but no significant differences for Na^+ content were found after prolonged periods indicating the limited participation of CNGC3 in the long term Na^+ uptake (Gobert et al. 2006). AtCNGC10 complemented K^+ uptake mutants of *E. coli* and *S. cerevisiae*, as well as the *akt1* mutant of *Arabidopsis*. Transgenic plants overproducing AtCNGC10 grew better on K^+-limited medium, whereas antisense plants had 50% less K^+ than wild type plants. Compared with the wild-type, mature plants of antisense lines had altered K^+ and Na^+ concentrations in shoots and were more sensitive to salt stress. The shoots of these plants contained higher Na^+ contents and significantly higher Na^+/K^+ ratios compared with wild-type, whereas roots had higher K^+ contents and lower Na^+/K^+ ratios. Upon salt exposure, the Na^+ efflux was higher and the K^+ efflux was smaller in the antisense lines, indicating that AtCNGC10 might function as a channel providing Na^+ influx and K^+ efflux at the root/soil interface ((Guo et al. 2008) and references therein). However, lower Ca^{2+} influx and reduced Ca^{2+}

cytosolic activity in antisense *cngc10* lines have also been reported (Guo et al. 2010), and AtCNGC10 is known to affect a suite of developmental processes (Borsics et al. 2007). Consequently, the pleiotropic nature of *AtCNGC10* gene suppression complicates the assessment of its role in salt tolerance. Moreover, it should be noted that CNGCs conduct both monovalent and divalent cations, but in the presence of millimolar concentrations Ca^{2+} likely present in the apoplast, monovalent cation conductance through CNGCs is restricted (Leng et al. 2002). On the other hand, treatment of plants with membrane permeable cGMP, which presumably affects salt tolerance by modulating CNGC channel activity, also alters the transcription of many genes, particularly of those encoding transporter proteins (Maathuis 2006). Considering the increasing evidence that CNGCs are components of Ca^{2+} and cyclic nucleotide-dependent signaling pathways, an indirect signaling role of CNGC on K^+ and Na^+ transport processes mediated by proteins other than CNGC themselves cannot be ruled out.

3.2 Sodium Efflux and Long-Distance Transport; the SOS System

Genetic and biochemical evidence indicates that the transport protein SOS1 catalyzes electroneutral Na^+/H^+ exchange at the plasma membrane (Qiu et al. 2002; Shi et al. 2002). SOS1 appears to be highly specific for Na^+ and does not transport other monovalent cations, such as K^+ or Li^+ (Quintero et al. 2002; Shi et al. 2002). Based on the expression pattern of *SOS1* and the characterization of *sos1* mutants in *Arabidopsis*, *Thellungiella*, and tomato plants, it has been suggested that SOS1 controls both Na^+ efflux in the root and long-distance Na^+ transport via xylem to partition this ion among root and shoot (Shi et al. 2002; Oh et al. 2009; Olias et al. 2009a, b). Despite substantial efflux of Na^+ across the plasma membrane of root cells, the net flux of Na^+ is unidirectional from soil to roots and then to the shoot, except for limited recirculation via the phloem (Tester and Davenport 2003). Once Na^+ has entered the root endodermis, which represents a barrier to ions (Peng et al. 2004; Fernandez-Garcia et al. 2009), it is translocated to aerial parts via xylem following the movement of water in the evapotranspiration stream. The Na^+ concentrations in the xylem sap of plants subjected to salinity stress appear to be in the low-millimolar range (7–70 mM) (Munns 1985; Shi et al. 2002; Olias et al. 2009a, b), not very different from cytosolic concentrations (Tester and Davenport 2003). Since mature xylem elements are part of the apoplast and the membrane electrical potential is negative inside xylem parenchyma cells, the loading of Na^+ into the xylem is thought to be active, i.e., mediated by secondary transporters dissipating the pre-established electrochemical gradient of H^+ between the xylem parenchyma cells and xylem elements (Tester and Davenport 2003). In *Arabidopsis* roots, a *SOS1:GUS* reporter gene was preferentially expressed in the epidermal cells of the root tip, shifting to the pericycle and parenchyma cells bordering xylem vessels in the mature root. In stem and petiole, expression was also maximal in parenchyma cells at the xylem/symplast boundary. The strong contribution of SOS1 to the halotolerance of *T. salsuginea* (salt cress), a halophyte that is a close relative of *Arabidopsis*, was demonstrated using post-transcriptional gene silencing (Oh et al. 2009). Suppressed plants showed faster Na^+ accumulation and slower removal of Na^+ from the root. Coherent with a role of SOS1 is the loading of Na^+ into the xylem, Na^+ strongly accumulated inside the pericycle of *sos1*-deficient plants, while Na^+ was confined in vacuoles of epidermis and cortex cells in the wild-type. In tomato, transgenic plants with reduced expression of *SOS1* showed enhanced sensitivity to salinity that correlated with higher accumulation of Na^+ in leaves and roots, but lower contents in stems of silenced plants under salt stress. Lower net Na^+ flux was observed in the xylem sap of suppressed lines. The ability of tomato plants to retain Na^+ in the stems, thus preventing Na^+ from reaching the photosynthetic tissues, was largely dependent on the function of SOS1 (Olias et al. 2009a, b). Thus, SOS1 is currently thought to function loading Na^+ into the xylem for controlled delivery to the shoot. In photosynthetic tissues that need to be protected from Na^+ toxicity, SOS1 could reduce the net rate of Na^+ unloading. Fine-tuning of net ion uptake by roots, xylem loading for long-distance distribution, ion retrieval by HKT proteins at specific plant tissues, and vacuolar compartmentation along the plant axis and in older leaves, would all be required to avert Na^+ toxicity (Pardo et al. 2006; Olias et al. 2009a, b).

The partially redundant protein kinase complexes formed by proteins SOS2/SOS3 and SOS2/SCaBP8 are major regulators of SOS1 activity (see below). *Arabidopsis* mutants lacking any of the SOS proteins manifest symptoms of K^+ deficiency on culture medium with low K^+ ($<200\ \mu M\ K^+$), implying that the SOS pathway assists K^+ acquisition (Zhu et al. 1998). SOS1 is the only known target of the SOS2/SOS3 kinase complex and SOS1 has been shown to mediate Na^+/H^+ exchange but to lack any detectable K^+ transport capacity, both in its basal state and after activation by the SOS2/SOS3 kinase (Qiu et al. 2002, 2004; Quintero et al. 2002; Shi et al. 2002;). High Na^+ concentration in the soil solution is thought to disturb K^+ acquisition by competing with binding sites in transport systems that mediate K^+ uptake (Hasegawa et al. 2000), but the K^+-deficiency phenotype of *sos* mutants is perplexing and raises the question of how Na^+ impinges on K^+ nutrition beyond the competition for common transporters. Two independent observations suggest that the K^+-deficient phenotype of *sos* mutants is secondary and just a consequence of disturbed Na^+ homeostasis. Firstly, *Thellungiella* plants with suppressed *SOS1* expression showed significantly higher Na^+ contents in cells of the pericycle and the stele, measured with the Na^+-specific probe CoroNa-Green. Micro X-ray analyses confirmed higher Na^+ content in the root stele and further identified xylem parenchyma cells as the primary point of Na^+ accumulation in the mutant root. The Na^+/K^+ ratio in cells surrounding the roots vasculature was 12-fold higher in mutant plants compared to wild-type, which is coherent with the stated role of SOS1 in loading Na^+ into the xylem (Oh et al. 2009). Secondly, the K^+-uptake capacity (membrane permeability) of the *sos1* cortical root cells measured electrophysiologically was normal in control conditions but strongly reduced in the presence of NaCl. Introducing 10 mM NaCl into the cytoplasm of patch-clamped whole root cells completely abrogated the activity of the K^+ channel AKT1 (Qi and Spalding 2004), which is known to contribute to physiological K^+ uptake (see Sect. 2.1). These results, suggest that AKT1 appears to be a target of Na^+ toxicity in *sos1* plants, resulting in poor growth due to impaired K^+ uptake. However, these findings could not be confirmed (Volkov and Amtmann 2006). In any case, the low-K^+ sensitivity phenotype of the *Arabidopsis sos1* mutant disappears in hydroponic culture with media formulated to be essentially free of Na^+, whereas as little as 5 mM Na^+ gives rise to impaired growth in 1 mM K^+ (Lacal et al. unpublished).

Significant progress has been made in understanding the regulation of SOS1. Na^+/H^+ exchange by SOS1 was found in plasma membrane vesicles from salt-stressed *Arabidopsis* plants but not in control plants, indicating that SOS1 activity was induced by salinity (Qiu et al. 2002). This up-regulation of SOS1 activity occurs both at the level of mRNA stability and through protein phosphorylation that regulates the specific rate of ion exchange (Quintero et al. 2002; Martinez-Atienza et al. 2007; Chung et al. 2008). Key intermediaries in the regulation of SOS1 are the protein kinase SOS2 (also known as CIPK24) and its associated Ca^{2+}-sensor protein SOS3 (CBL4) (Fig. 2). *Arabidopsis* plants deficient in either *SOS2* or *SOS3* share the salt-sensitive and K^+-deficiency phenotype of *sos1* plants (Zhu 2000). SOS2 is a serine/threonine protein kinase belonging to the SnRK3 family (also known as PKSs or CIPKs) (Kolukisaoglu et al. 2004). SOS3 belong to the recoverin-like family of small Ca^{2+}-binding proteins denoted as SCaBPs or CBLs (Gong et al. 2004; Kolukisaoglu et al. 2004). Upon Ca^{2+} binding, SOS3 undergoes dimerization and hydrophobic domains are exposed (Sanchez-Barrena et al. 2005). Changes in surface properties facilitate interaction with the FISL/NAF domain of SOS2 to relief the kinase from autoinhibition by its C-terminal domain, thereby transmitting the Ca^{2+} signal elicited by salt stress. Without SOS3, the N-terminal catalytic domain of SOS2 binds to the FISL/NAF motif and this intramolecular interaction locks the kinase in a "closed" state (Guo et al. 2001). SOS3 opens a cavity in SOS2 and binds to the autoinhibitory FISL/NAF motif, relieving the kinase from auto-inhibition (Sanchez-Barrena et al. 2007). Adjacent to the SOS3-binding domain of SOS2 is the PPI domain enabling the interaction of the protein kinase with ABI2, a type-2C protein phosphatase that is a negative intermediary in ABA signaling (Ohta et al. 2003). The interaction of SOS3 with the FISL/NAF domain partially occludes the PPI motif and in vitro competition experiments demonstrated that SOS3 and ABI2 do compete for binding to SOS2. Thus, the binding of ABI2 and SOS3 to SOS2 is mutually exclusive and kinase activation by SOS3 is only possible when the SOS2 kinase is not interacting with ABI2. Further studies are required to establish whether interaction between SOS2 and ABI2 is constitutive, keeping the kinase in a sequestered, inactive state. Such a regulatory mechanism would be reminiscent of the role that PP2C phosphatases play on the regulation of SnRK2 kinases in ABA signaling (Pardo 2010). In the ABA-dependent signaling pathway, PP2C protein phosphatases bind to and counteract the phosphorylation of the activation loop in SnRK2 protein kinases, keeping them inactive (Fujii et al. 2009; Umezawa et al. 2009; Vlad et al. 2009). This interaction is constant in the absence of ABA. The locking of ABA into the receptor proteins PYR/PYL/RCAR

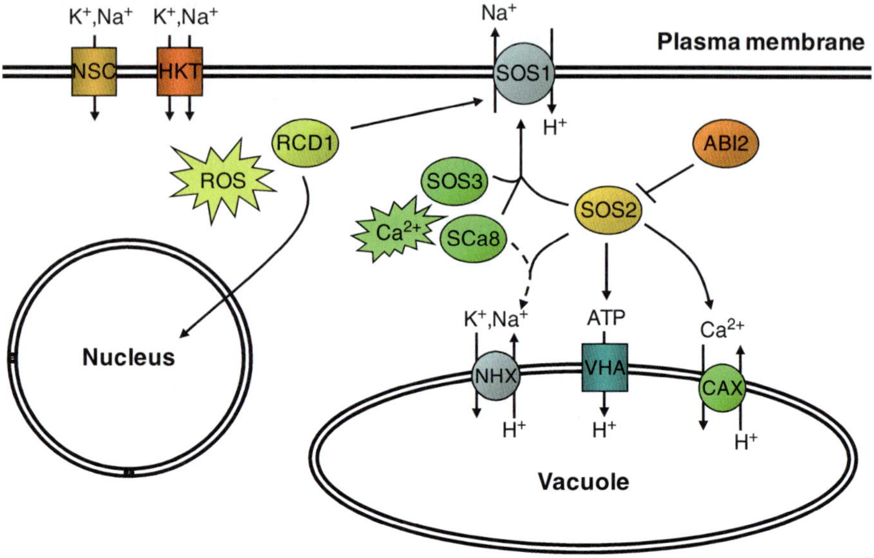

Fig. 2 Regulation of Na^+ homeostasis. A hypothetical plant cell is depicted showing relevant Na^+ transport processes and regulatory events. Na^+ ions enter the cell through, at least, nonselective cation channels (NSC) and HKT proteins at the plasma membrane. By an as yet unknown mechanism, Na^+ entry and/or accumulation is transduced into ROS and Ca^{2+} signaling which, in turn, regulate the activity of specific intermediaries such as RCD1 and SCaBP/CBL proteins, respectively. RCD1 interacts with SOS1 and shuttles between cytoplasm and nucleus where it is thought the control gene expression related to oxidative stress. The protein kinase SOS2 plays a focal role in the regulation of Na^+ and Ca^{2+} transport. SOS3 and SCaBP8/CBL10 (SCa8) recruit SOS2 to phosphorylate SOS1 and activate Na^+ efflux. SCaBP8 may also play a role in recruiting SOS2 to the tonoplast where the kinase regulates NHX proteins, albeit indirectly. SOS2, without ancillary SCaBP/CBL proteins, interact with and activate the V-ATPase (VHA) and CAX antiporters at the tonoplast. ABI2 plays a negative role in these regulatory events by inhibiting SOS2

creates a binding surface for the active site of PP2Cs, thereby relieving SnRK2s from inhibition (Ma et al. 2009b; Park et al. 2009). In turn, the activated SnRK2 phosphorylates downstream targets, including ABA-responsive transcription factors (ABF) to start transcriptional responses (Fujii et al. 2009).

Although the interaction between SOS2 and ABI2 may represent a cross-talk point between ABA signaling and the SOS pathway for ion homeostasis under salinity stress, there is no direct evidence of the regulation of the SOS proteins by ABA. Moreover, *sos1*, *sos2* and *sos3* mutants are specifically defective in Na^+ tolerance but not in ABA responses (Ohta et al. 2003). The dominant negative protein phosphatase ABI2 encoded by the *abi2-1* allele, which brings about ABA insensitivity rather than hypersensitivity, cannot interact with SOS2 and imparts greater tolerance to salt shock, suggesting that dissociation of the ABI2/SOS2 complex is a requisite for progression of the salt-stress signal. Whether or not activation of SOS1 is a signaling output of the stress phenotype associated with the *abi2-1* allele is presently unknown. Another unresolved issue is how Ca^{2+}-binding by SOS3 regulates interaction with SOS2. The structure of the SOS2/SOS3 complex showed two calcium atoms at domains EF1 and EF4 of SOS3 (Sanchez-Barrena et al. 2007). In contrast, the structure of Ca^{2+}-bound SOS3 suggested that all four Ca^{2+}-binding EF sites were occupied with Ca^{2+} (Sanchez-Barrena et al. 2005). Earlier reports showed that the interaction between SOS3 and the FISL/NAF domain of SOS2 was not dependent on supplemental Ca^{2+} (Guo et al. 2001). These conflicting data make it difficult to rationalize the role that stress-induced Ca^{2+} signaling plays in the mechanistic workings of the SOS system.

The rate of Na^+/H^+ exchange of SOS1 is controlled through protein phosphorylation by the SOS2/SOS3 and SOS2/SCaBP8(CBL10) kinase complexes (Qiu et al. 2002; Quintero et al. 2002; Martinez-Atienza et al. 2007) (Fig. 2). SOS1 activity in *Arabidopsis* plasma membrane vesicles could not be induced by salt stress in *sos2* or *sos3* mutants (Qiu et al. 2002). However, addition of a constitutively active, SOS3-independent form of SOS2 to plasma membrane vesicles isolated from *sos2* or *sos3* plants, but not from *sos1*, restored SOS1 activity. The SOS2/SOS3 kinase complex phosphorylates SOS1 at the C terminus and stimulates its Na^+ extrusion activity by releasing the exchanger from autoinhibition by its C-terminal part (Quintero et al. 2002; Pardo et al. 2005). Besides kinase activation, SOS3 was also shown to recruit SOS2 to the plasma membrane to achieve efficient interaction with SOS1 (Quintero et al. 2002). N-myristoylation of SOS3 is necessary for its functionality *in planta* (Ishitani et al. 2000). Nonetheless, the evaluation in Arabidopsis of an array of hyperactive mutant variants of SOS2 suggested that the kinase may reach the plant plasma membrane in the absence of a functional SOS3 protein, presumably as a result of interactions with alternative SCaBP/CBL proteins (Qiu et al. 2004). Indeed, recent reports have shown that the alternative protein kinase module SOS2/SCaBP8 does regulate SOS1 preferentially in the shoot, whereas the SOS2/SOS3 module would be most active in the root (Quan et al. 2007). Accordingly, *sos3 scabp8* and *sos2 scabp8* mutants display a phenotype similar to a *sos2*, which is more sensitive to salt than either *sos3* or *scabp8* alone. There is, however, evidence that SOS3 and SCaBP8 proteins are not distinct only in their specific expression pattern but also in functionality. Overexpression of SCaBP8 in the *sos3* mutant partially rescued the salt-sensitive phenotype, whereas overexpression of SOS3 failed to complement the *scabp8* mutant (Quan et al. 2007). Moreover, SOS2 phosphorylates SCaBP8 but not SOS3 (Lin et al. 2009). This phosphorylation was induced by salt stress, occurred at the membrane, stabilized the SCaBP8-SOS2 interaction, and enhanced the Na^+/H^+ exchange activity of SOS1. Also worth noting is that SCaBP8 attaches to membranes by a process that involves an N-terminal hydrophobic domain (Quan et al. 2007) that is unique among SCaBP-CBL proteins which most often use fatty-acyl modifications to interact with various plant cell membranes (Batistic et al. 2010). Lastly, SOS2(CIPK24) is also recruited to the plasma membrane by CBL1, although the salt-sensitivity of the *Arabidopsis cbl1* mutant is only moderate compared to *sos* mutants and the target(s) of SOS2/CBL1 remains unknown (Albrecht et al. 2003; Waadt et al. 2008).

Besides the reported localization of SCaBP8 in the plasma membrane of *Arabidopsis* and in yeast cells, in agreement with a role in the regulation of SOS1 (Quan et al. 2007), the SOS2/SCaBP8 complex (synonymous of CIPK24/CBL10) was also localized in large endosomal vesicles and in the tonoplast (Waadt et al. 2008; Batistic et al. 2010), and the *cbl10* (*scabp8*) mutant showed a Na^+ content that, unlike the SOS mutants, was significantly lower than in the wild-type under either normal or high-salt conditions (Kim et al. 2007). Moreover, the *sos2* mutant had reduced tonoplast Na^+/H^+-exchange activity originating from NHX proteins (Qiu et al. 2004). By contrast, tonoplast Na^+/H^+-exchange activity was significantly higher in the *sos1* mutant and unchanged in *sos3* relative to the wild-type. Unexpectedly, vacuolar cation/proton antiport activity in the *cbl10* mutant also remains unchanged, raising the possibility that the altered Na^+ contents in the *cbl10* mutant is not a consequence of direct regulation of NHX exchangers (Jiang, Quintero and Pardo, unpublished). The SOS2 protein is also known to interact with and regulate other ion transporters in the vacuoles, namely the peripheral regulatory subunits B1 and B2 of the V-ATPase that helps establish the proton gradient across the tonoplast (Batelli et al. 2007), and the $Ca2^+/H^+$ antiporter CAX1 which plays a role in intracellular $Ca2^+$ homeostasis (Cheng et al. 2004). These regulatory events are independent of CBL/SCaBP proteins and do not involve protein phosphorylation by SOS2 of the target proteins. Hence, it appears that SOS2 is a multi-functional protein kinase that coordinately regulates different aspects of Na^+ and $Ca2^+$ homeostasis for salt tolerance by interacting with distinct SCaBP-CBL calcium sensors and key transport proteins (Fig. 2).

As mentioned earlier, the transcript level of *SOS1* is up-regulated by salt stress by a mechanism that is partly dependent on SOS2 and SOS3 (Shi et al. 2000, 2003). A major component of the regulation of mRNA abundance appears to be post-transcriptional because salinity-induced mRNA accumulation was observed even when transcription of a *SOS1* cDNA was driven by the constitutive CaMV 35S promoter (Shi et al. 2003). The *SOS1* transcript is unstable and rapidly degrades with a half-life of approximately 10 min in the absence of stress (Chung et al. 2008). Salinity, dehydration, and H_2O_2 treatment increase the stability of SOS1 mRNA, suggesting that stress-induced mRNA stability is mediated by ROS. Accordingly, NADPH oxidase is also involved in the upregulation of SOS1 mRNA stability, presumably through the control of extracellular ROS production. The *cis*-element required for SOS1 mRNA instability resides in a 500-bp region at the 3′ end of the *SOS1* mRNA.

The link between SOS1 and ROS is further substantiated by the interaction of the SOS1 cytosolic C-terminal with RCD1 (Katiyar-Agarwal et al. 2006). The *Arabidopsis* protein RCD1 (radical-induced cell death) is a key transcriptional regulator of several ROS, hormonal, and abiotic stress-related responses (Jaspers et al. 2010). The *Arabidopsis* mutant *rcd1* is sensitive to externally applied ozone and superoxide but resistant to methyl-viologen, which generates ROS intracellularly in the chloroplasts (Katiyar-Agarwal et al. 2006). Like *rcd1* mutants, *sos1* mutant plants show an altered sensitivity to oxidative stresses (Katiyar-Agarwal et al. 2006; Chung et al. 2008). Without stress treatment, RCD1 is localized in the nucleus, but under high salt or oxidative stress RCD1 becomes nuclear-cytoplasmic and interacts with SOS1 (Katiyar-Agarwal et al. 2006). RCD1 also interacts with the transcription factors STO and DREB2A, which have been implicated in salt-stress tolerance (Belles-Boix et al. 2000). Accordingly, the *rcd1* mutation causes a decrease in salt tolerance and enhances the salt-stress sensitivity of *sos1* mutant plants. In addition to ion toxicity, salinity stress leads to the accumulation of high levels of ROS that disturb cellular redox homeostasis and lead to oxidative injuries (Miller et al. 2009). In salt-treated plants, the *sos1* and *rcd1* mutants accumulated higher levels of ROS compared with the wild-type (Katiyar-Agarwal et al. 2006). The lack of an additive effect between *sos1* and *rcd1* mutations on oxidative-stress tolerance indicate that SOS1 functions in the same pathway with RCD1 in oxidative-stress responses and that the role of SOS1 in oxidative stress is likely mediated by RCD1. However, the biochemical consequence of RCD1–SOS1 interaction in either protein is not known. Likely, the interaction affects oxidative-stress signaling, because the expression of ROS scavenging-related genes was altered by both *rcd1* and *sos1* mutants (Katiyar-Agarwal et al. 2006). These findings reveal a surprising function of a plasma membrane Na^+/H^+ antiporter in oxidative-stress tolerance and evoke the cross-talk between ion- and ROS-detoxification pathways (Miller et al. 2009).

A connection of SOS2 with ROS signaling has also been found (Verslues et al. 2007). SOS2 physically interacts with the H_2O_2 signaling protein nucleoside diphosphate kinase 2 (NDPK2) and with catalases 2 and 3. A *sos2 ndpk2* double mutant did not accumulate H_2O_2 in response to salt stress, suggesting that it is altered signaling rather than H_2O_2 toxicity alone that is responsible for the increased salt sensitivity of the *sos2 ndpk2* double mutant relative to single mutants. The interaction of SOS2 with NDPK2 occurred at the FISL motif, the same protein domain required for the interaction with SOS3 and adjacent to the interaction domain with the protein phosphatase ABI2. Interaction with SOS2 inhibited NDPK2 histidine autophosphorylation, indicating modulation of NDPK2 activity and, in turn, of its downstream targets, the H_2O_2-responsive mitogen-activated protein kinases AtMPK3 and AtMPK6. The *enh1* mutation causes enhanced accumulation of ROS, particularly under salt stress, and enhances the salt sensitivity of the *sos3* mutant (Zhu et al. 2007). The *ENH1* gene encodes a chloroplast-localized rubredoxin-like protein that shows greatest sequence similarity to rubredoxin proteins that play a role in superoxide detoxification. Since *sos2* but not *sos3* mutants show increased sensitivity to oxidative stress and the *enh1* mutation does not enhance *sos2* phenotypes, it appears that ENH1 also functions in the SOS2-dependent ROS signaling.

3.3 Transport at the Tonoplast for Na^+ Sequestration

In saline environments, the sequestration of ions that are potentially damaging to cellular metabolism (e.g., Cl^-, Na^+) provide an efficient mechanism for ion detoxification while maintaining high K^+/Na^+ ratios in the cytosol. Moreover, the accrual of ions inside the vacuole provides osmoticum for improved water uptake.

Biochemical and physiological data suggest a key role for Na^+/H^+ antiporters in regulating cytoplasmic Na^+ concentration and vacuolar sequestration (Blumwald 2000). In spite of the large number of putative cation/proton antiporters identified in plants by genomic sequencing (Maser et al. 2001) (see also http://wardlab.cbs.umn.edu/rice/; http://plantst.genomics.purdue.edu/), only the NHX family has been functionally linked to Na^+ sequestration into vacuoles (Blumwald 2000; Rodriguez-Rosales et al. 2009). Plant and fungal NHX antiporters are phylogenetically related to NHE subfamily of mammalian Na^+/H^+ exchangers and are present as multiple isoforms in all plant genomes sequenced to date (Rodriguez-Rosales et al. 2009). On the basis of protein sequence similarity, the NHE/NHX subfamily can be classified into two major groups that have been named Plasma Membrane (PM) and IntraCellular (IC) according to their subcellular localization (Brett et al. 2005). The PM group is exclusively present in animal cells whereas members of the IC group can be found in animals, plants and fungi. All plant NHXs characterized to date are assigned to the IC group and can be further divided into two main classes, class-I and class-II (Brett et al. 2005; Pardo et al. 2006; Rodriguez-Rosales et al. 2009). In *Arabidopsis*, isoforms AtNHX1–4 constitute the class-I category whereas AtNHX5 and 6 belong to class-II (Yokoi et al. 2002). The NHX gene family of rice has five members; OsNHX1–4 constitute the class-I, and OsNHX5 is the only class-II member (Pardo et al. 2006). All NHX proteins of class-I characterized to date are localized in the vacuolar membrane (Pardo et al. 2006; Rodriguez-Rosales et al. 2009). Class-II members are found in endosomal vesicles of plants and are grouped in the same clade than homologous proteins with various endosomal localizations present in animals and fungi (Brett et al. 2005; Pardo et al. 2006; Rodriguez-Rosales et al. 2009)). Each NHX class appears to have distinct ion selectivities too. Vacuolar exchangers of class-I catalyze Na^+/H^+ and K^+/H^+ exchange (reviewed in (Rodriguez-Rosales et al. 2009; Jiang et al. 2010)). The only class-II NHX antiporter of plants whose activity has been analyzed in detail is LeNHX2 of tomato. LeNHX2 colocalizes with prevacuolar and Golgi markers, affected the accumulation of K^+ but not Na^+ in intracellular compartments of transgenic plants and yeast, and purified LeNHX2 catalyzed K^+/H^+ exchange but only minor Na^+/H^+ exchange in reconstituted in proteoliposomes (Venema et al. 2003; Rodriguez-Rosales et al. 2008). Unlike AtNHX1 (class-I) that transports Na^+ and K^+ indistinguishably, the K^+/H^+ antiport activity of LeNHX2 is inhibited by low concentrations of Na^+ (Jiang et al. 2010). Presumably, the localization of class-II NHX proteins in endosomal compartments other than the vacuole imposes restrictions on ion selectivity and they may have a preference for K^+ over Na^+ to avoid poisoning of the endomembrane system (Hernandez et al. 2009; Rodriguez-Rosales et al. 2009). In fact, the dual affinity of vacuolar isoforms (class-I) implies that under standard growth conditions even class-I NHX proteins will mediate the accumulation of K^+ into vacuoles, as recently demonstrated in transgenic tomato overexpressing AtNHX1 (Leidi et al. 2010). Vacuoles are the main reservoirs of K^+ and ensure adequate supply to the cytosol (Leigh and Wyn Jones 1984). Cytosolic K^+ concentration declines only when vacuolar K^+ concentration decreases to values around 20 mM (Leigh and Wyn Jones 1984). Electrophysiological studies indicate the suitability of tonoplast cation/proton antiporters of the NHE/NHX subfamily for the homeostatic regulation of cytosolic K^+ when cells are provided with surplus concentrations of K^+ (Walker et al. 1996). Contrary to general belief, salt tolerance mediated by vacuolar NHX proteins is not strictly related to Na^+ accumulation, and reduced Na^+ content and increased K^+ content are equally often observed (reviewed in (Rodriguez-Rosales et al. 2009; Jiang et al. 2010). Evidence suggests that these proteins function in osmotic adjustment via K^+ or Na^+ accumulation, in accordance with their ion specificity and gene induction pattern. In addition, salt tolerance may be induced by other mechanisms involving indirect effects on vesicle trafficking via endosomal pH regulation by NHX proteins (Hernandez et al. 2009; Rodriguez-Rosales et al. 2009).

Very little is known about the signaling pathways regulating the expression and function of plant NHX proteins. Animal NHE transporters are controlled through distinct signaling networks that converge in the C-terminal regulatory domain. Modifications in the C terminus by phosphorylation and interaction with regulatory proteins, including the Ca^{2+}-binding proteins calmodulin and CHP (calcineurin B homologous), control transport activity by altering the affinity of the pore domain for intracellular protons (Putney et al. 2002). Similar mechanisms are beginning to emerge shedding light on the regulation of plant NHX exchangers. The activity of the *Arabidopsis* AtNHX1 is regulated by the binding of a calmodulin-like protein, AtCaM15, to its C terminus (Yamaguchi et al. 2005). The binding of AtCaM15 to AtNHX1 was Ca^{2+}- and pH-dependent and modified the Na^+/K^+ selectivity of the antiporter. AtCaM15 lowered the V_{max} of the Na^+/H^+ exchange activity, whereas K^+/H^+ exchange activity was not significantly altered. The interaction of AtCaM15 and the C terminus of AtNHX1 has been reported to occur in the vacuolar lumen (Yamaguchi et al. 2005), but there exist contradictory data on the topology of AtNHX1. Yamaguchi et al. (2003) reported a protein topology of AtNHX1 according to which the transmembrane domains (TMs) closer to the C terminus would have the opposite orientation relative to the corresponding TMs of human NHE. Hence, the AtNHX1 C-terminal domain would then be inside the vacuolar lumen, contrary to animal NHEs in which the regulatory C-terminal domain is cytosolic. By contrast, Sato and Sakaguchi (2005) concluded that AtNHX1 has the same topology than human NHE1 and that the C-terminal part of the plant protein is cytosolic. The reason for this discrepancy is unclear but the latter authors argued that in the study of Yamaguchi et al. (2003) the first TM of AtNHX1 was erroneously assigned as equivalent to the TM-1 of animal NHEs when it should be considered equivalent to TM-2. Intriguingly, AtCaM15, which interacts with the C terminus of AtNHX1, was localized in the vacuolar lumen suggesting the presence of signaling elements acting within the vacuole that could regulate AtNHX1 (Yamaguchi et al. 2005). On the other hand, proteomic analyses of tonoplast proteins have identified phospho-peptides derived from the C termini of NHX-like proteins of several species (Arabidopsis, (Whiteman et al. 2008b); rice, (Whiteman et al. 2008a); barley, (Endler et al. 2009)). These findings strongly suggest that the C termini of NHX proteins are cytosolic since ATP, as the substrate of protein kinases, is not localized inside the vacuole (Endler et al. 2009). In other words, the C terminus of

NHX proteins must be cytosolic to become phosphorylated by protein kinases. The protein kinase(s) targeting NHX proteins remain to be identified. As described in Sect. 3.2, NHX activity is compromised in the *sos2* mutant of *Arabidopsis* (Qiu et al. 2004). However, no phosphorylation of AtNHX1 by SOS2 was observed. Still, NHX proteins could be regulated through phosphorylation-independent, protein–protein interaction. Direct SOS2 stimulation of transport activity in the absence of a corresponding in vitro phosphorylation of the transport protein has been reported for CAX1, a tonoplast $Ca2^+/H^+$ exchanger in *Arabidopsis* (Cheng et al. 2004). Alternatively, SOS2 may indirectly affect the activity of NHX exchangers through the regulation of V-ATPase activity, which is also reduced in the *sos2* mutant (Batelli et al. 2007).

4 Concluding Remarks

Potassium and Na^+ homeostasis is achieved by the coordinate function of transporters that operate at different membranes in the various cell types that compose plant tissues. Plants are endowed with a large suite of K^+ and Na^+ transporters to secure K^+ supply to all cells and to avoid high concentrations of Na^+ in their cytoplasm. Besides constitutive transport capacities, the coordinate regulation of various transport systems is of chief importance to achieve ion homeostasis in widely varying environments. Regulation may take place at different levels that include transcriptional modulation of the genes encoding the transport systems, activation of the transporters themselves, and protein targeting to the different cell membranes. Plants can sense K^+ and Na^+ availability and they respond to different ionic stress conditions through sensors and signal cascades, although the mechanistic details remain largely unknown. The largest gap in our knowledge comprises the physicochemical clues that sensor proteins perceive and respond to. Current evidence indicates that sensing of the K^+ status involves, at least partly, monitoring of the membrane potential rather than cytosolic K^+ activity. Linking changes in membrane potential to know the generation of secondary signals transducers (ROS, Ca^{2+}) and to archetypical cascades of protein phosphorylation–dephosphorylation will be of prime importance. By analogy, it is tempting to speculate that the ionic component of salinity might also be perceived by early membrane depolarization caused by Na^+ influx. The systematic comparison of salt-tolerant and salt-sensitive species that are closely related phylogenetically, such as *T. halophyla* and *A. thaliana*, combined with the strength of molecular genetics, holds promise in the identification of principal components of stress tolerance.

Acknowledgments The authors apologize to the many authors whose original contributions could not be cited herein due to space constraints. This work was supported by grants BIO2009-08641 and CSD2007-00057 from MICINN (Spain) to JMP and grants 08696/PI/08 from Fundación Séneca of Región de Murcia (Spain) and AGL2009-08140 from MICINN of Spain to FR.

References

Ahn SJ, Shin R, Schachtman DP (2004) Expression of KT/KUP genes in Arabidopsis and the role of root hairs in K^+ uptake. Plant Physiol 134:1135–1145

Albrecht V, Weinl S, Blazevic D, D'Angelo C, Batistic O, Kolukisaoglu U, Bock R, Schulz B, Harter K, Kudla J (2003) The calcium sensor CBL1 integrates plant responses to abiotic stresses. Plant J 36:457–470

Ali R, Zielinski RE, Berkowitz GA (2006) Expression of plant cyclic nucleotide-gated cation channels in yeast. J Exp Bot 57:125–138

Anderson JA, Huprikar SS, Kochian LV, Lucas WJ, Gaber RF (1992) Functional expression of a probable Arabidopsis thaliana potassium channel in Saccharomyces cerevisiae. Proc Natl Acad Sci USA 89:3736–3740

Arazi T, Sunkar R, Kaplan B, Fromm H (1999) A tobacco plasma membrane calmodulin-binding transporter confers Ni^{2+} tolerance and Pb^{2+} hypersensitivity in transgenic plants. Plant J 20:171–182

Armengaud P, Breitling R, Amtmann A (2004) The potassium-dependent transcriptome of Arabidopsis reveals a prominent role of jasmonic acid in nutrient signaling. Plant Physiol 136:2556–2576

Balague C, Lin B, Alcon C, Flottes G, Malmstrom S, Kohler C, Neuhaus G, Pelletier G, Gaymard F, Roby D (2003) HLM1, an essential signaling component in the hypersensitive response, is a member of the cyclic nucleotide-gated channel ion channel family. Plant Cell 15:365–379

Bañuelos MA, Garciadeblas B, Cubero B, Rodriguez-Navarro A (2002) Inventory and functional characterization of the HAK potassium transporters of rice. Plant Physiol 130:784–795

Batelli G, Verslues PE, Agius F, Qiu Q, Fujii H, Pan S, Schumaker KS, Grillo S, Zhu JK (2007) SOS2 promotes salt tolerance in part by interacting with the vacuolar H^+-ATPase and upregulating its transport activity. Mol Cell Biol 27:7781–7790

Batistic O, Waadt R, Steinhorst L, Held K, Kudla J (2010) CBL-mediated targeting of CIPKs facilitates the decoding of calcium signals emanating from distinct cellular stores. Plant J 61:211–222

Belles-Boix E, Babiychuk E, Van Montagu M, Inze D, Kushnir S (2000) CEO1, a new protein from Arabidopsis thaliana, protects yeast against oxidative damage. FEBS Lett 482:19–24

Berthomieu P, Conejero G, Nublat A, Brackenbury WJ, Lambert C, Savio C, Uozumi N, Oiki S, Yamada K, Cellier F (2003) Functional analysis of AtHKT1 in Arabidopsis shows that Na+ recirculation by the phloem is crucial for salt tolerance. EMBO J 22:2004–2014

Blumwald E (2000) Sodium transport and salt tolerance in plants. Curr Opin Cell Biol 12:431–434

Borsics T, Webb D, Andeme-Ondzighi C, Staehelin LA, Christopher DA (2007) The cyclic nucleotide-gated calmodulin-binding channel AtCNGC10 localizes to the plasma membrane and influences numerous growth responses and starch accumulation in Arabidopsis thaliana. Planta 225:563–573

Brett CL, Donowitz M, Rao R (2005) Evolutionary origins of eukaryotic sodium/proton exchangers. Am J Physiol Cell Physiol 288:C223–C239

Buschmann PH, Vaidyanathan R, Gassmann W, Schroeder JI (2000) Enhancement of Na^+ uptake currents, time-dependent inward-rectifying K^+ channel currents, and K^+ channel transcripts by K^+ starvation in wheat root cells. Plant Physiol 122:1387–1398

Byrt CS, Platten JD, Spielmeyer W, James RA, Lagudah ES, Dennis ES, Tester M, Munns R (2007) HKT1;5-like cation transporters linked to Na^+ exclusion loci in wheat, Nax2 and Kna1. Plant Physiol 143:1918–1928

Cellier F, Conéjéro G, Ricaud L, Luu DT, Lepetit M, Gosti F, Casse F (2004) Characterization of *AtCHX17*, a member of the cation/H^+ exchangers, CHX family, from *Arabidopsis thaliana* suggests a role in K^+ homeostasis. Plant J 39:834–846

Cheng NH, Pittman JK, Zhu JK, Hirschi KD (2004) The protein kinase SOS2 activates the Arabidopsis H^+/Ca^{2+} antiporter CAX1 to integrate calcium transport and salt tolerance. J Biol Chem 279:2922–2926

Cheong YH, Pandey GK, Grant JJ, Batistic O, Li L, Kim BG, Lee SC, Kudla J, Luan S (2007) Two calcineurin B-like calcium sensors, interacting with protein kinase CIPK23, regulate leaf transpiration and root potassium uptake in Arabidopsis. Plant J 52:223–239

Chung JS, Zhu JK, Bressan RA, Hasegawa PM, Shi H (2008) Reactive oxygen species mediate Na+-induced SOS1 mRNA stability in Arabidopsis. Plant J 53:554–565

Davenport R (2002) Glutamate receptors in plants. Ann Bot 90:549–557

Davies C, Shin R, Liu W, Thomas MR, Schachtman DP (2006) Transporters expressed during grape berry (Vitis vinifera L.) development are associated with an increase in berry size and berry potassium accumulation. J Exp Bot 57:3209–3216

de Nadal E, Clotet J, Posas F, Serrano R, Gomez N, Arino J (1998) The yeast halotolerance determinant Hal3p is an inhibitory subunit of the Ppz1p Ser/Thr protein phosphatase. Proc Natl Acad Sci USA 95:7357–7362

Demidchik V, Maathuis FJM (2007) Physiological roles of nonselective cation channels in plants: from salt stress to signalling and development. New Phytol 175:387–404

Donaldson L, Ludidi N, Knight MR, Gehring C, Denby K (2004) Salt and osmotic stress cause rapid increases in Arabidopsis thaliana cGMP levels. FEBS Lett 569:317–320

Duby G, Hosy E, Fizames C, Alcon C, Costa A, Sentenac H, Thibaud JB (2008) AtKC1, a conditionally targeted Shaker-type subunit, regulates the activity of plant K^+ channels. Plant J 53:115–123

Dvořak J, Noaman MM, Goyal S, Gorham J (1994) Enhancement of the salt tolerance of Triticum turgidum L. by the Kna1 locus transferred from the Triticum aestivum L. chromosome 4D by homoeologous recombination. Theor Appl Genet 87:872–877

Elumalai RP, Nagpal P, Reed JW (2002) A mutation in the Arabidopsis KT2/KUP2 potassium transporter gene affects shoot cell expansion. Plant Cell 14:119–131

Endler A, Reiland S, Gerrits B, Schmidt UG, Baginsky S, Martinoia E (2009) In vivo phosphorylation sites of barley tonoplast proteins identified by a phosphoproteomic approach. Proteomics 9:310–321

Epstein E, Rains DW, Elzam OE (1963) Resolution of dual mechanisms of potassium absorption by barley roots. Proc Natl Acad Sci USA 49:684–692

Essah PA, Davenport R, Tester M (2003) Sodium influx and accumulation in Arabidopsis. Plant Physiol 133:307–318

Fernandez-Garcia N, Lopez-Perez L, Hernandez M, Olmos E (2009) Role of phi cells and the endodermis under salt stress in Brassica oleracea. New Phytol 181:347–360

Forment J, Mulet JM, Vicente O, Serrano R (2002) The yeast SR protein kinase Sky1p modulates salt tolerance, membrane potential and the Trk1, 2 potassium transporter. Biochim Biophys Acta 1565:36–40

Fujii H, Chinnusamy V, Rodrigues A, Rubio S, Antoni R, Park SY, Cutler SR, Sheen J, Rodriguez PL, Zhu JK (2009) In vitro reconstitution of an abscisic acid signalling pathway. Nature 462:660–664

Fulgenzi FR, Peralta ML, Mangano S, Danna CH, Vallejo AJ, Puigdomenech P, Santa-Maria GE (2008) The ionic environment controls the contribution of the barley HvHAK1 transporter to potassium acquisition. Plant Physiol 147(1):252–262

Garciadeblas B, Senn ME, Banuelos MA, Rodriguez-Navarro A (2003) Sodium transport and HKT transporters: the rice model. Plant J 34:788–801

Garciadeblas B, Barrero-Gil J, Benito B, Rodriguez-Navarro A (2007) Potassium transport systems in the moss Physcomitrella patens: pphak1 plants reveal the complexity of potassium uptake. Plant J 52:1080–1093

Gierth M, Maser P, Schroeder JI (2005) The potassium transporter AtHAK5 functions in K^+ deprivation-induced high-affinity K^+ uptake and AKT1 K^+ channel contribution to K^+ uptake kinetics in Arabidopsis roots. Plant Physiol 137:1105–1114

Glass A (1976) Regulation of potassium absorption in barley roots. An allosteric model. Plant Physiol 58:33–37

Glass ADM (1978) Regulation of potassium influx into intact roots of barley by internal potassium levels. Can J Bot Rev Can Bot 56:1759–1764

Gobert A, Park G, Amtmann A, Sanders D, Maathuis FJM (2006) Arabidopsis thaliana cyclic nucleotide gated channel 3 forms a non-selective ion transporter involved in germination and cation transport. J Exp Bot 57:791–800

Golldack D, Quigley F, Michalowski CB, Kamasani UR, Bohnert HJ (2003) Salinity stress-tolerant and -sensitive rice (Oryza sativa L.) regulate AKT1-type potassium channel transcripts differently. Plant Mol Biol 51:71–81

Gong DM, Guo Y, Schumaker KS, Zhu JK (2004) The SOS3 family of calcium sensors and SOS2 family of protein kinases in Arabidopsis. Plant Physiol 134:919–926

Guo Y, Halfter U, Ishitani M, Zhu JK (2001) Molecular characterization of functional domains in the protein kinase SOS2 that is required for plant salt tolerance. Plant Cell 13:1383–1399

Guo KM, Babourina O, Christopher DA, Borsics T, Rengel Z (2008) The cyclic nucleotide-gated channel, AtCNGC10, influences salt tolerance in Arabidopsis. Physiol Plant 134:499–507

Guo KM, Babourina O, Christopher DA, Borsic T, Rengel Z (2010) The cyclic nucleotide-gated channel AtCNGC10 transports Ca^{2+} and Mg^{2+} in Arabidopsis. Physiol Plant. doi:10.1111/j.1399-3054.2010.01366.x

Haro R, Sainz L, Rubio F, Rodriguez-Navarro A (1999) Cloning of two genes encoding potassium transporters in Neurospora crassa and expression of the corresponging cDNAs in Saccharomyces cerevisiae. Mol Microbiol 31:511–520

Haro R, Banuelos MA, Rodriguez-Navarro A (2010) High-affinity sodium uptake in land plants. Plant Cell Physiol 51:68–79

Hasegawa PM, Bressan RA, Zhu JK, Bohnert HJ (2000) Plant cellular and molecular responses to high salinity. Annu Rev Plant Physiol Plant Mol Biol 51:463–499

Hernandez A, Jiang X, Cubero B, Nieto PM, Bressan RA, Hasegawa PM, Pardo JM (2009) Mutants of the Arabidopsis thaliana Cation/H^+ antiporter AtNHX1 conferring increased salt tolerance in yeast: the endosome/prevacuolar compartment is a target for salt toxicity. J Biol Chem 284:14276–14285

Hirsch RE, Lewis BD, Spalding EP, Sussman MR (1998) A role for the AKT1 potassium channel in plant nutrition. Science 280:918–921

Honsbein A, Sokolovski S, Grefen C, Campanoni P, Pratelli R, Paneque M, Chen Z, Johansson I, Blatt MR (2009) A tripartite SNARE-K^+ channel complex mediates in channel-dependent K+ nutrition in Arabidopsis. Plant Cell 21:2859–2877

Horie T, Yoshida K, Nakayama H, Yamada K, Oiki S, Shinmyo A (2001) Two types of HKT transporters with different properties of Na^+ and K^+ transport in Oryza sativa. Plant J 27:129–138

Horie T, Costa A, Kim TH, Han MJ, Horie R, Leung HY, Miyao A, Hirochika H, An G, Schroeder JI (2007) Rice OsHKT2;1 transporter mediates large Na^+ influx component into K^+-starved roots for growth. EMBO J 26:3003–3014

Horie T, Hauser F, Schroeder JI (2009) HKT transporter-mediated salinity resistance mechanisms in Arabidopsis and monocot crop plants. Trends Plant Sci 14:660–668

Hua BG, Mercier RW, Leng Q, Berkowitz GA (2003) Plants do it differently. A new basis for potassium/sodium selectivity in the pore of an ion channel. Plant Physiol 132:1353–1361

Huang SB, Spielmeyer W, Lagudah ES, James RA, Platten JD, Dennis ES, Munns R (2006) A sodium transporter (HKT7) is a candidate for Nax1, a gene for salt tolerance in durum wheat. Plant Physiol 142:1718–1727

Ishitani M, Liu JP, Halfter U, Kim CS, Shi WM, Zhu JK (2000) SOS3 function in plant salt tolerance requires N-myristoylation and calcium binding. Plant Cell 12:1667–1677

Ivashikina N, Becker D, Ache P, Meyerhoff O, Felle HH, Hedrich R (2001) K^+ channel profile and electrical properties of Arabidopsis root hairs. FEBS Lett 508:463–469

Jaquinod M, Villiers F, Kieffer-Jaquinod S, Hugouvieu V, Bruley C, Garin J, Bourguignon J (2007) A proteomics dissection of Arabidopsis thaliana vacuoles isolated from cell culture. Mol Cell Proteomics 6:394–412

Jaspers P, Brosche M, Overmyer K, Kangasjarvi J (2010) The transcription factor interacting protein RCD1 contains a novel conserved domain. Plant Signal Behav 5:78–80

Jiang X-Y, Leidi EO, Pardo JM (2010) How do vacuolar NHX exchangers function in plant salt tolerance? Plant Signal Behav

Jung JY, Shin R, Schachtman DP (2009) Ethylene mediates response and tolerance to potassium deprivation in Arabidopsis. Plant Cell 21:607–621

Kaddour R, Narsi N, Mïrah S, Berthomieu P, Lachaal M (2009) Comparative effect of potassium on K and Na uptake and transport in two accessions of Arabidopsis thaliana during salinity stress. CR Biol 332:784–794

Kaplan B, Sherman T, Fromm H (2007) Cyclic nucleotide-gated channels in plants. FEBS Lett 581:2237–2246

Katiyar-Agarwal S, Zhu J, Kim K, Agarwal M, Fu X, Huang A, Zhu JK (2006) The plasma membrane Na^+/H^+ antiporter SOS1 interacts with RCD1 and functions in oxidative stress tolerance in Arabidopsis. Proc Natl Acad Sci USA 103:18816–18821

Kim B-G, Waadt R, Cheong YH, Pandey GK, Dominguez-Solis JR, Schültke S, Lee SC, Kudla J, Luan S (2007) The calcium sensor CBL10 mediates salt tolerance by regulating ion homeostasis in Arabidopsis. Plant J 52:473–484

Kim MJ, Ciani S, Schachtman DP (2010) A peroxidase contributes to ROS production during Arabidopsis root response to potassium deficiency. Mol Plant 3(2):420–427

Köhler C, Merkle T, Neuhaus G (1999) Characterisation of a novel gene family of putative cyclic nucleotide- and calmodulin-regulated ion channels in Arabidopsis thaliana. Plant J 18:97–104

Kolukisaoglu U, Weinl S, Blazevic D, Batistic O, Kudla J (2004) Calcium sensors and their interacting protein kinases: genomics of the Arabidopsis and rice CBL-CIPK signaling networks. Plant Physiol 134:43–58

Lagarde D, Basset M, Lepetit M, Conejero G, Gaymard F, Astruc S, Grignon C (1996) Tissue-specific expression of Arabidopsis AKT1 gene is consistent with a role in K^+ nutrition. Plant J 9:195–203

Laurie S, Feeney KA, Maathuis FJM, Heard PJ, Brown SJ, Leigh RA (2002) A role for HKT1 in sodium uptake by wheat roots. Plant J 32:139–149

Lebaudy A, Hosy E, Simonneau T, Sentenac H, Thibaud JB, Dreyer I (2008) Heteromeric K+ channels in plants. Plant J 54:1076–1082

Lee SC, Lan WZ, Kim BG, Li L, Cheong YH, Pandey GK, Lu G, Buchanan BB, Luan S (2007) A protein phosphorylation/dephosphorylation network regulates a plant potassium channel. Proc Natl Acad Sci USA 104:15959–15964

Leidi EO, Barragan V, Rubio L, El-Hamdaoui A, Ruiz T, Cubero B, Fernandez JA, Bressan RA, Mike HP, Quintero FJ (2010) The AtNHX1 exchanger mediates potassium compartmentation in vacuoles of transgenic tomato. Plant J 61:495–506

Leigh RA, Wyn Jones RG (1984) A hypothesis relating critical potassium concentrations for growth to the distribution and functions of this ion in the plant cell. New Phytol 97:1–13

Lejay L, Wirth J, Pervent M, Cross JM-F, Tillard P, Gojon A (2008) Oxidative pentose phosphate pathway-dependent sugar sensing as a mechanism for regulation of root ion transporters by photosynthesis. Plant Physiol 146:2036–2053

Leng Q, Mercier RW, Hua BG, Fromm H, Berkowitz GA (2002) Electrophysiological analysis of cloned cyclic nucleotide-gated ion channels. Plant Physiol 128:400–410

Li J, Yang H, Ann Peer W, Richter G, Blakeslee J, Bandyopadhyay A, Titapiwantakun B, Undurraga S, Khodakovskaya M, Richards EL, Krizek B, Murphy AS, Gilroy S, Gaxiola R (2005a) Arabidopsis H+-PPase AVP1 regulates auxin-mediated organ development. Science 310:121–125

Li X, Borsics T, Harrington H, Christopher D (2005b) Arabidopsis AtCNGC10 rescues potassium channel mutants of E. coli, yeast and Arabidopsis and is regulated by calcium/calmodulin and cyclic GMP in E. coli. Funct Plant Biol 32:643–653

Li L, Kim BG, Cheong YH, Pandey GK, Luan S (2006) A Ca2+ signaling pathway regulates a K^+ channel for low-K response in Arabidopsis. Proc Natl Acad Sci USA 103:12625–12630

Lin H, Yang Y, Quan R, Mendoza I, Wu Y, Du W, Zhao S, Schumaker KS, Pardo JM, Guo Y (2009) Phosphorylation of SOS3-LIKE CALCIUM BINDING PROTEIN8 by SOS2 protein kinase stabilizes their protein complex and regulates salt tolerance in Arabidopsis. Plant Cell 21:1607–1619

Luan S, Lan W, Chul Lee S (2009) Potassium nutrition, sodium toxicity, and calcium signaling: connections through the CBL-CIPK network. Curr Opin Plant Biol 12:339–346

Ma W, Qi Z, Smigel A, Walker RK, Verma R, Berkowitz GA (2009a) Ca^{2+}, cAMP, and transduction of non-self perception during plant immune responses. Proc Natl Acad Sci USA 106:20995–21000

Ma Y, Szostkiewicz I, Korte A, Moes D, Yang Y, Christmann A, Grill E (2009b) Regulators of PP2C phosphatase activity function as abscisic acid sensors. Science 324:1064–1068

Maathuis FJ (2006) The role of monovalent cation transporters in plant responses to salinity. J Exp Bot 57:1137–1147

Maathuis FJM, Sanders D (1993) Energization of potassium uptake in Arabidopsis thaliana. Planta 191:302–307

Maathuis FJ, Sanders D (2001) Sodium uptake in Arabidopsis roots is regulated by cyclic nucleotides. Plant Physiol 127:1617–1625

Madrid R, Gomez MJ, Ramos J, Rodriguez-Navarro A (1998) Ectopic potassium uptake in trk1 trk2 mutants of Saccharomyces cerevisiae correlates with a highly hyperpolarized membrane potential. J Biol Chem 273:14838–14844

Mangano S, Silberstein S, Santa-Maria GE (2008) Point mutations in the barley HvHAK1 potassium transporter lead to improved K^+-nutrition and enhanced resistance to salt stress. FEBS Lett 582:3922–3928

Martinez-Atienza J, Jiang XY, Garciadeblas B, Mendoza I, Zhu JK, Pardo JM, Quintero FJ (2007) Conservation of the salt overly sensitive pathway in rice. Plant Physiol 143:1001–1012

Martinez-Cordero MA, Martinez V, Rubio F (2004) Cloning and functional characterization of the high-affinity K + transporter HAK1 of pepper. Plant Mol Biol 56:413–421

Martinez-Cordero MA, Martinez V, Rubio F (2005) High-affinity K^+ uptake in pepper plants. J Exp Bot 56:1553–1562

Maser P, Thomine S, Schroeder JI, Ward JM, Hirschi K, Sze H, Talke IN, Amtmann A, Maathuis FJM, Sanders D, Harper JF, Tchieu J, Gribskov M, Persans MW, Salt DE, Kim SA, Guerinot ML (2001) Phylogenetic relationships within cation transporter families of Arabidopsis. Plant Physiol 126:1646–1667

Mendoza I, Rubio F, Rodriguez-Navarro A, Pardo JM (1994) The protein phosphatase calcineurin is essential for Nacl tolerance of Saccharomyces cerevisiae. J Biol Chem 269:8792–8796

Miller G, Suzuki N, Ciftci-Yilmaz S, Mittler R (2009) Reactive oxygen species homeostasis and signalling during drought and salinity stresses. Plant Cell Environ 33:453–467

Mulet JM, Leube MP, Kron SJ, Rios G, Fink GR, Serrano R (1999) A novel mechanism of ion homeostasis and salt tolerance in yeast: the Hal4 and Hal5 protein kinases modulate the Trk1-Trk2 potassium transporter. Mol Cell Biol 19:3328–3337

Munns R (1985) Na+, K+ and Cl- in xylem sap flowing to shoots of NaCl-treated barley. J Exp Bot 36:1032–1042

Munns R, Tester M (2008) Mechanisms of salinity tolerance. Annu Rev Plant Biol 59:651–681

Munns R, James RA, Lauchli A (2006) Approaches to increasing the salt tolerance of wheat and other cereals. J Exp Bot 57:1025–1043

Munson AM, Haydon DH, Love SL, Fell GL, Palanivel VR, Rosenwald AG (2004) Yeast ARL1 encodes a regulator of K+ influx. J Cell Sci 117:2309–2320

Nieves-Cordones M, Martinez-Cordero MA, Martinez V, Rubio F (2007) An NH_4^+-sensitive component dominates high-affinity K^+ uptake in tomato plants. Plant Sci 172:273–280

Nieves-Cordones M, Miller A, Alemán F, Martinez V, Rubio F (2008) A putative role for the plasma membrane potential in the control of the expression of the gene encoding the tomato high-affinity potassium transporter HAK5. Plant Mol Biol 68:521–532

Nieves-Cordones M, Aleman F, Martinez V, Rubio F (2009) The Arabidopsis thaliana HAK5 K+ transporter is required for plant growth and K^+ acquisition from low K^+ solutions under saline conditions. Mol Plant 3(2):326–333

Oh D-H, Leidi E, Zhang Q, Hwang S-M, Li Y, Quintero FJ, Jiang X, D'Urzo MP, Lee SY, Zhao Y, Bahk JD, Bressan RA, Yun D-J, Pardo JM, Bohnert HJ (2009) Loss of halophytism by interference with SOS1 expression. Plant Physiol 151:210–222

Ohta M, Guo Y, Halfter U, Zhu JK (2003) A novel domain in the protein kinase SOS2 mediates interaction with the protein phosphatase 2C ABI2. Proc Natl Acad Sci USA 100:11771–11776

Olias R, Eljakaoui Z, Li J, Alvarez De Morales P, Marin-Manzano MC, Pardo JM, Belver A (2009a) The plasma membrane Na^+/H^+ antiporter SOS1 is essential for salt tolerance in tomato and affects the partitioning of Na^+ between plant organs. Plant Cell Environ 32:904–916

Olias R, Eljakaoui Z, Pardo JM, Belver A (2009b) The Na^+/H^+ exchanger SOS1 controls extrusion and distribution of Na^+ in tomato plants under salinity conditions. Plant Signal Behav 4:973–976

Pandey GK, Cheong YH, Kim BG, Grant JJ, Li LG, Luan S (2007) CIPK9: a calcium sensor-interacting protein kinase required for low-potassium tolerance in Arabidopsis. Cell Res 17:411–421

Pardo JM (2010) Biotechnology of water and salinity stress tolerance. Curr Opin Biotech 21(2):185–196

Pardo JM, Martinez-Atienza J, Quintero FJ (2005) Function of the SOS system on the salinity tolerance of plants. Comp Biochem Physiol A Physiol 141:S341

Pardo JM, Cubero B, Leidi EO, Quintero FJ (2006) Alkali cation exchangers: roles in cellular homeostasis and stress tolerance. J Exp Bot 57:1181–1199

Park S-Y, Fung P, Nishimura N, Jensen DR, Fujii H, Zhao Y, Lumba S, Santiago J, Rodrigues A, T-fF C, Alfred SE, Bonetta D, Finkelstein R, Provart NJ, Desveaux D, Rodriguez PL, McCourt P, Zhu J-K, Schroeder JI, Volkman BF, Cutler SR (2009) Abscisic acid inhibits type 2C protein phosphatases via the PYR/PYL family of START proteins. Science 324:1068–1071

Peng Y-H, Zhu Y-F, Mao Y-Q, Wang S-M, Su W-A, Tang Z-C (2004) Alkali grass resists salt stress through high [K^+] and an endodermis barrier to Na^+. J Exp Bot 55:939–949

Perez-Valle J, Jenkins H, Merchan S, Montiel V, Ramos J, Sharma S, Serrano R, Yenush L (2007) Key role for intracellular K^+ and protein kinases Sat4/Hal4 and Hal5 in the plasma membrane stabilization of yeast nutrient transporters. Mol Cell Biol 27:5725–5736

Putney LK, Denker SP, Barber DL (2002) The changing face of the Na^+/H^+ exchanger, NHE1: structure, regulation, and cellular actions. Annu Rev Pharmacol Toxicol 42:527–552

Qi Z, Spalding EP (2004) Protection of plasma membrane K^+ transport by the salt overly sensitive1 Na^+-H^+ antiporter during salinity stress. Plant Physiol 136:2548–2555

Qi Z, Hampton CR, Shin R, Barkla BJ, White PJ, Schachtman DP (2008) The high affinity K+ transporter AtHAK5 plays a physiological role in planta at very low K^+ concentrations and provides a caesium uptake pathway in Arabidopsis. J Exp Bot 59:595–607

Qiu QS, Guo Y, Dietrich MA, Schumaker KS, Zhu JK (2002) Regulation of SOS1, a plasma membrane Na^+/H^+ exchanger in Arabidopsis thaliana, by SOS2 and SOS3. Proc Natl Acad Sci USA 99:8436–8441

Qiu QS, Guo Y, Quintero FJ, Pardo JM, Schumaker KS, Zhu JK (2004) Regulation of vacuolar Na+/H+ exchange in Arabidopsis thaliana by the salt-overly-sensitive (SOS) pathway. J Biol Chem 279:207–215

Quan RD, Lin HX, Mendoza I, Zhang YG, Cao WH, Yang YQ, Shang M, Chen SY, Pardo JM, Guo Y (2007) SCABP8/CBL10, a putative calcium sensor, interacts with the protein kinase SOS2 to protect Arabidopsis shoots from salt stress. Plant Cell 19:1415–1431

Quintero FJ, Blatt MR (1997) A new family of K^+ transporters from Arabidopsis that are conserved across phyla. FEBS Lett 415:206–211

Quintero FJ, Ohta M, Shi HZ, Zhu JK, Pardo JM (2002) Reconstitution in yeast of the Arabidopsis SOS signaling pathway for Na^+ homeostasis. Proc Natl Acad Sci USA 99:9061–9066

Raghothama KG (2000) Phosphate transport and signaling. Curr Opin Plant Biol 3:182–187

Ramos J, Haro R, Rodriguez-Navarro A (1990) Regulation of potassium fluxes in Saccharomyces cerevisiae. Biochim Biophys Acta 1029:211–217

Reintanz B, Szyroki A, Ivashikina N, Ache P, Godde M, Becker D, Palme K, Hedrich R (2002) AtKC1, a silent Arabidopsis potassium channel alpha -subunit modulates root hair K^+ influx. Proc Natl Acad Sci USA 99:4079–4084

Remans T, Nacry P, Pervent M, Girin T, Tillard P, Lepetit M, Gojon A (2006) A central role for the nitrate transporter NRT2.1 in the integrated morphological and physiological responses of the root system to nitrogen limitation in Arabidopsis. Plant Physiol 140:909–921

Ren Z-H, Gao J-P, Li L-G, Cai X-L, Huang W, Chao D-Y, Zhu M-Z, Wang Z-Y, Luan S, Lin H-X (2005) A rice quantitative trait locus for salt tolerance encodes a sodium transporter. Nat Genet 37:1141–1146

Rigas S, Debrosses G, Haralampidis K, Vicente-Agullo F, Feldmann K, Grabov A, Dolan L, Hatzopoulos P (2001) Trh1 encodes a potassium transporter required for tip growth in arabidopsis root hairs. Plant Cell 13:139–151

Rodriguez-Navarro A (2000) Potassium transport in fungi and plants. Biochim Biophys Acta-Rev Biomembr 1469:1–30

Rodriguez-Navarro A, Rubio F (2006) High-affinity potassium and sodium transport systems in plants. J Exp Bot 57:1149–1160

Rodriguez-Navarro A, Blatt MR, Slayman CL (1986) A potassium-proton symport in Neurospora crassa. J Gen Physiol 87:649–674

Rodriguez-Rosales MP, Jiang X, Galvez FJ, Aranda MN, Cubero B, Venema K (2008) Overexpression of the tomato K^+/H^+ antiporter LeNHX2 confers salt tolerance by improving potassium compartmentalization. New Phytol 179:366–377

Rodriguez-Rosales MP, Galvez FJ, Huertas R, Aranda MN, Baghour M, Cagnac O, Venema K (2009) Plant NHX cation/proton antiporters. Plant Signal Behav 4:265–276

Ruan Y-L, Llewellyn DJ, Furbank RT (2001) The control of single-celled cotton fiber elongation by developmentally reversible gating of plasmodesmata and coordinated expression of sucrose and K^+ transporters and expansion. Plant Cell 13:47–60

Rubio F, Gassmann W, Schroeder JI (1995) Sodium-driven potassium uptake by the plant potassium transporter HKT1 and mutations conferring salt tolerance. Science 270:1660–1663

Rubio F, Santa-Maria GE, Rodriguez-Navarro A (2000) Cloning of Arabidopsis and barley cDNAs encoding HAK potassium transporters in root and shoot cells. Physiol Plant 109:34–43

Rubio F, Flores P, Navarro J, Martinez V (2003) Effects of Ca^{2+}, K^+ and cGMP on Na^+ uptake in pepper plants. Plant Sci 165:1043–1050

Rubio F, Nieves-Cordones M, Aleman F, Martinez V (2008) Relative contribution of AtHAK5 and AtAKT1 to K^+ uptake in the high-affinity range of concentrations. Physiol Plant 134:598–608

Rubio F, Alemán F, Nieves-Cordones M, Vicente M (2010) Studies on Arabidopsis *athak5*, *atakt1* double mutants disclose the range of concentrations at which AtHAK5, AtAKT1 and unknown systems mediate K^+ uptake. Physiol Plant 138(2):220–228

Sanchez-Barrena MJ, Martinez-Ripoll M, Zhu JK, Albert A (2005) The structure of the Arabidopsis thaliana SOS3: molecular mechanism of sensing calcium for salt stress response. J Mol Biol 345:1253–1264

Sanchez-Barrena MJ, Fujii H, Angulo I, Martinez-Ripoll M, Zhu JK, Albert A (2007) The structure of the C-terminal domain of the protein kinase AtSOS2 bound to the calcium sensor AtSOS3. Mol Cell 26:427–435

Santa-María GE, Rubio F, Dubcovsky J, Rodriguez-Navarro A (1997) The HAK1 gene of barley is a member of a large gene family and encodes a high-affinity potassium transporter. Plant Cell 9:2281–2289

Sato Y, Sakaguchi M (2005) Topogenic properties of transmembrane segments of Arabidopsis thaliana NHX1 reveal a common topology model of the Na^+/H^+ exchanger family. J Biochem 138:425–431

Schachtman DP, Schroeder JI (1994) Structure and transport mechanism of a high-affinity potassium uptake transporter from higher plants. Nature 370:655–658

Schuurink RC, Shartzer SF, Fath A, Jones RL (1998) Characterization of a calmodulin-binding transporter from the plasma membrane of barley aleurone. Proc Natl Acad Sci USA 95:1944–1949

Senn ME, Rubio F, Banuelos MA, Rodriguez-Navarro A (2001) Comparative functional features of plant potassium HvHAK1 and HvHAK2 transporters. J Biol Chem 276:44563–44569

Sentenac H, Bonneaud N, Minet M, Lacroute F, Salmon JM, Gaymard F, Grignon C (1992) Cloning and expression in yeast of a plant potassium ion transport system. Science 256:663–665

Shabala S, Pang JY, Zhou MX, Shabala L, Cuin TA, Nick P, Wegner LH (2009) Electrical signalling and cytokinins mediate effects of light and root cutting on ion uptake in intact plants. Plant Cell Environ 32:194–207

Shi HZ, Ishitani M, Kim CS, Zhu JK (2000) The Arabidopsis thaliana salt tolerance gene SOS1 encodes a putative Na^+/H^+ antiporter. Proc Natl Acad Sci USA 97:6896–6901

Shi HZ, Quintero FJ, Pardo JM, Zhu JK (2002) The putative plasma membrane Na+/H+ antiporter SOS1 controls long-distance Na^+ transport in plants. Plant Cell 14:465–477

Shi HZ, Lee BH, Wu SJ, Zhu JK (2003) Overexpression of a plasma membrane Na^+/H^+ antiporter gene improves salt tolerance in Arabidopsis thaliana. Nat Biotech 21:81–85

Shin R, Schachtman DP (2004) Hydrogen peroxide mediates plant root cell response to nutrient deprivation. Proc Natl Acad Sci USA 101:8827–8832

Shin R, Berg RH, Schachtman DP (2005) Reactive oxygen species and root hairs in Arabidopsis root response to nitrogen, phosphorus and potassium deficiency. Plant Cell Physiol 46:1350–1357

Shin R, Burch AY, Huppert KA, Tiwari SB, Murphy AS, Guilfoyle TJ, Schachtman DP (2007) The Arabidopsis transcription factor MYB77 modulates auxin signal transduction. Plant Cell 19:2440–2453

Spalding EP, Hirsch RE, Lewis DR, Qi Z, Sussman MR, Lewis BD (1999) Potassium uptake supporting plant growth in the absence of AKT1 channel activity: Inhibition by ammonium and stimulation by sodium. J Gen Physiol 113:909–918

Su H, Balderas E, Vera-Estrella R, Golldack D, Quigley F, Zhao CS, Pantoja O, Bohnert JH (2003) Expression of the cation transporter McHKT1 in a halophyte. Plant Mol Biol 52:967–980

Sunarpi HT, Motoda J, Kubo M, Yang H, Yoda K, Horie R, Chan WY, Leung HY, Hattori K, Konomi M, Osumi M, Yamagami M, Schroeder JI, Uozumi N (2005) Enhanced salt tolerance mediated by AtHKT1 transporter-induced Na^+ unloading from xylem vessels to xylem parenchyma cells. Plant J 44:928–938

Talke IN, Blaudez D, Maathuis FJM, Sanders D (2003) CNGCs: prime targets of plant cyclic nucleotide signalling? Trends Plant Sci 8:286–293

Tester M, Davenport R (2003) Na^+ tolerance and Na^+ transport in higher plants. Ann Bot 91:503–527

Umezawa T, Sugiyama N, Mizoguchi M, Hayashi S, Myouga F, Yamaguchi-Shinozaki K, Ishihama Y, Hirayama T, Shinozaki K (2009) Type 2C protein phosphatases directly regulate abscisic acid-activated protein kinases in Arabidopsis. Proc Natl Acad Sci USA 106:17588–17593

Venema K, Belver A, Marin-Manzano MC, Rodriguez-Rosales MP, Donaire JP (2003) A novel intracellular K^+/H^+ antiporter related to Na^+/H^+ antiporters is important for K^+ ion homeostasis in plants. J Biol Chem 278:22453–22459

Verslues PE, Batelli G, Grillo S, Agius F, Kim YS, Zhu J, Agarwal M, Katiyar-Agarwal S, Zhu JK (2007) Interaction of SOS2 with nucleoside diphosphate kinase 2 and catalases reveals a point

of connection between salt stress and H_2O_2 signaling in Arabidopsis thaliana. Mol Cell Biol 27:7771–7780
Vicente-Agullo F, Rigas S, Desbrosses G, Dolan L, Hatzopoulos P, Grabov A (2004) Potassium carrier TRH1 is required for auxin transport in Arabidopsis roots. Plant J 40:523–535
Vlad F, Rubio S, Rodrigues A, Sirichandra C, Belin C, Robert N, Leung J, Rodriguez PL, Lauriere C, Merlot S (2009) Protein phosphatases 2C regulate the activation of the Snf1-related kinase OST1 by abscisic acid in Arabidopsis. Plant Cell. doi:10.1105/tpc.109.069179
Volkov V, Amtmann A (2006) Thellungiella halophila, a salt-tolerant relative of Arabidopsis thaliana, has specific root ion-channel features supporting K^+/Na^+ homeostasis under salinity stress. Plant J 48:342–353
Waadt R, Schmidt LK, Lohse M, Hashimoto K, Bock R, Kudla J (2008) Multicolor bimolecular fluorescence complementation reveals simultaneous formation of alternative CBL/CIPK complexes in planta. Plant J 56:505–516
Walker DJ, Leigh RA, Miller AJ (1996) Potassium homeostasis in vacuolate plant cells. Proc Natl Acad Sci USA 93:10510–10514
Wang Y, Wu W-H (2010) Plant sensing and signaling in response to K^+-deficiency. Mol Plant 3:280–287
Wang YH, Garvin DF, Kochian LV (2001) Nitrate-induced genes in tomato roots. Array analysis reveals novel genes that may play a role in nitrogen nutrition. Plant Physiol 127:345–359
Wang B, Davenport RJ, Volkov V, Amtmann A (2006) Low unidirectional sodium influx into root cells restricts net sodium accumulation in Thellungiella halophila, a salt-tolerant relative of Arabidopsis thaliana. J Exp Bot 57:1161–1170
Ward JM, Mäser P, Schroeder JI (2009) Plant ion channels: gene families, physiology, and functional genomics analyses. Annu Rev Physiol 71:59–82
Whiteman S-A, Nühse TS, Ashford DA, Sanders D, Maathuis FJ (2008a) A proteomic and phosphoproteomic analysis of Oryza sativa plasma membrane and vacuolar membrane. Plant J 56:146–156
Whiteman S-A, Serazetdinova L, Jones AME, Sanders D, Rathjen J, Peck SC, Maathuis FJM (2008b) Identification of novel proteins and phosphorylation sites in a tonoplast enriched membrane fraction of Arabidopsis thaliana. Proteomics 8:3536–3547
Wyn-Jones RG, Gorham J (1986) The potential for enhancing the salt tolerance of wheat and other important crop plants. Outlook Agric 15:33–39
Xu J, Li HD, Chen LQ, Wang Y, Liu LL, He L, Wu WH (2006) A protein kinase, interacting with two calcineurin B-like proteins, regulates K^+ transporter AKT1 in Arabidopsis. Cell 125:1347–1360
Yamaguchi T, Aharon GS, Sottosanto JB, Blumwald E (2005) Vacuolar Na^+/H^+ antiporter cation selectivity is regulated by calmodulin from within the vacuole in a Ca^{2+}- and pH-dependent manner. Proc Natl Acad Sci USA 44:16107–16112
Yamaguchi T, Apse MP, Shi H, Blumwald E (2003) Topological analysis of a plant vacuolar Na^+/H^+ antiporter reveals a luminal C terminus that regulates antiporter cation selectivity. Proc Natl Acad Sci USA 14:12510–12515
Yao X, Horie T, Xue S, Leung H-Y, Katsuhara M, Brodsky DE, Wu Y, Schroeder JI (2010) Differential sodium and potassium transport selectivities of the rice OsHKT2;1 and OsHKT2;2 transporters in plant cells. Plant Physiol 152:341–355
Yenush L, Mulet JM, Arino J, Serrano R (2002) The Ppz protein phosphatases are key regulators of K+ and pH homeostasis: implications for salt tolerance, cell wall integrity and cell cycle progression. EMBO J 21:920–929
Yokoi S, Quintero FJ, Cubero B, Ruiz MT, Bressan RA, Hasegawa PM, Pardo JM (2002) Differential expression and function of Arabidopsis thaliana NHX Na^+/H^+ antiporters in the salt stress response. Plant J 30:529–539
Zhao J, Cheng NH, Motes CM, Blancaflor EB, Moore M, Gonzales N, Padmanaban S, Sze H, Ward JM, Hirschi KD (2008) AtCHX13 is a plasma membrane K^+ transporter. Plant Physiol 148:796–807

Zhu JK (2000) Genetic analysis of plant salt tolerance using arabidopsis. Plant Physiol 124:941–948

Zhu JK, Liu JP, Xiong LM (1998) Genetic analysis of salt tolerance in Arabidopsis: evidence for a critical role of potassium nutrition. Plant Cell 10:1181–1191

Zhu J, Fu X, Koo YD, Zhu JK, Jenney FE Jr, Adams MWW, Zhu Y, Shi H, Yun DJ, Hasegawa PM, Bressan RA (2007) An enhancer mutant of Arabidopsis salt overly sensitive 3 mediates both ion homeostasis and the oxidative stress response. Mol Cell Biol 27:5214–5224

Iron Transport and Signaling in Plants

S. Thomine and V. Lanquar

Abstract Iron is an essential micronutrient for most living organisms. Paradoxically, although iron is abundant in many soils, iron availability is very often limiting for plant growth. In addition, iron is potentially highly toxic to cells. Therefore, iron homeostasis needs to be tightly regulated. This chapter focuses on the iron transport pathways dedicated to iron uptake, distribution and sequestration in plants, and the processes that regulate their activities. Nongraminaceous and graminaceous plant species acquire iron from the soil through two distinct strategies based on iron reduction and iron chelation, respectively. We describe the molecular mechanisms underlying these strategies and the factors responsible for their up-regulation under iron deficiency. The acquisition of iron by plants is regulated at several levels by local and systemic signals. The systemic signaling pathway appears to integrate multiple inputs from hormonal signals, diurnal regulation, and the plant nutritional demand.

1 Introduction

Iron (Fe) is an essential micronutrient for most living organisms. Cell metabolism uses Fe for its ability to transfer electrons by shuttling from its reduced state Fe(II) and its oxidized state Fe(III). Iron is a major player in electron transfer chains in mitochondrial respiration and chloroplast photosynthesis. Although some enzymes, such as Fe-dependent superoxide dismutase, use molecular Fe as a cofactor directly, most proteins use Fe-containing cofactors, which can be classified under two main

S. Thomine (✉)
Institut des Sciences du Végétal CNRS, Avenue de la Terrasse, 91198 Gif-sur-Yvette cedex, France
e-mail: Sebastien.Thomine@isv.cnrs-gif.fr

V. Lanquar
Department of Plant Biology, Carnegie Institution, 260 Panama Street, Stanford, CA 94305, USA
e-mail: vlanquar@stanford.edu

M. Geisler and K. Venema (eds.), *Transporters and Pumps in Plant Signaling*, Signaling and Communication in Plants 7,
DOI 10.1007/978-3-642-14369-4_4,

forms, namely heme Fe, which is found in cytochromes or hemoglobin and Fe sulfur clusters, which are used by numerous biosynthetic enzymes. Photosystem I requires 12 iron atoms arranged in three Fe_4S_4 clusters for its function.

Paradoxically, although Fe is abundant in the earth crust (4.2%) and in many soils (2–60%), Fe availability is very often limiting for the growth of living organisms at all levels. Iron fertilization in oceans has shown that this element is a major limitation of plankton growth in many conditions (Buesseler et al. 2004; Mills et al. 2004). Iron deficiency is thought to limit agricultural yields on as much as 30% of arable lands. Most importantly, dietary Fe deficiency is affecting about one billion human beings on earth, leading to asthenia, increased sensitivity to diseases, and in some cases, death. All living organisms have evolved very efficient and sometimes sophisticated strategies to acquire Fe. Iron acquisition is often the object of a fierce ecological competition. More specifically, pathogenic bacteria have evolved strategies to withdraw Fe from their host, while hosts attempt to reduce invasion by limiting Fe availability is a widespread resistance mechanism to bacterial invasion (Expert 1999; Goetz et al. 2002). The mechanisms involved in Fe acquisition are often costly for the organism and need to be controlled with respect to Fe availability.

However, Fe ability to change its oxidation states in physiological conditions also makes this element potentially highly toxic to cells. In the presence of oxygen or reactive oxygen species (ROS), Fe becomes involved in the Fenton chemistry, which leads to the generation of highly toxic ROS, superoxide anions (O_2^-), and hydroxyl radicals (HO°). In turn, ROS generated by the Fenton reactions will react with all cell components: proteins, leading to nonfunctional oxidized proteins, lipids, leading to membrane degradation, and DNA, leading to genotoxic stress, cancer, and mutations. As Fe is both necessary for many facets of cell metabolism and potentially toxic, Fe availability must be highly controlled by local and long-distance signaling networks. Whether Fe itself may act as a regulator of cell fate and development is not very well documented. Nevertheless, reports indicate abnormal cell differentiation in a plant mutant impaired in Fe uptake (Henriques et al. 2002) and a role for NGAL lipocalin, a protein which binds Fe complexed to small organic ligands, as a morphogen acting through apoptosis in animals (Yang et al. 2002). This chapter focuses on the Fe transport pathways dedicated to Fe uptake, distribution and sequestration, and the processes that regulate their activities. We will briefly recapitulate the current knowledge on Fe transport processes and their regulation in yeast and mammals as references and we will describe in more detail the mechanisms of Fe transport and their regulation in plant cells.

2 Fe Transport and Signaling in Yeast and Mammalian Cells

2.1 A Brief Overview of Fe Transport and Signaling in Yeast

Because yeast is an easy-to-handle model organism amenable to molecular genetics, knowledge on Fe transport and regulation in these fungi is very advanced. Yeast provides one of the most complete pictures on the way Fe transport may be regulated. Yeast Fe uptake requires a reduction step from FeIII to FeII achieved by transmembrane NADPH-dependent ferric reductases (FRE1 and FRE2). Subsequently, Fe is taken up via two main pathways. The high-affinity pathway is a protein complex including a Cu-dependent Fe oxydase (FET3) facing the extracellular medium, which reoxidizes FeII to FeIII, and a Fe^{3+} transporter Ftr1p, which carries Fe into the yeast cytosol (Askwith et al. 1994; Stearman et al. 1996; Kwok et al. 2006). The two steps are coupled and allow high-affinity and high-specificity Fe uptake. Fet4p, a divalent metal permease, constitutes the low-affinity Fe uptake system which is less specific (Dix et al. 1994). In yeast, the vacuole is the main site for Fe sequestration and storage. Import of Fe into the vacuole by the Ccc1p iron transporter buffers cytosolic Fe concentration (Li et al. 2001). When Fe is scarce, two Fe transport systems are induced and allow the retrieval of Fe from the vacuolar stores: Smf3p, a divalent metal transporter belonging to the NRAMP family (Portnoy et al. 2000) and a complex between the Cu-dependent Fe^{2+} oxidase Fet5p and the Fe^{3+} transporter Fth1p, analogous to the Fet3p/Ftr1p complex at the plasma membrane (Spizzo et al. 1997; Urbanowski and Piper 1999). Iron acquisition and remobilization is controlled by a pair of transcription factors, Aft1p and Aft2p, regulating overlapping but not identical sets of genes (Yamaguchi-Iwai et al. 1995, 1996; Blaiseau et al. 2001; Rutherford et al. 2003). Aft1p binds to the promoter of Fe-deficiency responsive genes under Fe starvation (Yamaguchi-Iwai et al. 1996). The activity of Aft1p appears to be regulated mostly through the regulation of its cellular localization. The inhibition of Aft1p under Fe-sufficient conditions requires FeS cluster biosynthesis in mitochondria and their export to the cytosol (Chen et al. 2004; Rutherford et al. 2005). Recent reports indicate that an heterodimeric complex between Grx3p or Grx4p (glutaredoxin 3 or 4) and Fra2p would be responsible for the sensing of cytosolic FeS cluster mediating the inhibition of Aft1p activity under Fe sufficient conditions (Ojeda et al. 2006; Kumanovics et al. 2008; Li et al. 2009).

2.2 *A Brief Overview of Fe Transport and Signaling in Mammals*

In mammals, the uptake of Fe from the intestine is achieved by three types of transporters present in enterocytes. DCT1/NRAMP2/DMT1 (Divalent Metal Transporter 1) and HCP1 (Heme-Carrier Protein) take up iron as Fe^{2+} and as heme Fe, respectively, at the apical face of enterocytes facing the lumen of the guts (Gunshin et al. 1997; Fleming and Andrews 1998; Shayeghi et al. 2005). A complex between Ferroportin (FPN/IREG) and haephestin acting as exporter of Fe^{2+} and Cu-dependent Fe oxydase, respectively, is responsible for the release of Fe^{3+} in the plasma at the basal face of enterocytes (McKie et al. 2000). FPN also mediates Fe release from other organs storing Fe such as the liver. Iron circulates as Fe^{3+} bound to transferrin and is taken up into cells by transferrin receptor (TfR)-mediated endocytosis and subsequently exported from endosomes by NRAMP2/DCT1/DMT1.

Iron homeostasis is controlled at different levels. At the cellular level, the messenger RNA of the major Fe homeostasis proteins such as the TfR, FPN/IREG, and DMT1 contain IRE (Iron Response Element) in their 3′ or 5′ untranslated sequences (Hentze and Kuhn 1996). When cellular Fe level is low, IRP (IRE binding protein) bind IRE and stabilize mRNAs with IRE in their 3′UTR or inhibit the translation of mRNA with IRE in their 5′UTR. There are two IRPs, IRP1 and IRP2. The binding of IRP1 to IRE is dependent on the availability of FeS clusters. When FeS clusters are bound to IRP1, IRP1 is active as a cytosolic aconitase. IRP1 becomes able to bind IRE and regulate Fe homeostasis only in the absence of FeS clusters. IRP2 regulates Fe homeostasis in response to oxidative stress. IRP2 is degraded when it is oxidized. If free Fe is present in the cell and the Fenton reaction with oxygen is initiated, the degradation of IRP2 inhibits the mechanisms that make Fe available. At the organism level, hepcidin, a circulating peptide hormone, regulates transferrin-bound Fe levels in the plasma (Nemeth and Ganz 2006). Hepcidin is produced when Tf Fe levels are elevated or under pathogen attack, indicating that limiting circulating Fe levels, available to pathogens, is a strategy to reduce pathogen invasion in mammals. Hepcidin acts directly by binding and inducing the degradation of FPN/IREG (Nemeth et al. 2004). Degradation of FPN/IREG prevents the efflux of Fe stored in enterocytes, hepatocytes, or spleen cells to the plasma. Hence in mammals, Fe homeostasis is controlled at the cellular level by the availability of FeS clusters and at the level of the organism by a specific hormone, hepcidin.

3 Fe Transport Systems in Plant Cells

3.1 *Fe Uptake*

Although very abundant in soils, Fe is poorly soluble and found predominantly under Fe hydroxide forms. In neutral or alkaline soils, the availability of ferric Fe is strongly reduced. While plants need around 10^{-8} M Fe, the solubility of Fe^{3+} fluctuates from 10^{-17} M at pH 7 to 10^{-6} M at pH 3.3 (Marschner 1997; Fox and Guerinot 1998; Hell and Stephan 2003). To cope with the low availability of Fe in soils, graminaceous and nongraminaceous plants have evolved two distinct strategies to efficiently take up Fe from the soil. Both strategies are directed at making Fe available, a process called Fe mobilization.

3.1.1 Strategy I

Nongraminaceous monocots and all dicots, including *Arabidopsis thaliana,* activate the strategy I or reduction-based strategy of Fe uptake upon Fe deficiency. Iron uptake through strategy I requires at least three steps: first, an acidification of the root hair zone through the extrusion of protons to solubilize Fe chelates; second, the activation of ferric reductase activity; and third, the induction of a high-affinity ferrous Fe transport system (Fig. 1).

In many species including tomato, cucumber, or Arabidopsis, Fe deficiency is associated with an increased acidification of the rhizophere (Marschner 1997; Dell'Orto et al. 2000). This acidification step was proposed to be mediated by plasma membrane ATP-dependent proton pumps (Fig. 1, H^+-ATPase). Arabidopsis genome encodes 12 H^+-ATPases. AHA1, one of the most abundant isoforms present at the plasma membrane, is not affected by Fe deficiency, suggesting a role in basal function independent of Fe deficiency. However, among the 12 H^+-ATPase isoforms, *AHA2* and *AHA7* are up-regulated by Fe deficiency. The measurement of net proton efflux on different *aha* mutants indicates that AHA2 is the main form responsible for the root hair zone acidification under Fe deficiency (Santi and Schmidt 2009). In addition, AHA7 is necessary for the induction of the root hair

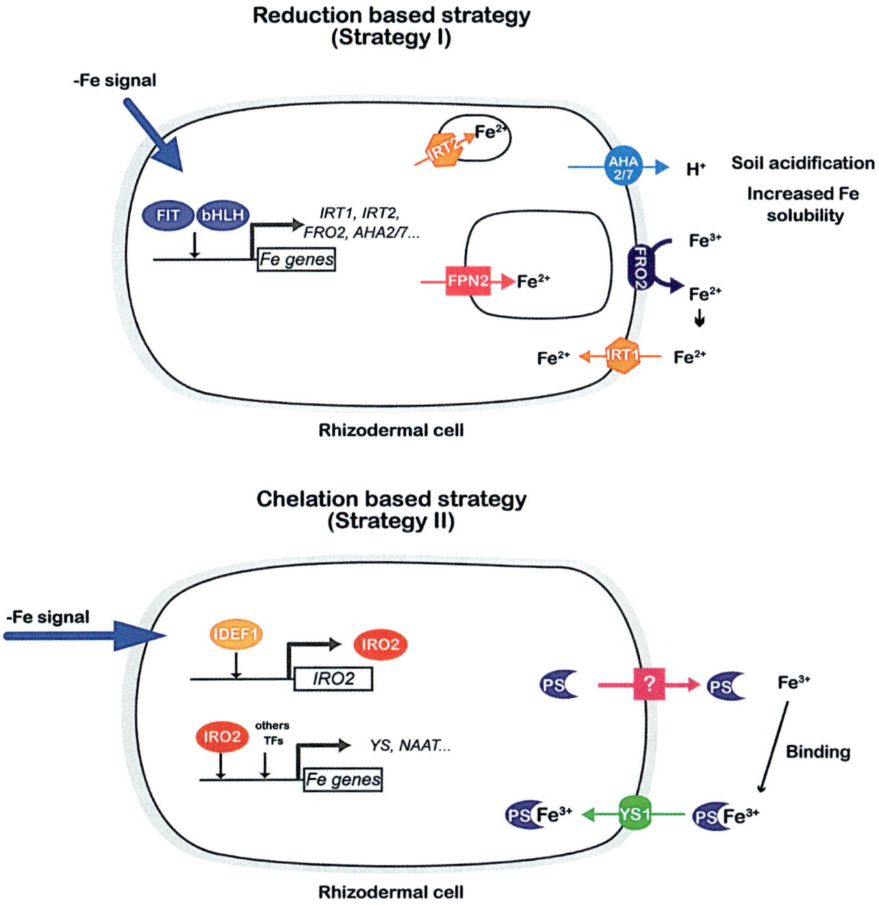

Fig. 1 Iron acquisition mechanisms and their transcriptional regulation in strategy I and strategy II plants

formation in response to Fe deficiency. The counterpart in cucumber involved in the Fe deficiency response would be the gene *CsHA1* (Santi et al. 2005).

The solubilization of the FeIII chelates is followed by the reduction of ferric Fe to ferrous Fe performed by a ferric-chelate reductase (Fig. 1). The characterization of the EMS mutants, *frd1-1* and *frd1-3*, which fail to activate ferric chelate reductase activity (Yi and Guerinot 1996), combined with search for genes homologous to yeast Fe ferric-reductase and a human NADPH oxidase, led to the identification of the *AtFRO2* gene (Robinson et al. 1999). *AtFRO2* (Ferric oxidase–reductase) encodes a plasma membrane ferric-chelate reductase-containing FAD- and NADPH-binding sites and is primarily expressed in roots in response to Fe deficiency (Robinson et al. 1999; Mukherjee et al. 2006). *Atfro2* loss-of-function mutants do not survive to mature stage unless high concentrations of exogenous Fe are supplied. Among the eight members identified in Arabidopsis, AtFRO6 has also been localized at the plasma membrane but its main expression in shoots suggests that *AtFRO6* has probably no contribution in the reduction of Fe in the rhizoderm (Wu et al. 2005; Feng et al. 2006; Mukherjee et al. 2006; Jeong et al. 2008). Homologues of the *AtFRO* genes have been identified in tomato, cucumber, and pea (Waters et al. 2002; Li et al. 2004; Waters et al. 2007).

In the third step, the reduction of ferric chelates to ferrous Fe is followed by Fe^{2+} uptake (Fig. 1). IRT1 (Iron-regulated transporter 1), the Fe^{2+} high-affinity uptake transporter, was identified by screening an Arabidopsis cDNA library in *fet3fet4*, a yeast mutant deficient in Fe uptake (Eide et al. 1996). Although the uptake affinity is highest for Fe^{2+} (Km ~10 μM), AtIRT1 has a broad metal selectivity (Mn^{2+}, Cd^{2+}, and Zn^{2+}) (Korshunova et al. 1999; Rogers et al. 2000). *AtIRT1* mRNA and proteins are detectable in root epidermis only in Fe-deficiency conditions and the protein is localized at the plasma membrane. *irt1* mutants display a strong chlorotic phenotype and do not survive after the stage of 4–6 leaves (Henriques et al. 2002; Varotto et al. 2002; Vert et al. 2002). Uptake experiments of Fe^{55} reveal that under Fe deficiency, the major root entrance of Fe^{2+} is performed through AtIRT1 (Vert et al. 2002). Nevertheless, *irt1* knockout mutant phenotypes are rescued by generous Fe fertilization. This indicates that, in addition to high-affinity uptake pathway by AtIRT1, Fe may enter plant cells through one or several additional low-affinity uptake pathways, which remain to be identified.

3.1.2 Strategy II

The strategy II, used by graminaceous species, which include the main cereal crops, is a chelation-based strategy. It involves the secretion in the rhizophere of low molecular weight molecules, named phytosiderophores (PS), which bind Fe^{3+} (Fig. 1). PS belong to the mugineic acid (MA) family, first identified in 1978 in barley (Takemoto et al. 1978). Until now, nine different MAs have been identified in different grasses (Romheld and Marschner 1986; Ueno et al. 2007). MAs are synthesized enzymatically from L-methionine with intermediate steps catalyzed by SAM synthetase, nicotianamine synthase (NAS), and nicotianamine aminotransferase

(NAAT) forming *S*-adenosyl methionine (SAM), nicotianamine (NA), and deoxymugineic acid, respectively (Mori 1999). The expression of the genes encoding these enzymes is increased during Fe deficiency, consistent with a need to enhance Fe acquisition (Negishi et al. 2002). The production of MA is specific for graminaceous species, but all plants are able to produce NA, which is important for the transport and sequestration of metals in plants (Curie et al. 2009). The mechanism of PS release in the soil is not known. However, different pathways involving transporters, anion channels, or exocytose have been proposed (Nishizawa and Mori 1987; Sakaguchi et al. 1999; Negishi et al. 2002).

When PS are released in the rhizophere, they bind to Fe^{3+} and the complex is reimported into root cells through transporters (Fig. 1). The first Fe(III)-MA transporter was identified by characterization of the *yellow-stripe 1* (*ys1*) maize mutant. This mutant displays interveinal chlorosis, characteristic of Fe-deficiency symptoms, due to a failure in Fe–MA uptake (Von Wiren et al. 1994; Curie et al. 2001). The *ZmYS1* gene encodes a plasma membrane transporter, whose expression in roots and shoots increases during Fe starvation (Curie et al. 2001; Roberts et al. 2004). Fe–NA transport is electrogenic and pH-dependent indicating that ZmYS1 is a proton –Fe–MA cotransporter (Schaaf et al. 2004). In the rice genome, 18 putative yellow-stripe 1 like genes (*OsYSLs*) have been identified (Koike et al. 2004). Among them, *OsYSL15* presented highly induced expression in the epidermis and stele of Fe-deficient roots. The protein is localized at the plasma membrane and *OsYSL15*-silenced plants display an early growth arrest rescued by Fe resupply (Inoue et al. 2009; Lee et al. 2009). *OsYSL15* induces current in the presence of FeIII-DMA, suggesting an electrogenic transport as shown for *ZmYS1*. OsYSL15 and HvYS1 likely represent the primary Fe–PS uptake transporter, corresponding to maize YS1, in rice and barley roots, respectively (Murata et al. 2006; Inoue et al. 2009; Lee et al. 2009).

The clear distinction between the two different Fe uptake strategies evolved by grasses and nongrasses needs to be slightly revisited. Recent data indicate that, in rice, both mechanisms coexist. Rice plants produce FeIII–PS complexes, although in lower quantities compared to other grasses, and can also take up ferrous Fe through OsIRT1 and OsIRT2 transporters (Ishimaru et al. 2006). A mutant with a nonfunctional NAAT enzyme, thus unable to use the strategy II, is rescued by the addition of EDTA–Fe and displays up-regulation Fe^{2+} uptake systems (OsIRT1, OsIRT2) (Cheng et al. 2007). The rice genome contains nonfunctional copies of ferric chelate reductase genes. Rice has likely acclimated to submerged growth conditions, which favor the presence of Fe^{2+} over Fe^{3+} and decrease the need for a ferric reductase activity. Nevertheless, expression of a functional recombinant ferric chelate reductase in rice improves tolerance to Fe deficiency (Ishimaru et al. 2007).

3.2 *Intracellular Fe Distribution*

Cytosolic-free Fe concentrations are believed to be extremely low. However, the total Fe concentration in cells grown in complete medium reaches the micromolar range, suggesting efficient compartmentation of the metal (Outten and O'Halloran 2001; Finney and O'Halloran 2003; Fig. 2). The removal of Fe from the cytosol contributes to at least two aims: the storage under harmless forms and the delivery to the locations where Fe–heme or Fe–S clusters are synthesized.

3.2.1 Vacuole

As in yeast, the plant vacuole is a location for metal storage (Pich et al. 2001; Lanquar et al. 2010). In particular, the endodermal vacuole of Arabidopsis embryo contains Fe, which is remobilized during germination providing autonomy to the seedling for a few days (Lanquar et al. 2005; Kim et al. 2006; Roschzttardtz et al. 2009).

Iron is loaded into the vacuole by AtVIT1, a homologue of the Ccc1p yeast iron/manganese transporter (Fig. 2). *AtVIT1* is expressed in shoots and roots of adult plants but is particularly important during the seed formation. Metal mapping with

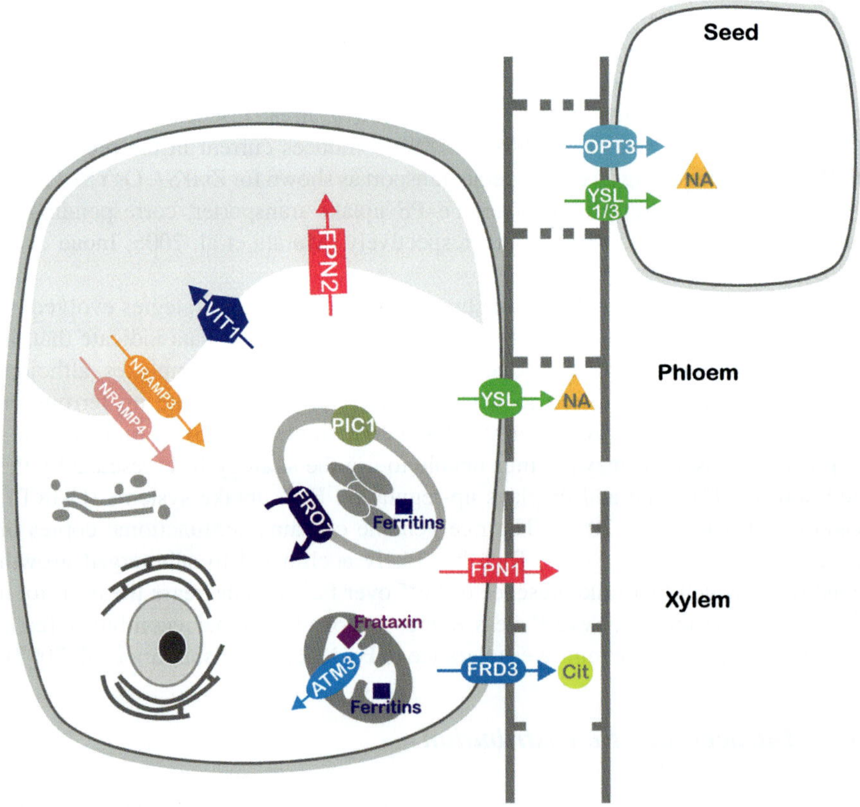

Fig. 2 Transporters involved in intracellular and interorgan iron distribution in Arabidopsis

X-ray microtomography and Fe staining with PERL DAB of *vit1* mutant embryos show that Fe does not locate in the vacuoles of the endodermal cells in the provasculature of the cotyledons and embryonic axis, as the wild type, but is concentrated in the vacuoles of subepidermal cells of cotyledons instead (Kim et al. 2006; Roschzttardtz et al. 2009). *vit1* loss of function mutant displays a chlorotic phenotype during germination on alkaline soil implying that the adequate location of Fe is required for proper plant development in these conditions (Kim et al. 2006). Unlike many components of the Fe transport pathway, *AtVIT1* expression is not regulated by Fe (Kim et al. 2006).

The ferroportin FPN2 transporter (or IREG2), a plant homologue of mammalian IREG1, is involved in the transport of nickel and cobalt (Schaaf et al. 2006; Morrissey et al. 2009). A role for FPN2 in Fe transport into vacuoles of epidermal and cortical root cells has also been proposed (Fig. 2), buffering the influx of metals into the cytoplasm by vacuolar sequestration. Although *FPN2* transcripts are detected in Fe-sufficient root, their accumulation is increased under Fe-deficient conditions. *fpn2* mutants display delayed or reduced Fe-deficiency response implying that failure to sequester Fe into vacuoles leads to alterations in the perception of Fe deficiency (Morrissey et al. 2009).

During germination, AtNRAMP3 and AtNRAMP4 redundantly mediate the retrieval of Fe from the vacuole (Lanquar et al. 2005) (Fig. 2). Although the Fe content in *nramp3nramp4* seeds is not altered, the seedlings display a strong chlorotic phenotype when germinated on low iron. Electron microscopy coupled to the electron spectroscopic imaging as well as Perl/DAB staining showed that, in *nramp3nramp4* double mutant, Fe remains trapped inside the vacuoles of endodermal cells and is not remobilized during germination (Lanquar et al. 2005; Roschzttardtz et al. 2009). The analysis of the *nramp3nramp4* double mutant contributed to the identification of the vacuole as an important Fe storage compartment in seeds. In seedlings, *AtNRAMP3* and *AtNRAMP4* transcripts are detected in vascular tissues. AtNRAMP3 and AtNRAMP4 proteins are regulated at the transcriptional level by Fe deficiency (Thomine et al. 2000, 2003; Lanquar et al. 2005). Although their involvement in Fe mobilization in the germinating seeds is clear, a function in Fe transport in the adult plant remains to be characterized.

3.2.2 Chloroplast

Heme synthesis and Fe–S cluster assembly take place in the chloroplast and about 90% of the leaf Fe is located in this organelle (Shingles et al. 2002; Abdel-Ghany et al. 2005). In the stroma, Fe is caged by ferritins under a harmless and available form (Briat et al. 2009).

The mechanisms of Fe influx into chloroplasts are not yet elucidated but uptake experiments on isolated vesicles from pea chloroplastic inner membrane suggest the involvement of a Fe^{2+} uniporter. The influx of Fe^{2+} would be driven by the electrical gradient (Shingles et al. 2002). In agreement with a mechanism involving Fe^{2+} import into the chloroplast, the ferric-chelate reductase AtFRO7 is targeted to the chloroplast envelope (Jeong et al. 2008) (Fig. 2). Seedlings lacking *AtFRO7* do not survive under Fe-limiting conditions. The *fro7* mutant photosynthetic activity is affected and the Fe concentration is reduced by one third in *fro7* chloroplasts.

AtPIC1 (Permease In Chloroplast 1) displays homology with cyanobacterial permeases and is located in the inner membrane of Arabidoposis chloroplasts. The phenotype of *PIC1* loss of function mutants clearly indicates an involvement of the PIC1 protein in photosynthesis, chloroplasts morphogenesis, and metal homeostasis (Duy et al. 2007). Expression of *AtPIC1* rescues the growth defect of *fet3fet4* on low Fe media. These results suggest that PIC1 may be involved in Fe uptake into chloroplasts (Fig. 2).

3.2.3 Mitochondria

Together with plastids, mitochondria are one of the locations of Fe–S cluster biogenesis. Moreover, many mitochondrial enzymes from the electron transport chain use Fe as a cofactor. The mechanisms of Fe import into mitochondria are still unknown. So far, the only transport system characterized at the molecular level involved in Fe homeostasis in plant mitochondria is AtATM3 (Fig. 2).

ATM proteins are half-molecule ABC transporters. The yeast homologue of AtATM3, Atm1p, is involved in the export of Fe–S clusters out of mitochondria. Atm1Δ mutant displays a *petite* phenotype, reflecting a respiration defect, associated with an Fe overaccumulation into mitochondria and a constitutive expression of the plasma membrane Fe uptake systems (Kispal et al. 1997, 1999). The expression of the *AtATM3* in Atm1Δ restores a wild-type phenotype (Chen et al. 2007). *starik*, a null *AtATM3* mutant, is a dwarf and chlorotic plant and shows Fe overaccumulation in mitochondria similar to the phenotype observed in Atm1Δ (Kushnir et al. 2001). *AtATM3* mRNA are detected in all plant tissues. AtATM3 is thus probably involved in constitutive export of Fe–S clusters from mitochondria. AtATM1 and AtATM2, two close homologues of ATM3, are also targeted to mitochondria. However, their involvement in Fe homeostasis is not clear yet (Chen et al. 2007). A recent report indicates that ATM3 functions in the export of pterin precursors from mitochondria and that its role in Fe homeostasis in mitochondria is likely indirect (Teschner et al. 2010).

3.2.4 Other Compartments

Recently, AtIRT2, a close homologue of AtIRT1, has been proposed to sequester the Fe accumulated after activation of the Fe deficiency response into vesicles. *AtIRT2* encodes a membrane protein localized in intracellular vesicles, which complements the Fe uptake defect of the *fet3fet4* yeast mutant. *AtIRT2* expression is increased under Fe deficiency (Vert et al. 2001). In contrast to the *irt1* knockout plant mutant, the absence of *AtIRT2* does not lead to a chlorotic phenotype and *irt1* chlorosis is not rescued by the overexpression of AtIRT2 (Vert et al. 2002, 2009). Like *AtIRT1* and *AtFRO2*, *AtIRT2* is under the control of *FIT* and the transcripts are detected in the rhizoderm. *AtIRT1* transcripts overaccumulate in *AtIRT2* over-expressing lines. It was proposed that increased Fe sequestration into vesicles leads to cytosolic Fe depletion, leading to *AtIRT1* up-regulation (Vert et al. 2009).

3.3 Long-Distance Transport

Iron circulates through the plant and is distributed to the different tissues. The detection of Fe in xylem and phloem indicates that bot systems are necessary for adequate Fe supply to the plant.

3.3.1 Xylem Transport

In the xylem sap, Fe is associated to organic acids and more specifically to citrate (Cataldo et al. 1988; Durrett et al. 2007; Larbi et al. 2010; Rellan-Alvarez et al. 2010). The *frd3* mutant was isolated in a screen for altered ferric chelate reductase activity (Yi and Guerinot 1996; Rogers and Guerinot 2002). The protein, which belongs to the multidrug and toxin efflux family (MATE), is detected at the plasma membrane of pericycle and vascular parenchyma cells. Recently, the FRD3 (ferric reductase defective 3) protein was characterized as a citrate efflux transporter, responsible for citrate loading in the xylem (Fig. 2). Xylem citrate levels are decreased in *frd3* and, in conditions of Fe excess, Fe overaccumulates in *frd3* roots, strengthening the evidence that Fe moves through the xylem as a ferric–citrate complex (Durrett et al. 2007).

FPN1 is one of the two plant homologues of mammalian IREG1. In analogy with the function of IREG1in the release of Fe to circulating plasma, FPN1 could be to a certain extent implicated in the loading of Fe into the xylem for root-to-shoot delivery (Fig. 2). The protein is localized at the plasma membrane and *FPN1* gene is expressed in vascular tissues. *FPN1* is primarily involved in cobalt transport but may also be able to transport Fe (Morrissey et al. 2009). *FPN1* is not regulated in response to Fe and the *fpn1* mutant shows chlorosis although its Fe content is not altered. Determining the precise role of FPN1 in plant Fe homeostasis will require further investigations.

3.3.2 Phloem Transport

In the phloem, Fe is associated with NA and in many species the Fe concentration in phloem is higher than in the xylem (Marschner 1997). The best candidates for loading and retrieval of Fe–NA or NA into and from the phloem are the YSL genes (YS1-like protein), a subfamily of OligoPeptide Transporters (OPT, Fig. 2).

Eight *AtYSL* genes were identified in Arabidopsis by homology with *ZmYS1*. Although it was shown in heterologous systems without ambiguity that ZmYS1 and OsYSL2 are able to transport Fe–NA or Fe–MA, the complementation of the *fet3fet4* yeast mutant by *AtYSL2* is controversial (DiDonato et al. 2004; Koike et al. 2004; Schaaf et al. 2005). *AtYSL1*, *AtYSL2*, and *AtYSL3* expression is up-regulated in the presence of Fe. Metal content analysis of *ysl* single or double mutants suggests that these genes are involved in the lateral transport of Fe–NA (Waters et al. 2006). In addition, *ysl1ysl3* double mutants exhibit interveinal chlorosis and reduced fertility. *AtYSL1 and AtYSL3* are required for Fe loading in seeds as well as for the mobilization of metals during leaf senescence (Le Jean et al. 2005; Waters et al. 2006). The rice OsYSL2, which is expressed in the phloem and developing seeds, is necessary for Fe transport in shoots and Fe supply to seeds (Ishimaru et al. 2010).

AtOPT3, a member of the OPT family (as YSLs), is probably involved in Fe transport (Wintz et al. 2003). *AtOPT3* is up-regulated by Fe starvation. AtOPT3 restores the growth of the *fet3fet4* yeast mutant, independently of the presence of NA. *AtOPT3* is preferentially expressed in vascular tissues and seed funiculus. A null mutation of *OPT3* is embryo lethal. An *OPT3* knock-down mutant displays Fe overaccumulation in root and shoot vascular tissues and a constitutive up-regulation of root Fe-deficiency responses, indicating a wrong perception of the Fe status. *opt3* knock-down mutant also exhibit a defect in Fe loading in seeds (Stacey et al. 2008). The Fe form transported by OPT3 is not known, but considering that other OPT members from yeast transport tetra and penta-peptides, OPT3 likely transports Fe chelates to peptides or peptide-like molecules.

4 Mechanisms of the Regulation of Fe Uptake in Plants

Under Fe starvation, the Fe uptake mechanisms are clearly regulated at the transcriptional level. In strategy I plants, such as Arabidopsis, tomato, or pea, the transcription of the ferric chelate reductase and IRT1 genes is strongly up-regulated under Fe-deficient conditions. In strategy II plants such as rice, maize or barley, the transcription of the genes of the phytosiderophore biosynthetic pathway and phytosiderophore uptake transporters are also strongly up-regulated under Fe deficiency. During the last decade, the first transcriptional regulators involved in the Fe-deficiency responses of strategy I and strategy II plants have been identified. This represents an important step in the understanding of the signaling mechanisms that regulate Fe transport. In addition, in strategy I plants such as Arabidopsis, the Fe uptake pathway appears to be regulated at the post-transcriptional level. In contrast, no evidence was found in plants for a control of mRNA stability or translation, analogous to the IRE/IRP system described in mammals (Arnaud et al. 2007).

4.1 Transcriptional Control of Fe Uptake in Strategy I Plants

The identification of the gene affected by the *fer* mutation in tomato represented the first breakthrough in the identification of the transcriptional regulators of Fe uptake in strategy I plants. The *fer* tomato mutant shows symptoms of Fe deficiency such as chlorosis even when it is fed with sufficient Fe (Ling et al. 1996). This mutant fails to activate Fe-deficiency responses in roots including the up-regulation of ferric chelate reductase activity and LeIRT1 transcription, rhizophere acidification as well as the increase in root hair density. The FER gene encodes a transcription factor of the bHLH family (Ling et al. 2002). The FER protein localizes in the nucleus (Brumbarova and Bauer 2005). FER is expressed in the root tissues where Fe uptake and Fe transfer to the central cylinder take place (Ling et al. 2002; Brumbarova and Bauer 2005). FER protein and mRNA accumulation is downregulated under generous Fe nutrition and is up-regulated in the tomato *chloronerva* mutant which displays constitutive activation of Fe-deficiency responses (Brumbarova and Bauer 2005).

The identification of *FER* in tomato allowed the investigation of *FIT*, its orthologue in *Arabidopsis thaliana*. FIT was initially named FIT1 or FRU1 and corresponds to bHLH29 of the 1b group of Arabidopsis bHLH transcription factors (Colangelo and Guerinot 2004; Jakoby et al. 2004; Yuan et al. 2005). *fit* knockout mutants are strongly chlorotic and do not set seeds unless supplemented with generous Fe nutrition (Colangelo and Guerinot 2004). *FIT* is expressed in the epidermal and cortical cell layers of the domain of Arabidopsis roots where Fe uptake takes place. *FIT* transcription is induced upon Fe deficiency (Colangelo and Guerinot 2004; Jakoby et al. 2004). Interestingly FIT controls at least partly its own transcriptional activation, indicating that it is part of an amplification loop regulating Fe-deficiency responses. Comparative microarray analyses of the transcriptional response to Fe deficiency between wild-type and *fit* mutant roots provided an overview of the genes that are directly or indirectly controlled by FIT (Colangelo and Guerinot 2004). In addition to *AtFRO2*, this study identified a number of transporters that are involved in Fe-deficiency responses whose expression is mis-regulated in the *fit* mutant: ZIP9, IRT2, IRT1, and FPN2 which may be directly involved in Fe transport (Vert et al. 2002; Schaaf et al. 2006; Morrissey et al. 2009; Vert et al. 2009) but also transporters for other metals such as COPT2, a Cu transporter (Sancenon et al. 2003), NRAMP1, the Mn uptake system (Cailliatte et al. 2010), and HMA3 and MTP3 responsible for vacuolar sequestration of Cd and Zn, respectively (Arrivault et al. 2006; Morel et al. 2009). In addition, the induction of AHA7, a proton pumping ATPase, and AtPDR9, an ABC transporter homologous to NtPDR3, of yet uncharacterized function, are also under the control of FIT. *AtAHA7*, *AtIRT1*, and *AtFRO2* are under the control of FIT and display the same expression pattern in Fe-deficient roots: they are probably the main players involved in the Strategy I response to Fe deficiency (Vert et al. 2003; Santi and Schmidt 2009). Genome wide analysis of gene expression indicates an orchestrated remodeling of metal homeostasis by FIT in Fe-deficient roots. However, AtNRAMP3 and AtNRAMP4 vacuolar Fe transporters, although upregulated in Fe-deficient roots, are not under the control of FIT (Colangelo and Guerinot 2004). Likewise, FER was shown to control only a subset of Fe-regulated transporters in tomato roots: although LeIRT1, LeIRT2, LeNRAMP1, and LeNRAMP3 are all up-regulated under Fe deficiency, only the up-regulation of LeIRT1 and LeNRAMP1 is controlled by the FER bHLH transcription factor (Bereczky et al. 2003).

FIT is necessary for the induction of the Fe uptake system in root; however, its overexpression is not sufficient to trigger strong constitutive upregulation of Fe-deficiency responses (Jakoby et al. 2004; Yuan et al. 2005). bHLH transcription factors often function as heterodimers. Within group 1b of Arabidopsis bHLH transcription factors, a subgroup containing bHLH38, 39, 100, and 101 was found to be upregulated under Fe deficiency (Wang et al. 2007). Knocking out these bHLH genes does not impact Fe-deficiency responses (Wang et al. 2007; Yuan et al. 2008); however, bHLH38 and 39 are able to interact with *AtIRT1* and *AtFRO2* promoters in yeast and with FIT protein in plant cells (Yuan et al. 2008). Furthermore, although individual overexpression of FIT, bHLH38, or bHLH39 leads to no or marginal upregulation or Fe-deficiency responses (Jakoby et al. 2004; Yuan et al. 2005), double overexpression of FIT with bHLH38 or bHLH39 triggers robust constitutive upregulation of Fe uptake and generates plants with higher Fe content in shoots (Yuan et al. 2008). These results are consistent with a model in which FIT/bHLH38 or FIT/bHLH39 complexes stimulate the transcription of Fe acquisition genes under Fe deficiency (Fig. 1). Upstream regulators of these complexes are still to be discovered. POPEYE (PYE), another bHLH transcription factor upregulated in root pericycle cells in response to Fe deficiency was recently identified (Long et al. 2010). PYE controls genes involved inter- and intracellular Fe transport as well as in root and root hair development. Accordingly, *pye* mutants show exacerbated chlorosis and altered root development under Fe starvation.

4.2 Post-transcriptional Control of Fe Acquisition Mechanisms in Strategy I Plants

In addition to the transcriptional regulation, attempts to overexpress components of the strategy I Fe uptake system revealed post-transcriptional regulations. Ectopic overexpression of *AtIRT1* or *AtFRO2* gene leads to high accumulation of *AtIRT1* and *AtFRO2* transcripts in roots and shoots irrespective of the Fe nutrition regime (Connolly et al. 2002; Connolly et al. 2003). However, the increase of the ferric chelate reductase activity in *AtFRO2* overexpressing plants is observed only under Fe deficiency (Connolly et al. 2003). This indicates a downregulation of the amount or activity of this enzyme under Fe-sufficient conditions. In the case of AtIRT1, protein accumulation in overexpressing plants is observed only in roots of Fe-deficient plants (Connolly et al. 2002). It was proposed that the protein is rapidly degraded in shoots and in Fe-sufficient roots. Accordingly, mutation of two lysine residues that are putative targets for conjugation with ubiquitin leads to a stabilization of the protein in the shoots and to a lesser extent in the roots of Arabidopsis plants overexpressing the mutated version (Kerkeb et al. 2008). The rapid degradation of AtIRT1 protein in the presence of Fe was proposed to be important to protect plants from Fe overload.

The possibility that the stability or the translation of mRNA of Fe-responsive genes is regulated by secondary RNA structures in the 5′ or 3′ UTR similar to IRE (iron-regulatory elements) characterized in mammalian cells has been addressed (Arnaud et al. 2007). The disruption of one of the cytosolic aconitases, which correspond to mammalian IRP1 (IRE binding protein 1), did not alter plant responses to Fe. More specifically it did not affect the induction of ferritin Fe storage proteins under Fe-sufficient conditions. In addition, IRE homologous sequences could not be detected in the 5′ or 3′ UTR of Fe-regulated genes in Arabidopsis. The possibility remains that secondary RNA structure with no homology to mammalian IRE exist in these genes, and that they are recognized by proteins which are not homologous to mammalian IRPs.

4.3 Transcriptional Control of Fe Uptake in Strategy II Plants

To investigate the mechanisms of transcriptional regulation by the Fe status in strategy II plants, Kobayashi and collaborators dissected the promoter of barley IDS2 gene, which is highly and specifically induced under Fe deficiency (Kobayashi et al. 2003). A combination of promoter deletion and promoter scanning using the β-glucuronidase gene as a reporter allowed the identification of two distinct Iron-Deficiency Elements (IDE) in the promoter of IDS2: IDE1 (ATCAAG-CATGCTTCTTGC) and IDE2 (TTGAACGGCAAGTTTCACGCTGTCACT). IDE1 were found in the promoters of several genes that are regulated by Fe availability in barley and rice: nicotianamine aminotransferase (HvNAAT)-A, HvNAAT-B, nicotianamine synthase (HvNAS1), HvIDS3, OsNAS1, OsNAS2, and OsIRT1. Interestingly, IDE1 and IDE2 are functional in tobacco and Arabidopsis even though they are strategy I plants (Kobayashi et al. 2005). Moreover, IDE1 is present in the promoters of *AtIRT1* and *AtFRO2* in Arabidopsis and in the promoter of NtPDR3, an Fe-deficiency inducible ABC transporter from *Nicotiana plumbaginifolia* (Ducos et al. 2005; Kobayashi et al. 2007b).

One hybrid screens identified two transcription factors from rice, IDEF1 and IDEF2, which bind specifically IDE1 and IDE2, respectively. IDEF1 and IDEF2 belong to the ABI3/VP1 and NAC families of transcription factors, respectively (Kobayashi et al. 2007a; Ogo et al. 2008). *IDEF1* and *IDEF2* are expressed in both rice roots and shoots, irrespective of the Fe status of the plant (Kobayashi et al. 2007a, 2010; Ogo et al. 2008). Overexpression of IDEF1 is sufficient to induce the transcription of target genes such as *OsIRT1* and *OsIRO2* encoding an Fe transporter and a bHLH transcription factor, respectively (Kobayashi et al. 2007a). This suggests a posttranscriptional regulation of IDEF1 proteins dependent on Fe nutrition. IDEF1 overexpressing plants exhibit increased tolerance to Fe deficiency while knock-down plants exhibit higher sensitivity to Fe deficiency. IDEF1 mediates a biphasic response to Fe deficiency (Kobayashi et al. 2009). In the first phase, IDEF1 mediates the up-regulation of Fe acquisition and utilization genes such as *OsIRO2*, *OsYSL15*, *OsIRT1*, *OsYSL2*, *OsNAS1*, *OsNAS2*, *OsNAS3*, and *OsDMAS1* and subsequently it mediates the up-regulation of genes encoding late embryogenesis abundant proteins in vegetative organs. Silencing of IDEF2 NAC transcription factor leads to aberrant Fe homeostasis in rice (Ogo et al. 2008). Notably, the up-regulation of the Fe nicotianamine transporter OsYSL2 under Fe deficiency is strongly decreased in rice plant with decreased IDEF2 expression.

Among the genes that are up-regulated in Fe-deficient rice, *OsIRO2* encodes a transcription factor from the bHLH family (Ogo et al. 2006). OsIRO2 promoter contains IDE1, and OsIRO2 expression is enhanced in IDEF1 overexpressing rice, suggesting that OsIRO2 is a target of IDEF1. In turn, OsIRO2 regulates directly or indirectly 59 genes induced by Fe deficiency, including many genes involved in Fe acquisition and utilization (Ogo et al. 2007). Genes which contain OsIRO2-binding DNA motif (ACCACGTGGTTTT) in their promoters and which are mis-regulated in OsIRO2-silenced plants, such as OsNAS1, OsFDH, and OsAPT1, are likely direct targets of this transcription factor. Interestingly, OsIRO2 also regulates a set of transcription factors including NAC transcription factor distinct from IDEF2 (Ogo et al. 2007). This suggests a model in which the transcriptional regulation of the Fe-deficiency response in rice is mediated by a complex cascade of transcription factors (Fig. 1). Silencing OsIRO2 leads to higher sensitivity to Fe deficiency. In contrast overexpression of OsIRO2 does not improve tolerance to Fe deficiency (Ogo et al. 2007). This situation is reminiscent of the results obtained with FIT bHLH transcription factor in strategy I plants. It suggests that additional components are needed to achieve constitutive expression of Fe acquisition and utilization genes in rice. IDEF2 may represent one such component as many promoters containing binding elements for OsIRO2 also contain IDE2 (Ogo et al. 2008).

The characterization of *cis-* and *trans-*acting factors involved in gene regulation in response to Fe deficiency opens many questions. Is the transcriptional cascade conserved between strategy I and strategy II plants? FIT/FER and OsIRO2 are bHLH transcription factors of the same group that are up-regulated under Fe deficiency and play major roles in the regulation of Fe acquisition. However, FIT/FER expression is restricted to the root tissues that perform Fe acquisition, whereas OsIRO2 is expressed in both roots and shoots. What are the upstream factors controlling FER/FIT expression? The identification of IDEF1 and IDEF2, which are not regulated by Fe at the transcriptional level, raises the question of the post-transcriptional regulation events that could mediate their activation under Fe-limiting conditions and their relationships to Fe-sensing mechanisms. In both strategy I and strategy II plants, the transcriptional regulation is the main regulatory level in response to Fe deficiency. This is similar to the yeast response to Fe deprivation and perhaps more distant to the situation in mammalian cells where posttranscriptional regulation plays a central role. The Fe acquisition system of strategy I plant combines regulation at both levels of transcription and protein stability. Is the stability of protein-involved Fe acquisition in strategy II plants is also regulated by Fe availability?

5 Local Versus Long-Distance Regulation

Evidence for a dual regulation of Fe acquisition in strategy I plants by a shoot-derived systemic signal and a local signal could be obtained from split-root experiments on Arabidopsis (Vert et al. 2003). *AtIRT1* and *AtFRO2* mRNA expressions were analyzed in a system where half of the roots are in a Fe-deficient media and half of the roots are in the presence of Fe. *AtIRT1* and *AtFRO2* mRNA levels are downregulated in the half-part of the roots where Fe is lacking, whereas in the half-part which was in Fe-sufficient media, the mRNA levels are strongly upregulated. AtIRT1 protein levels and *AtIRT2* transcript levels followed the same regulation pattern (Vert et al. 2003, 2009). The plant locally downregulates the expression of the genes in the roots where the Fe is completely lacking and the Fe-deficient status of the shoots leads to upregulation of the expression of the Fe uptake system in the roots where Fe is present. Thus, Arabidopsis is able to assess its own shoot Fe status and to sense the root local Fe concentration (Fig. 3). Whether the long-distance regulation is mediated by a positive or a repressive

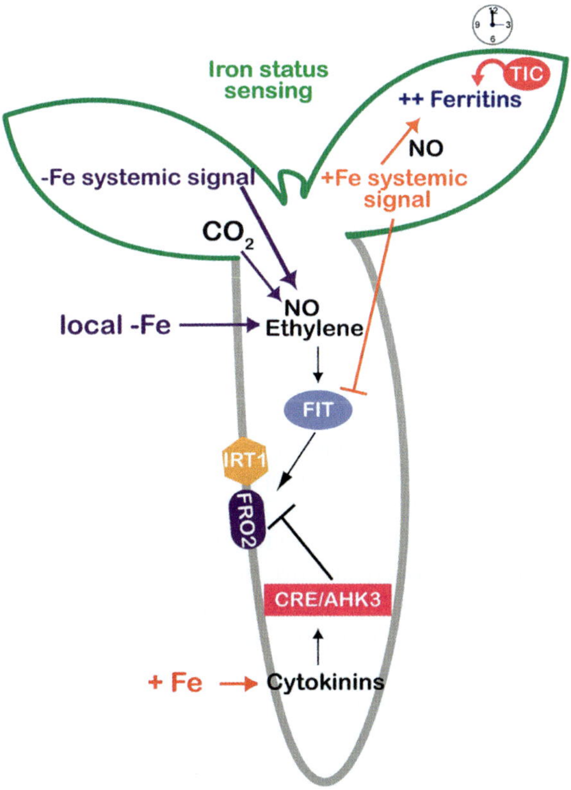

Fig. 3 Systemic regulation of iron acquisition mechanisms in strategy I plants

signal is not known (Vert et al. 2003). By contrast, other Fe-regulated transporters, such as NRAMP3 and NRAMP4, do not seem to be dually controlled by local and systemic signals (Vert et al. 2009). AtIRT1, *AtIRT2*, and *AtFRO2* are under the control of the FIT transcription factor, whereas NRAMP3 and NRAMP4 are not, suggesting that FIT may be a component of the systemic signaling pathway (Fig. 3).

Analysis of Fe-deficiency responses in plants perturbed in Fe cellular compartmentalization indicates local sensing of cytosolic Fe levels. NRAMP3 mediates the release of Fe from the vacuole (Thomine et al. 2003); IRT2 and FPN2 are involved in Fe sequestration in vesicles and vacuoles, respectively (Morrissey et al. 2009; Vert et al. 2009). Loss of NRAMP3 function or overexpression of IRT2 exacerbates responses to Fe deficiency (Thomine et al. 2003; Vert et al. 2009). Conversely, overexpression of NRAMP3 or loss of FPN2 function attenuates responses to Fe deficiency, such as *AtIRT1* expression or ferric chelate reductase activity (Thomine et al. 2003; Morrissey et al. 2009). Thus, the balance between Fe sequestration and mobilization locally modulates Fe-deficiency responses: the root cells respond to increased Fe sequestration by upregulation of Fe-deficiency responses and vice versa.

Mutants impaired in the regulation of Fe-deficiency responses constitute useful tools to investigate the signaling pathways regulating Fe acquisition. Some mutants, such as *fer* in tomato and *fit* in Arabidopsis, do not turn on the Fe-deficiency responses under Fe starvation (see Sect. 4.1). By contrast, other mutants, such as *dgl* and *brz* in pea, *chln* in tomato, and *frd3* and *opt3* in Arabidopsis, are unable to turn off the Fe-deficiency responses under Fe-sufficient conditions.

The *brz* mutant exhibits bronze-spotted necrotic leaves on account of an excessive Fe accumulation (Welch and Larue 1990). The root Fe concentration of the mutant is similar to the parental control genotype but more Fe is translocated to the shoots. Under Fe-sufficient conditions, *brz* constitutively expresses Fe-deficiency responses: acidification of the soil, expression of ferric-chelate reductase gene (*FRO1*), and higher Fe uptake (Welch and Larue 1990; Waters et al. 2002). The *dgl* mutant also accumulates excessive Fe levels in its vegetative tissues, manifested as brown-degenerated leaves. Like *brz*, *dgl* constitutively expresses *FRO1* (Waters et al. 2002). Shoot/root reciprocal grafting revealed that the genotype of the shoots controls the *brz* or *dgl* phenotype. This indicates that a signal transmitted to the roots is involved in the Fe-deficiency response (Fig. 3) (Grusak et al. 1990; Grusak and Pezeshgi 1996). Moreover, proper regulation of *FRO1* is maintained in the shoots of both mutants (Waters et al. 2002). The two mutants are not allelic, suggesting that two signals are involved in the ability to sense the Fe status (Grusak and Pezeshgi 1996). Alternatively, the two *loci* could encode distinct components of the same signaling pathway. The double grafting of control and *dgl* shoots on a control rootstock still leads to constitutive activity of the ferric-chelate reductase. Thus, it was proposed that the *dgl* mutant shoots transmit a signaling compound which promotes Fe-deficiency responses (Grusak and Pezeshgi 1996). The genes impaired by the *brz* and *dgl* mutations remain to be identified.

Like *brz* and *dgl*, the tomato mutant *chloronerva* (*chln*) exhibits constitutive upregulation of the Fe-deficiency responses, associated with Fe over accumulation in shoots and roots. In contrast with *brz* and *dgl*, *chln* phenotype is manifested as interveinal chlorosis, a symptom of Fe deficiency. Grafting the *chloronerva* mutant onto wild type or vice versa normalizes the mutant phenotype. This indicates that a transportable substance may move from the wild-type root stock or scion to the mutant organs and correct the mutant phenotype (Stephan and Grun 1989; Becker et al. 1995). *CHLN* encodes a nicotianamine synthase and *chln* lacks nicotianamine synthase enzymatic activity (Ling et al. 1999). Exogenous application of nicotianamine suppressed the chlorotic phenotype (Stephan and Grun 1989). Due to the absence of NA, Fe precipitates in *chln* shoots and roots as insoluble ferric phosphate, which is not accessible for nutrition (Becker et al. 1995). NA synthesis is thus necessary for Fe availability and transit in the plant (see Sect. 3.3.2) (Ling et al. 1999; Takahashi et al. 2003). Unavailability of Fe in *chln* accounts for the upregulation of Fe acquisition system in this mutant and its chlorotic phenotype (Ling et al. 1996). The *fer* gene is epistatic over the *chln* gene. This indicates that the long-distance signaling targets the transcription of the Fe acquisition system through this bHLH transcription factor.

The Arabidopsis *frd3* mutant also displays constitutive up-regulation of Fe-deficiency responses (Yi and Guerinot 1996). In the case of *frd3*, grafting experiments revealed that the genotype of the roots controls the phenotype (Green and Rogers 2004). In agreement, the FRD3 protein is expressed in the pericycle cells of the roots. *frd3* defects could be traced back to a defect in citrate secretion into the xylem (Durrett et al. 2007). This impairs both the loading of Fe in the xylem and Fe reuptake by shoot cells. The phenotype of *frd3* reflects the importance of root-to-shoot Fe transport for the adequate sensing of Fe (see Sect. 3.3.1). As in the case of *chln*, decreased availability of Fe accounts for the constitutive upregulation of the Fe-deficiency response.

The investigation of mutants such as *frd3*, *opt3*, *bzl*, or *dgl* and the results obtained from wild-type plants indicate that the plant integrates the intracellular Fe status and the local Fe concentration to acclimate its deficiency responses. In addition, the Fe must be either correctly localized in cells or under an adequate speciation-chelation form to be properly sensed. The molecular bases of the systemic signal and of the local Fe sensing remain to be determined.

6 Hormonal Signals

Hormones are good candidates to be part of the systemic signal triggering the responses to Fe deficiency in roots when shoots are Fe deficient. As a matter of fact, some morphological and physiological responses to Fe deficiency can be mimicked or inhibited by ethylene, auxin, cytokinins, and ABA addition (Fig. 3) (Schmidt et al. 2000). The key question is then to discriminate between a tangible involvement of the hormone in Fe signaling and a mere crosstalk, where the hormone is an element of the morphological response.

6.1 Ethylene

Fe deficiency is associated with morphological changes in the subapical root zone such as a tissue swelling associated with an increased number of root hairs and the development of transfer cells in the outer root cell layers (Landsberg 1986). The application of ethylene or 1-aminocyclopropane-1-carboxilic acid (ACC), an ethylene precursor, on seedlings mimics some of these morphological changes, such as the increased root hair density. Reciprocally, the inhibition of ethylene production or perception has the opposite effect triggering a reduction of root hair density (Romera and Alcantara 1994; Romera et al. 1999).

A production of ethylene occurs after 3–9 days of Fe deficiency depending on the plant species (Romera et al. 1999; Li and Li 2004). Whether the ethylene production is concomitant with ferric-chelate reductase activation or follows it is still not clear (Romera et al. 1999; Li and Li 2004). Phenotypical analysis of mutants altered in the ethylene biosynthetic pathway or ethylene perception indicates that ethylene is necessary for the formation of root hairs in response to Fe deficiency but does not affect the activity of the ferric-chelate reductase (Schmidt et al. 2000; Li and Li 2004; Lucena et al. 2006). However, some authors proposed that different ethylene pathways mediate subapical root hair formation and ferric-chelate reductase/Fe transport induction (Lucena et al. 2006). The alternate ethylene pathway mediating ferric-chelate reductase up-regulation remains to be elucidated.

In Arabidopsis and tomato, *IRT1*, *FRO2*, and *FIT/FER* expression under Fe-deficient condition is increased by ethylene. In Fe-sufficient conditions, this hormone does not affect the expression of these genes (Lucena et al. 2006). Concordant results were obtained in cucumber: up-regulation of *CsFRO1* and *CsIRT1* transcript levels is attenuated when Fe-deficient plants are treated with an ethylene inhibitor and enhanced by the addition of ACC. Moreover, among the two H^+-ATPases CsHA1 and CsHA2, only CsHA1, the isoform responding to Fe deficiency, is regulated in response to ACC or an ethylene inhibitor (Waters et al. 2007). Based on these results, ethylene would act as an enhancer of Fe-deficiency responses (Fig. 3). As ethylene does not affect the expression of Fe-acquisition genes in Fe-sufficient conditions, it is possible that Fe acts as an inhibitor of the ethylene response, overriding the effect of ethylene. The absence of ferric reductase regulation in the tomato *fer* mutant after supply of ACC indicates that ethylene action could be mediated through the FIT/FER transcription factor. The identification of an ethylene-responsive element in FIT promoter supports the notion of ethylene being one component of the Fe-deficiency response (Lucena et al. 2006).

Although one study reports that the inhibition of ethylene synthesis or ethylene action reduces the uptake of Fe–PS in Barley, the data obtained so far in Strategy II plants do not suggest an important role for ethylene in responses to Fe deficiency (Romera et al. 2006; Welch et al. 1997). It would be interesting to investigate the effect of ethylene in rice, in which elements of Fe acquisition via strategy I and II coexist.

6.2 *Cytokinins*

Many developmental and cellular processes including germination, maintenance of meristems, specification of the root vascular system, inhibition of lateral root formation, and leaf senescence are controlled by cytokinins (CK) (Perilli et al. 2010). Moreover, CK inhibit phosphate, nitrogen, and sulfate uptake as well as responses to deficiencies in these nutrients. In a screen for signals regulating Supp., Fe-deficiency responses, Seguela and collaborators found that CK is the most potent hormone to inhibit the transcription of *AtIRT1* (Seguela et al. 2008).

CK repress *AtIRT11, AtFRO2*, and *FIT* gene expression independently of the Fe status. Even though *AtFRO2* and *AtIRT1* mRNA levels are expressed at low levels in the *fit* knockout mutant, the addition of CK further decreased their amount, indicating that the repression is not mediated by the *FIT* transcription factor. The Fe-deficiency responsive genes *AtNRAMP3* and *AtNRAMP4* are not repressed by CK, suggesting that this hormone specifically targets components of the root Fe uptake system. The repression of *AtIRT1* requires the presence of the CRE1/AHK3, two CK receptors, which upon hormone binding are phosphorylated and transfer a signal to downstream elements (Seguela et al. 2008). Accordingly, the resupply of Fe, which is known to trigger a rapid decrease of *AtIRT1* mRNA levels and to turn off Fe-deficiency response, transiently upregulates genes involved in CK synthesis and signaling.

CK inhibit root development. *AtIRT1* expression is also repressed by other treatments which reduce root growth (Seguela et al. 2008). Thus, the root growth rate appears to regulate the high-affinity Fe uptake system independently of the Fe status. This suggests that the components of the Fe-deficiency response are negatively regulated by CK to match the plant nutritional demand (Fig. 3).

6.3 *Nitric Oxide*

Nitric oxide (NO) is a small diffusible signaling molecule existing in different redox states. NO regulates Fe metabolism in plants and in animals. The nitric oxide radical (NO•) binds Fe with high affinity and the various redox states depend on the pH and redox potential of the cell. Both responses to Fe deficiency and Fe excess involve NO signaling but whether the same NO species or the same production pathway are involved is not clearly established yet.

The supply of NO donors, such as sodium nitroprusside (SNP) or *S*-nitrosoglutathione (GSNO), to Fe-deficient maize seedlings alleviates their chlorotic symptoms (Graziano et al. 2002). Although the Fe content of deficient plants treated or not with NO remains the same, the chlorophyll levels and transcript abundance of rubisco large subunit and of the D1 protein of photosystem II are restored (Graziano et al. 2002). Conversely, the application of NO scavengers, which eliminate endogenous NO, on Fe-sufficient plants triggers Fe-deficiency symptoms. The addition of NO in *ys1* and *ys3* maize mutants eliminates their interveinal chlorosis

(Graziano et al. 2002). As the Fe concentration in plants is not modified by NO application, it seems probable that NO renders Fe more available, either by remobilizing it from an inaccessible pool, or by improving its root to shoot translocation.

NO alleviates Fe-deficiency chlorotic symptoms not only in monocots but in dicots as well. Iron-deficient tomato plants produce ~two times more NO in their roots than Fe-replete plants. NO production in the shoots remains identical regardless of the Fe status (Graziano and Lamattina 2007). NO is generated in the root epidermis within hours after transfer to Fe-deficient conditions. In addition, NO can trigger the morphological changes associated with Fe deficiency, such as an increase in the number of root hairs. Addition of NO scavenger prevents the upregulation of *Fer*, *LeFRO1*, and *LeIRT1* under Fe deficiency, and reduces the number of root hairs (Graziano and Lamattina 2007). NO production triggered by Cd treatment upregulates *AtIRT1* expression in Arabidopsis (Besson-Bard et al. 2009). Thus, NO is involved in the physiological and the morphological responses to Fe deficiency in Strategy I plants. The enzymatic pathway leading to endogenous NO production in the root epidermis of Fe-deficient plants is still not clear. A nitrate reductase tomato mutant displays reduced ability to induce the expression of *Fer*, *LeFRO1*, and *LeIRT1* during Fe starvation, suggesting the involvement of nitrate reductase in NO production under Fe deficiency.

The treatment of the *fer* mutant with GSNO reverts the chlorotic phenotype although it does not restore the expression of *LeIRT1* and *LeFRO2*, indicating that the transcription factor is necessary to trigger root responses to Fe deficiency. Moreover, two putative consensus *S*-nitrosylation motifs were found in the FER protein that may represent NO-regulation sites (Graziano and Lamattina 2007). An interrelation between ethylene and NO signaling at this step is possible.

Elevation of ambient CO_2 concentration alleviates the symptoms of Fe deficiency in tomato plants grown on the poorly available hydrous Fe(III) oxydes as the only Fe source (Jin et al. 2009). Elevation of ambient CO_2 enhances Fe-deficiency responses: increased subapical root hair number and apical root swelling, higher proton extrusion and ferric-chelate reductase activity, and upregulation of *LeIRT1*, *LeFRO1*, and *Fer* genes. The effect of ambient CO_2 on Fe-deficiency responses correlates with NO production. A model in which elevation of ambient CO_2 enhances Fe-deficiency response via NO production is likely. CO_2 elevation induces NO production in Fe-sufficient plants as well. However, the expression of *LeIRT1*, *LeFRO1*, and *Fer* genes is not affected in these conditions. This indicates that NO is not the only component triggering the Fe-deficiency response (Graziano and Lamattina 2007; Jin et al. 2009). How the CO_2 induces the production of NO is not yet established. NO is a component of the Fe-deficiency response: it contributes to Fe availability either by an increased Fe acquisition from the soil, by remobilization of some internal Fe pool, or by a more efficient Fe utilization by the plant (Fig. 3).

As mentioned above, NO also takes part in the response to Fe excess. Ferritins buffer Fe excess to prevent against oxidative stress. The addition of high concentration of Fe on Arabidopsis cell suspension triggers NO production in chloroplasts.

The formation of NO in plastids is followed by the accumulation of *AtFer1* and *AtFer4* transcripts, enabling the storage of Fe under an innocuous form (Fig. 3) (Murgia et al. 2002; Arnaud et al. 2006; Martin et al. 2009). What is the source of the production of the NO in the plastid is not clear yet. Inversely, the application of NO scavengers impairs *AtFer1* mRNA accumulation. Analysis of the *AtFER1* promoter sequence identified an Fe-dependent regulatory sequence (IDRS) involved in the transcriptional repression of *AtFER1* when cellular Fe concentrations are low (Murgia et al. 2002). NO would act downstream of Fe, relieving the effect on the repressor through a pathway that involves a PP2A-type protein phosphatase and 26S proteasome-dependent degradation of the repressor (Murgia et al. 2002; Arnaud et al. 2006). Thus, in response to excessive Fe supply, NO enables the transcription of *AtFER1* as part of a mechanism to prevent oxidative stress.

Frataxin is a mitochondrial protein present in many organisms from mammals to plants (Fig. 2). This protein is proposed to play a role in Fe–S biogenesis or Fe–S insertion into proteins (Busi et al. 2006). In yeast, a frataxin loss of function mutant displays oxidative stress symptoms probably related to excessive Fe accumulation in mitochondria (Radisky et al. 1999). In Arabidopsis, one gene encoding frataxin was identified. An Arabidopsis frataxin knock-down mutant accumulates more Fe in mitochondria and plastids, which correlates with higher root hair density and an increased accumulation of *AtFer1* and *AtFer4* transcripts (Martin et al. 2009). This mutant produces higher levels of NO in the roots. As in chloroplasts, the authors propose that NO is part of the mechanism to prevent Fe-induced oxidative stress (Martin et al. 2009).

7 Diurnal Regulation and Control by the Circadian Clock

As observed for some macronutrient uptake systems, Fe-deficiency responses in graminaceous or nongraminaceous plants are diurnally regulated (Gazzarrini et al. 1999). The observation that heterozygous *irt1* +/− mutant flowers earlier in short days than wild-type plants, whereas in long days, the heterozygous is phenotypically identical to the wild type, led to the examination of *AtIRT1* and *AtFRO2* diurnal regulation (Vert et al. 2003). In Fe-sufficient conditions, the abundance of *AtIRT1* and *AtFRO2* mRNA is maximal during the light period and decreases during the dark phase. In Fe-deficient plants, these two genes are not light-regulated, indicating that the Fe nutritional status overrides the diurnal rhythm (Vert et al. 2003). *AtAHA7* expression displays the same diurnal regulation pattern as *AtIRT1* and *AtFRO2*. Supp. by contrast, *AtAHA2* and *AtFIT1* mRNA levels increase during the light period but this regulation is maintained under Fe-deficiency (Santi and Schmidt 2009).

In Fe-deficient barley, phytosiderophore secretion peaks about 2 h after the beginning of the light period. Gene expression analysis in Fe-deficient barley roots using a rice microarray allowed the identification of ~50 genes that undergo diurnal regulation (Negishi et al. 2002). The closer investigation of five of these genes, which are putatively involved in the secretion of MAs, revealed that their RNA abundance increased during the dark period and started to decrease after the beginning of the light phase (Negishi et al. 2002). Interestingly, this regulation is different from the one observed for the high-affinity uptake system of dicots (Vert et al. 2003; Santi and Schmidt 2009).

The diurnal regulation of Fe acquisition genes in strategy I and II plants may be due either to their sensitivity to light or to their dependence on the circadian clock. The first hint that Fe homeostasis may be under the control of the circadian oscillator came from a genetic screen for deregulated expression of the ferritin gene *AtFER1*. This screen identified mutants in the TIC gene (Duc et al. 2009). TIC is a nuclear protein regulating the plant circadian clock, which affects more particularly genes expressed in the evening (Ding et al. 2007). *AtFER1* is highly expressed in Fe-deficient *tic2* knock-out mutants, indicating that the TIC protein is involved in the down-regulation of *AtFer1* (Fig. 3). Several genes involved in the response to Fe excess, *AtFER1, AtFER3*, *AtFER4*, and *AtAPX1,* are under the control of the circadian clock and *AtFER3*, *AtFER4*, and *AtAPX1* are derepressed in a *tic2* mutant as well (Duc et al. 2009). Even though its leaf Fe content is similar to the wild type, the *tic2* mutant displays internerval chlorosis, which is rescued by generous Fe supply. *tic2* is also more sensitive to Fe excess. Despite its Fe-deficient phenotype, the components of the deficiency response are not upregulated in *tic2*, suggesting that the Fe concentration is still correctly sensed. Upon Fe overload, *AtFER1* expression does not depend on TIC, and the TIC-dependent regulation of *AtFER1* is independent of the IDRS element. As for the diurnal regulation of *AtIRT1* and *AtFRO2*, the Fe nutritional status overrides the clock regulation (Vert et al. 2003; Duc et al. 2009).

8 Conclusion

The molecular mechanisms of Fe acquisition in strategy I and II plants have been elucidated. The transcription factors responsible for their up-regulation under Fe deficiency have been identified. The acquisition of Fe by plants is regulated at several levels by local and systemic long-distance signals. The systemic signaling pathway appears to integrate multiple inputs from hormonal signals, diurnal regulation, and nutritional demand (Fig. 3). However, several important elements are still to be discovered. First, the nature of the systemic signal transmitted by Fe-deficient shoots which activate Fe uptake in roots remains elusive. Is a specific hormone, analogous to hepcidin in mammals, involved? In mammals, hepcidin establishes a strong connection between Fe homeostasis and inflammatory responses (Nemeth and Ganz 2006). Fe homeostasis and pathogen resistance are also connected in plants (Expert 1999; Dellagi et al. 2005, 2009; Segond et al. 2009). It will be interesting to find out whether the systemic signal mediating Fe-deficiency responses is also mobilized during pathogen attack. Second, knowledge about the nature of the Fe-sensing mechanisms in plants is missing. Mammalian and yeast cells sense Fe through specific proteins that bind FeS clusters, such as cytosolic aconitase and glutaredoxins. It will be interesting to find out if plant cells use a similar pathway to sense their Fe status. It will be interesting also to find out if similar sensing mechanisms are used for local and systemic signaling.

Many transporters which control intracellular Fe distribution have been identified. However, key processes such as the uptake of Fe in plastids and mitochondria need to be fully elucidated. The signals that control Fe intracellular distribution are unknown. It will be of major interest to discover in which cell compartment Fe is sensed.

Our current knowledge of the mechanisms of Fe distribution between plant organs is fragmentary. Clear evidence points to nicotianamine and citrate as important ligands controlling Fe distribution (Ling et al. 1999; Rogers and Guerinot 2002; Takahashi et al. 2003; Klatte et al. 2009). The YSL transporters are good candidates for the transport of NA and NA–Fe between plant cells; however, the exact role of the different YSL is still unclear. As for intracellular distribution, the signals regulating Fe distribution between organs are unknown. It is striking that, in contrast with transporters involved in Fe acquisition, Arabidopsis YSL genes that have been studied are up-regulated in the presence of Fe (Le Jean et al. 2005; Waters et al. 2006).

Although many components of Fe homeostasis have been identified in the recent years, important discoveries on the mechanisms of Fe sensing, Fe signaling, and Fe trafficking within and between plant cells and organs are still to be made.

References

Abdel-Ghany SE, Muller-Moule P, Niyogi KK, Pilon M, Shikanai T (2005) Two P-type ATPases are required for copper delivery in *Arabidopsis thaliana* chloroplasts. Plant Cell 17:1233–1251

Arnaud N, Murgia I, Boucherez J, Briat JF, Cellier F, Gaymard F (2006) An iron-induced nitric oxide burst precedes ubiquitin-dependent protein degradation for Arabidopsis AtFer1 ferritin gene expression. J Biol Chem 281:23579–23588

Arnaud N, Ravet K, Borlotti A, Touraine B, Boucherez J, Fizames C, Briat JF, Cellier F, Gaymard F (2007) The iron-responsive element (IRE)/iron-regulatory protein 1 (IRP1)-cytosolic aconitase iron-regulatory switch does not operate in plants. Biochem J 405:523–531

Arrivault S, Senger T, Kramer U (2006) The Arabidopsis metal tolerance protein AtMTP3 maintains metal homeostasis by mediating Zn exclusion from the shoot under Fe deficiency and Zn oversupply. Plant J 46:861–879

Askwith C, Eide D, Vanho A, Bernard PS, Li LT, Daviskaplan S, Sipe DM, Kaplan J (1994) The Fet3 gene of S-cerevisiae encodes a multicopper oxidase required for ferrous iron uptake. Cell 76:403–410

Becker R, Fritz E, Manteuffel R (1995) Subcellular localization and characterization of excessive iron in the nicotianamine-less tomato mutant chloronerva. Plant Physiol 108:269–275

Bereczky Z, Wang HY, Schubert V, Ganal M, Bauer P (2003) Differential regulation of nramp and irt metal transporter genes in wild type and iron uptake mutants of tomato. J Biol Chem 278:24697–24704

Besson-Bard A, Gravot A, Richaud P, Auroy P, Duc C, Gaymard F, Taconnat L, Renou JP, Pugin A, Wendehenne D (2009) Nitric oxide contributes to cadmium toxicity in Arabidopsis by promoting cadmium accumulation in roots and by up-regulating genes related to iron uptake. Plant Physiol 149:1302–1315

Blaiseau PL, Lesuisse E, Camadro JM (2001) Aft2p, a novel iron-regulated transcription activator that modulates, with Aft1p, intracellular iron use and resistance to oxidative stress in yeast. J Biol Chem 276:34221–34226

Briat JF, Duc C, Ravet K, Gaymard F (2009) Ferritins and iron storage in plants. Biochim Biophys Acta [Epub ahead of print]

Brumbarova T, Bauer P (2005) Iron-mediated control of the basic helix-loop-helix protein FER, a regulator of iron uptake in tomato. Plant Physiol 137:1018–1026

Buesseler KO, Andrews JE, Pike SM, Charette MA (2004) The effects of iron fertilization on carbon sequestration in the southern ocean. Science 304:414–417

Busi MV, Maliandi MV, Valdez H, Clemente M, Zabaleta EJ, Araya A, Gomez-Casati DF (2006) Deficiency of *Arabidopsis thaliana* frataxin alters activity of mitochondrial Fe-S proteins and induces oxidative stress. Plant J 48:873–882

Cailliatte R, Schikora A, Briat JF, Mari S, Curie C (2010) High-affinity manganese uptake by the metal transporter NRAMP1 Is essential for Arabidopsis growth in low manganese conditions. Plant Cell 22:904–917

Cataldo DA, McFadden KM, Garland TR, Wildung RE (1988) Organic constituents and complexation of nickel(II), iron(III), cadmium(II), and plutonium(IV) in soybean xylem exudates. Plant Physiol 86:734–739

Chen OS, Crisp RJ, Valachovic M, Bard M, Winge DR, Kaplan J (2004) Transcription of the yeast iron regulon does not respond directly to iron but rather to iron-sulfur cluster biosynthesis. J Biol Chem 279:29513–29518

Chen S, Sanchez-Fernandez R, Lyver ER, Dancis A, Rea PA (2007) Functional characterization of AtATM1, AtATM2, and AtATM3, a subfamily of Arabidopsis half-molecule ATP-binding cassette transporters implicated in iron homeostasis. J Biol Chem 282:21561–21571

Cheng L, Wang F, Shou H, Huang F, Zheng L, He F, Li J, Zhao FJ, Ueno D, Ma JF, Wu P (2007) Mutation in nicotianamine aminotransferase stimulated the Fe(II) acquisition system and led to iron accumulation in rice. Plant Physiol 145:1647–1657

Colangelo EP, Guerinot ML (2004) The essential basic helix-loop-helix protein FIT1 is required for the iron deficiency response. Plant Cell 16:3400–3412

Connolly EL, Fett JP, Guerinot ML (2002) Expression of the IRT1 metal transporter is controlled by metals at the levels of transcript and protein accumulation. Plant Cell 14:1347–1357

Connolly EL, Campbell NH, Grotz N, Prichard CL, Guerinot ML (2003) Overexpression of the FRO2 ferric chelate reductase confers tolerance to growth on low iron and uncovers posttranscriptional control. Plant Physiol 133:1102–1110

Curie C, Panaviene Z, Loulergue C, Dellaporta SL, Briat JF, Walker EL (2001) Maize yellow stripe1 encodes a membrane protein directly involved in Fe(III) uptake. Nature 409: 346–349

Curie C, Cassin G, Couch D, Divol F, Higuchi K, Le Jean M, Misson J, Schikora A, Czernic P, Mari S (2009) Metal movement within the plant: contribution of nicotianamine and yellow stripe 1-like transporters. Ann Bot 103:1–11

Dellagi A, Rigault M, Segond D, Roux C, Kraepiel Y, Cellier F, Briat JF, Gaymard F, Expert D (2005) Siderophore-mediated upregulation of Arabidopsis ferritin expression in response to Erwinia chrysanthemi infection. Plant J 43:262–272

Dellagi A, Segond D, Rigault M, Fagard M, Simon C, Saindrenan P, Expert D (2009) Microbial siderophores exert a subtle role in Arabidopsis during infection by manipulating the immune response and the iron status. Plant Physiol 150:1687–1696

Dell'Orto M, Santi S, De Nisi P, Cesco S, Varanini Z, Zocchi G, Pinton R (2000) Development of Fe-deficiency responses in cucumber (Cucumis sativus L.) roots: involvement of plasma membrane H(+)-ATPase activity. J Exp Bot 51:695–701

DiDonato RJ Jr, Roberts LA, Sanderson T, Eisley RB, Walker EL (2004) Arabidopsis yellow stripe-like2 (YSL2): a metal-regulated gene encoding a plasma membrane transporter of nicotianamine-metal complexes. Plant J 39:403–414

Ding Z, Millar AJ, Davis AM, Davis SJ (2007) Time for coffee encodes a nuclear regulator in the *Arabidopsis thaliana* circadian clock. Plant Cell 19:1522–1536

Dix DR, Bridgham JT, Broderius MA, Byersdorfer CA, Eide DJ (1994) The Fet4 gene encodes the low affinity Fe(Ii) transport protein of Saccharomyces Cerevisiae. J Biol Chem 269: 26092–26099

Duc C, Cellier F, Lobreaux S, Briat JF, Gaymard F (2009) Regulation of iron homeostasis in *Arabidopsis thaliana* by the clock regulator time for coffee. J Biol Chem 284:36271–36281

Ducos E, Fraysse S, Boutry M (2005) NtPDR3, an iron-deficiency inducible ABC transporter in Nicotiana tabacum. FEBS Lett 579:6791–6795

Durrett TP, Gassmann W, Rogers EE (2007) The FRD3-mediated efflux of citrate into the root vasculature is necessary for efficient iron translocation. Plant Physiol 144:197–205

Duy D, Wanner G, Meda AR, von Wiren N, Soll J, Philippar K (2007) PIC1, an ancient permease in Arabidopsis chloroplasts, mediates iron transport. Plant Cell 19:986–1006

Eide D, Broderius M, Fett J, Guerinot ML (1996) A novel iron-regulated metal transporter from plants identified by functional expression in yeast. Proc Natl Acad Sci USA 93:5624–5628

Expert D (1999) Withholding and exchanging iron: interactions between Erwinia spp. and their plant hosts. Annu Rev Phytopathol 37:307–334

Feng H, An F, Zhang S, Ji Z, Ling HQ, Zuo J (2006) Light-regulated, tissue-specific, and cell differentiation-specific expression of the Arabidopsis Fe(III)-chelate reductase gene AtFRO6. Plant Physiol 140:1345–1354

Finney LA, O'Halloran TV (2003) Transition metal speciation in the cell: insights from the chemistry of metal ion receptors. Science 300:931–936

Fleming MD, Andrews NC (1998) Mammalian iron transport: an unexpected link between metal homeostasis and host defense. J Lab Clin Med 132:464–468

Fox TC, Guerinot ML (1998) Molecular biology of cation transport in plants. Annu Rev Plant Physiol Plant Mol Biol 49:669–696

Gazzarrini S, Lejay L, Gojon A, Ninnemann O, Frommer WB, von Wiren N (1999) Three functional transporters for constitutive, diurnally regulated, and starvation-induced uptake of ammonium into Arabidopsis roots. Plant Cell 11:937–948

Goetz DH, Holmes MA, Borregaard N, Bluhm ME, Raymond KN, Strong RK (2002) The neutrophil lipocalin NGAL is a bacteriostatic agent that interferes with siderophore-mediated iron acquisition. Mol Cell 10:1033–1043

Graziano M, Lamattina L (2007) Nitric oxide accumulation is required for molecular and physiological responses to iron deficiency in tomato roots. Plant J 52:949–960

Graziano M, Beligni MV, Lamattina L (2002) Nitric oxide improves internal iron availability in plants. Plant Physiol 130:1852–1859

Green LS, Rogers EE (2004) FRD3 controls iron localization in Arabidopsis. Plant Physiol 136:2523–2531

Grusak MA, Pezeshgi S (1996) Shoot-to-root signal transmission regulates root Fe(III) reductase activity in the dgl mutant of pea. Plant Physiol 110:329–334

Grusak MA, Welch RM, Kochian LV (1990) Physiological characterization of a single-gene mutant of Pisum Sativum exhibiting excess iron accumulation. 1. root iron reduction and iron uptake. Plant Physiol 93:976–981

Gunshin H, Mackenzie B, Berger UV, Gunshin Y, Romero MF, Boron WF, Nussberger S, Gollan JL, Hediger MA (1997) Cloning and characterization of a mammalian proton-coupled metal-ion transporter. Nature 388:482–488

Hell R, Stephan UW (2003) Iron uptake, trafficking and homeostasis in plants. Planta 216:541–551

Henriques R, Jasik J, Klein M, Martinoia E, Feller U, Schell J, Pais MS, Koncz C (2002) Knockout of Arabidopsis metal transporter gene IRT1 results in iron deficiency accompanied by cell differentiation defects. Plant Mol Biol 50:587–597

Hentze MW, Kuhn LC (1996) Molecular control of vertebrate iron metabolism: mRNA-based regulatory circuits operated by iron, nitric oxide, and oxidative stress. Proc Natl Acad Sci USA 93:8175–8182

Inoue H, Kobayashi T, Nozoye T, Takahashi M, Kakei Y, Suzuki K, Nakazono M, Nakanishi H, Mori S, Nishizawa NK (2009) Rice OsYSL15 is an iron-regulated iron(III)-deoxymugineic acid transporter expressed in the roots and is essential for iron uptake in early growth of the seedlings. J Biol Chem 284:3470–3479

Ishimaru Y, Suzuki M, Tsukamoto T, Suzuki K, Nakazono M, Kobayashi T, Wada Y, Watanabe S, Matsuhashi S, Takahashi M, Nakanishi H, Mori S, Nishizawa NK (2006) Rice plants take up iron as an Fe3+-phytosiderophore and as Fe2+. Plant J 45:335–346

Ishimaru Y, Kim S, Tsukamoto T, Oki H, Kobayashi T, Watanabe S, Matsuhashi S, Takahashi M, Nakanishi H, Mori S, Nishizawa NK (2007) Mutational reconstructed ferric chelate reductase confers enhanced tolerance in rice to iron deficiency in calcareous soil. Proc Natl Acad Sci USA 104:7373–7378

Ishimaru Y, Masuda H, Bashir K, Inoue H, Tsukamoto T, Takahashi M, Nakanishi H, Aoki N, Hirose T, Ohsugi R, Nishizawa NK (2010) Rice metal-nicotianamine transporter, OsYSL2, is required for long distance transport of iron and manganese. Plant J 62:379–390

Jakoby M, Wang HY, Reidt W, Weisshaar B, Bauer P (2004) FRU (BHLH029) is required for induction of iron mobilization genes in *Arabidopsis thaliana*. FEBS Lett 577:528–534

Jeong J, Cohu C, Kerkeb L, Pilon M, Connolly EL, Guerinot ML (2008) Chloroplast Fe(III) chelate reductase activity is essential for seedling viability under iron limiting conditions. Proc Natl Acad Sci USA 105:10619–10624

Jin CW, Du ST, Chen WW, Li GX, Zhang YS, Zheng SJ (2009) Elevated carbon dioxide improves plant iron nutrition through enhancing the iron-deficiency-induced responses under iron-limited conditions in tomato. Plant Physiol 150:272–280

Kerkeb L, Mukherjee I, Chatterjee I, Lahner B, Salt DE, Connolly EL (2008) Iron-induced turnover of the Arabidopsis iron-regulated transporter1 metal transporter requires lysine residues. Plant Physiol 146:1964–1973

Kim SA, Punshon T, Lanzirotti A, Li L, Alonso JM, Ecker JR, Kaplan J, Guerinot ML (2006) Localization of iron in Arabidopsis seed requires the vacuolar membrane transporter VIT1. Science 314:1295–1298

Kispal G, Csere P, Guiard B, Lill R (1997) The ABC transporter Atm1p is required for mitochondrial iron homeostasis. FEBS Lett 418:346–350

Kispal G, Csere P, Prohl C, Lill R (1999) The mitochondrial proteins Atm1p and Nfs1p are essential for biogenesis of cytosolic Fe/S proteins. EMBO J 18:3981–3989

Klatte M, Schuler M, Wirtz M, Fink-Straube C, Hell R, Bauer P (2009) The analysis of Arabidopsis nicotianamine synthase mutants reveals functions for nicotianamine in seed iron loading and iron deficiency responses. Plant Physiol 150:257–271

Kobayashi T, Nakayama Y, Itai RN, Nakanishi H, Yoshihara T, Mori S, Nishizawa NK (2003) Identification of novel cis-acting elements, IDE1 and IDE2, of the barley IDS2 gene promoter conferring iron-deficiency-inducible, root-specific expression in heterogeneous tobacco plants. Plant J 36:780–793

Kobayashi T, Suzuki M, Inoue H, Itai RN, Takahashi M, Nakanishi H, Mori S, Nishizawa NK (2005) Expression of iron-acquisition-related genes in iron-deficient rice is co-ordinately induced by partially conserved iron-deficiency-responsive elements. J Exp Bot 56:1305–1316

Kobayashi T, Ogo Y, Itai RN, Nakanishi H, Takahashi M, Mori S, Nishizawa NK (2007a) The transcription factor IDEF1 regulates the response to and tolerance of iron deficiency in plants. Proc Natl Acad Sci USA 104:19150–19155

Kobayashi T, Yoshihara T, Itai RN, Nakanishi H, Takahashi M, Mori S, Nishizawa NK (2007b) Promoter analysis of iron-deficiency-inducible barley IDS3 gene in Arabidopsis and tobacco plants. Plant Physiol Biochem 45:262–269

Kobayashi T, Itai RN, Ogo Y, Kakei Y, Nakanishi H, Takahashi M, Nishizawa NK (2009) The rice transcription factor IDEF1 is essential for the early response to iron deficiency, and induces vegetative expression of late embryogenesis abundant genes. Plant J 60:948–961

Kobayashi T, Ogo Y, Aung MS, Nozoye T, Itai RN, Nakanishi H, Yamakawa T, Nishizawa NK (2010) The spatial expression and regulation of transcription factors IDEF1 and IDEF2. Ann Bot 105(7):1109–1117. doi:10.1093/aob/mcq002

Koike S, Inoue H, Mizuno D, Takahashi M, Nakanishi H, Mori S, Nishizawa NK (2004) OsYSL2 is a rice metal-nicotianamine transporter that is regulated by iron and expressed in the phloem. Plant J 39:415–424

Korshunova YO, Eide D, Clark WG, Guerinot ML, Pakrasi HB (1999) The *IRT1* protein from *Arabidopsis thaliana* is a metal transporter with a broad substrate range. Plant Mol Biol 40:37–44

Kumanovics A, Chen OS, Li L, Bagley D, Adkins EM, Lin H, Dingra NN, Outten CE, Keller G, Winge D, Ward DM, Kaplan J (2008) Identification of FRA1 and FRA2 as genes involved in regulating the yeast iron regulon in response to decreased mitochondrial iron-sulfur cluster synthesis. J Biol Chem 283:10276–10286

Kushnir S, Babiychuk E, Storozhenko S, Davey MW, Papenbrock J, De Rycke R, Engler G, Stephan UW, Lange H, Kispal G, Lill R, Van Montagu M (2001) A mutation of the mitochondrial ABC transporter Sta1 leads to dwarfism and chlorosis in the Arabidopsis mutant starik. Plant Cell 13:89–100

Kwok EY, Severance S, Kosman DJ (2006) Evidence for iron channeling in the Fet3p-Ftr1p high-affinity iron uptake complex in the yeast plasma membrane. Biochemistry 45:6317–6327

Landsberg EC (1986) Function of rhizodermal transfer cells in the Fe stress response mechanism of Capsicum-annuum-L. Plant Physiol 82:511–517

Lanquar V, Lelievre F, Bolte S, Hames C, Alcon C, Neumann D, Vansuyt G, Curie C, Schroder A, Kramer U, Barbier-Brygoo H, Thomine S (2005) Mobilization of vacuolar iron by AtNRAMP3 and AtNRAMP4 is essential for seed germination on low iron. EMBO J 24:4041–4051

Lanquar V, Ramos MS, Lelievre F, Barbier-Brygoo H, Krieger-Liszkay A, Kramer U, Thomine S (2010) Export of vacuolar manganese by AtNRAMP3 and AtNRAMP4 is required for optimal photosynthesis and growth under manganese deficiency. Plant Physiol 152:1986–1999

Larbi A, Morales F, Abadia A, Abadia J (2010) Changes in iron and organic acid concentrations in xylem sap and apoplastic fluid of iron-deficient Beta vulgaris plants in response to iron resupply. J Plant Physiol 167:255–260

Le Jean M, Schikora A, Mari S, Briat JF, Curie C (2005) A loss-of-function mutation in AtYSL1 reveals its role in iron and nicotianamine seed loading. Plant J 44:769–782

Lee S, Chiecko JC, Kim SA, Walker EL, Lee Y, Guerinot ML, An G (2009) Disruption of OsYSL15 leads to iron inefficiency in rice plants. Plant Physiol 150:786–800

Li XX, Li CJ (2004) Is ethylene involved in regulation of root ferric reductase activity of dicotyledonous species under iron deficiency? Plant Soil 261:147–153

Li L, Chen OS, McVey Ward D, Kaplan J (2001) CCC1 is a transporter that mediates vacuolar iron storage in yeast. J Biol Chem 276:29515–29519

Li L, Cheng X, Ling HQ (2004) Isolation and characterization of Fe(III)-chelate reductase gene LeFRO1 in tomato. Plant Mol Biol 54:125–136

Li H, Mapolelo DT, Dingra NN, Naik SG, Lees NS, Hoffman BM, Riggs-Gelasco PJ, Huynh BH, Johnson MK, Outten CE (2009) The yeast iron regulatory proteins Grx3/4 and Fra2 form heterodimeric complexes containing a [2Fe-2S] cluster with cysteinyl and histidyl ligation. Biochemistry 48:9569–9581

Ling HQ, Pich A, Scholz G, Ganal MW (1996) Genetic analysis of two tomato mutants affected in the regulation of iron metabolism. Mol Gen Genet 252:87–92

Ling HQ, Koch G, Baumlein H, Ganal MW (1999) Map-based cloning of chloronerva, a gene involved in iron uptake of higher plants encoding nicotianamine synthase. Proc Natl Acad Sci USA 96:7098–7103

Ling HQ, Bauer P, Bereczky Z, Keller B, Ganal M (2002) The tomato fer gene encoding a bHLH protein controls iron-uptake responses in roots. Proc Natl Acad Sci USA 99:13938–13943

Long TA, Tsukagoshi H, Busch W, Lahner B, Salt DE, Benfey PN (2010) The bHLH Transcription Factor POPEYE Regulates Response to Iron Deficiency in Arabidopsis Roots. Plant Cell (Epub ahead of print)

Lucena C, Waters BM, Romera FJ, Garcia MJ, Morales M, Alcantara E, Perez-Vicente R (2006) Ethylene could influence ferric reductase, iron transporter, and H+-ATPase gene expression by affecting FER (or FER-like) gene activity. J Exp Bot 57:4145–4154

Marschner H (1997) Mineral nutrition of higher plants, 2nd edn. Academic Press, London

Martin M, Colman MJ, Gomez-Casati DF, Lamattina L, Zabaleta EJ (2009) Nitric oxide accumulation is required to protect against iron-mediated oxidative stress in frataxin-deficient Arabidopsis plants. FEBS Lett 583:542–548

McKie AT, Marciani P, Rolfs A, Brennan K, Wehr K, Barrow D, Miret S, Bomford A, Peters TJ, Farzaneh F, Hediger MA, Hentze MW, Simpson RJ (2000) A novel duodenal iron-regulated transporter, IREG1, implicated in the basolateral transfer of iron to the circulation. Mol Cell 5:299–309

Mills MM, Ridame C, Davey M, La Roche J, Geider RJ (2004) Iron and phosphorus co-limit nitrogen fixation in the eastern tropical North Atlantic. Nature 429:292–294

Morel M, Crouzet J, Gravot A, Auroy P, Leonhardt N, Vavasseur A, Richaud P (2009) AtHMA3, a P1B-ATPase allowing Cd/Zn/Co/Pb vacuolar storage in Arabidopsis. Plant Physiol 149:894–904

Mori S (1999) Iron acquisition by plants. Curr Opin Plant Biol 2:250–253

Morrissey J, Baxter IR, Lee J, Li L, Lahner B, Grotz N, Kaplan J, Salt DE, Guerinot ML (2009) The ferroportin metal efflux proteins function in iron and cobalt homeostasis in Arabidopsis. Plant Cell 21:3326–3338

Mukherjee I, Campbell NH, Ash JS, Connolly EL (2006) Expression profiling of the Arabidopsis ferric chelate reductase (FRO) gene family reveals differential regulation by iron and copper. Planta 223:1178–1190

Murata Y, Ma JF, Yamaji N, Ueno D, Nomoto K, Iwashita T (2006) A specific transporter for iron (III)-phytosiderophore in barley roots. Plant J 46:563–572

Murgia I, Delledonne M, Soave C (2002) Nitric oxide mediates iron-induced ferritin accumulation in Arabidopsis. Plant J 30:521–528

Negishi T, Nakanishi H, Yazaki J, Kishimoto N, Fujii F, Shimbo K, Yamamoto K, Sakata K, Sasaki T, Kikuchi S, Mori S, Nishizawa NK (2002) cDNA microarray analysis of gene expression during Fe-deficiency stress in barley suggests that polar transport of vesicles is implicated in phytosiderophore secretion in Fe-deficient barley roots. Plant J 30:83–94

Nemeth E, Ganz T (2006) Regulation of iron metabolism by hepcidin. Annu Rev Nutr 26:323–342

Nemeth E, Tuttle MS, Powelson J, Vaughn MB, Donovan A, Ward DM, Ganz T, Kaplan J (2004) Hepcidin regulates cellular iron efflux by binding to ferroportin and inducing its internalization. Science 306:2090–2093

Nishizawa N, Mori S (1987) The particular vesicle appearing in barley root cells and its relation to mugineic acid secretion. J Plant Nutr 10:1013–1020

Ogo Y, Itai RN, Nakanishi H, Inoue H, Kobayashi T, Suzuki M, Takahashi M, Mori S, Nishizawa NK (2006) Isolation and characterization of IRO2, a novel iron-regulated bHLH transcription factor in graminaceous plants. J Exp Bot 57:2867–2878

Ogo Y, Itai RN, Nakanishi H, Kobayashi T, Takahashi M, Mori S, Nishizawa NK (2007) The rice bHLH protein OsIRO2 is an essential regulator of the genes involved in Fe uptake under Fe-deficient conditions. Plant J 51:366–377

Ogo Y, Kobayashi T, Nakanishi Itai R, Nakanishi H, Kakei Y, Takahashi M, Toki S, Mori S, Nishizawa NK (2008) A novel NAC transcription factor, IDEF2, that recognizes the iron deficiency-responsive element 2 regulates the genes involved in iron homeostasis in plants. J Biol Chem 283:13407–13417

Ojeda L, Keller G, Muhlenhoff U, Rutherford JC, Lill R, Winge DR (2006) Role of glutaredoxin-3 and glutaredoxin-4 in the iron regulation of the Aft1 transcriptional activator in Saccharomyces cerevisiae. J Biol Chem 281:17661–17669

Outten CE, O'Halloran TV (2001) Femtomolar sensitivity of metalloregulatory proteins controlling zinc homeostasis. Science 292:2488–2492

Perilli S, Moubayidin L, Sabatini S (2010) The molecular basis of cytokinin function. Curr Opin Plant Biol 13:21–26

Pich A, Manteuffel R, Hillmer S, Scholz G, Schmidt W (2001) Fe homeostasis in plant cells: does nicotianamine play multiple roles in the regulation of cytoplasmic Fe concentration? Planta 213:967–976

Portnoy ME, Liu XF, Culotta VC (2000) Saccharomyces cerevisiae expresses three functionally distinct homologues of the nramp family of metal transporters. Mol Cell Biol 20:7893–7902

Radisky DC, Babcock MC, Kaplan J (1999) The yeast frataxin homologue mediates mitochondrial iron efflux. Evidence for a mitochondrial iron cycle. J Biol Chem 274:4497–4499

Rellan-Alvarez R, Giner-Martinez-Sierra J, Orduna J, Orera I, Rodriguez-Castrillon JA, Garcia-Alonso JI, Abadia J, Alvarez-Fernandez A (2010) Identification of a tri-iron(III), tri-citrate complex in the xylem sap of iron-deficient tomato resupplied with iron: new insights into plant iron long-distance transport. Plant Cell Physiol 51:91–102

Roberts LA, Pierson AJ, Panaviene Z, Walker EL (2004) Yellow stripe1. Expanded roles for the maize iron-phytosiderophore transporter. Plant Physiol 135:112–120

Robinson NJ, Procter CM, Connolly EL, Guerinot ML (1999) A ferric-chelate reductase for iron uptake from soils. Nature 397:694–697

Rogers EE, Guerinot ML (2002) FRD3, a member of the multidrug and toxin efflux family, controls iron deficiency responses in *Arabidopsis*. Plant Cell 14:1787–1799

Rogers EE, Eide DJ, Guerinot ML (2000) Altered selectivity in an *Arabidopsis* metal transporter. Proc Natl Acad Sci USA 97:12356–12360

Romera FJ, Alcantara E (1994) Iron-deficiency stress responses in cucumber (Cucumis sativus L.) roots (a possible role for ethylene?). Plant Physiol 105:1133–1138

Romera FJ, Alcantara E, De la Guardia MD (1999) Ethylene production by Fe-deficient roots and its involvement in the regulation of Fe-deficiency stress responses by strategy I plants. Ann Bot 83:51–55

Romera FJ, Lucena C, Alcantara E (2006) Plant hormones influencing iron uptake in plants. In: Barton LL, Abadia J (eds) Iron nutrition in plants and rhizopheric microorganisms. Springer, Netherlands, pp 251–278

Romheld V, Marschner H (1986) Evidence for a specific uptake system for iron phytosiderophores in roots of grasses. Plant Physiol 80:175–180

Roschzttardtz H, Conejero G, Curie C, Mari S (2009) Identification of the endodermal vacuole as the iron storage compartment in the Arabidopsis embryo. Plant Physiol 151:1329–1338

Rutherford JC, Jaron S, Winge DR (2003) Aftlp and Aft2p mediate iron-responsive gene expression in yeast through related promoter elements. J Biol Chem 278:27636–27643

Rutherford JC, Ojeda L, Balk J, Muhlenhoff U, Lill R, Winge DR (2005) Activation of the iron regulon by the yeast Aft1/Aft2 transcription factors depends on mitochondrial but not cytosolic iron-sulfur protein biogenesis. J Biol Chem 280:10135–10140

Sakaguchi T, Nishizawa N, Nakanishi H, Yoshimura E, Mori S (1999) The role of potassium in the secretion of mugineic acids family phytosiderophores from iron-deficient barley roots. Plant Soil 215:221–227

Sancenon V, Puig S, Mira H, Thiele DJ, Penarrubia L (2003) Identification of a copper transporter family in *Arabidopsis thaliana*. Plant Mol Biol 51:577–587

Santi S, Schmidt W (2009) Dissecting iron deficiency-induced proton extrusion in Arabidopsis roots. New Phytol 183:1072–1084

Santi S, Cesco S, Varanini Z, Pinton R (2005) Two plasma membrane H(+)-ATPase genes are differentially expressed in iron-deficient cucumber plants. Plant Physiol Biochem 43:287–292

Schaaf G, Ludewig U, Erenoglu BE, Mori S, Kitahara T, von Wiren N (2004) ZmYS1 functions as a proton-coupled symporter for phytosiderophore- and nicotianamine-chelated metals. J Biol Chem 279:9091–9096

Schaaf G, Schikora A, Haberle J, Vert G, Ludewig U, Briat JF, Curie C, von Wiren N (2005) A putative function for the arabidopsis Fe-Phytosiderophore transporter homolog AtYSL2 in Fe and Zn homeostasis. Plant Cell Physiol 46:762–774

Schaaf G, Honsbein A, Meda AR, Kirchner S, Wipf D, von Wiren N (2006) AtIREG2 encodes a tonoplast transport protein involved in iron-dependent nickel detoxification in *Arabidopsis thaliana* roots. J Biol Chem 281:25532–25540

Schmidt W, Tittel J, Schikora A (2000) Role of hormones in the induction of iron deficiency responses in Arabidopsis roots. Plant Physiol 122:1109–1118

Segond D, Dellagi A, Lanquar V, Rigault M, Patrit O, Thomine S, Expert D (2009) NRAMP genes function in *Arabidopsis thaliana* resistance to Erwinia chrysanthemi infection. Plant J 58:195–207

Seguela M, Briat JF, Vert G, Curie C (2008) Cytokinins negatively regulate the root iron uptake machinery in Arabidopsis through a growth-dependent pathway. Plant J 55:289–300

Shayeghi M, Latunde-Dada GO, Oakhill JS, Laftah AH, Takeuchi K, Halliday N, Khan Y, Warley A, McCann FE, Hider RC, Frazer DM, Anderson GJ, Vulpe CD, Simpson RJ, McKie AT (2005) Identification of an intestinal heme transporter. Cell 122:789–801

Shingles R, North M, McCarty RE (2002) Ferrous ion transport across chloroplast inner envelope membranes. Plant Physiol 128:1022–1030

Spizzo T, Byersdorfer C, Duesterhoeft S, Eide D (1997) The yeast FET5 gene encodes a FET3-related multicopper oxidase implicated in iron transport. Mol Gen Genet 256:547–556

Stacey MG, Patel A, McClain WE, Mathieu M, Remley M, Rogers EE, Gassmann W, Blevins DG, Stacey G (2008) The Arabidopsis AtOPT3 protein functions in metal homeostasis and movement of iron to developing seeds. Plant Physiol 146:589–601

Stearman R, Yuan DS, Yamaguchi-Iwai Y, Klausner RD, Dancis A (1996) A permease-oxidase complex involved in high-affinity iron uptake in yeast. Science 271:1552–1557

Stephan UW, Grun M (1989) Physiological disorders of the nicotianamine-auxotroph tomato mutant chloronerva at diffent levels of iron nutrition.II. Iron deficiency response and heavy metal metabolism. Biochem Physiol Pflanz 185:189–200

Takahashi M, Terada Y, Nakai I, Nakanishi H, Yoshimura E, Mori S, Nishizawa NK (2003) Role of nicotianamine in the intracellular delivery of metals and plant reproductive development. Plant Cell 15:1263–1280

Takemoto T, Nomoto K, Fushiya S, Ouchi R, Kusano G, Hikino H, Takagi S, Matuura Y, Kakudo M (1978) Structure of mugineic acid, a new amino acid possessing an iron-chelating activity from roots washing of water-cultured Hordeum vulgare L. Proc Jpn Acad B Phys Biol Sci 54:469–473

Teschner J, Lachmann N, Schulze J, Geisler M, Selbach K, Santamaria-Araujo J, Balk J, Mendel RR, Bittner F (2010) A novel role for Arabidopsis mitochondrial ABC transporter ATM3 in molybdenum cofactor biosynthesis. Plant Cell 22:468-480

Thomine S, Wang R, Ward JM, Crawford NM, Schroeder JI (2000) Cadmium and iron transport by members of a plant transporter gene family in *Arabidopsis* with homology to *NRAMP* genes. Proc Nat Acad Sci USA 97:4991–4996

Thomine S, Lelièvre F, Debarbieux E, Schroeder JI, Barbier-Brygoo H (2003) AtNRAMP3, a multispecific vacuolar metal transporter involved in plant responses to iron deficiency. Plant J 34:685–695

Ueno D, Rombola AD, Iwashita T, Nomoto K, Ma JF (2007) Identification of two novel phytosiderophores secreted by perennial grasses. New Phytol 174:304–310

Urbanowski JL, Piper RC (1999) The iron transporter Fth1p forms a complex with the Fet5 iron oxidase and resides on the vacuolar membrane. J Biol Chem 274:38061–38070

Varotto C, Maiwald D, Pesaresi P, Jahns P, Salamini F, Leister D (2002) The metal ion transporter IRT1 is necessary for iron homeostasis and efficient photosynthesis in *Arabidopsis thaliana*. Plant J 31:589–599

Vert G, Briat JF, Curie C (2001) *Arabidopsis IRT2* gene encodes a root-periphery iron transporter. Plant J 26:181–189

Vert G, Grotz N, Dedaldechamp F, Gaymard F, Guerinot ML, Briat JF, Curie C (2002) IRT1, an *Arabidopsis* transporter essential for iron uptake from the soil and for plant growth. Plant Cell 14:1223–1233

Vert GA, Briat JF, Curie C (2003) Dual regulation of the Arabidopsis high-affinity root iron uptake system by local and long-distance signals. Plant Physiol 132:796–804

Vert G, Barberon M, Zelazny E, Seguela M, Briat JF, Curie C (2009) Arabidopsis IRT2 cooperates with the high-affinity iron uptake system to maintain iron homeostasis in root epidermal cells. Planta 229:1171–1179

Von Wiren N, Mori S, Marschner H, Romheld V (1994) Iron inefficiency in maize mutant ys1 (Zea mays L. cv yellow-stripe) Is caused by a defect in uptake of iron phytosiderophores. Plant Physiol 106:71–77

Wang HY, Klatte M, Jakoby M, Baumlein H, Weisshaar B, Bauer P (2007) Iron deficiency-mediated stress regulation of four subgroup Ib BHLH genes in *Arabidopsis thaliana*. Planta 226:897–908

Waters BM, Blevins DG, Eide DJ (2002) Characterization of FRO1, a pea ferric-chelate reductase involved in root iron acquisition. Plant Physiol 129:85–94

Waters BM, Chu HH, Didonato RJ, Roberts LA, Eisley RB, Lahner B, Salt DE, Walker EL (2006) Mutations in Arabidopsis yellow stripe-like1 and yellow stripe-like3 reveal their roles in metal ion homeostasis and loading of metal ions in seeds. Plant Physiol 141:1446–1458

Waters BM, Lucena C, Romera FJ, Jester GG, Wynn AN, Rojas CL, Alcantara E, Perez-Vicente R (2007) Ethylene involvement in the regulation of the H(+)-ATPase CsHA1 gene and of the new isolated ferric reductase CsFRO1 and iron transporter CsIRT1 genes in cucumber plants. Plant Physiol Biochem 45:293–301

Welch RM, Larue TA (1990) Physiological characteristics of Fe accumulation in the; bronze' mutant of *Pisum sativum* L., cv; Sparkle' E107 (brz brz). Plant Physiol 93:723–729

Welch RM, Norvell WA, Gesuwan P, Schaefer S (1997) Possible role of root-ethylene in Fe(III)-phytometallophore uptake in strategy II species. Plant Soil 196:229–232

Wintz H, Fox T, Wu YY, Feng V, Chen W, Chang HS, Zhu T, Vulpe C (2003) Expression profiles of *Arabidopsis thaliana* in mineral deficiencies reveal novel transporters involved in metal homeostasis. J Biol Chem 278:47644–47653

Wu H, Li L, Du J, Yuan Y, Cheng X, Ling HQ (2005) Molecular and biochemical characterization of the Fe(III) chelate reductase gene family in *Arabidopsis thaliana*. Plant Cell Physiol 46:1505–1514

Yamaguchi-Iwai Y, Dancis A, Klausner RD (1995) AFT1: a mediator of iron regulated transcriptional control in Saccharomyces cerevisiae. EMBO J 14:1231–1239

Yamaguchi-Iwai Y, Stearman R, Dancis A, Klausner RD (1996) Iron-regulated DNA binding by the AFT1 protein controls the iron regulon in yeast. EMBO J 15:3377–3384

Yang J, Goetz D, Li JY, Wang W, Mori K, Setlik D, Du T, Erdjument-Bromage H, Tempst P, Strong R, Barasch J (2002) An iron delivery pathway mediated by a lipocalin. Mol Cell 10:1045–1056

Yi Y, Guerinot ML (1996) Genetic evidence that induction of root Fe(III) chelate reductase activity is necessary for iron uptake under iron deficiency. Plant J 10:835–844

Yuan YX, Zhang J, Wang DW, Ling HQ (2005) AtbHLH29 of *Arabidopsis thaliana* is a functional ortholog of tomato FER involved in controlling iron acquisition in strategy I plants. Cell Res 15:613–621

Yuan Y, Wu H, Wang N, Li J, Zhao W, Du J, Wang D, Ling HQ (2008) FIT interacts with AtbHLH38 and AtbHLH39 in regulating iron uptake gene expression for iron homeostasis in Arabidopsis. Cell Res 18:385–397

Ca^{2+} Pumps and Ca^{2+} Antiporters in Plant Development

Jon K. Pittman, Maria Cristina Bonza, and Maria Ida De Michelis

Abstract Calcium (Ca^{2+}) efflux transporters remove Ca^{2+} from the cytosol of the cell either by transporting it out of the cell across the plasma membrane or into internal organelles. These transporters, which include Ca^{2+}-ATPases and Ca^{2+}/H^+ antiporters, have a critical role in preventing Ca^{2+} toxicity, maintaining cytosolic Ca^{2+} at a low resting level, and transferring Ca^{2+} to specific cellular locations where it is required. Many genes encoding plant Ca^{2+}-ATPases and Ca^{2+}/H^+ antiporters have now been identified and characterised to elucidate their biochemical and genetic features. Furthermore, the use of gene knockouts has begun to provide evidence for an involvement of these Ca^{2+} transporters in Ca^{2+}-signaling networks and in various aspects of plant development.

1 Introduction

Calcium (Ca^{2+}) is an essential nutrient and an important component in cellular signaling. As a nutrient Ca^{2+} is critical for biochemical reactions within the mitochondria and chloroplast, it has structural roles within the cell wall, is involved in membrane dynamics and it has been implicated in mediating or regulating programmed cell death (White and Broadley 2003; Hirschi 2004). Ca^{2+} has been demonstrated to mediate cellular-signaling events in plants in response to a variety of environmental and developmental situations (Sanders et al. 2002; Hetherington and Brownlee 2004; McAinsh and Pittman 2009). Elevations in cytosolic Ca^{2+}

J.K. Pittman (✉)
Faculty of Life Sciences, University of Manchester, Michael Smith Building, Oxford Road, Manchester M13 9PT, UK
e-mail: jon.pittman@manchester.ac.uk

M.C. Bonza and M.I. De Michelis
Dipartimento di Biologia "L. Gorini", Università di Milano, via G. Celoria 26 20133 Milano, Italy
e-mail: cristina.bonza@unimi.it; mariaida.demichelis@unimi.it

M. Geisler and K. Venema (eds.), *Transporters and Pumps in Plant Signaling*, Signaling and Communication in Plants 7,
DOI 10.1007/978-3-642-14369-4_5,

($[Ca^{2+}]_{cyt}$) have been observed in response to a wide variety of environmental stimuli including salt stress, drought, cold and oxidative stress (Knight et al. 1996; Kiegle et al. 2000; Evans et al. 2005). The characteristics of a $[Ca^{2+}]_{cyt}$ elevation following each stimuli appear to be specific such that the spatial and temporal dynamics of the Ca^{2+} elevation can be regarded as a "Ca^{2+} signature" for a given stimulus (McAinsh and Pittman 2009). It is thought that this stimulus-specific nature of $[Ca^{2+}]_{cyt}$ allows the cell to distinguish between endogenous and/or environmental stimuli and then mediate the appropriate downstream response. In support of this, a number of studies have demonstrated that alteration in a Ca^{2+} signature can indeed give rise to a different downstream response (Allen et al. 2001; Miwa et al. 2006). However, the exact mechanisms which generate and regulate the $[Ca^{2+}]_{cyt}$ elevations are not fully understood.

Ca^{2+} is very toxic when it accumulates at a high concentration within the cell, with the cytosol particularly sensitive to excess Ca^{2+}. Excess Ca^{2+} will interact with phosphate to form insoluble complexes which will interfere with the role of phosphate as an energy donor. High concentrations of Ca^{2+} will also out-compete other cations such as Mg^{2+}, which function as co-factors for essential enzymes and proteins. Active Ca^{2+} efflux mechanisms to rapidly remove excess Ca^{2+} from the cytosol either by transporting it out of the cell or by sequestering it into internal stores are therefore essential and developed early in evolution (Case et al. 2007). These transporters allow the maintenance of $[Ca^{2+}]_{cyt}$ at submicromolar concentrations. The other essential "housekeeping" function of Ca^{2+} efflux transporters is to regulate the partitioning of Ca^{2+} into the various organelles where it is required and maintain Ca^{2+} at the optimal concentration in these compartments. A large concentration gradient of Ca^{2+} therefore exists between the cytosol and most internal compartments, and between the cytosol and apoplast. The $[Ca^{2+}]_{cyt}$ concentration in many cell types is approximately 100 nM, whereas apoplastic and vacuolar Ca^{2+} can reach 1–10 mM. Such a Ca^{2+} gradient will allow the rapid mobilisation of Ca^{2+} into the cytosol, which in combination with the Ca^{2+} efflux pathways for maintaining low $[Ca^{2+}]_{cyt}$, makes Ca^{2+} a suitable compound to function as a cellular messenger. These housekeeping functions of Ca^{2+} efflux transporters are therefore instrumental in the generation of $[Ca^{2+}]_{cyt}$ signals.

The opening of Ca^{2+} permeable influx channels at the plasma membrane and endomembranes in response to a stimulus will release Ca^{2+} into the cytosol and cause the generation of a Ca^{2+} spike. Ca^{2+} efflux activity will then lead to a reduction in a Ca^{2+} spike as $[Ca^{2+}]_{cyt}$ concentration is reduced and returned to the resting concentration. However, it is possible that Ca^{2+} efflux transporters may play a more specific role in modulating the dynamics of a Ca^{2+} signature which is distinct to their function of merely providing tolerance to high concentrations of $[Ca^{2+}]_{cyt}$ (McAinsh and Pittman 2009; Dodd et al. 2010). It may be hypothesised that the activation of different Ca^{2+} efflux transporter isoforms or the differential regulation of an individual efflux transporter will give rise to differing Ca^{2+} transport kinetics which will return $[Ca^{2+}]_{cyt}$ to resting concentration at different rates and will therefore be able to impact on the characteristic of a transient Ca^{2+} elevation. Evidence of a direct signaling role for Ca^{2+} efflux transporters is beginning to emerge and in this chapter we will review the possible involvement of these Ca^{2+} transporters in specific cell-signaling pathways.

The removal of Ca^{2+} from the cytosol is against an electrochemical potential gradient for Ca^{2+} and is therefore an energy-dependent process. Most eukaryotes have two main mechanisms of cellular Ca^{2+} efflux: a direct energised high-affinity pathway mediated by Ca^{2+}-ATPases and a secondary energised lower-affinity Ca^{2+} exchanger pathway which is energised by the counter exchange of another ion, either Na^+ or H^+. Plants also have two types of Ca^{2+} efflux transporters, ATP-dependent Ca^{2+} pumps (Sze et al. 2000; Boursiac and Harper 2007) and Ca^{2+}/H^+ antiporters that utilise the proton motive force (Shigaki and Hirschi 2006) (Fig. 1). The genes which encode both transporter types have been identified in plants. This has enabled detailed investigations into the functions and mechanisms of these Ca^{2+} transporters through the use of reverse genetics and heterologous expression. This chapter will review our current understanding of plant Ca^{2+} pumps and Ca^{2+}/H^+ antiporters, including the biochemical and regulatory characteristics of the transporters,

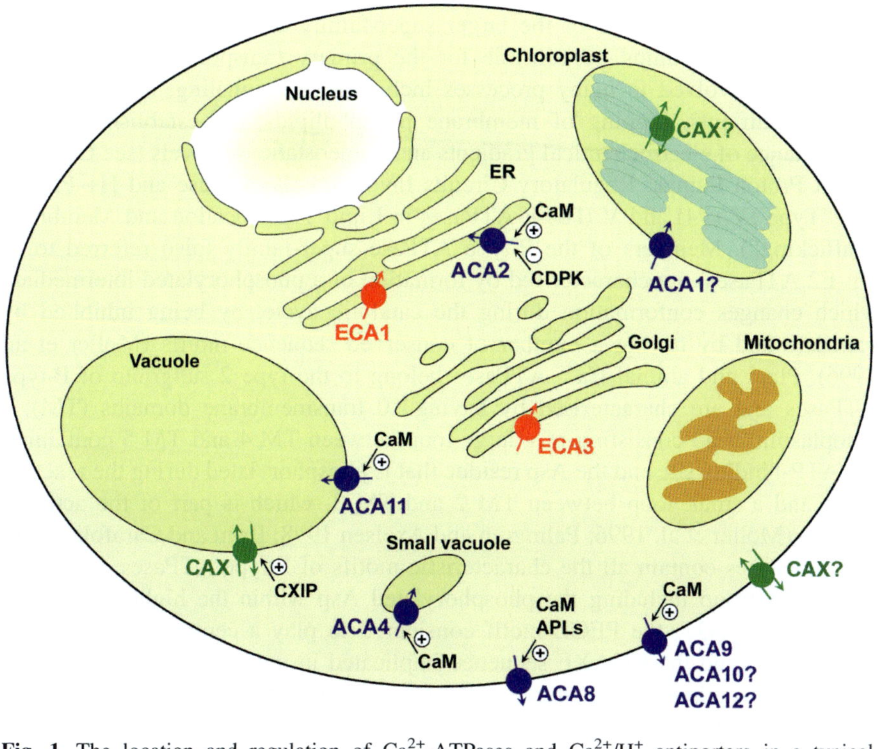

Fig. 1 The location and regulation of Ca^{2+}-ATPases and Ca^{2+}/H^+ antiporters in a typical Arabidopsis cell. ACA-type Ca^{2+}-ATPases are shown in *blue*, ECA-type Ca^{2+}-ATPases are shown in *red* and CAX Ca^{2+}/H^+ antiporters are shown in *green*. Some Ca^{2+} transport regulators are shown including calmodulin (CaM), acidic phospholipids (APLs), Ca^{2+}-dependent protein kinase (CDPK) and CAX-interacting protein (CXIP). The *plus* symbol indicates activation of transport by the regulator, and the *minus* symbol indicates inhibition of transport by the regulator

and will describe what we currently know about the roles of these Ca^{2+} transporters in specific aspects of plant development.

2 Ca^{2+}-ATPases

Although the first report on ATP-dependent Ca^{2+} transport in plant membrane vesicles dates back to the late seventies (Gross and Marmè 1978), the first attempts to identify and characterise plant Ca^{2+}-ATPases in microsomes and/or purified membrane fractions isolated from different plant materials lead to conflicting results, in terms of biochemical characteristics, membrane location and regulatory properties. Only by the late eighties it became clear that plant cells contain two major types of Ca^{2+}-ATPases, one type resembling those present in the sarcoplasmic/endoplasmic reticulum (SERCA, sarco-endoplasmic reticulum Ca^{2+}-ATPase) and another type resembling those in the plasma membrane (PMCA, plasma membrane Ca^{2+}-ATPase) of animal cells.

These are all members of the larger super-family of P-type ATPases. These ATP-consuming pumps, responsible for the primary transport of charged substrates, are involved in many processes including cell signaling, the movement of micronutrients, flipping of membrane phospholipids, the establishment and maintenance of electrochemical gradients and homeostatic ion levels (see Chapters "Plant Proton Pumps: Regulatory Circuits Involving H+-ATPase and H+-PPase" and "Type IV (P4) and V (P5) P-ATPases in Lipid Translocation and Membrane Trafficking"). Members of the P-type ATPase super-family (also referred to as E1–E2 ATPases) are characterised by formation of a phosphorylated intermediate which changes conformation during the catalytic cycle, by being inhibited by vanadate and by having a number of conserved sequence motifs (Møller et al. 1996). Plant and animal Ca^{2+}-ATPases belong to the type 2 subgroup of P-type ATPases and are characterised by having 10 transmembrane domains (TM), a cytoplasmic head consisting of a large loop between TM 4 and TM 5 containing the ATP-binding site and the Asp residue that is phosphorylated during the reaction cycle and a small loop between TM 2 and TM 3, which is part of the actuator domain (Møller et al. 1996; Palmgren and Axelsen 1998; Brini and Carafoli 2009). Ca^{2+}-ATPases contain all the characteristic motifs of P-type ATPases and of the type 2 subgroup including the phosphorylated Asp within the highly conserved sequence DKTGT, the PEGL motif considered to play a central role in energy transduction and the KGAXE sequence implicated in nucleotide binding (Møller et al. 1996; Axelsen and Palmgren 1998; Brini and Carafoli 2009).

In contrast to what is observed in animal cells, in plant cells each Ca^{2+} pump type is not necessarily restricted to one particular organelle or membrane, so that a purified membrane fraction may not be safely used to biochemically characterise a single Ca^{2+} pump type and, on the other hand, intracellular localisation cannot be deduced from biochemical characteristics. Only by exploiting a few biochemical traits peculiar of one Ca^{2+}-ATPase type (see below) is it possible to highlight the contribution of each type of Ca^{2+}-ATPase to Ca^{2+} pumping in a certain membrane fraction, as nicely shown by Hwang et al. (1997) in carrot suspension-cultured cells.

Several genes encoding plant Ca^{2+}-ATPases have been cloned since the early nineties, and phylogenetic analysis has shown that they group either with animal SERCA in the 2A subgroup of P-type ATPases or with PMCA in the 2B subgroup (Geisler et al. 2000a; Baxter et al. 2003). Given their multiple intracellular localisation, they have been named, respectively, ECA (for ER-type Ca^{2+}-ATPase) and ACA (for auto-inhibited Ca^{2+}-ATPase) (Geisler et al. 2000a). Availability of cDNA has allowed heterologous expression of single isoforms in yeast. In particular, a yeast strain (K616) devoid of endogenous Ca^{2+}-ATPases has proven very useful both for analysis of possible physiological function and for biochemical characterization of individual plant Ca^{2+}-ATPase isoforms (Sze et al. 2000). Only work clearly inherent for one Ca^{2+}-ATPase type, or isoform, will be expressly described below. For earlier work the reader should refer to previous reviews (Askerlund and Sommarin 1996; Evans and Williams 1998).

2.1 *Characteristics of Type 2A Ca^{2+}-ATPases (ECAs)*

Genome sequencing has shown that plant cells possess different isoforms of ECA, which share a number of structural and functional similarities. They are ca. 115 kDa proteins with more than 50% identity to animal SERCA. Their common biochemical features include a strong preference for ATP as a nucleoside triphosphate substrate, high affinity for Ca^{2+} (a $K_{0.5}$ value as low as 30 nM free Ca^{2+} has been reported for the Arabidopsis isoform ECA1, Liang and Sze 1998), and inhibition by cyclopiazonic acid, a specific inhibitor of SERCA (Brini and Carafoli 2009). Notably, plant ECAs differ from SERCA in the sensitivity to thapsigargin, a plant-derived sesquiterpene lactone, which inhibits SERCA activity at subnanomolar concentrations (Brini and Carafoli 2009). Plant ECAs are either insensitive to thapsigargin up to micromolar concentrations (Liang and Sze 1998) or show a strongly reduced sensibility to this inhibitor (IC_{50} values between 90 and 700 nM, Ordenes et al. 2002; Johnson et al. 2009). The thapsigargin-binding site is localised between the third and the eighth TM of SERCA (Brini and Carafoli 2009); although two key residues for thapsigargin coordination are fully conserved in animal and plant pumps, it is conceivable that in plant isoforms amino acid substitutions within the thapsigargin-binding pocket sterically alter thapsigargin access (Johnson et al. 2009).

Another peculiar feature of plant ECAs is their ability to transport, beside Ca^{2+}, also Mn^{2+} and Zn^{2+} ions, as shown by their ability to complement the phenotype of yeast strains K616 and *pmr1*: these strains, lacking the Golgi pump PMR1 which is thought to remove these cations from the cytosol and prevent toxicity (Lapinskas et al. 1995), are unable to grow in the presence of millimolar concentrations of Mn^{2+} or Zn^{2+} (Liang et al. 1997; Wu et al. 2002; Li et al. 2008; Mills et al. 2008; Johnson et al. 2009).

The transport mechanism of plant ECAs has not been investigated. SERCA is known to transport two Ca^{2+} ions per catalytic cycle, in obligatory exchange with two or three H^+. The two Ca^{2+} ions are coordinated by six residues in TM 4, 5, 6 and 8, three of which are involved also in H^+ translocation (Obara et al. 2005; Brini and Carafoli 2009). Since all of these residues are fully conserved in all the ECA isoforms cloned so far, it is likely that also plant ECAs transport two Ca^{2+} ions per catalytic cycle, in exchange with protons.

As mentioned above, plant genomes encode different ECA isoforms: e.g. four isoforms are present in Arabidopsis and three in rice. They are divided in two clusters based on sequence alignments and intron positions (Pittman et al. 1999; Baxter et al. 2003). In Arabidopsis all of them are expressed in all major organs (Mills et al. 2008) but only ECA1 and ECA3 have been characterised in detail. ECA1 was found localised at the ER (Liang et al. 1997), while ECA3 was found associated with Golgi and post-Golgi vesicles (Li et al. 2008; Mills et al. 2008) (Fig. 1). Indeed, Ca^{2+}-pumping activities with characteristics coherent with being driven by ECAs have been observed in ER-enriched fractions from various plant materials as well as in membrane vesicles from the Golgi apparatus of pea and cauliflower (Askerlund and Sommarin 1996; Hwang et al. 1997; Evans and Williams 1998; Sze et al. 2000; Ordenes et al. 2002). However, other locations are possible for other ECA isoforms: an antibody raised against the tomato ECA LCA1, recognised proteins associated with the tonoplast, the plasma membrane (PM) and the nuclear envelope (Ferrol and Bennett 1996; Downie et al. 1998).

Nothing is known about the regulation of ECAs in plants, except for a unique ECA isoform that has been cloned from maize. CAP1 has a C-terminal extension containing a calmodulin (CaM)-binding site, and its activity is, albeit moderately and with low affinity, stimulated by CaM similar to that of type 2B Ca^{2+}-ATPases (Subbaiah and Sachs 2000).

2.2 Characteristics of Type 2B Ca^{2+}-ATPases (ACAs)

The prototype of the type 2B Ca^{2+}-ATPases is the erythrocyte Ca^{2+} pump (PMCA isoform 4b) which has been among the first CaM targets identified (Carafoli 1991; Brini and Carafoli 2009). Type 2B Ca^{2+}-ATPases share a number of structural and functional features with type 2A Ca^{2+}-ATPases, but have unique regulatory properties due to the presence of an extended cytosolic domain whose auto-inhibitory action can be suppressed by binding of CaM. Localisation of this terminal regulatory domain is the main structural difference between plant and animal members of the type 2B group of Ca^{2+}-ATPases. In fact, while in PMCA this domain is localised in the extended C-terminus of the protein (Carafoli 1991; Brini and Carafoli 2009), in plant ACA isoforms, which have a very short cytosolic C-terminus, the auto-inhibitory, CaM-binding domain is localised in the extended cytosolic N-terminus preceding the first TM (Geisler et al. 2000a; Baxter et al. 2003; Kabala and Klobus 2005; Boursiac and Harper 2007). In contrast to animal cells, in which PMCA isoforms are localised exclusively at the PM, plant cells have ACAs localised at different membranes, including PM, tonoplast, ER and possibly chloroplast envelope (Geisler et al. 2000a; Kabala and Klobus 2005; Boursiac and Harper 2007) (Fig. 1); these are encoded by as many as 10 genes in Arabidopsis and 11 in rice (Baxter et al. 2003). These isoforms have subtle differences which will be discussed below, but share a number of characteristics.

Sequence alignment shows that only three of the six residues that define SERCA's two Ca^{2+}-binding sites are conserved in ACAs, as in the animal members of the type 2B subgroup of Ca^{2+}-ATPases (Brini and Carafoli 2009). Coherently, PMCA has been shown to catalyse transport of one Ca^{2+} per catalytic cycle, in exchange for one or two H^+ (Carafoli 1991; Brini and Carafoli 2009). The stoichiometry of plant ACAs has not been determined, but values of $K_{0.5}$ for free Ca^{2+} in the micromolar range have been determined with no sign of cooperativity (Sze et al. 2000; Bonza et al. 2001; Meneghelli et al. 2008), in agreement with one Ca^{2+} per ATP stoichiometry. The mechanism of transport has been analysed only for PM-localised ACAs of radish and Arabidopsis, which have been shown to catalyse a Ca^{2+}/H^+ exchange of unknown stoichiometry (Rasi-Caldogno et al. 1987; Luoni et al. 2000).

ACAs are unique among P-type ATPases for being able to use ITP or GTP as an alternative nucleoside-triphosphate substrate to ATP. This characteristic has been very useful specially for characterisation of the PM isoforms in native membrane vesicles, since it allows the avoidance of interference by the much more abundant H^+-ATPase which is highly specific for ATP (Williams et al. 1990; Carnelli et al. 1992; Hwang et al. 1997). ACAs are also particularly sensitive to inhibition by fluorescein derivatives which are known to bind to the nucleotide-binding domain (De Michelis et al. 1993; Askerlund and Sommarin 1996; Geisler et al. 2000a; Bonza et al. 2004).

ACAs are activated by CaM which determines both an increase of V_{max} and a decrease of the $K_{0.5}$ for free Ca^{2+} (Rasi-Caldogno et al. 1993; Hwang et al. 2000a; Sze et al. 2000; Bonza et al. 2001; Meneghelli et al. 2008). Activation is due to direct binding of CaM to a site localised in the N-terminus of the protein (Malmström et al. 1997; Harper et al. 1998; Bonza et al. 2000; Geisler et al. 2000b; Hwang et al. 2000a; Lee et al. 2007). This characteristic has been widely used for enzyme purification through CaM-affinity chromatography (Dieter and Marmè 1981; Askerlund and Sommarin 1996; Evans and Williams 1998; Sze et al. 2000).

The auto-inhibitory role of the N-terminus of ACAs has been shown both by controlled proteolysis and by analysis of N-deleted mutants: proteolytic cleavage of an N-terminal 9–15 kDa fragment or genetic deletion of the first 67–80 amino acids generates constitutively active proteins, which cannot be further activated by CaM (Rasi-Caldogno et al. 1993, 1995; Harper et al. 1998; Olbe and Sommarin 1998; Hwang et al. 2000a; Malmström et al. 2000; Sze et al. 2000; Luoni et al. 2004; Bækgaard et al. 2006). Notably, only constitutively active ACA mutants are able to complement the phenotype of the yeast K616 strain making it able to grow in extremely low Ca^{2+} concentrations (Harper et al. 1998; Chung et al. 2000; Geisler et al. 2000b; Hwang et al. 2000a; Sze et al. 2000; Bonza et al. 2004; Schiøtt and Palmgren 2005; Bækgaard et al. 2006; Lee et al. 2007).

The CaM-binding and the auto-inhibitory domains are partially overlapping, as shown by the ability of 20–25 amino acids long peptides reproducing the CaM-binding site to inhibit the activity of N-truncated pumps (Hwang et al. 2000a; Malmström et al. 2000; Luoni et al. 2004) as well as by analysis of single-point mutants (Curran et al. 2000; Bækgaard et al. 2006). However, the auto-inhibitory domain probably extends beyond the CaM-binding site; in fact, (1) a peptide reproducing the entire N-terminal extension of Arabidopsis isoform ACA8 (aa 1–116) is about 10-fold more efficient than that reproducing the CaM-binding site (aa 41–63) in inhibiting the activity of the N-truncated pump (Luoni et al. 2004) and (2) single-point mutations in the N-terminus downstream of the CaM-binding site of Arabidopsis isoform ACA2 generate a deregulated enzyme (Curran et al. 2000). Moreover, in a helical wheel projection of the CaM-binding site, residues involved in auto-inhibition cluster on one face of the α-helix, while residues involved in CaM-binding are scattered all around the α-helix (Bækgaard et al. 2006).

In contrast with vertebrates in which a gene family encodes a single CaM isoform, or with fungi and yeast that have a single CaM-coding gene, plants contain different isoforms of CaM, very similar to CaMs from other organisms, as well as CaM-like proteins with CaM activity ("divergent CaMs") with different ability to activate target proteins (Snedden and Fromm 1998; Zielinski 1998). ACAs are activated by divergent CaMs as much as by canonical CaM isoforms, but their affinity for divergent isoforms is lower than that of canonical CaMs (Lee et al. 2000; Yamniuk and Vogel 2004; Luoni et al. 2006).

While the terminal regulatory domain of both animal and plant members of type 2B Ca^{2+}-ATPases has been described in detail, much less is known of how its auto-inhibitory action is exerted. In Arabidopsis isoforms ACA2 and ACA8 or in PMCA4b, mutation of a conserved Asp residue, localised at the cytoplasmic end of TM 2, generates a deregulated pump which is almost insensitive to further activation by CaM (Curran et al. 2000; Bredeston and Adamo 2004; Fusca et al. 2009). However, a detailed characterisation of the PMCA4b mutant suggests that this mutation does not displace the auto-inhibitory domain, but rather directly changes accessibility of Ca^{2+} to the transmembrane domain (Bredeston and Adamo 2004; Corradi and Adamo 2007).

Cross-linking of PMCA with a peptide corresponding to the extended CaM-binding site allowed the identification of two putative sites of intramolecular interaction within the cytoplasmic head containing the catalytic domain: one is localised in the big cytoplasmic loop connecting TM 4 and 5 and the other in the small cytoplasmic loop connecting TM 2 and 3 (Carafoli 1991; Brini and Carafoli 2009). The latter region, which is part of the actuator domain, is highly conserved between the members of the type 2B Ca^{2+}-ATPases group, and a peptide reproducing this region of Arabidopsis isoform ACA8 has been shown to interact with ACA8 N-terminus in pull-down experiments (Luoni et al. 2004). Site-directed mutagenesis of this region of ACA8 has shown that mutation to Ala of any of the six tested conserved acidic residues originates a partially deregulated enzyme with higher basal activity in the absence of added CaM (Fusca et al. 2009). Although none of these single point mutants is fully deregulated, these results point out the relevance of acidic residues in this region of ACA8 in auto-inhibition; for at least one of these mutants evidence has been presented that it loosens the interaction between the N-terminus and the catalytic head making the CaM-binding site more accessible (Fusca et al. 2009). Since Ala scanning mutagenesis of the N-terminus of plant isoforms of type 2B Ca^{2+}-ATPases reveals the involvement of basic residues in the auto-inhibitory domain (Curran et al. 2000; Bækgaard et al. 2006), it is possible that the negative charge conferred by acidic residues to the surface area of the small cytoplasmic loop favours and/or stabilises its auto-inhibitory interaction with the N-terminal auto-inhibitory domain (Fusca et al. 2009).

The activity of PMCA is regulated also by acidic phospholipids, which activate the enzyme through a complex mechanism, only partially overlapping that of CaM (Carafoli 1991; Brini and Carafoli 2009). This has been shown to be the case at least for PM-localised isoforms of plant ACAs (Bonza et al. 2001; Meneghelli et al. 2008). As for PMCA, acidic phospholipids such as phosphatidylserine or phosphatidylinositol-4P activate Arabidopsis isoform ACA8 via two distinct mechanisms, involving their binding to different sites: acidic phospholipids binding to a site in the protein N-terminus, overlapping the auto-inhibitory and CaM-binding domain, stimulates ACA8 activity similar to CaM or to cleavage of the N-terminus, while binding to a second, as yet unidentified, site further stimulates ACA8 activity by lowering its $K_{0.5}$ for free Ca^{2+} (Meneghelli et al. 2008).

ACAs, as their animal counterpart PMCA (Carafoli 1991; Brini and Carafoli 2009), can be regulated also by phosphorylation. Two Ser residues in the N-terminus of the Brassica isoform BCA1 are phosphorylated in vitro by protein kinase C. One of these Ser residues is within the CaM-binding domain but no effect of phosphorylation on CaM binding was observed (Malmström et al. 2000). Similarly, in vitro phosphorylation of a Ser residue of Arabidopsis isoform ACA2 just downstream the CaM-binding site by a calcium-dependent protein kinase does not disrupt CaM binding. However, it severely inhibits CaM-stimulated enzyme activity, suggesting that pump activity can be finely tuned by two different Ca^{2+}-signaling pathways (Hwang et al. 2000b).

The different isoforms of ACA are divided in four clusters, based on intron number and position and on sequence alignment (Baxter et al. 2003). Current evidence suggests a correlation between clusters and intracellular localization. Cluster 1 comprises three Arabidopsis isoforms: ACA2, that is localised at the ER (Hong et al. 1999), ACA7, of which nothing is known to date and ACA1; the latter has been originally proposed to be localised at the plastid inner envelope (Huang et al. 1993, 1994), but no Ca^{2+}-ATPase activity has been detected in this membrane (Roh et al. 1998) and it has been found associated with the ER in a proteomic survey of Arabidopsis organelles (Dunkley et al. 2006). Cluster 2 comprises Arabidopsis isoforms ACA4 and ACA11 and Brassica isoform BCA1, which are all localised at the vacuolar membrane (Malmström et al. 1997; Geisler et al. 2000b; Malmström et al. 2000; Lee et al. 2007). Cluster 4 comprises Arabidopsis isoforms ACA8, ACA9 and ACA10: isoforms ACA8 and ACA9 have been shown to be localised at the PM (Bonza et al. 2000; Schiøtt et al. 2004), and the same localization is likely for ACA10, given the ability of ACA8 to complement the phenotype of a ACA10 knockout mutant (George et al. 2008). Nothing is known about intracellular localization of members of cluster 3, but a PM localization has been proposed based on the ability of Arabidopsis isoform ACA12 to rescue a pollen defect in a ACA9 knockout mutant (Boursiac and Harper 2007). In Arabidopsis all of the isoforms are expressed, at least in some cell type, and most cells express multiple isoforms of two or three different clusters. However, only few isoforms have been analysed in detail (Boursiac and Harper 2007).

Members of clusters 1 and 2 are very similar both at the gene level (five or six introns in conserved positions) and at the protein level (in Arabidopsis, 1014–1030 aa proteins more than 60% identical to each other). Members of cluster 4 share a complex gene structure with as many as 33 introns, and a molecular mass slightly higher than that of the other clusters (the three Arabidopsis isoforms are 1069–1086 aa proteins approximately 70% identical to each other) mainly due to a longer N-terminus domain. Members of cluster 3 are unique in having intron-less genes; moreover, they are markedly divergent in the putative CaM-binding region (Baxter et al. 2003; Kabala and Klobus 2005; Boursiac and Harper 2007).

Overall, the most variable region is the N-terminus, suggesting that the different ACA isoforms may have different regulatory properties. While stimulation of Ca^{2+}-ATPase activity by exogenous CaM can be easily observed in tonoplast or ER vesicles, stimulation of Ca^{2+}-ATPase activity by exogenous CaM in isolated PM vesicles is low or even undetectable unless endogenous CaM is removed by extensive washing with strong Ca^{2+} chelators (Robinson et al. 1988; Williams et al. 1990; De Michelis et al. 1993; Rasi-Caldogno et al. 1993; Kurosaki and Kaburaki 1994; Dainese et al. 1997; Olbe et al. 1997). Moreover, the PM Ca^{2+}-ATPase efficiently binds CaM in overlay experiments (Rasi-Caldogno et al. 1995; Askerlund 1997; Dainese et al. 1997; Bonza et al. 1998; Olbe and Sommarin 1998; Bonza et al. 2000) and tightly binds to CaM-agarose (Bonza et al. 1998; Bonza et al. 2000). These results suggest that PM-localised ACA isoforms may have a higher affinity for CaM than those in other membranes (Robinson et al. 1988; Williams et al. 1990; Rasi-Caldogno et al. 1993; Kurosaki and Kaburaki 1994; Dainese et al. 1997; Olbe et al. 1997). The interaction with CaM of Arabidopsis PM isoform ACA8 has been analysed in detail by different experimental approaches, leading to estimate a K_d value of about 30 nM (Bækgaard et al. 2006; Luoni et al. 2006). This value is definitely lower than that reported for BCA1, a tonoplast isoform of Brassica (110 nM, Yamniuk and Vogel 2004), but rather similar to that reported for ACA2, an ER isoform of Arabidopsis (30–40 nM, Harper et al. 1998; Hwang et al. 2000a; Hwang et al. 2000b). The behaviour of PM-localised ACA isoforms might thus reflect, more than a higher affinity for CaM, a higher stability of the complex. In fact, analysis of the kinetics of interaction between the N-terminus of ACA8 and CaM by surface plasmon resonance spectroscopy shows that association is rapid, but dissociation is slow, setting ACA8 in the group of slow-response type 2B Ca^{2+}-ATPases (Bækgaard et al. 2006; Luoni et al. 2006).

2.3 Physiological Role of Ca^{2+}-ATPases

Three main experimental approaches have been used in the attempt of defining the physiological role of Ca^{2+}-ATPase: (1) analysis of gene expression in different cell types and in different physiological states; (2) a pharmacological approach exploiting differential sensitivity to inhibitors such as cyclopiazonic acid or fluorescein derivatives of ECAs and ACAs, and (3) phenotypic characterisation of knockout mutants of single Ca^{2+}-ATPase isoforms. Each of these approaches has its pitfalls: (1) up or downregulation of gene expression per se does not allow us to draw any solid conclusion, especially when, as in most cases, it is not shown to determine a change in protein level and/or enzyme activity; (2) sensitivity to inhibitors, even when sufficiently specific, does not allow to discriminate between different isoforms of one Ca^{2+}-ATPase group, and (3) knockout mutants of single Ca^{2+}-ATPase isoforms often do not have any evident phenotype due to redundancy; moreover when a phenotype is observed it may be an indirect, rather than straightforward, consequence of the lack of the protein, which might influence other processes such as development. Nevertheless, the integrated and critical use of the above-mentioned approaches has allowed researchers to ascertain the involvement of Ca^{2+} -ATPases, or of a specific subgroup, or even of a single isoform in some developmental process as well as in the response to external stimuli.

Analysis of T-DNA insertional mutants of Arabidopsis *ECA1* and *ECA3* has shown that the ability of ECAs to transport Mn^{2+} is important both in Mn^{2+} nutrition and in tolerance to Mn^{2+} stress. When grown on standard nutrient medium the *eca1* knockout mutant is indistinguishable from the wild type, but it clearly shows reduced root growth and toxicity symptoms when exposed to 0.5 mM Mn^{2+}; the overall Mn^{2+}content of the plants is similar to that of the wild type, indicating that toxicity is due to altered Mn^{2+} compartmentation due to lack of the ER-localised ECA1 (Wu et al. 2002). A similar phenotype was observed for knockout mutants of *ECA3*, an isoform localised at Golgi and post-Golgi vesicles (Li et al. 2008), thus, detoxification of Mn^{2+} would require sequestration both in the ER and in Golgi. On the other hand, Mills et al. (2008) found that an *eca3* knockout mutant was not more sensitive than the wild type to Mn^{2+} toxicity, but rather required higher Mn^{2+} concentrations for growth, suggesting that ECA3 may play an essential role also in Mn^{2+} nutrition. Both *eca1* and *eca3* knockout mutants have slightly altered Ca^{2+} sensitivity (Wu et al. 2002; Li et al. 2008), and the *eca3* mutant showed increased protein secretion, suggesting that alteration of Ca^{2+} and/or Mn^{2+} homeostasis in the endosomes may perturb protein secretion (Li et al. 2008).

A role in gibberellin signaling in aleurone tissue has been suggested for OsECA1, an isoform of rice, based on the finding that its expression is rapidly induced by gibberellin and that its over-expression bypasses the need for gibberellin to activate an α-amylase gene (Chen et al. 1997); however, the available data are not enough to define which role this ECA isoform actually plays in the process.

For ACA pumps, a definite phenotype has been described only for knockout mutants of two members of cluster 4 in Arabidopsis. ACA9 is a plasma membrane-localised isoform, primarily expressed in the pollen. Its knockout mutants display reduced growth of pollen tubes with high frequency of aborted fertilisation leading to a threefold reduction in seed set (Schiøtt et al. 2004) (Fig 2a). While a large amount of evidence indicates that Ca^{2+} signaling plays an important role in the regulation of pollen tube growth and involves Ca^{2+} influx through plasma membrane channels, the role played by ACA9 has not been clarified yet: it might be required to prevent a toxic accumulation of Ca^{2+} in the cytoplasm, or, more

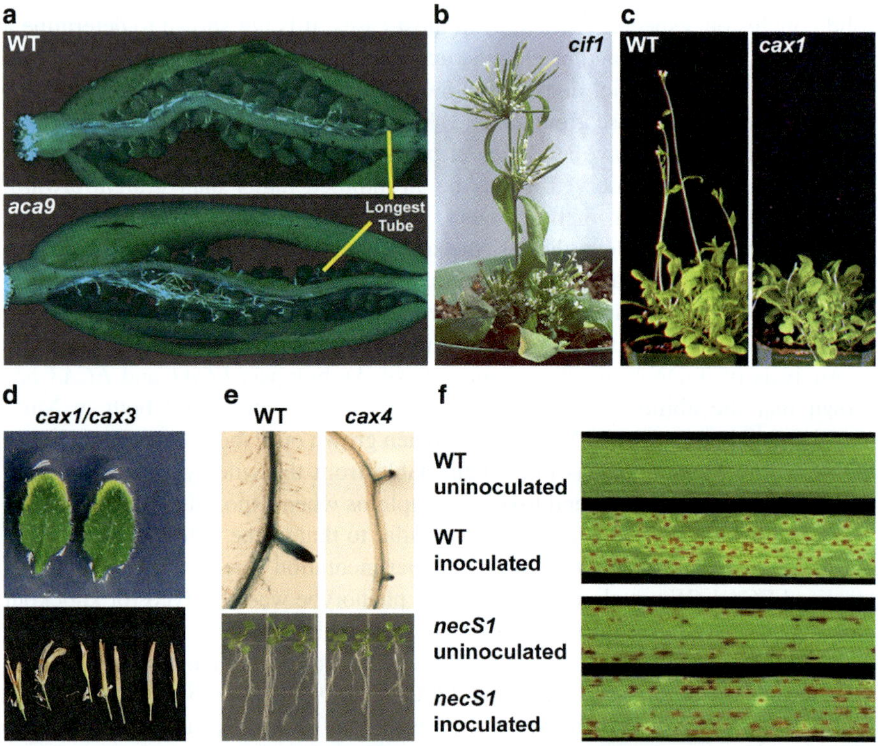

Fig. 2 Phenotypes of ACA and CAX mutant plants. (**a**) Pollen tube defects of the *aca9* Arabidopsis mutant compared to wild type (WT). Pistils are shown stained with aniline blue (from Schiøtt et al. 2004, copyright 2004, National Academy of Sciences, USA). (**b**) Inflorescence phenotype of the *cif1* mutant Arabidopsis, which is a mutant of *ACA10* (George et al. 2008). (**c**) Delayed bolting of the inflorescence stem in the Arabidopsis *cax1* mutant compared to WT (from Cheng et al. 2003, www.plantcell.org, copyright American Society of Plant Biologist). (**d**) Leaf and seed pod phenotypes of the *cax1/cax3* double mutant Arabidopsis (from Cheng et al. 2005, www.plantphysiol.org, copyright American Society of Plant Biologist). (**e**) Auxin sensitivity of *cax4* mutant Arabidopsis following treatment with 50 nM IAA. Top panels: change in IAA content in *cax4* as indicated by the *DR5*::*GUS* reporter; bottom panels: reduced primary root growth in *cax4* (from Mei et al. 2009, reproduced with permission from John Wiley and Sons). (**f**) Formation of the hypersensitive response and resistance to stem rust disease in the *necS1* barley mutant, which is a mutant of *HvCAX1*, compared to the WT cultivar (from Zhang et al. 2009, reproduced with kind permission from Springer Science+Business Media)

intriguingly, it might be involved in regulating the dynamics of the signaling Ca^{2+} waves (Schiøtt et al. 2004; Boursiac and Harper 2007).

In contrast to *ACA9*, *ACA8* and *ACA10* are expressed throughout the plant during vegetative and reproductive growth (Schiøtt and Palmgren 2005; Cerana et al. 2006; George et al. 2008). Nevertheless, knockout mutants of *ACA8* or *ACA10* do not have an altered phenotype, with a noticeable exception: when *ACA10* is deleted in a genotype containing a naturally occurring dominant allele of an unlinked gene (*CIF2*, encoding an unknown protein), this causes altered adult vegetative development and the formation of floral clusters (George et al. 2008) (Fig. 2b). Intriguingly, knockout of *ACA8* in the same genotype does not produce phenotypic alterations, but *ACA8* over-expression complements the phenotype of the *aca10* knockout mutant. The unique role of ACA10 in development probably reflects subtle differences in expression patterns, protein levels or regulatory properties between ACA10 and ACA8, which can be overcome or mitigated by over-expression (George et al. 2008). There is also the intriguing possibility that the CIF2 protein may encode a novel ACA10 regulator.

The expression of *ACA8*, but not that of *ACA10*, is up-regulated by cold stress (Schiøtt and Palmgren 2005) which is known to determine large transient changes in $[Ca^{2+}]_{cyt}$ concentration (Knight et al. 1996). This up-regulation may lead to an increase in the capacity of the transporter to extrude Ca^{2+} to the apoplast, similar to that observed upon cold acclimation in winter rye leaves (Puhakainen et al. 1999).

A possible role for type 2B Ca^{2+}-ATPases of small vacuoles (ACA4, Geisler et al. 2000b) and of the ER (ACA2, Anil et al. 2008) in salt tolerance is suggested by their ability to improve salt tolerance in yeast. Salt exposure increases the expression of *ACA4* in Arabidopsis plantlets (Geisler et al. 2000b) as well as of different isoforms of ECA or ACA in barley roots (Garciadeblas et al. 2001), but no evidence for a salt stress-induced increase of protein level and/or activity is available for higher plants. Recently, an ACA isoform localised at small vacuoles of the moss *Physcomitrella patens* was nicely shown to be involved in the moss response to salt stress: its knockout mutants exhibit enhanced susceptibility to salt stress and sustained elevation of cytosolic-free Ca^{2+} concentration in response to salt treatment (Qudeimat et al. 2008).

The high sensitivity of ACAs to inhibition by fluorescein derivatives such as erythrosin B and eosin Y (De Michelis et al. 1993) has allowed their use in vivo to selectively inhibit ACA-mediated Ca^{2+} fluxes and to determine their involvement in Ca^{2+} homeostasis under different conditions (Felle et al. 1992; Beffagna et al. 2000; Romani et al. 2004). By such a method it was shown that in *Egeria densa* leaves ABA rapidly (within 10 min) induces an increase in the activity of a PM-localised ACA; interestingly, due to the Ca^{2+}/H^+ exchange catalysed by this pump, this activation also determines a decrease in net proton extrusion and of the value of cytoplasmic pH (Beffagna et al. 2000). A role of PM-localised ACAs in ABA signaling is suggested also by the ABA-induced stimulation of expression of *ACA8* and *ACA9* in Arabidopsis seedlings, which at least in the case of ACA8 has been shown to determine an increase in protein level in the PM (Cerana et al. 2006). However, the relatively slow induction of *ACA8* and *ACA9* mRNAs (within 2–4 h) and accumulation of ACA8 protein at the PM (within 8 h) suggest that these events may be involved, rather than in short-term effects of ABA on $[Ca^{2+}]_{cyt}$ concentration, in a mechanism for desensitisation of the cell to ABA signal when, under stress condition, ABA is a constant presence.

The integrated use of the in vivo pharmacological approach with biochemical analysis of the PM has allowed us to demonstrate the involvement of a PM-localised ACA in the short-term response of Arabidopsis cultured cells to elicitors such as oligogalacturonides (Romani et al. 2004), which are known to determine a rapid and transient increase of cytoplasmic pH (Lamb and Dixon 1997; Grant et al. 2000). After an immediate stimulation of Ca^{2+} influx, oligogalacturonides stimulate eosin Y-sensitive Ca^{2+} efflux: such a stimulation can be ascribed to an increased formation of the active Ca^{2+}–CaM–Ca^{2+}–ATPase complex and is essential to turn the oxidative burst off (Romani et al. 2004).

3 Ca^{2+}/H^+ Antiporters

Low-affinity, high-capacity Ca^{2+} sequestration into plant vacuoles mediated by Ca^{2+}/H^+ antiport activity has been identified in many plant species and is likely to be the predominant ubiquitous pathway for vacuolar Ca^{2+} loading (Schumaker and Sze 1985; Blumwald and Poole 1986; Bush and Sze 1986) (Fig. 1). Ca^{2+}/H^+ antiport activity is encoded by cation/H^+ exchanger (CAX) genes. These are members of the larger CaCA (Ca^{2+}/cation antiporter) superfamily of transporters which also includes the related animal Na^+/Ca^{2+} exchanger (NCX) gene family (Cai and Lytton 2004). CAX genes are present in prokaryotes, fungi, invertebrates and lower vertebrates in addition to plants but absent from higher animals (Shigaki et al. 2006). All of the characterised CAX proteins including those from non-plant species share common structural features. They are approximately 400 amino acids in length with 11 TM domains separated by a negatively charged cytosolic loop of unknown function called the acidic motif between TM 6 and 7, and a long hydrophilic N-terminal tail, predicted to be cytosolic, that has been found to have regulatory function in some plant Ca^{2+}/H^+ antiporters (Shigaki and Hirschi 2006). Such a topology is only predicted for plant CAX proteins but experimentally determined topology mapping of the yeast Ca^{2+}/H^+ antiporter confirms this structure (Segarra and Thomas 2008).

Structure–function studies have begun to identify some of the potential substrate specificity determinants of the plant CAX proteins. A nine-amino acid region called the Ca^{2+} domain was mapped between TM 1 and 2 of Arabidopsis CAX1, and has been shown to be essential for CAX1 Ca^{2+} transport (Shigaki et al. 2001). Two highly conserved sequence repeats present on loops between TM 3 and 4 (c-1 repeat region) and between TM 8 and 9 (c-2 repeat region) have been proposed to function as cation selectivity filters and specific residues within these regions are required for Ca^{2+} transport (Kamiya and Maeshima 2004; Shigaki et al. 2005).

Plant CAX genes were first identified in Arabidopsis through the use of a yeast complementation screen (Hirschi et al. 1996) and subsequently characterised in other species including mung bean and rice (Ueoka-Nakanishi et al. 1999; Kamiya and Maeshima 2004). Genome sequencing and transcriptomic analyses suggest that CAX genes are ubiquitous in plants (Shigaki et al. 2006). When it has been tested, all the plant CAX genes characterised so far appear to encode proteins with tonoplast-localised Ca^{2+}/H^+ antiport activity (Shigaki and Hirschi 2006). Although some CAX proteins encode transporters with high Ca^{2+} specificity, others have a much broader specificity and can transport a range of divalent cations, such as Mn^{2+} and Cd^{2+} in addition to Ca^{2+} (Hirschi et al. 2000). Recent phylogenetic analysis suggests that plant CAX genes can be divided into two phylogenetic subgroups, Type 1-A and Type 1-B (Shigaki et al. 2006). For example, the six Arabidopsis CAX genes are split between the two groups. This phylogenetic variation may indicate functional variation and indeed CAX2, CAX5 and CAX6 in the Type 1-B group have broader substrate specificities compared to CAX1, CAX3 and CAX4 in the Type 1-A group. Arabidopsis also possesses five other related genes (originally named *CAX7–CAX11*) but which have subsequently been shown to be more similar to some animal NCX genes and have been placed in a separate phylogenetic group called cation/Ca^{2+} exchanger (CCX) (Shigaki et al. 2006). However, as yet there is no direct evidence of any Ca^{2+}/H^+ antiport activity mediated by CCX proteins.

There is some evidence for Ca^{2+}/H^+ antiport activity at other membrane locations including the plasma membrane (Kasai and Muto 1990) and the chloroplast thylakoid membrane (Ettinger et al. 1999) although the genes encoding non-tonoplast Ca^{2+}/H^+ antiporters have not yet been identified (Fig. 1). A CAX protein from soybean involved in Na^+ tolerance may be plasma membrane localised but no Ca^{2+} transport activity for GmCAX1 has been determined (Luo et al. 2005) so plasma membrane Ca^{2+}/H^+ antiport activity performed by a CAX protein awaits definitive identification.

3.1 Biochemical and Regulatory Properties of Ca^{2+}/H^+ Antiporters

Biochemical studies using purified tonoplast fractions from a variety of plant species have demonstrated that proton-coupled Ca^{2+} antiport activity is dependent on ΔpH and has a $K_{0.5}$ for Ca^{2+} in the range of 4–67 μM and V_{max} values of 3–34 nmol Ca^{2+} mg protein^{-1} min^{-1} (Schumaker and Sze 1985; Blumwald and Poole 1986; Bush and Sze 1986; Blackford et al. 1990; DuPont et al. 1990; Ueoka-Nakanishi et al. 1999), suggesting that compared to Ca^{2+} pumps this is a low-affinity Ca^{2+} transport pathway but with a high capacity for transport. This is indicative of Ca^{2+}/H^+ antiporters playing a major role in Ca^{2+} sequestration when $[Ca^{2+}]_{cyt}$ concentrations are high, a role that has been confirmed for the yeast vacuolar Ca^{2+}/H^+ antiporter (Miseta et al. 1999). The estimated stoichiometry of

Ca^{2+}/H^+ antiport is $1Ca^{2+}/3H^+$ (Blackford et al. 1990) which indicates that this protein would be able to efficiently transport Ca^{2+} into the vacuole by utilising the proton motive force. This proton gradient across the tonoplast is generated by the V-type H^+-ATPase (V-ATPase) and the vacuolar H^+-pyrophosphatase (V-PPase). The activity of these H^+ pumps is therefore critical for Ca^{2+}/H^+ antiport activity as demonstrated by inhibitor studies and genetic analysis. V-ATPase inhibitors such as bafilomycin significantly inhibit Ca^{2+}/H^+ antiport (Hirschi et al. 1996; Ueoka-Nakanishi et al. 1999). Likewise, when tonoplast-specific V-ATPase activity is deleted in an Arabidopsis *vha-a2 vha-a3* double knockout mutant, Ca^{2+} content in the plant leaves are significantly reduced (Krebs et al. 2010). In contrast, when V-PPase activity is over-expressed, Ca^{2+}/H^+ antiport activity is similarly increased (Gaxiola et al. 2001). Even more intriguingly, manipulation of Ca^{2+}/H^+ antiport activity either through knockout or overexpression appears to alter V-ATPase activity, suggesting that there may be some feedback between the Ca^{2+}/H^+ antiporter and the primary energising V-ATPase (Barkla et al. 2008).

The identification of CAX genes has allowed the characterisation of individual Ca^{2+}/H^+ antiporter isoforms in comparison to the purified plant fractions that may contain a heterogeneous population of Ca^{2+} transport proteins. Heterologous expression studies of some CAX proteins in yeast, such as CAX1 from Arabidopsis and VCAX1 from mung bean have confirmed that they show biochemical characteristics similar to the activity determined from plants (Hirschi et al. 1996; Ueoka-Nakanishi et al. 2000). For example, CAX1 expressed and purified from yeast shows ΔpH-dependent Ca^{2+}/H^+ antiport activity with a K_m for Ca^{2+} of 10–15 μM (Hirschi et al. 1996) and optimal activity at pH 7.5 (Pittman et al. 2005). This is equivalent to the characteristics of the Ca^{2+}/H^+ antiporter solubilised and reconstituted from oat root (Schumaker and Sze 1990). In contrast Arabidopsis CAX2 has a lower affinity for Ca^{2+} than CAX1 and a broader cation selectivity (Edmond et al. 2009), indicative of its role as a transition metal transporter and not a major component in Ca^{2+} homeostasis (Pittman et al. 2004).

The yeast heterologous expression system has been particularly useful for plant Ca^{2+}/H^+ antiporter analysis, not only for characterising Ca^{2+} transport kinetics and determining substrate specificity, but for gaining insights into the regulatory mechanisms of these transporters. The yeast-based analysis has indicated that rather than being constitutively active, plant Ca^{2+}/H^+ antiporters are regulated by a post-translational mechanism analogous to that of the ACA-type Ca^{2+} pumps. Analysis of Arabidopsis CAX1 found that Ca^{2+} transport activity can be modulated via a regulatory domain at the N-terminus of the protein (Pittman and Hirschi 2001; Pittman et al. 2002a). The *vcx1 pmc1* Ca^{2+} hypersensitive yeast mutant cannot grow on high Ca^{2+} media due to impaired Ca^{2+} tolerance because of the lack of the vacuolar Ca^{2+}-ATPase PMC1 and the Ca^{2+}/H^+ antiporter VCX1. When CAX1 is expressed in this mutant, the yeast remains Ca^{2+} hypersensitive due to weak CAX1 Ca^{2+}/H^+ antiport activity. However, N-terminal truncation of CAX1 activates Ca^{2+} antiport activity and allows the yeast to grow on high calcium (Pittman and Hirschi 2001), equivalent to the situation with the N-terminally cleaved ACAs in yeast. Protein expression and tonoplast localisation of both CAX variants are equivalent.

This and other evidence indicate that the inability of full-length CAX1 to efficiently transport Ca^{2+} is due to auto-inhibition mediated by a domain on the cytosolic N-terminal tail which interacts with another N-terminal region of CAX1 (Pittman et al. 2002a). Regulatory proteins are thought to activate CAX1 by interaction with the N-terminal tail, similar to CaM stimulation of ACA pumps, although CaM does not appear to be an activator of CAX1. Through a yeast-screening approach, a number of CAX-interacting proteins (CXIP) have been identified that interact with the N-terminus of CAX1 and regulate its activity (Cheng and Hirschi 2003; Cheng et al. 2004a, b). These include CXIP3, which is identical to FKBP15-2, a member of the FKBP-type immunophilin family of proteins that may be involved in protein folding (Harrar et al. 2001), CXIP4 which is a novel plant-specific protein of unknown function and CXIP5, which is identical to the Ser/Thr kinase SOS2 (also called CIPK24) which is known as an activator of the Na^+ transporter SOS1 (Zhu 2003). It has yet to be confirmed whether SOS2 regulates CAX1 by phosphorylation, however, a candidate Ser residue (Ser-25) within the N-terminal domain that may become phosphorylated has been identified through mutagenesis (Pittman et al. 2002a). If Ser-25 is mutated to Asp to mimic constitutive phosphorylation, CAX1 is activated, but when mutated to Ala no activation occurs. Furthermore, SOS2 is unable to activate the S25A-CAX1 mutant (Cheng et al. 2004b).

In addition to the identified CXIP proteins, another Arabidopsis CAX protein, CAX3, has been identified as a regulator of CAX1. CAX3 is the most closely related CAX gene to CAX1. Although it has the potential to transport Ca^{2+} when a putative Ca^{2+} specificity domain is mutated (Shigaki et al. 2001), either full-length or N-terminal-truncated CAX3 does not show any Ca^{2+} transport activity when expressed in yeast. When full-length CAX1 and CAX3 are co-expressed in yeast, Ca^{2+} transport activity is induced (Zhao et al. 2009b). Activation of Ca^{2+}/H^+ antiport activity is mediated by direct protein–protein interaction through the formation of a heteromeric CAX1–CAX3 protein complex (Zhao et al. 2009a, b). These studies confirmed that such a complex can form in plant cells as well as in the yeast system, and that CAX1 and CAX3 can also form homomeric complexes. Oligomerization is common for many transport proteins. Mung bean VCAX1 has also been shown to form homomeric complexes (Ueoka-Nakanishi et al. 2000) and it is possible that all CAX proteins may assemble in either homomeric or heteromeric forms. We hypothesise that under particular conditions, such as in response to a specific stress stimulus CAX1 and CAX3 interact to form a complex with altered Ca^{2+} transport activity. For example, the formation of the CAX1–CAX3 complex in Arabidopsis tissue was identified following salt stress (Zhao et al. 2009b). The mechanisms of this interaction and the functional relevance of the heteromeric CAX complex require further study.

It is now appreciated that the Arabidopsis CAX1 cDNA originally identified in the yeast Ca^{2+} sensitivity suppression screen (Hirschi et al. 1996) was a truncated partial-length cDNA which lacked the regulatory domain, thus allowing it to efficiently transport Ca^{2+} when expressed in the yeast. In the plant, native CAX1 only appears to exist as the full-length protein and will therefore require the various modes of activation which have been identified through the protein interaction studies. Other Ca^{2+}/H^+ antiporters from Arabidopsis, from other plant species and from algae appear to share this mode of regulation (Pittman et al. 2002b, 2004, 2009; Kamiya et al. 2005). However, this regulatory characteristic does vary between plant CAX proteins. Full-length mung bean VCAX1 has detectable Ca^{2+} transport activity in yeast although this is increased in the N-terminal-truncated VCAX1 variant (Pittman et al. 2002b). Furthermore, full-length CAX proteins from tomato (Edmond et al. 2009) and the monocot plant *Puccinellia* (Liu et al. 2009) have high Ca^{2+}/H^+ antiport activity in yeast. Interestingly, the transport activity of the N-terminal-truncated PutCAX1 from *Puccinellia* is significantly reduced compared to the full-length version. In all this suggests a high degree of interspecies variation. As these regulatory mechanisms of the Ca^{2+}/H^+ antiporters have been elucidated principally through yeast studies, there will be a need to demonstrate these mechanisms in vivo. There is some evidence that the N-terminal regulatory domain of Arabidopsis CAX1 has equivalent functional characteristics in a plant cell as in yeast (Mei et al. 2007), however, the relevance of these regulatory mechanisms to the plant is still not clear.

3.2 Roles of Ca^{2+}/H^+ Antiporters in Ca^{2+} Signaling and Plant Development

While significant advances have been made to our understanding of the biochemical and regulatory properties of the plant Ca^{2+}/H^+ antiporters using yeast heterologous expression, we have less detail in our understanding of the physiological functions of these transporters in plants. Much of our current insight into the roles of the Arabidopsis CAX proteins in the plant has come from gene-knockout studies. Knockouts of all the Arabidopsis CAX genes provide rather subtle phenotypes. This is likely due to the redundancy inherent in plant Ca^{2+} homeostatic mechanisms. Not only are there six CAX genes but there are many Ca^{2+}-ATPases that can perform similar Ca^{2+} transport functions. This is illustrated with the *cax1* knockout line in which various CAX genes, particularly *CAX3* and *CAX4*, are upregulated when *CAX1* is lacking (Cheng et al. 2003). In addition, the vacuolar Ca^{2+}-ATPase *ACA4* is upregulated and CaM-stimulated tonoplast Ca^{2+}-ATPase activity is increased as the plant attempts to compensate for the lack of *CAX1*. Despite this redundancy, the *cax1* plants have a reduction in vacuolar Ca^{2+}/H^+ antiport activity of approximately 50% compared to wild type and exhibit a number of interesting phenotypes (Catalá et al. 2003; Cheng et al. 2003). There are alterations to the development of the Arabidopsis plant. Bolting of the *cax1* inflorescence stem is delayed by 5–7 days compared to wild-type plants such that 27 days after germination, the primary inflorescence stem of the *cax1-1* line was reduced in length by approximately 70% (Cheng et al. 2003) (Fig. 2c). The number and total length of branching shoots, and the number and length of lateral roots were also reduced in the *cax1* plants. When *CAX3* is also deleted in Arabidopsis to give a *cax1 cax3* double knockout, the plants are significantly developmentally impaired (Cheng et al. 2005). Soil-grown plants are dwarfed, have necrotic lesions at the leaf tips and shoot apex, and have a significant reduction in seed pod size and seed number (Fig. 2d). An obvious question is, what is the primary cause of these morphological phenotypes, is it simply due to impaired Ca^{2+} homeostasis or due to an impact on Ca^{2+} signaling?

The phenotypes of the *cax* knockout plants can be compared and contrasted with those phenotypes gained from *CAX1* enhanced or ectopic expression. When the deregulated, N-terminal-truncated version of CAX1 (sCAX1) is expressed in tobacco under the control of the strong 35S promoter, the tobacco plants show a variety of phenotypes including stunted growth, apical burning and ion sensitivity (Hirschi 1999). The 35S::sCAX1 plants have increased total Ca^{2+} content due to increased vacuolar Ca^{2+} accumulation, and all of the phenotypes are reversed by application of exogenous Ca^{2+}. This suggests that the phenotypes are due to Ca^{2+} deficiency-like symptoms as the increased and deregulated sequestration of Ca^{2+} into the vacuole due to enhanced Ca^{2+}/H^{+} antiport activity is disrupting the normal partitioning of Ca^{2+} to particular tissues and at the cellular level, to particular organelles where it is required. Ectopic expression of 35S::sCAX1 in tomato similarly led to impaired development of the plant including increased incidence of the soft fruit Ca^{2+}-deficiency disorder blossom-end rot (Park et al. 2005), further suggesting that impaired Ca^{2+} nutrition is a cause of these phenotypes. It is therefore quite likely that the impaired development of the *cax* knockout plants is similarly due to impaired Ca^{2+} nutrition. However, we must also consider the possibility that other ion imbalances in the *cax* plants may give rise to altered plant development. Deletion of *CAX1* and *CAX3* lowers activity of both V-ATPase and the plasma membrane H^{+}-ATPase (Cheng et al. 2005; Zhao et al. 2008) which could lead to alteration in cellular pH. In addition, reduction in Ca^{2+}/H^{+} antiport activity will cause reduced vacuolar H^{+} "leak" which may also impact on cellular pH. One possible scenario is therefore that alteration in cellular pH homeostasis through CAX gene deletion may also impact on plant development. Future studies which examine pH levels in these mutants could be quite informative.

In addition to developmental phenotypes, the Arabidopsis *cax* mutants exhibit altered hormonal responses. Exogenous application of auxins such as IAA leads to inhibition of root growth in plants. The *cax1* mutant plants treated with exogenous IAA were found to be partially resistant to its effects such that the *cax1* primary roots were longer than the wild-type roots following IAA treatment (Cheng et al. 2003). Whether there is a link between alteration in auxin response by *cax1* and the observed alteration in lateral root development is still unclear. This response to auxin appears specific to the *cax1* mutant and is not shared by *cax3*. In contrast, the *cax4* mutant does display an altered root growth response to exogenous auxin, but this is quite different to that exhibited by *cax1*. The *cax4* mutant has increased sensitivity to IAA with an increased reduction in primary root length compared to wild type following IAA treatment (Mei et al. 2009) (Fig. 2e). Furthermore, the *DR5::GUS* auxin reporter was found to be insensitive to auxin when expressed in the *cax4* mutant, suggesting that IAA content is reduced in the root tips of *cax4* mutants. One possibility is that an increase in $[Ca^{2+}]_{cyt}$ in *cax4* due to reduced Ca^{2+} sequestration may give rise to altered auxin homeostasis in the root. The AGC-type kinase PINOID regulates the PIN auxin efflux carrier and PINOID has been suggested to be regulated by Ca^{2+} (Benjamins et al. 2003), thus alteration in Ca^{2+} homeostasis may impact on root auxin gradients. However, this hypothesis does not explain the differences between *cax1* and *cax4* with respect to auxin. Differences in expression between the two genes may partly explain the different responses as *CAX1* is principally shoot/leaf expressed, while *CAX4* is mainly root expressed (Cheng et al. 2005; Mei et al. 2009).

While phenotypes from Arabidopsis *cax* mutants clearly demonstrate that vacuolar Ca^{2+}/H^{+} antiporters are critical for mediating cellular Ca^{2+} tolerance and Ca^{2+} homeostasis (Cheng et al. 2005), some of the knockout mutant phenotypes may indicate a direct involvement of these transporters in specific Ca^{2+}-signaling responses, particularly in response to abiotic and biotic stress. Both *cax1* and *cax3* have increased sensitivity to ABA as exhibited by reduced germination of the mutant seeds in responses to increasing concentrations of exogenous hormone (Zhao et al. 2008). The *cax3* Arabidopsis seedlings also have increased sensitivity to salt stress compared to wild-type and *cax1* seedlings (Zhao et al. 2008), while the *cax1* knockout has increased tolerance to freezing following cold acclimation (Catalá et al. 2003). Various cold-response genes are induced following *CAX1* deletion, suggesting that CAX1 is a negative regulator of the cold-acclimation response. A particularly interesting phenotype which has recently been uncovered suggests an involvement of Ca^{2+}/H^{+} antiport activity in pathogen response. The barley mutant *necS1*, in which *HvCAX1* is mutated, has increased resistance to the stem rust pathogen *Puccinia graminis* compared to the wild-type and stem rust-susceptible background cultivar (Fig. 2f). Mutation of the *HvCAX1* gene is likely to be the cause of this disease resistance as silencing of *HvCAX1* lead to an increase in the hypersensitive response, a form of programmed cell death of the cells infected by the pathogen, and thus a decrease in pathogen infection (Zhang et al. 2009).

Most biotic and abiotic stresses including pathogen infection, cold and salinity give rise to elevation of $[Ca^{2+}]_{cyt}$ which can often be discerned as Ca^{2+} oscillations with specific dynamic characteristics (Knight et al. 1996; Grant et al. 2000; Kiegle et al. 2000). Ca^{2+}/H^{+} antiporters may be directly involved in modulating the $[Ca^{2+}]_{cyt}$ elevations following stress induction to generate a specific Ca^{2+} signature and may therefore be required for the correct transduction of the signal. The ability of the Ca^{2+}/H^{+} antiporters to be regulated by multiple modes of action is consistent with this potential role as a Ca^{2+} modulator. As yet evidence for a direct role of Ca^{2+}/H^{+} antiporters in controlling Ca^{2+} signals is still lacking. Analysis of the Arabidopsis *det3* mutant, which has a 60% reduction in V-ATPase activity, demonstrated that guard cell abiotic stress-induced $[Ca^{2+}]_{cyt}$ elevations no longer oscillate as in wild type (Allen et al. 2000). Furthermore, these perturbed oscillations inhibited stomatal closure. Indirect inhibition of Ca^{2+}/H^{+} antiport activity due to inhibition of its primary H^{+} pump energiser could be the cause for these disrupted stress-induced $[Ca^{2+}]_{cyt}$ oscillations.

4 Summary and Future Perspectives

Studies of model plants such as Arabidopsis and rice have revealed the full extent of the Ca^{2+} efflux transporter repertoire. Arabidopsis has 14 Ca^{2+} pumps and six Ca^{2+}/H^+ antiporters plus the possibility of other transporters providing Ca^{2+} efflux transport activity, such as the type 1 P-type ATPase HMA1 (Moreno et al. 2008). Here, we have summarised how genetic and biochemical analyses have begun to allow us to infer the physiological roles of some of these transporters. However, there is still a long way to go and many hurdles to overcome before we can identify the specific functions of each Ca^{2+} transporter isoform. The large variety of Ca^{2+} transporters will make it difficult to define the physiological role played by each isoform of a transporter group, particularly as each cell type expresses different Ca^{2+}-ATPases and Ca^{2+}/H^+ antiporters. Furthermore, even each membrane type can be endowed with different Ca^+ transporters belonging to the ECA, ACA or CAX group, such as ECA1 and ACA2 at the ER, and ACA11 and CAX1 at the tonoplast. In addition, it appears that different transporter types may be involved in the same process, for example, expression of both ACA8 and CAX1 is upregulated by cold stress, suggesting shared involvement in freezing tolerance (Catalá et al. 2003; Schiøtt and Palmgren 2005). Consequently it is often difficult to determine a clear cut phenotype in a single gene knockout mutant. To date, the involvement of a specific subgroup, or even a single isoform of Ca^{2+}-ATPase or Ca^{2+}/H^+ antiporter in a number of physiological processes has been demonstrated, but in most cases the exact role they play has not been demonstrated. A key challenge for future research will therefore be to determine how each of the Ca^{2+} transporters are delineated and how they overlap or interact to mediate maintenance of low $[Ca^{2+}]_{cyt}$ and its re-establishment in response to specific environmental and developmental stimuli.

While the biochemical characteristics shared by each type of Ca^{2+} efflux transporter (ECA, ACA, and CAX) have been described in sufficient detail, much less is known about the subtle differences which may exist between different isoforms of the same Ca^{2+} transporter type. For example, there appear to be differences in the Ca^{2+} transport kinetics and regulatory characteristics between Ca^{2+}/H^+ antiporter isoforms (Pittman et al. 2002b; Edmond et al. 2009) but these variations require better analysis. Variation between Ca^{2+}-ATPase isoforms may include variable sensitivity to inhibitors, such as with the possible different sensitivity to thapsigargin by members of the two clusters of ECA (Johnson et al. 2009). Subtle differences in regulatory properties between closely related transporters may also exist, such as differential regulation by phosphorylation. For example, while the CaM-binding site is highly conserved at least in members of each cluster of ACA, the surrounding region, in which phosphorylation has been shown to occur (Hwang et al. 2000b; Malmström et al. 2000), is much more variable. Thus, single ACA isoforms may differ from each other in the degree of autoinhibition and in the response to different regulatory mechanisms. From this point of view, members of ACA cluster 3 (ACA12 and ACA13), that have very low or undetectable expression level in basal conditions, but are dramatically induced upon exposure of a specific stress, such as in response to pathogens (Boursiac and Harper 2007), are particularly intriguing. Not only is their N-terminal CaM-binding and auto-inhibitory region highly divergent, but they also have variation within TM 2 where an Asp residue conserved in all other ACAs is instead an Asn residue. In different ACA isoforms mutation of Asp at this region to Asn generates a deregulated pump almost insensitive to further activation by CaM (Curran et al. 2000; Bredeston and Adamo 2004; Fusca et al. 2009). It is thus possible that some Ca^{2+} pumps such as ACA12 and ACA13 are either deregulated or differently regulated pumps, whose activity might be required only under specific circumstances (Boursiac and Harper 2007).

Future work will hopefully shed some light onto the relevance of these regulatory and kinetic variations. One possibility is that variation in the kinetics of Ca^{2+} transport by a specific ECA, ACA or CAX transporter isoform may allow the generation of a cytosolic Ca^{2+} signal with specific dynamics. The involvement of a particular Ca^{2+} transporter isoform in a Ca^{2+} signaling event may therefore partly explain how a cell is able to generate different Ca^{2+} signals in response to different stimuli. However, clearer evidence is required to demonstrate the direct role of a Ca^{2+}-ATPase or Ca^{2+}/H^+ antiporter in modulating a cytosolic Ca^{2+} signal.

Acknowledgements We thank Jim Connorton (University of Manchester) for comments to the manuscript and are very grateful to Bob Sharrock (Montana State University) for providing the *cif1* picture.

References

Allen GJ, Chu SP, Schumacher K, Shimazaki CT, Vafeados D, Kemper A, Hawke SD, Tallman G, Tsien RY, Harper JF, Chory J, Schroeder JI (2000) Alteration of stimulus-specific guard cell calcium oscillations and stomatal closing in Arabidopsis *det3* mutant. Science 289:2338–2342

Allen GJ, Chu SP, Harrington CL, Schumacher K, Hoffman T, Tang YY, Grill E, Schroeder JI (2001) A defined range of guard cell calcium oscillation parameters encodes stomatal movements. Nature 411:1053–1057

Anil VS, Rajkumar P, Kumar P, Mathew MK (2008) A plant Ca^{2+} pump, ACA2, relieves salt hypersensitivity in yeast. Modulation of cytosolic calcium signature and activation of adaptive Na^+ homeostasis. J Biol Chem 283:3497–3506

Askerlund P (1997) Calmodulin-stimulated Ca^{2+}-ATPases in the vacuolar and plasma membranes in cauliflower. Plant Physiol 114:999–1007

Askerlund P, Sommarin M (1996) Calcium efflux transporters in higher plants. In: Smallwood M, Knox JP, Bowles DJ (eds) Membranes: specialized functions in plants. BIOS Scientific Publishers, Oxford, pp 281–299

Axelsen KB, Palmgren MG (1998) Evolution of substrate specificities in the P-type ATPase superfamily. J Mol Evol 46:84–101

Bækgaard L, Luoni L, De Michelis MI, Palmgren MG (2006) The plant plasma membrane Ca^{2+} pump ACA8 contains overlapping as well as physically separated autoinhibitory and calmodulin-binding domains. J Biol Chem 281:1058–1065

Barkla BJ, Hirschi KD, Pittman JK (2008) Exchangers man the pumps: functional interplay between proton pumps and proton-coupled Ca^{2+} exchangers. Plant Signal Behav 3:354–356

Baxter I, Tchieu J, Sussman MR, Boutry M, Palmgren MG, Gribskov M, Harper JF, Axelsen KB (2003) Genomic comparison of P-type ATPase ion pumps in Arabidopsis and rice. Plant Physiol 132:618–628

Beffagna N, Romani G, Sforza MC (2000) H^+ fluxes at plasmalemma level: in vivo evidence for a significant contribution of the $Ca2^+$-ATPase and for the involvement of its activity in the abscisic acid-induced changes in *Egeria densa* leaves. Plant Biol 2:168–175

Benjamins R, Ampudia CSG, Hooykaas PJJ, Offringa R (2003) PINOID-mediated signaling involves calcium-binding proteins. Plant Physiol 132:1623–1630

Blackford S, Rea PA, Sanders D (1990) Voltage sensitivity of H^+/Ca^{2+} antiport in higher-plant tonoplast suggests a role in vacuolar calcium accumulation. J Biol Chem 265:9617–9620

Blumwald E, Poole RJ (1986) Kinetics of Ca^{2+}/H^+ antiport in isolated tonoplast vesicles from storage tissue of *Beta vulgaris* L. Plant Physiol 80:727–731

Bonza C, Carnelli A, De Michelis MI, Rasi-Caldogno F (1998) Purification of the plasma membrane Ca^{2+}-ATPase from radish seedlings by calmodulin-agarose affinity chromatography. Plant Physiol 116:845–851

Bonza MC, Morandini P, Luoni L, Geisler M, Palmgren MG, De Michelis MI (2000) At-ACA8 encodes a plasma membrane-localized calcium-ATPase of Arabidopsis with a calmodulin-binding domain at the N terminus. Plant Physiol 123:1495–1505

Bonza MC, Luoni L, De Michelis MI (2001) Stimulation of plant plasma membrane Ca^{2+}-ATPase activity by acidic phospholipids. Physiol Plant 112:315–320

Bonza MC, Luoni L, De Michelis MI (2004) Functional expression in yeast of an N-deleted form of At-ACA8, a plasma membrane Ca^{2+}-ATPase of *Arabidopsis thaliana*, and characterization of a hyperactive mutant. Planta 218:814–823

Boursiac Y, Harper JF (2007) The origin and function of calmodulin regulated Ca^{2+} pumps in plants. J Bioenerg Biomembr 39:409–414

Bredeston LM, Adamo HP (2004) Loss of autoinhibition of the plasma membrane Ca^{2+} pump by substitution of aspartic 170 by asparagine – activation of plasma membrane calcium ATPase 4 without disruption of the interaction between the catalytic core and the C-terminal regulatory domain. J Biol Chem 279:41619–41625

Brini M, Carafoli E (2009) Calcium pumps in health and disease. Physiol Rev 89:1341–1378

Bush DR, Sze H (1986) Calcium-transport in tonoplast and endoplasmic-reticulum vesicles isolated from cultured carrot cells. Plant Physiol 80:549–555

Cai XJ, Lytton J (2004) The cation/Ca^{2+} exchanger superfamily: phylogenetic analysis and structural implications. Mol Biol Evol 21:1692–1703

Carafoli E (1991) Calcium pump of the plasma membrane. Physiol Rev 71:129–153

Carnelli A, De Michelis MI, Rasi-Caldogno F (1992) Plasma-membrane Ca^{2+}-ATPase of radish seedlings.1. Biochemical-characterization using ITP as a substrate. Plant Physiol 98:1196–1201

Case RM, Eisner D, Gurney A, Jones O, Muallem S, Verkhratsky A (2007) Evolution of calcium homeostasis: from birth of the first cell to an omnipresent signalling system. Cell Calcium 42:345–350

Catalá R, Santos E, Alonso JM, Ecker JR, Martínez-Zapater JM, Salinas J (2003) Mutations in the Ca^{2+}/H^+ transporter *CAX1* increase *CBF/DREB1* expression and the cold-acclimation response in Arabidopsis. Plant Cell 15:2940–2951

Cerana M, Bonza MC, Harris R, Sanders D, De Michelis MI (2006) Abscisic acid stimulates the expression of two isoforms of plasma membrane Ca^{2+}-ATPase in *Arabidopsis thaliana* seedlings. Plant Biol 8:572–578

Chen XF, Chang MC, Wang BY, Wu R (1997) Cloning of a Ca^{2+}-ATPase gene and the role of cytosolic Ca^{2+} in the gibberellin-dependent signaling pathway in aleurone cells. Plant J 11:363–371

Cheng NH, Hirschi KD (2003) Cloning and characterization of CXIP1, a novel PICOT domain-containing *Arabidopsis* protein that associates with CAX1. J Biol Chem 278:6503–6509

Cheng NH, Pittman JK, Barkla BJ, Shigaki T, Hirschi KD (2003) The Arabidopsis *cax1* mutant exhibits impaired ion homeostasis, development, and hormonal responses and reveals interplay among vacuolar transporters. Plant Cell 15:347–364

Cheng NH, Liu JZ, Nelson RS, Hirschi KD (2004a) Characterization of CXIP4, a novel *Arabidopsis* protein that activates the H^+/Ca^{2+} antiporter, CAX1. FEBS Lett 559:99–106

Cheng NH, Pittman JK, Zhu JK, Hirschi KD (2004b) The protein kinase SOS2 activates the *Arabidopsis* H^+/Ca^{2+} antiporter CAX1 to integrate calcium transport and salt tolerance. J Biol Chem 279:2922–2926

Cheng NH, Pittman JK, Shigaki T, Lachmansingh J, LeClere S, Lahner B, Salt DE, Hirschi KD (2005) Functional association of Arabidopsis CAX1 and CAX3 is required for normal growth and ion homeostasis. Plant Physiol 138:2048–2060

Chung WS, Lee SH, Kim JC, Heo WD, Kim MC, Park CY, Park HC, Lim CO, Kim WB, Harper JF, Cho MJ (2000) Identification of a calmodulin-regulated soybean Ca^{2+}-ATPase (SCA1) that is located in the plasma membrane. Plant Cell 12:1393–1407

Corradi GR, Adamo HP (2007) Intramolecular fluorescence resonance energy transfer between fused autofluorescent proteins reveals rearrangements of the N- and C-terminal segments of the plasma membrane Ca^{2+} pump involved in the activation. J Biol Chem 282:35440–35448

Curran AC, Hwang I, Corbin J, Martinez S, Rayle D, Sze H, Harper JF (2000) Autoinhibition of a calmodulin-dependent calcium pump involves a structure in the stalk that connects the transmembrane domain to the ATPase catalytic domain. J Biol Chem 275:30301–30308

Dainese P, James P, Baldan B, Carafoli E (1997) Subcellular and tissue distribution, partial purification, and sequencing of calmodulin-stimulated Ca^{2+}-transporting ATPases from barley (*Hordeum vulgare* L) and tobacco (*Nicotiana tabacum*). Eur J Biochem 244:31–38

De Michelis MI, Carnelli A, Rasi-Caldogno F (1993) The Ca^{2+} pump of the plasma-membrane of *Arabidopsis thaliana* – characteristics and sensitivity to fluorescein derivatives. Bot Acta 106:20–25

Dieter P, Marmè D (1981) A calmodulin dependent microsomal ATPase from corn (*Zea mays* L.). FEBS Lett 125:245–248

Dodd AN, Kudla J, Sanders D (2010) The language of calcium signaling. Annu Rev Plant Biol 61:593–620

Downie L, Priddle J, Hawes C, Evans DE (1998) A calcium pump at the higher plant nuclear envelope? FEBS Lett 429:44–48

Dunkley TPJ, Hester S, Shadforth IP, Runions J, Weimar T, Hanton SL, Griffin JL, Bessant C, Brandizzi F, Hawes C, Watson RB, Dupree P, Lilley KS (2006) Mapping the *Arabidopsis* organelle proteome. Proc Natl Acad Sci USA 103:6518–6523

DuPont FM, Bush DS, Windle JJ, Jones RL (1990) Calcium and proton transport in membrane-vesicles from barley roots. Plant Physiol 94:179–188

Edmond C, Shigaki T, Ewert S, Nelson M, Connorton J, Chalova V, Noordally Z, Pittman JK (2009) Comparative analysis of CAX2-like cation transporters indicates functional and regulatory diversity. Biochem J 418:145–154

Ettinger WF, Clear AM, Fanning KJ, Peck ML (1999) Identification of a Ca^{2+}/H^+ antiport in the plant chloroplast thylakoid membrane. Plant Physiol 119:1379–1385

Evans DE, Williams LE (1998) P-type calcium ATPases in higher plants – biochemical, molecular and functional properties. Biochim Biophys Acta 1376:1–25

Evans NH, McAinsh MR, Hetherington AM, Knight MR (2005) ROS perception in *Arabidopsis thaliana*: the ozone-induced calcium response. Plant J 41:615–626

Felle HH, Tretyn A, Wagner G (1992) The role of the plasma-membrane Ca^{2+}-ATPase in Ca^{2+} homeostasis in *Sinapis alba* root hairs. Planta 188:306–313

Ferrol N, Bennett AB (1996) A single gene may encode differentially localized Ca^{2+}-ATPases in tomato. Plant Cell 8:1159–1169

Fusca T, Bonza MC, Luoni L, Meneghelli S, Marrano CA, De Michelis MI (2009) Single point mutations in the small cytoplasmic loop of ACA8, a plasma membrane Ca^{2+}-ATPase of *Arabidopsis thaliana*, generate partially deregulated pumps. J Biol Chem 284:30881–30888

Garciadeblas B, Benito B, Rodriguez-Navarro A (2001) Plant cells express several stress calcium ATPases but apparently no sodium ATPase. Plant Soil 235:181–192

Gaxiola RA, Li JS, Undurraga S, Dang LM, Allen GJ, Alper SL, Fink GR (2001) Drought- and salt-tolerant plants result from overexpression of the AVP1 H^+-pump. Proc Natl Acad Sci USA 98:11444–11449

Geisler M, Axelsen KB, Harper JF, Palmgren MG (2000a) Molecular aspects of higher plant P-type Ca^{2+}-ATPases. Biochim Biophys Acta 1465:52–78

Geisler M, Frangne N, Gomès E, Martinoia E, Palmgren MG (2000b) The ACA4 gene of Arabidopsis encodes a vacuolar membrane calcium pump that improves salt tolerance in yeast. Plant Physiol 124:1814–1827

George L, Romanowsky SM, Harper JF, Sharrock RA (2008) The ACA10 Ca^{2+}-ATPase regulates adult vegetative development and inflorescence architecture in Arabidopsis. Plant Physiol 146:716–728

Grant M, Brown I, Adams S, Knight M, Ainslie A, Mansfield J (2000) The *RPM1* plant disease resistance gene facilitates a rapid and sustained increase in cytosolic calcium that is necessary for the oxidative burst and hypersensitive cell death. Plant J 23:441–450

Gross J, Marmè D (1978) ATP-dependent Ca^{2+}-uptake into plant membrane vesicles. Proc Natl Acad Sci USA 75:1232–1236

Harper JF, Hong BM, Hwang ID, Guo HQ, Stoddard R, Huang JF, Palmgren MG, Sze H (1998) A novel calmodulin-regulated Ca^{2+}-ATPase (ACA2) from *Arabidopsis* with an N-terminal autoinhibitory domain. J Biol Chem 273:1099–1106

Harrar Y, Bellini C, Faure JD (2001) FKBPs: at the crossroads of folding and transduction. Trends Plant Sci 6:426–431

Hetherington AM, Brownlee C (2004) The generation of Ca^{2+} signals in plants. Annu Rev Plant Biol 55:401–427

Hirschi KD (1999) Expression of Arabidopsis *CAX1* in tobacco: altered calcium homeostasis and increased stress sensitivity. Plant Cell 11:2113–2122

Hirschi KD (2004) The calcium conundrum. Both versatile nutrient and specific signal. Plant Physiol 136:2438–2442

Hirschi KD, Zhen RG, Cunningham KW, Rea PA, Fink GR (1996) CAX1, an H^+/Ca^{2+} antiporter from *Arabidopsis*. Proc Natl Acad Sci USA 93:8782–8786

Hirschi KD, Korenkov VD, Wilganowski NL, Wagner GJ (2000) Expression of Arabidopsis CAX2 in tobacco. Altered metal accumulation and increased manganese tolerance. Plant Physiol 124:125–133

Hong BM, Ichida A, Wang YW, Gens JS, Pickard BC, Harper JF (1999) Identification of a calmodulin-regulated Ca^{2+}-ATPase in the endoplasmic reticulum. Plant Physiol 119:1165–1175

Huang LQ, Berkelman T, Franklin AE, Hoffman NE (1993) Characterization of a gene encoding a Ca^{2+}-ATPase-like protein in the plastid envelope. Proc Natl Acad Sci USA 90:10066–10070

Huang LQ, Berkelman T, Franklin AE, Hoffman NE (1994) Errata corrige. Proc Natl Acad Sci USA 91:9664

Hwang I, Ratterman DM, Sze H (1997) Distinction between endoplasmic reticulum-type and plasma membrane-type Ca2+ pumps. Plant Physiol 113:535–548

Hwang I, Harper JF, Liang F, Sze H (2000a) Calmodulin activation of an endoplasmic reticulum-located calcium pump involves an interaction with the N-terminal autoinhibitory domain. Plant Physiol 122:157–167

Hwang I, Sze H, Harper JF (2000b) A calcium-dependent protein kinase can inhibit a calmodulin-stimulated Ca^{2+} pump (ACA2) located in the endoplasmic reticulum of *Arabidopsis*. Proc Natl Acad Sci USA 97:6224–6229

Johnson NA, Liu FL, Weeks PD, Hentzen AE, Kruse HP, Parker JJ, Laursen M, Nissen P, Costa CJ, Gatto C (2009) A tomato ER-type Ca^{2+}-ATPase, LCA1, has a low thapsigargin-sensitivity and can transport manganese. Arch Biochem Biophys 481:157–168

Kabala K, Klobus GY (2005) Plant Ca^{2+}-ATPases. Acta Physiol Plant 27:559–574

Kamiya T, Maeshima M (2004) Residues in internal repeats of the rice cation/H^+ exchanger are involved in the transport and selection of cations. J Biol Chem 279:812–819

Kamiya T, Akahori T, Maeshima M (2005) Expression profile of the genes for rice cation/H^+ exchanger family and functional analysis in yeast. Plant Cell Physiol 46:1735–1740

Kasai M, Muto S (1990) Ca^{2+} pump and Ca^{2+}/H^+ antiporter in plasma membrane vesicles isolated by aqueous two-phase partitioning from corn leaves. J Membr Biol 114:133–142

Kiegle E, Moore CA, Haseloff J, Tester MA, Knight MR (2000) Cell-type-specific calcium responses to drought, salt and cold in the *Arabidopsis* root. Plant J 23:267–278

Knight H, Trewavas AJ, Knight MR (1996) Cold calcium signaling in Arabidopsis involves two cellular pools and a change in calcium signature after acclimation. Plant Cell 8:489–503

Krebs M, Beyhl D, Gorlich E, Al-Rasheid KAS, Marten I, Stierhof Y-D, Hedrich R, Schumacher K (2010) Arabidopsis V-ATPase activity at the tonoplast is required for efficient nutrient storage but not for sodium accumulation. Proc Natl Acad Sci USA 107:3251–3256

Kurosaki F, Kaburaki H (1994) Calmodulin-dependency of a Ca^{2+}-pump at the plasma membrane of cultured carrot cells. Plant Sci 104:23–30

Lamb C, Dixon RA (1997) The oxidative burst in plant disease resistance. Annu Rev Plant Physiol Plant Mol Biol 48:251–275

Lapinskas PJ, Cunningham KW, Liu XF, Fink GR, Culotta C (1995) Mutations in *PMR1* suppress oxidative damage in yeast-cells lacking superoxide-dismutase. Mol Cell Biol 15:1382–1388

Lee SH, Johnson JD, Walsh MP, Van Lierop JE, Sutherland C, Xu AD, Snedden WA, Kosk-Kosicka D, Fromm H, Narayanan N, Cho MJ (2000) Differential regulation of Ca^{2+}/calmodulin-dependent enzymes by plant calmodulin isoforms and free Ca^{2+} concentration. Biochem J 350:299–306

Lee SM, Kim HS, Han HJ, Moon BC, Kim CY, Harper JF, Chung WS (2007) Identification of a calmodulin-regulated autoinhibited Ca^{2+}-ATPase (ACA11) that is localized to vacuole membranes in *Arabidopsis*. FEBS Lett 581:3943–3949

Li X, Chanroj S, Wu Z, Romanowsky SM, Harper JF, Sze H (2008) A distinct endosomal Ca^{2+}/Mn^{2+} pump affects root growth through the secretory process. Plant Physiol 147: 1675–1689

Liang F, Sze H (1998) A high-affinity Ca^{2+} pump, ECA1, from the endoplasmic reticulum is inhibited by cyclopiazonic acid but not by thapsigargin. Plant Physiol 118:817–825

Liang F, Cunningham KW, Harper JF, Sze H (1997) ECA1 complements yeast mutants defective in Ca^{2+} pumps and encodes an endoplasmic reticulum-type Ca^{2+}-ATPase in *Arabidopsis thaliana*. Proc Natl Acad Sci USA 94:8579–8584

Liu H, Zhang XX, Takano T, Liu SK (2009) Characterization of a PutCAX1 gene from *Puccinellia tenuiflora* that confers Ca^{2+} and Ba^{2+} tolerance in yeast. Biochem Biophys Res Commun 383:392–396

Luo GZ, Wang HW, Huang J, Tian AG, Wang YJ, Zhang JS, Chen SY (2005) A putative plasma membrane cation/proton antiporter from soybean confers salt tolerance in *Arabidopsis*. Plant Mol Biol 59:809–820

Luoni L, Bonza MC, De Michelis MI (2000) H^+/Ca^{2+} exchange driven by the plasma membrane Ca^{2+}-ATPase of *Arabidopsis thaliana* reconstituted in proteoliposomes after calmodulin-affinity purification. FEBS Lett 482:225–230

Luoni L, Meneghelli S, Bonza MC, De Michelis MI (2004) Auto-inhibition of *Arabidopsis thaliana* plasma membrane Ca^{2+}-ATPase involves an interaction of the N-terminus with the small cytoplasmic loop. FEBS Lett 574:20–24

Luoni L, Bonza MC, De Michelis MI (2006) Calmodulin/Ca^{2+}-ATPase interaction at the *Arabidopsis thaliana* plasma membrane is dependent on calmodulin isoform showing isoform-specific Ca^{2+} dependencies. Physiol Plant 126:175–186

Malmström S, Askerlund P, Palmgren MG (1997) A calmodulin-stimulated Ca^{2+}-ATPase from plant vacuolar membranes with a putative regulatory domain at its N-terminus. FEBS Lett 400:324–328

Malmström S, Åkerlund HE, Askerlund P (2000) Regulatory role of the N terminus of the vacuolar calcium-ATPase in cauliflower. Plant Physiol 122:517–526

McAinsh MR, Pittman JK (2009) Shaping the calcium signature. New Phytol 181:275–294

Mei H, Zhao J, Pittman JK, Lachmansingh J, Park S, Hirschi KD (2007) *In planta* regulation of the *Arabidopsis* Ca^{2+}/H^+ antiporter CAX1. J Exp Bot 58:3419–3427

Mei H, Cheng NH, Zhao J, Park P, Escareno RA, Pittman JK, Hirschi KD (2009) Root development under metal stress in *Arabidopsis thaliana* requires the H^+/cation antiporter CAX4. New Phytol 183:95–105

Meneghelli S, Fusca T, Luoni L, De Michelis MI (2008) Dual mechanism of activation of plant plasma membrane Ca^{2+}-ATPase by acidic phospholipids: evidence for a phospholipid binding site which overlaps the calmodulin-binding site. Mol Membr Biol 25:539–546

Mills RF, Doherty ML, López-Marqués RL, Weimar T, Dupree P, Palmgren MG, Pittman JK, Williams LE (2008) ECA3, a Golgi-localized P_{2A}-type ATPase, plays a crucial role in manganese nutrition in Arabidopsis. Plant Physiol 146:116–128

Miseta A, Kellermayer R, Aiello DP, Fu LW, Bedwell DM (1999) The vacuolar Ca^{2+}/H^+ exchanger Vcx1p/Hum1p tightly controls cytosolic Ca^{2+} levels in *S. cerevisiae*. FEBS Lett 451:132–136

Miwa H, Sun J, Oldroyd GED, Downie JA (2006) Analysis of calcium spiking using a cameleon calcium sensor reveals that nodulation gene expression is regulated by calcium spike number and the developmental status of the cell. Plant J 48:883–894

Møller JV, Juul B, le Maire M (1996) Structural organization, ion transport, and energy transduction of P-type ATPases. Biochim Biophys Acta 1286:1–51

Moreno I, Norambuena L, Maturana D, Toro M, Vergara C, Orellana A, Zurita-Silva A, Ordenes VR (2008) AtHMA1 is a thapsigargin-sensitive Ca^{2+}/heavy metal pump. J Biol Chem 283:9633–9641

Obara K, Miyashita N, Xu C, Toyoshima L, Sugita Y, Inesi G, Toyoshima C (2005) Structural role of countertransport revealed in Ca^{2+} pump crystal structure in the absence of Ca^{2+}. Proc Natl Acad Sci USA 102:14489–14496

Olbe M, Sommarin M (1998) The spinach plasma membrane Ca^{2+} pump is a 120-kDa polypeptide regulated by calmodulin-binding to a terminal region. Physiol Plant 103:35–44

Olbe M, Widell S, Sommarin M (1997) Active calcium transporters in isolated membranes of wheat root cells. J Exp Bot 48:1767–1777

Ordenes VR, Reyes FC, Wolff D, Orellana A (2002) A thapsigargin-sensitive Ca^{2+} pump is present in the pea Golgi apparatus membrane. Plant Physiol 129:1820–1828

Palmgren MG, Axelsen KB (1998) Evolution of P-type ATPases. Biochim Biophys Acta 1365:37–45

Park S, Cheng NH, Pittman JK, Yoo KS, Park J, Smith RH, Hirschi KD (2005) Increased calcium levels and prolonged shelf life in tomatoes expressing Arabidopsis H^+/Ca^{2+} transporters. Plant Physiol 139:1194–1206

Pittman JK, Hirschi KD (2001) Regulation of CAX1, an Arabidopsis Ca^{2+}/H^+ antiporter. Identification of an N-terminal autoinhibitory domain. Plant Physiol 127:1020–1029

Pittman JK, Mills RF, O'Connor CD, Williams LE (1999) Two additional type IIA Ca^{2+}-ATPases are expressed in *Arabidopsis thaliana*: evidence that type IIA sub-groups exist. Gene 236:137–147

Pittman JK, Shigaki T, Cheng NH, Hirschi KD (2002a) Mechanism of N-terminal autoinhibition in the *Arabidopsis* Ca^{2+}/H^+ antiporter CAX1. J Biol Chem 277:26452–26459

Pittman JK, Sreevidya CS, Shigaki T, Ueoka-Nakanishi H, Hirschi KD (2002b) Distinct N-terminal regulatory domains of Ca^{2+}/H^+ antiporters. Plant Physiol 130:1054–1062

Pittman JK, Shigaki T, Marshall JL, Morris JL, Cheng NH, Hirschi KD (2004) Functional and regulatory analysis of the *Arabidopsis thaliana* CAX2 cation transporter. Plant Mol Biol 56:959–971

Pittman JK, Shigaki T, Hirschi KD (2005) Evidence of differential pH regulation of the *Arabidopsis* vacuolar Ca^{2+}/H^+ antiporters CAX1 and CAX2. FEBS Lett 579:2648–2656

Pittman JK, Edmond C, Sunderland PA, Bray CM (2009) A cation-regulated and proton gradient-dependent cation transporter from *Chlamydomonas reinhardtii* has a role in calcium and sodium homeostasis. J Biol Chem 284:525–533

Puhakainen T, Pihakaski-Maunsbach K, Widell S, Sommarin M (1999) Cold acclimation enhances the activity of plasma membrane Ca^{2+} ATPase in winter rye leaves. Plant Physiol Biochem 37:231–239

Qudeimat E, Faltusz AMC, Wheeler G, Lang D, Brownlee C, Reski R, Frank W (2008) A P-IIB-type Ca^{2+}-ATPase is essential for stress adaptation in *Physcomitrella patens*. Proc Natl Acad Sci USA 105:19555–19560

Rasi-Caldogno F, Pugliarello MC, De Michelis MI (1987) The Ca^{2+}-transport ATPase of plant plasma membrane catalyzes a nH^+/Ca^{2+} exchange. Plant Physiol 83:994–1000

Rasi-Caldogno F, Carnelli A, De Michelis MI (1993) Controlled proteolysis activates the plasma-membrane Ca^{2+} pump of higher-plants - a comparison with the effect of calmodulin in plasma-membrane from radish seedlings. Plant Physiol 103:385–390

Rasi-Caldogno F, Carnelli A, De Michelis MI (1995) Identification of the plasma membrane Ca^{2+}-ATPase and of its autoinhibitory domain. Plant Physiol 108:105–113

Robinson C, Larsson C, Buckhout TJ (1988) Identification of a calmodulin stimulated (Ca^{2+} + Mg^{2+}) ATPase in a plasma membrane fraction isolated from maize (*Zea mays*). Physiol Plant 72:177–184

Roh MH, Shingles R, Cleveland MJ, McCarty RE (1998) Direct measurement of calcium transport across chloroplast inner-envelope vesicles. Plant Physiol 118:1447–1454

Romani G, Bonza MC, Filippini I, Cerana M, Beffagna N, De Michelis MI (2004) Involvement of the plasma membrane Ca^{2+}-ATPase in the short-term response of *Arabidopsis thaliana* cultured cells to oligogalacturonides. Plant Biol 6:192–200

Sanders D, Pelloux J, Brownlee C, Harper JF (2002) Calcium at the crossroads of signaling. Plant Cell 14:S401–S417

Schiøtt M, Palmgren MG (2005) Two plant Ca^{2+} pumps expressed in stomatal guard cells show opposite expression patterns during cold stress. Physiol Plant 124:278–283

Schiøtt M, Romanowsky SM, Bækgaard L, Jakobsen MK, Palmgren MG, Harper JF (2004) A plant plasma membrane Ca^{2+} pump is required for normal pollen tube growth and fertilization. Proc Natl Acad Sci USA 101:9502–9507

Schumaker KS, Sze H (1985) A Ca^{2+}/H^+ antiport system driven by the proton electrochemical gradient of a tonoplast H^+-ATPase from oat roots. Plant Physiol 79:1111–1117

Schumaker KS, Sze H (1990) Solubilization and reconstitution of the oat root vacuolar H^+/Ca^{2+} exchanger. Plant Physiol 92:340–345

Segarra VA, Thomas L (2008) Topology mapping of the vacuolar Vcx1p Ca^{2+}/H^+ exchanger from *Saccharomyces cerevisiae*. Biochem J 414:133–141

Shigaki T, Hirschi KD (2006) Diverse functions and molecular properties emerging for CAX cation/H^+ exchangers in plants. Plant Biol 8:419–429

Shigaki T, Cheng NH, Pittman JK, Hirschi K (2001) Structural determinants of Ca^{2+} transport in the *Arabidopsis* H^+/Ca^{2+} antiporter CAX1. J Biol Chem 276:43152–43159

Shigaki T, Barkla BJ, Miranda-Vergara MC, Zhao J, Pantoja O, Hirschi KD (2005) Identification of a crucial histidine involved in metal transport activity in the Arabidopsis cation/H^+ exchanger CAX1. J Biol Chem 280:30136–30142

Shigaki T, Rees I, Nakhleh L, Hirschi KD (2006) Identification of three distinct phylogenetic groups of CAX cation/proton antiporters. J Mol Evol 63:815–825

Snedden WA, Fromm H (1998) Calmodulin, calmodulin-related proteins and plant responses to the environment. Trends Plant Sci 3:299–304

Subbaiah CC, Sachs MM (2000) Maize *cap1* encodes a novel SERCA-type calcium-ATPase with a calmodulin-binding domain. J Biol Chem 275:21678–21687

Sze H, Liang F, Hwang I, Curran AC, Harper JF (2000) Diversity and regulation of plant Ca^{2+} pumps: Insights from expression in yeast. Annu Rev Plant Physiol Plant Mol Biol 51:433–462

Ueoka-Nakanishi H, Nakanishi Y, Tanaka Y, Maeshima M (1999) Properties and molecular cloning of Ca^{2+}/H^{+} antiporter in the vacuolar membrane of mung bean. Eur J Biochem 262:417–425
Ueoka-Nakanishi H, Tsuchiya T, Sasaki M, Nakanishi Y, Cunningham KW, Maeshima M (2000) Functional expression of mung bean Ca^{2+}/H^{+} antiporter in yeast and its intracellular localization in the hypocotyl and tobacco cells. Eur J Biochem 267:3090–3098
White PJ, Broadley MR (2003) Calcium in plants. Ann Bot 92:487–511
Williams LE, Schueler SB, Briskin DP (1990) Further characterization of the red beet plasma-membrane Ca^{2+}-ATPase using GTP as an alternative substrate. Plant Physiol 92:747–754
Wu ZY, Liang F, Hong BM, Young JC, Sussman MR, Harper JF, Sze H (2002) An endoplasmic reticulum-bound Ca^{2+}/Mn^{2+} pump, ECA1, supports plant growth and confers tolerance to Mn^{2+} stress. Plant Physiol 130:128–137
Yamniuk AP, Vogel HJ (2004) Structurally homologous binding of plant calmodulin isoforms to the calmodulin-binding domain of vacuolar calcium-ATPase. J Biol Chem 279:7698–7707
Zhang L, Lavery L, Gill U, Gill K, Steffenson B, Yan GP, Chen XM, Kleinhofs A (2009) A cation/proton-exchanging protein is a candidate for the barley *NecS1* gene controlling necrosis and enhanced defense response to stem rust. Theor Appl Genet 118:385–397
Zhao J, Barkla BJ, Marshall J, Pittman JK, Hirschi KD (2008) The *Arabidopsis cax3* mutants display altered salt tolerance, pH sensitivity and reduced plasma membrane H^{+}-ATPase activity. Planta 227:659–669
Zhao J, Connorton JM, Guo YQ, Li XK, Shigaki T, Hirschi KD, Pittman JK (2009a) Functional Studies of Split *Arabidopsis* Ca^{2+}/H^{+} Exchangers. J Biol Chem 284:34075–34083
Zhao J, Shigaki T, Mei H, Guo YQ, Cheng NH, Hirschi KD (2009b) Interaction between Arabidopsis Ca^{2+}/H^{+} Exchangers CAX1 and CAX3. J Biol Chem 284:4605–4615
Zhu JK (2003) Regulation of ion homeostasis under salt stress. Curr Opin Plant Biol 6:441–445
Zielinski RE (1998) Calmodulin and calmodulin-binding proteins in plants. Annu Rev Plant Physiol Plant Mol Biol 49:697–725

Part III
Nutrient Transport

Nitrate Transporters and Root Architecture

Nick Chapman and Tony Miller

Abstract Nitrogen (N) is one of the most important limiting factors for plant growth and crop production. The root is the most important organ for acquring soil N that is available as NO_3^-, NH_4^+ or amino acids. Soil NO_3^- availability to roots is transient and the concentrations of NO_3^- can rapidly change in response to climatic factors. Stable soil surface aggregates facilitate a network of continuous and connected pores that can positively affect water flow to the root, and thus the delivery of dissolved NO_3^-. Within the root, NO_3^- uptake and transport are realised by NO_3^- transporters (NRTs). Uniquely, NRT1.1 is capable of functioning in both high- and low-affinity uptake and possesses an NO_3^- sensing and signaling capability, regulating other key players in NO_3^- uptake, transport and signaling. NRT expression and function are regulated by plant N status and can directly influence the root system architecture, due in part to an overlap with the developmentally important hormones auxin, ethylene, cytokinin and abscisic acid.

1 Introduction

Plants obtain the majority of their essential nutrient ions from the soil. These ions not only serve to promote healthy growth but also act as signaling molecules that regulate vital developmental processes. However, soils exhibit spatial and temporal variation in the availability of nutrient ions and are thus described as heterogeneous. Sessile by nature, plants need to sense changes (both local and bulk) in essential nutrient ions within their growth environment and respond by targeting specific changes in morphology and metabolism to maintain growth.

N. Chapman and T. Miller (✉)
Rothamsted Research, Harpenden, Hertfordshire AL5 2JQ, UK
e-mail: tony.miller@bbsrc.ac.uk

M. Geisler and K. Venema (eds.), *Transporters and Pumps in Plant Signaling*,
Signaling and Communication in Plants 7,
DOI 10.1007/978-3-642-14369-4_6,

Nitrogen (N) is the most important limiting factor for plant growth and crop production after light and water. The soil exhibits spatial and temporal heterogeneity in terms of N availability for plants and constant cycling between the different forms occurs (Fig. 1). In temperate climates, nitrate (NO_3^-) is the primary source of N due to it being more mobile in the soil than ammonium (NH_4^+) and will be the focus for this chapter. Within bulk soil, transpiration drives the delivery of water and dissolved nutrient ions to the root surface. NO_3^- availability to roots is therefore transient and the concentrations of NO_3^- in the soil can rapidly change in response to climatic factors, such as temperature and pH (Miller et al. 2007a, b). Stable soil surface aggregates facilitate a network of continuous and connected pores which can positively affect water flow to the root.

Much of our understanding of NO_3^- uptake, transport and sensing by higher plants has been accrued using the *Arabidopsis thaliana* model which has a short generation time, small size and modest growth requirements. The complete sequencing of its genome by the end of 2000 (119 Mb on 5 chromosomes) has provided a significant tool for the investigation of numerous physiological, biochemical, morphological and genetic processes involved in the development of higher plants (Rensink and Buell 2004; Feng and Mundy 2006). Most of this chapter will focus on evidence from this plant.

It is the aim of this chapter to describe the NO_3^- transporters (NRTs) and their influence on root system architecture (RSA). After briefly outlining the importance of NO_3^- at the whole plant level, the focus will turn to the root as the main organ for plant NO_3^- acquisition and how the RSA is important to function, and how this in

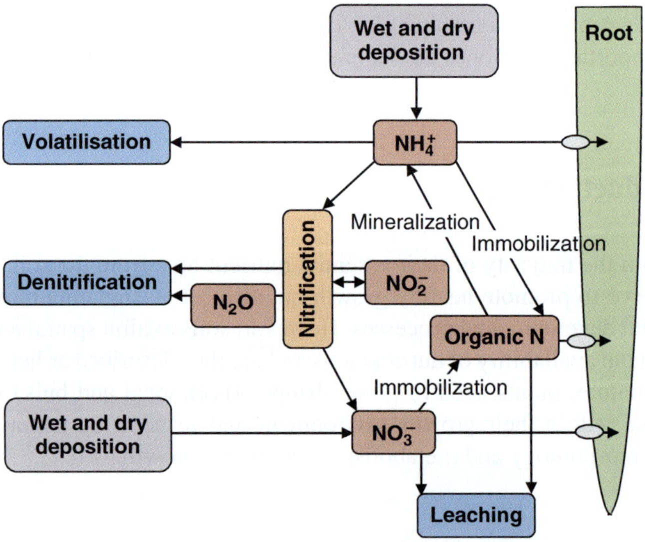

Fig. 1 A schematic representation of the cycling between the main N pools (*boxes*) and fluxes (*arrows*) within terrestrial ecosystems. N_2 fixation and animal input are not included here. Reproduced from Miller and Cramer (2005)

turn is regulated. The important NRT families will be illustrated and the mechanisms by which they are regulated will be discussed. The capability of certain NRTs to sense NO_3^- availability and signal to activate uptake will be described before we draw conclusions on the understanding of NO_3^- and RSA. Finally, we will provide an outlook on future research.

2 NO_3^- at the Whole Plant Level

Plants need N as a basic building block, therefore monitoring and adjusting uptake, storage and efflux is pivotal to survival and growth. Once uptake has occurred, NO_3^- undergoes several reductive steps within the cell, ultimately producing amino acids (Fig. 2). It is these amino acids which are used as the basic molecules for growth and development and these molecules may mediate feedback signaling but this will be discussed later.

The plant regulates how much N is taken up, stored, metabolised or lost. The balance between these processes determines the plant N status and this is important for controlling growth and development. The N status of the whole plant is indicated by the amount of stored N and this is reflected by the tissue NO_3^- concentration: a measure of what is accumulated in the cell vacuole. This chapter will focus on NO_3^- uptake and the root as the main organ for acquisition, but NO_3^- transport and assimilation in other tissues can influence uptake.

NRTs are combined with small peptide transporters (PTRs) in higher plants to form the NRT1/PTR family of transporters. NO_3^- uptake from the soil is achieved

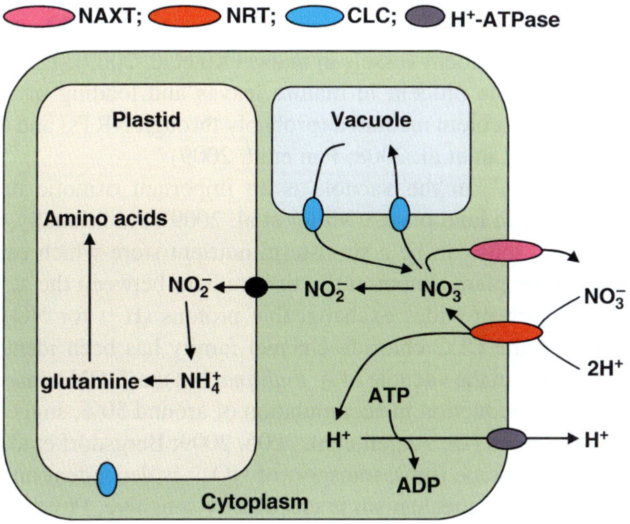

Fig. 2 Schematic representation of NO_3^- uptake and assimilation by a plant cell. Reproduced from Miller and Cramer (2005)

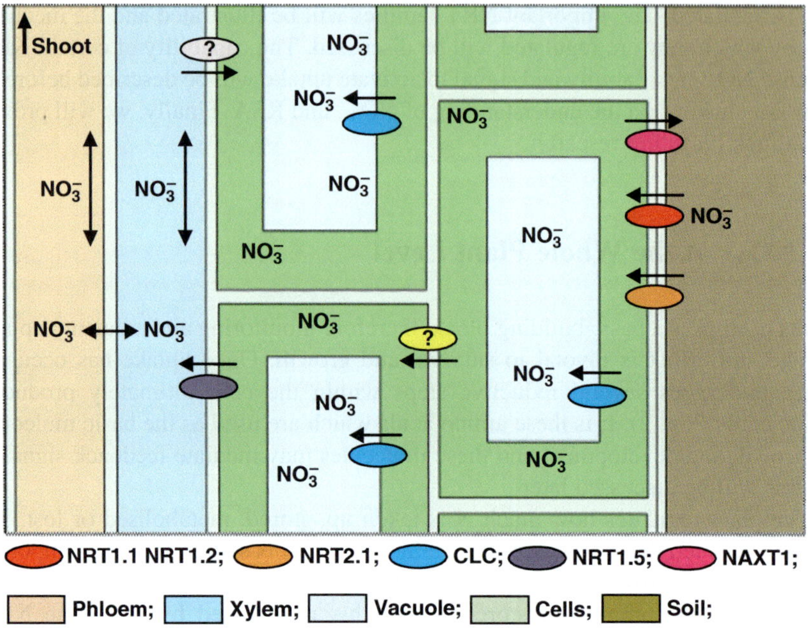

Fig. 3 A schematic representation of NO_3^- uptake, transport and efflux in the root

by NRTs in the root (Fig. 3) but N is needed in aerial tissues to support growth and development. The mechanism of N translocation from the root to the shoot has not been fully elucidated, but it is clear that the xylem and phloem vessels play a key role. Changes in root morphology correlate with NO_3^- levels within the xylem and NO_3^- translocation via phloem vessels in maize (Xu et al. 2009). There is evidence that NO_3^- loading of the phloem in mature leaves and loading of the xylem in mature roots are partly protein mediated (probably through NRTs) and are therefore likely to be regulated (Lin et al. 2008; Fan et al. 2009).

The storage of NO_3^- in the vacuole is an important osmotic driver of cell expansion and growth in land plants (Miller et al. 2009). Additionally, this internal localisation of NO_3^- ions can be a short-term nutrient store which can be readily remobilised within the plant. Importantly, the gradient between the acidic vacuole and alkaline cytoplasm provides exchangeable protons (H^+) for NO_3^- transport. A tonoplast NRT of the CLC chloride channel family has been identified for the accumulation of NO_3^- in the vacuole of *A. thaliana*, but the T-DNA insertion mutant *clca-1* only exhibits a reduction in accumulation of around 50%, suggesting further mechanisms are involved (De Angeli et al. 2006, 2009; Bergsdorf et al. 2009).

At the plasma membrane (pm), transport of NO_3^- is dependent on co-transport with two H^+ and thus the mechanism is electrically sensitive. Physical parameters such as cytosolic pH and membrane potential of the cell can also alter NO_3^- uptake (Miller and Smith 2008). At the cell membrane, a decreased electrical potential results in less available energy for transport, but also has been shown to alter the affinity of a transport protein for NO_3^- (Zhou et al. 1998).

While more is known about influx, pm NO_3^- efflux has also been found to be protein mediated, selective, passive and saturable. The efflux systems at both the tonoplast and the pm are regulated, proportional to whole tissue NO_3^- concentrations and NO_3^- inducible (Teyker et al. 1988; van der Leij et al. 1998). At the cortex of mature roots, pm efflux is achieved by NAXT1. This NRT1/PTR family member was identified using mass spectrometry and confirmed with a GFP reporter line (Segonzac et al. 2007).

To maintain the N status of the plant, NO_3^- transport throughout the growing plant must be regulated. In order to obtain NO_3^-, the root must find and take up NO_3^- from the environment within which it is growing. The RSA is modified in response to NO_3^- availability and it is the root that will now be the focus for the remainder of this chapter.

3 Physiology of the Root and NO_3^-

3.1 Root Development

Three major processes regulate RSA. First, cell division of initial cells at the primary root (PR) meristem enables indeterminate growth by adding new cells to the root. Second, lateral root (LR) formation increases the exploratory capacity of the root system. Third, root hair formation increases the total surface of each root component (Lopez-Bucio et al. 2003).

The mature PR comprises a single concentric layer of cells, surrounding the central vascular tissue. Root growth requires cell division in the root meristem, just behind the tip, and expansion in a zone just behind the apex. Turgor pressure from water influx into root cells provides the driving force for growth (Clark et al. 2003). The anticlinal division following the elongation of initial cells sustains each cell file. Daughter cells of the initials divide and differentiate into the specific root tissues.

The tips of each root possess a root cap (RC) which protects the root during penetration of the growth substrate (Driouich et al. 2007; Iijima et al. 2008). Root border cells are dead, detached cells from the RC found in the rhizosphere which reduce the frictional resistance of the soil to root penetration (Somasundaram et al. 2008). Their production and separation is regulated by phytohormones and environmental factors (Driouich et al. 2007).

Fine projections originating from epidermal cells, root hairs make up to 77% of the total root surface area of cultivated crops, and function in the uptake of nutrients and water, whilst aiding the anchorage of the plant in the soil (Bibikova and Gilroy 2002; Singh et al. 2008). It is widely accepted that the function of root hairs is to increase root surface area, enhancing uptake of plant growth resources, including

NO_3^-, from the soil. However, the soil water potential between root hairs has been shown to quickly reach that of the root (Segal et al. 2008). Thus, the tangential growth of root hairs serves to increase the diameter of the cylinder that is characterised by the root water potential in the most energy-efficient manner, augmenting the effective surface area of the root for the uptake of water and thus delivery of NO_3^- to the root surface (Wang et al. 2006; Segal et al. 2008). Root hairs are advantageous in the scavenging of less mobile nutrients, such as phosphate and ammonium, as they are able to penetrate soil particles and explore small pores within the soil.

3.2 The Influence of NO_3^- on RSA

In numerous plant species, LR development and elongation has been shown to be stimulated by local NO_3^- application (Drew and Saker 1975; Visser et al. 2008). In *A. thaliana*, vertical agar experiments demonstrated an increased LR density in response to high-NO_3^- patches (Zhang and Forde 1998). Conversely, seedlings grown on globally high-NO_3^- substrate exhibit suppressed LR development (Zhang et al. 1999). This inhibition occurs after emergence of the LR but before meristem activation, thus the elongation of older LRs is unaffected (Zhang and Forde 1998). These results indicate that the systemic status of the seedling is more important in determining root structure because in both experiments seedlings encountered locally high NO_3^- concentrations but produced different local root responses. Indeed, high shoot N status was implicated by the use of *A. thaliana* mutants that were defective in NO_3^- reductase activity. These plants cannot use NO_3^- as an N source, but still exhibited a local morphological response to high NO_3^- concentrations (Zhang and Forde 1998; Zhang et al. 1999).

PR length is known to increase with low bulk NO_3^- concentrations and decrease in high-NO_3^- supply (Linkohr et al. 2002). This represents a foraging response of the plant to search out NO_3^- within its growth environment (reviewed by De Kroon et al. 2009). PR growth is strongly inhibited when glutamate is sensed by the root tip. When wild type *A. thaliana* plants are supplied with NO_3^- and glutamate, the inhibitory effect is overridden (Walch-Liu and Forde 2008). Secondary root development in response to NO_3^- supply has not been studied.

3.3 Root Water Influences NO_3^- Uptake

The influence of local NO_3^- availability on root hydraulic properties may also adjust patterns of water uptake. As a function of this, net NO_3^- uptake increases likely due to increased delivery of dissolved NO_3^- to the root surface. Rapid localised changes in membrane hydraulic conductance have been observed, facilitated by aquaporin rearrangement, in response to locally high NO_3^- concentration

(Gorska et al. 2008). Although there is evidence that certain aquaporins demonstrate transport of uncharged NH_3 (Loque et al. 2005), it seems unlikely that the NO_3^- will pass through water channels. Although changing only a single amino acid residue within a mammalian aquaporin was shown to switch function to an anion channel (Liu et al. 2005); this topic is worthy of further investigation for the plant aquaporins and NRTs. Intriguingly, water transport has been demonstrated through some mammalian sodium cotransporters (Zeuthen et al. 1997) and this may be worthy of further investigation for plant NRTs. The NO_3^- and water response pathways could be integrated early during development of the plant (Deak and Malamy 2005). However, this will be discussed in more detail elsewhere within this book (see Chapter "Plant Aquaporins").

These targeted root behavioural responses to NO_3^- supply require the plant to monitor internal NO_3^- status, sense external NO_3^- availability and alter the RSA accordingly. The underlying signaling mechanisms for each of these processes, the components of which they are comprised, and their regulation, will be the focus for the remainder of this chapter.

4 Achieving Uptake: NRTs

Within the root, NO_3^- uptake and transport are realised by NRTs. The transport function of these membrane proteins are well known, but the underlying signaling networks have only recently been explained. This section will discuss the important NRT families.

4.1 High- and Low-affinity Transport Systems

Investigations into the influence of NO_3^- supply on plant physiology concluded that plants have developed three NO_3^- transport systems to cope with heterogeneous supply in the field (Crawford and Glass 1998). When NO_3^- is available at low concentrations (below 1 mM), uptake is achieved via two saturable high-affinity transport systems (HATS). The constitutive system (cHATS) is available when plants have been previously starved of NO_3^-, whereas the inducible system (iHATS) is stimulated by the presence of NO_3^-. The low-affinity transport system (LATS) achieves NO_3^- uptake at external NO_3^- concentrations above 1 mM (Crawford and Glass 1998). However, both types of HATS can contribute to NO_3^- uptake above 1 mM.

Historically, it was considered that the NRTs should be assigned to each of the transport systems based on functional characterisation of their uptake activity. However, it has now become clear that NRT function is more complex. Thus, this section will make reference to the HATS and LATS when describing the NRTs, but greater emphasis will be placed on gene family and complete functionality rather than simply uptake capability.

4.2 NRT1s

This family is believed to have important functionality in higher plants and this is reflected in the large numbers of NRT1 genes found in *A. thaliana* (53) and rice (80; Tsay et al. 2007). NRT1 proteins are comprised of 12 putative transmembrane-spanning domains with a large hydrophilic loop between the 6th and 7th transmembrane regions. This feature is consistent across all, and unique to, higher plants. The NRT1 family encompasses amino acid and PTRs and hence should be more correctly termed the NRT1/PTR family (Forde 2000; Orsel et al. 2002a, b).

Closely related members of the NRT1/PTR family have evolved distinct functions in plants. The analysis of *atptr* mutants has proved to be a powerful tool in understanding the function of these PTRs. By studying N levels in mutants supplied with dipeptides as a sole N source, AtPTR1 was shown to function in the uptake of peptides within the root. Using a similar approach, AtPTR5 was shown to demonstrate peptide transport activity from germinating pollen to seed development (Komarova et al. 2008). Interestingly, NRT1.1 has also been implicated in the germination of dormant seeds in response to N supply (Alboresi et al. 2005), suggesting that some members could have multiple and overlapping functions.

In 1978, *CHL1* (now known as *AtNRT1.1*) was the first *NRT1/PTR* gene to be identified in plants and confirmed by the use of a transferred DNA-tagged *A. thaliana* mutant (Doddema et al. 1978). The gene was shown to encode a proton-coupled NRT in the *Xenopus laevis* oocyte expression system (Tsay et al. 1993). Furthermore, AtNRT1.1 was found to possess a high- and low-affinity transport phase for the uptake of NO_3^- indicating that NRT1.1 is a dual-affinity NRT (Liu et al. 1999). The affinity of AtNRT1.1 to NO_3^- is regulated by the phosphorylation of a threonine residue (Liu and Tsay 2003; Tsay et al. 2007). The phosphorylated AtNRT1.1 functions as a high-affinity NRT, and the dephosphorylated AtNRT1.1 as a low-affinity transporter. Plants with defective *AtNRT1.1* expression have been shown to exhibit reduced response to NO_3^- patches (Remans et al. 2006a). In addition to NO_3^- uptake and a potential signaling role for NRT1.1, promoter-tagged fluorescent protein lines and immunolocalisation studies have been used to demonstrate the functional expression of AtNRT1.1 in guard cells. Mutant lines grown in the presence of NO_3^- exhibit reduced stomatal opening and are thus more drought tolerant compared to wild type. Therefore, NRT1.1 is required for the correct opening of the stomata and implying a key role for NO_3^- transport in guard cell function (Guo et al. 2003). This adds further weight to the overlap of the NO_3^- and water signaling pathways.

NRT1.1 demonstrates high-affinity NO_3^- transport when expressed in yeast (Martin et al. 2008). The *Brassica napus* transporter *BnNRT1-2* and *Os08g05910* and *Os10g40600* from rice have been suggested as orthologues of *AtNRT1.1* (Chen et al. 2008). Further research into their function is required to determine if these orthologues also demonstrate dual-affinity NO_3^- uptake. A further 11 AtNRT1s were studied but all exhibited low-affinity NO_3^- transport activity (Huang et al. 1999; Chiu et al. 2004; Almagro et al. 2008; Lin et al. 2008; Fan et al. 2009). The constitutively expressed AtNRT1.2 is located in the epidermis and functions in the cLATS (Huang et al. 1999). The rice equivalent to AtNRT1.2, OsNRT1 is also a root-epidermal low-affinity NRT (Lin et al. 2000). In the leaf petiole, AtNRT1.4 achieves low-affinity NO_3^- uptake (Chiu et al. 2004). A study using *X. laevis* oocytes demonstrated that NRT1.5 is a low-affinity, pH-dependent, bidirectional NRT. Localised to the plasma membrane of root pericycle cells in proximity to xylem vessels, NRT1.5 has been postulated to function in the xylem loading of NO_3^- (Lin et al. 2008). Interestingly, root-to-shoot transport of NO_3^- was not completely lost in the knock-out mutant suggesting there is an alternative mechanism involved in the xylem loading of NO_3^-. A similar approach determined that NRT1.6 is a low-affinity transporter that lacks the capacity to transport dipeptides and functions in the delivery of NO_3^- from maternal tissues to the developing embryo (Almagro et al. 2008). In aerial tissues, the NO_3^- transporter NRT1.7 is positioned in the phloem of the leaf minor vein and functions to transport NO_3^- from older leaves to younger ones. This led to the idea that NO_3^- itself can be remobilised, via the phloem, and this remobilisation is important to sustain growth (Fan et al. 2009).

4.3 NRT2s

There are seven *NRT2* genes in Arabidopsis. Located adjacently within one chromosome, *AtNRT2.1* and *AtNRT2.2* are involved solely in the HATS (Cerezo et al. 2001; Remans et al. 2006b; Chen et al. 2008). *AtNRT2.1* was shown to have a more important role in iHATS due to reduced iHATS expression in *nrt2.1* and *nrt2.2* mutant lines of 50–72% and 19%, respectively (Li et al. 2007). Both *AtNRT2.1* and *AtNRT2.2* are inducible and influence the RSA via NO_3^- uptake and sensing, although the effect of *AtNRT2.1* can be modified by exogenous sucrose application and light exposure (Lejay et al. 1999; Desnos 2008; Vidal and Gutierrez 2008).

AtNRT2.1 is located at the plasma membrane of root cortical and epidermal cells (Krapp et al. 1998; Chopin et al. 2007b). Whilst the monomeric form of the protein is involved in NO_3^- transport and is the most abundant form, there are other truncated forms that co-exist at the cell membrane (Wirth et al. 2007), but the functional activity of these forms in unknown. Functional NO_3^- transport requires the NAR2.1 protein to be expressed for plasma membrane targeting of the NRT2.1 monomer (see Sect. 4.4). An AtNRT2.1 orthologue is known to function in the HATS of wheat and the mRNA accumulates in the root (Yin et al. 2007). The NO_3^--induced TaNRT2 is located in the root and induced in response to both low and high concentrations of NO_3^-. Transcripts were undetected in plants grown under N limiting conditions or where NH_4^+ was the sole N source (Zhao et al. 2004).

Much evidence exists for the implication of *AtNRT2.1* in a NO_3^- transport-independent sensing role in LR initiation (Crawford and Glass 1998; Forde 2000;

Cerezo et al. 2001; Orsel et al. 2002a, b; Little et al. 2005; Remans et al. 2006a, b; Wirth et al. 2007). The study of *lin1*, an *AtNRT2.1* mutant line, suggested that AtNRT2.1 acts as a NO_3^- sensor or signal transducer. In the wild type, a high sucrose:NO_3^- ratio represses LR initiation, but the repression is removed in *lin1*. Indeed, this response of the lin1 mutant is observed in media without NO_3^- illustrating that this phenotype is independent of NO_3^- and indicative of a NO_3^- sensor or signal transducer function for *AtNRT2.1* (Little et al. 2005). The *lin1* mutant was selected from ethyl methanesulfonate-mutagenised seed and had a mutation in a single glycine residue that is likely to compromise the NO_3^- transport function (Little et al. 2005). Conversely, a different study demonstrated that the *atnrt2.1* gene knock-out mutant exhibited the opposite phenotype to *lin1*, a reduced LR initiation (Remans et al. 2006a, b). While this difference could be partly explained by differences in the growth conditions between studies, the results mean that the role of AtNRT2.1 in sensing NO_3^- remains elusive.

High expression levels of *AtNRT2.7* have been detected in reproductive organs and seeds, and it is believed that this vacuolar membrane transporter plays a role in seed NO_3^- accumulation (Chopin et al. 2007a). The vacuolar location is likely to be important to function as the CLC transporters are also implicated in vacuolar accumulation of NO_3^-. Interestingly, NRT1.1 has been implicated in the germination of dormant seeds and is known to regulate NRT2.1 expression. Therefore, it is not inconceivable to believe that NRT1.1 may regulate other NRT2 members. It is possible to postulate a regulatory interaction between NRT1.1 and NRT2.7, perhaps in response to the accumulation of NO_3^- to a certain threshold level, beyond which accumulation ceases and germination is initiated. This would imply a further sensing role for NRT1.1 itself or an intermediate molecule.

The *A. thaliana* mutant *atnrt2.1-1* possesses deletions of both NRT2.1 and NRT2.2 genes and exhibits suppression in up-regulation of the NO_3^- HATS in response to N starvation. This mutant was used to investigate RSA in response to low NO_3^- availability. An increase in the number of visible LRs was reported when wild type plants were transferred from 10 to 1/0.5 mM NO_3^- supply, whilst mean LR length increased when plants were transferred from 10 to 0.1/0.05 mM NO_3^-. The response of *atnrt2.1-1* to moderate NO_3^- limitation produced a RSA similar to the wild type response to severe NO_3^- stress. Indeed, this RSA response could reflect the reduced NO_3^- uptake measured in the mutant line, suggesting that uptake rate of NO_3^- could be more important than external NO_3^- concentration in influencing RSA. However, the *nrt2.1* mutant exhibited inhibited LR initiation in response to N limitation independent of the NO_3^- uptake and the inhibition persisted even when NO_3^- was added to the external medium. This is indicative of a direct stimulatory role for NRT2.1 in LR initiation and suggests that uptake alone is not responsible for RSA responses to N limitation (Orsel et al. 2004; Little et al. 2005; Remans et al. 2006a, b).

AtNRT2.1 expression rapidly increases during early vegetative growth, peaking just before floral stem emergence and decreases to minimal levels in flowering and silique-bearing plants. A series of experiments with altered N supply and source found that NO_3^- induced NRT2.1 expression, but amino acids (specifically glutamine) repressed expression. This provides evidence for a signaling role for glutamine regarding the regulation of NO_3^- uptake (Nazoa et al. 2003). In the same study, young roots did not demonstrate NRT2.1 expression despite exhibiting a similar rate of NO_3^- influx to older roots, suggesting that another high-affinity transporter functions in root tips (Nazoa et al. 2003). Indeed, NRT1.1 has been implicated in glutamine signaling at the root tip and is known to function in both HATS and LATS (Walch-Liu and Forde 2008).

4.4 NAR2s (NRT3)

The NAR2 (NRT3) proteins are required for *NRT2.1* function (Orsel et al. 2006; Chen et al. 2008). With just a single putative transmembrane spanning domain, NAR2 proteins seem to be necessary for targeting some NRT2 proteins to the plasma membrane. Indeed, NO_3^- elicited currents are only observed in *X. laevis* oocytes injected with both *CrNAR2* and *CrNRT2.1* mRNA, but not when injected with just one (Zhou et al. 2000). Later, the yeast split-ubiquitin system was used to confirm a direct interaction between the two proteins (Orsel et al. 2006). However, not all NAR2 proteins can form functional interactions with NRT2 proteins. For example, in barley, only HvNAR2.3 can generate an operational unit with HvNRT2.1 (Tong et al. 2005; Chen et al. 2008).

A. thaliana has two *NAR2* genes: *AtNAR2.1* (*AtNRT3.1*) and *AtNAR2.2* (AtNRT3.2; Chen et al. 2008). The former has been identified as important in the HATS (Okamoto et al. 2006; Orsel et al. 2006; Wirth et al. 2007). The *nar2.1* null mutant shows an extensive reduction of the HATS. Interestingly, the expression of cHATS in the *nrt2.1 nrt2.2* double mutant was only reduced by approximately a third of the reduction observed in the *nar2.1* null, suggesting that a further unidentified NRT2 is involved in the cHATS (Li et al. 2007).

5 Molecular Regulation of Nitrate Transporters

When considering the whole plant, net NO_3^- uptake is regulated by demand and the various uptake systems are induced by the presence of NO_3^- (Crawford and Glass 1998; Daniel-Vedele et al. 1998). However, the uptake systems can be negatively regulated by assimilatory products providing a mechanism for regulating net NO_3^- uptake related to whole plant N status (Muller and Touraine 1992). The levels of regulation will now be discussed.

5.1 Gene Expression

The regulatory mechanisms of long-distance NO_3^- transport within the plant remains largely unknown. However, the inducible HATS is feedback regulated relative to the plant demand for NO_3^-, and transcription of the *NRT* genes is feedback repressed by the secondary products of NO_3^- metabolism (Chen et al. 2008; Vidal and Gutierrez 2008).

AtNRT2.1 regulation has been comprehensively studied at the mRNA level. Up-regulation of *AtNRT2.1* expression at every level, from transport activity to promoter activation, is induced by NO_3^- itself and repressed by downstream N metabolites (Loque et al. 2003). AtNRT2.1 transcript accumulates at the epidermis and cortex of mature roots (Nazoa et al. 2003) and is greatly affected by several environmental factors. The expression of AtNRT2.1 is induced by NO_3^-, down-regulated by high N status via downstream N metabolites such as NH_4^+ and amino acids and positively regulated by sugars and light (Lejay et al. 1999; Zhou et al. 2000; Nazoa et al. 2003). The positive regulation of NRT2.1 by light is achieved indirectly via the reduced repression of NRT2.1 by NRT1.1, which is mediated by *LONG HYPOCOTYL5* (*HY5*) and *HY5 HOMOLOGUE* (HYH; Jonassen et al. 2008). NRT2.1 transcript levels positively correlate with NO_3^- HATS activity, suggesting that high-affinity NO_3^- uptake is affected by the transcriptional regulation of NRT2.1. As already stated, AtNAR2.1 expression closely parallels that of AtNRT2.1 and it too is repressed by negative feedback involving N metabolites (Krouk et al. 2006). NRT2.1 expression is up-regulated by NO_3^- starvation in wild type plants and by N limitation in a NO_3^- reductase-deficient mutant when grown on NO_3^- as the sole N source. Thus, NRT2.1 is feedback repressed by the downstream N metabolites of NO_3^- reduction. This is not the case for NRT1.1, which is not likely to be regulated by the presence of NO_3^- reductase, but by the N status of the plant (Lejay et al. 1999).

For *AtNRT2.1*, gene expression is regulated by a *cis*-acting 150bp element upstream of the promoter TATA box which is able to confer regulation to a minimal promoter (Girin et al. 2007). Split-root experiments demonstrate that NO_3^- activation occurs locally while metabolite-mediated repression is a function of whole plant N status. Even sucrose regulation of *NRT2.1* is mediated by this element, implying a potential interaction between N and C signaling and indeed several motifs have been identified within the region that correspond to the regulation of N and C status (Girin et al. 2007). Using a novel systems biology approach, a sub-network of genes regulated by the downstream metabolites of N was identified, with the master clock control gene *CCA1* involved in the regulation of central N metabolism enzymes (Gutierrez et al. 2007). This suggests that N signaling could influence the endogenous clock of the plant, representing a complex coordination of gene expression.

5.2 Post-translational Regulation of NRTs

The phosphorylation of AtNRT1.1 is controlled by the plant in response to changes in external NO_3^- concentrations encountered by the root (Liu and Tsay 2003). The ability of AtNRT1.1 to switch from a high-affinity to a low-affinity transporter is due to the dephosphorylation of the T101 residue. By using an uptake- and sensing-decoupled mutant, NRT1.1 has been shown to function as a NO_3^- sensor. A low-level primary response to NO_3^- is maintained in NRT1.1 via the phosphorylation of T101 by CIPK23 enabling NRT1.1 to sense a wide range of NO_3^- concentrations (Ho et al. 2009). Several NRT and NO_3^--regulated genes require the CBL-interacting protein kinase CIPK8 and a NIN-like protein NLP7 for complete induction by NO_3^-. Of particular importance, the CIPKs are likely to partake in a range of molecular responses to NO_3^-. Indeed, CIPK8 has been shown to transduce the NO_3^- signal in the LATS (Hu et al. 2009).

Whilst phosphorylation events are known to regulate activity of the dual-affinity transporter AtNRT1.1 in response to environmental cues, a similar mechanism has been suggested for the regulation of NRT2.1 (Liu and Tsay 2003). Indeed, a number of conserved protein kinase C recognition motifs are observed in the N- and C-terminal domains of HvNRT2.1 (Forde 2000). The presence of several different forms of the AtNRT2.1 protein in the plasma membrane, all of which are likely to rely on post-translational modification in response to environmental cues, suggest that each could have a specific function (Wirth et al. 2007). Interestingly, no rapid changes in abundance of AtNRT2.1 are detected when the plant is subjected to light, sucrose or N treatments that are known to strongly affect NRT2.1 transcript level and HATS activity (Wirth et al. 2007). Thus, it is likely that post-translational modification generates the different forms of NRT2.1 observed at the plasma membrane.

Sequence analysis of the NRT2s has identified some possible 14-3-3 regulatory sites; this is particularly interesting because of the role of these proteins in regulation of key N assimilatory enzymes. For example, the C terminus of the tobacco NRT2.1 gene has a perfect 14-3-3-binding consensus (Miller et al. 2007a, b).

5.3 *Overlap with Hormones*

RSA is in part regulated by plant hormones (reviewed by Rubio et al. 2009). Changes in the tissue concentrations and transport pattern of these hormones can modify root morphology, often in response to nutrient treatments (reviewed by Casson and Lindsey 2003; and Fukaki and Tasaka 2009). More recently, several studies using a systems approach has reproducibly identified gene expression patterns in response to NO_3^- and hormone signaling (Vidal and Gutierrez 2008; Krouk et al. 2009; Nero et al. 2009).

The auxin transport genes of the *PIN-FORMED* (*PIN*) family regulate RSA by locating the auxin to sites of growth. The establishment of auxin maxima at sites of cell division help to drive root growth. Aside from the developmentally important auxin and ethylene, abscisic acid (ABA) has been implicated in responses to water stress, mechanical impedance and LR formation in response to NO_3^- (Signora et al. 2001). Cytokinins (CKs) are known to interact with auxin to influence development of the RSA and this is thought to be regulated by nutrient uptake (Takei et al. 2001).

The complex relationship between hormones and root nutrient responses remains unresolved (Linkohr et al. 2002). An overlap between NO_3^- and auxin response pathways has been suggested, due to evidence obtained from auxin-resistant mutants (Zhang and Forde 1998; Zhang et al. 1999). However, a direct effect of auxin in root response to NO_3^- treatments was criticised (Linkohr et al. 2002; Bao et al. 2007). Indeed, auxin transport mutants displayed similar RSA phenotypes to the wild type when subjected to certain NO_3^- treatments. Although the effect of NO_3^- on LR primordia development in *A. thaliana* could not be directly attributed to localised auxin levels (Bao et al. 2007), a decrease in root auxin concentration has been considered to be important in the response of the inhibitory effect of high NO_3^- on root growth in maize (Tian et al. 2008). Furthermore, AtNRT1.1 has been shown to be regulated by auxin in *A. thaliana* root tips, where glutamate inhibition of PR growth occurs (Guo et al. 2002).

In addition, a recent study used Fluorescence-assisted Cell Sorting to characterise the transcriptome responses of five *A. thaliana* cell types to NO_3^- supply (Gifford et al. 2008). The results demonstrated localised gene responses and identified the NO_3^- repression of a microRNA specifically expressed in the LR cap and pericycle cells. The microRNA 167a/b expressed specifically where LR emergence occurs was repressed by supply of NO_3^-. The Auxin Response Factor (ARF) gene ARF8 is a target of miRNA 167a/b and is induced when NO_3^- is supplied and thus the emergence of initiated LRs is reduced. The miR167/ARF8 regulatory system has been shown to mediate NO_3^- signaling during LR development such that the microRNA 167a/b interacts with ARF8 to repress LR initiation in the presence of NO_3^- (Gifford et al. 2008). However, an inhibitor of glutamine synthesis prevented this induction of ARF8 suggesting that this regulatory mechanism for LR elongation is mediated by organic N signaling rather than NO_3^- (Gifford et al. 2008), although there still remains some doubt over the N source which is responsible for this regulatory mechanism (Gojon et al. 2009).

Obviously, auxin remains central to the regulation of RSA, but in particular to the inhibition of *A. thaliana* LRs in the systemic response to high NO_3^- concentration. Elevated NO_3^- results in an increased level of active CK in the root and CK has been shown to disrupt auxin levels in LR founder cells via altered *PIN* expression (Takei et al. 2001). Also, NO_3^- supplementation has implicated ABA in positively regulating the development of LRs (Signora et al. 2001).

In addition to auxin regulation of LR development, ethylene has been shown to modulate NRTs in order to regulate the development of LRs in response to NO_3^-. Two ethylene synthesis antagonists were shown to alleviate the high NO_3^--induced inhibition of LR growth. Furthermore, the ethylene mutant lines *etr1-3* and *ein2-1* demonstrated less reduction in LR length and number than wild type plants when grown on high NO_3^-. Crucially, the up-regulation of AtNRT1.1 and down-regulation of AtNRT2.1 were also instigated by ethylene synthesis precursors but expression of both transporters in *etr1-3* and *ein2-1* became insensitive to high NO_3^- concentrations (Tian et al. 2009). Thus, ethylene represses LR initiation and PR elongation by relocating the site of auxin biosynthesis to the root elongation

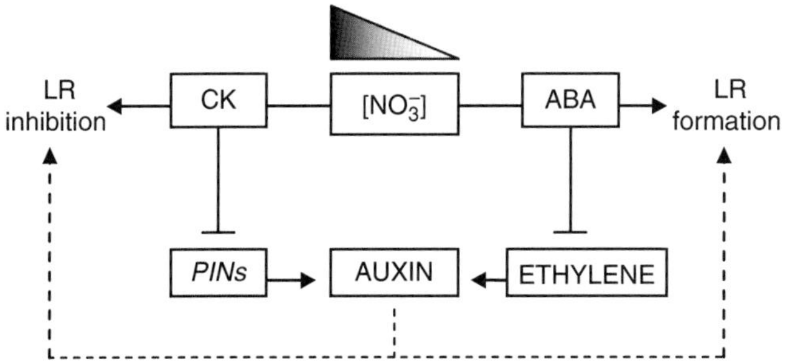

Fig. 4 A schematic representation of the key hormone interactions involved in LR responses to NO_3^- treatments

zone. However, ABA is known to repress ethylene production and application of ABA results in PR elongation and increased LR initiation.

The key interaction in this complex network of hormones is the antagonistic relationship between CK and auxin. Increased root CK, in response to increased NO_3^-, perturbs correct auxin distribution resulting in a failure to create the auxin maxima required for LR formation. It would seem that the increased ABA-dependent LR formation, in response to increased NO_3^- concentration, is capable of overriding the effect of CK. Thus, it could be speculated that CK-mediated repression of LR formation could be important in the systemic inhibition at high NO_3^- levels, and ABA-mediated alleviation of auxin repression is capable of overriding this inhibition to increase LR formation, perhaps in response to locally high NO_3^- (Fig. 4).

6 NO_3^- Signaling

One of the difficulties in assigning a signaling role for NRTs is to separate this function from the actual transport job of the protein. The first gene observed to play a regulatory role in the transport of NO_3^- was NRT1.1 (CHL1: Liu and Tsay 2003). In the NRT1.1 mutant line *chl1-5*, the expression of NRT2.1 is not repressed when plants are grown on a high N supply. Thus, NRT1.1 seemed to play an important regulatory role in NRT2.1 expression and the NO_3^- HATS (Munos et al. 2004). It was proposed that NRT2.1 is regulated by the activity of NRT1.1 rather than simply by its presence, where NRT2.1 expression is regulated by an N demand signal as a function of NRT1.1 NO_3^- uptake activity (Krouk et al. 2006).

The functional AtNRT2.1/NAR2.1 unit is known to be down-regulated by NO_3^- itself. This down-regulation is specifically triggered by AtNRT1.1 via a mechanism which is independent of the negative feedback exerted by downstream N metabolites (Munos et al. 2004; Krouk et al. 2006). AtNRT2.1 expression has been shown to increase rapidly following supply of NO_3^- to previously starved roots and declining as NO_3^- supply is maintained. NO_3^- accumulates at high levels in

NO_3^- reductase (NR) mutants due to defective metabolism of NO_3^-. AtNRT2.1 is likely to be induced by NO_3^- and repressed by downstream metabolites as increased transcript levels are observed in NR mutants. Indeed, glutamine has been shown to be important in down-regulating AtNRT2.1, NpNRT2.1 and HvNRT2 (Krapp et al. 1998; Nazoa et al. 2003; Remans et al. 2006a, b). In fact, amino acids and glutamine have been implicated in signaling for plant N demand via the phloem (Tillard et al. 1998). Interestingly, a high concentration of NO_3^- represses AtNRT2.1 expression in the presence of ammonium (Munos et al. 2004; Krouk et al. 2006). Significantly, the high-NO_3^- repression of AtNRT2.1 is relieved in the *chl1* mutant, indicating that AtNRT1.1 is required for this response (Munos et al. 2004).

The RSA responses described earlier in this chapter are the result of the activation of a signaling pathway which is triggered by NO_3^-. This is especially true of the elongation of LRs in response to a patch of locally high NO_3^-. Two genes are involved in signaling for this elongation morphology: *AtANR1* and *AtNRT1.1*. Encoding a putative MADS box transcription factor, *ANR1* is strongly expressed in the root tip along with *AtNRT1.1*. The involvement of *NRT1.1* in the signaling pathway was elucidated using split-root experiments with the *chl1* mutant line which showed a depleted LR elongation response in high NO_3^-, but exhibited wild type uptake levels. Interestingly, the mutation of *NRT1.1* reduces *ANR1* expression in the root tip indicative of a downstream role for *ANR1* relative to *NRT1.1*. The plasma membrane localisation of AtNRT1.1 at the root surface and its ability to switch between high- and low-affinity transport would make it ideal for sensing the heterogeneous external NO_3^- concentrations encountered by the proliferating root. Such a role was supported by the observation that glutamate-induced repression of PR growth is overcome in wild type plants grown on glutamate and NO_3^- medium, but not in the *chl1* mutant. This suggests that NRT1.1-dependent NO_3^- signaling antagonises the inhibitory effect of glutamate (Walch-Liu and Forde 2008; Wang et al. 2009).

PR growth is strongly inhibited when glutamate is sensed by the root tip. When wild type *A. thaliana* are supplied with NO_3^- and glutamate, the inhibitory effect is overridden. This is not the case for the *chl1* mutant, indicating that NO_3^- signaling to relieve PR growth inhibition is NRT1.1 dependent (Walch-Liu and Forde 2008). Significantly, when the *chl1* mutant line is complemented with wild type NRT1.1, inhibition of PR growth by glutamate is restored. If the mutant line is complemented with a non-phosphorylable NRT1.1T101A mutant protein, then the inhibitory effect is not restored. Since this phosphorylation mutant line can still transport NO_3^-, the ability of NRT1.1 to relieve the inhibition of PR growth by glutamate must be due to a specific signaling function of the phosphorylated form of NRT1.1 (Liu and Tsay 2003). Indeed, this was shown to be the case, with NRT1.1 demonstrating a NO_3^- sensing mechanism which regulates the phosphorylation status of the T101 residue (Ho et al. 2009). The phosphorylation of the T101 residue was shown to be mediated by CIPK23 and occurred during high-affinity binding of NO_3^-. With higher external NO_3^- concentrations, low-affinity binding occurs, repressing the phosphorylation of the T101 residue (Fig. 5).

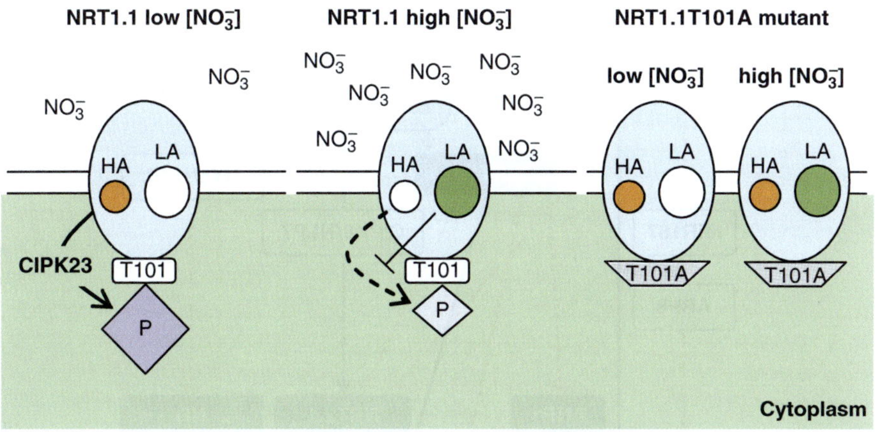

Fig. 5 Schematic representation of NRT1.1 NO_3^- sensing mediated by CIPK23 (redrawn from Ho et al. 2009). The *blue ovals* represent NRT1.1 at the plasma membrane. *Small circles* indicate the high affinity (HA) binding site, filled *orange* when NO_3^- is bound. *Large circles* represent the low affinity (LA) binding site, filled *green* when NO_3^- is bound. *P* denotes phosphorylation of the T101 residue and the *purple* colour gradient represents the level of $NRT1.1^{T101}$ phosphorylation. Reproduced from Ho et al. (2009)

Some NO_3^- assimilation genes are stimulated by light activation of phytohormones. *HY5* and *HYH* have been shown to be important for high NR activity in red and blue light via the NR promoter NIA2 (Jonassen et al. 2008). More recently, HY5 and HYH were shown to inhibit NRT1.1 across a range of light treatments and consistently throughout several tissue types (Jonassen et al. 2009). Thus, there exists a small signaling pathway between HY5/HYH and NRT1.1 which causes the final steps of NO_3^- reduction to coincide with a reduction in NRT1.1 activity. This could be in response to a threshold NO_3^- level beyond which the plant needs to cease uptake and promote reduction.

Phospholipase D (PLD) produces phosphatidic acid (PA) which functions as a lipid messenger implicated in cell growth and proliferation in response to N stress. Over-expressing and knock-out *A. thaliana* lines for the membrane-associated PLDε demonstrated altered RSA and bioaccumulation responses to N treatments, which corresponded to PA levels in the plants. It was postulated that lipid signaling could provide a mechanism to connect NRT sensing of N status to changes in RSA responses (Hong et al. 2009).

There are several direct and indirect NO_3^- signaling mechanisms which can achieve changes in RSA (Fig. 6). As RSA has such a behaviourally important function in achieving nutrient uptake for survival, we should not be surprised by the complex interactions involved in producing RSA in response to altered NO_3^- availability and demand. Whilst our knowledge of the signaling processes is developing (reviewed by Krouk et al. 2010), there is a long way to go to fully understand the interactions and mediatory players in generating RSA in response to NO_3^-.

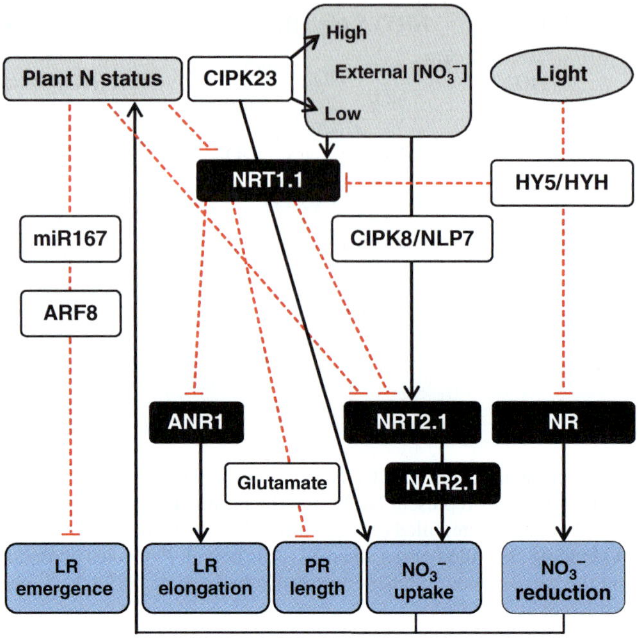

Fig. 6 Schematic representation demonstrating the regulation of gene (*black*) expression by external input (*grey*) to execute certain RSA characteristics (*blue*). *White boxes* represent intermediary interactions and *red lines* represent a regulatory step. Based on Gojon et al. (2009)

7 Conclusions

This chapter has outlined the function and contribution of several NRTs to NO_3^- uptake and transport within the root of a growing plant. Our understanding of NRT transport function is greatly developed with NO_3^- transport characterised for many tissues and developmental stages. Yet it is the important underlying signaling mechanisms which remain largely unknown.

The transport function of a few members of the NRT1 family has been described in detail for various tissue types, including root, xylem, leaf and embryonic, and across higher plants. Most members of the NRT1 family function in low-affinity transport, apart from the very important NRT1.1, which has dual affinity, and is the best characterised NRT in terms of a sensing and signaling role. Indeed, our understanding of the phosphorylation mechanism regulating the cabability of NRT1.1 to switch between LATS and HATS for NO_3^-, and the sensing of external NO_3^- concentration which drives this switch, has greatly increased (Ho et al. 2009). We also have a better understanding of the ability of NRT1.1 to regulate NRT2.1 and ultimately the RSA (reviewed by Gojon et al. 2009). However, there are many potential interactions for NRT1.1 with other NRTs during plant development to regulate plant N status which remains uncharacterised. For example, the ability of NRT1.1 to sense NO_3^- could be implicated in several growth-regulating processes including the progression from seed dormancy to embryonic growth, or in the indirect regulation of NO_3^- reduction. If NRT1.1 is capable of regulating NRT2.1 and ANR1, then the regulation of other key components should not be ruled out.

The transport function of many NRT2 family members has been well characterised. NRT2.1 and NRT2.2 function in the HATS of NO_3^- uptake, and in the case of the important NRT2.1, it is known that several forms exist at the plasma membrane but that NO_3^- uptake requires the presence of NAR2.1 (Chopin et al. 2007b). While evidence has been shown to implicate NRT2.1 in a transport-independent signaling role in the regulation of LR proliferation, little of the molecular signaling mechanisms involved have been characterised (Wirth et al. 2007). If we are to fully understand the interplay between external NO_3^- and RSA, then the cellular regulation of NRT2.1 in terms of its ability to sense NO_3^- is required.

It is clear that through our understanding of the effect of external NO_3^- on RSA, mediated by the NRT families, it is possible to explain how the plant regulates N status and thus growth and development. The metabolic processes undertaken within plant cells after NO_3^- uptake are well characterised, and the importance of certain metabolites has been described (Miller and Cramer 2005). It is likely that these processes will remain central to explaining NRT function at the whole plant level and the influence they have on RSA.

We know that hormonal regulation of plant root development is influenced by plant N status. And indeed much evidence has been described in this chapter for the overlap of NO_3^- signaling with auxin, ethylene, ABA and CK pathways in the generation of RSA (Signora et al. 2001; Takei et al. 2004; Tian et al. 2009). Yet there remains uncertainty over the exact mechanisms underlying these interactions, and further characterisation will be essential to complete the story.

8 Outlook

Much of the knowledge we possess related to NRTs has been obtained in the Petri dish. These experiments provide an ideal means of quickly elucidating gene expression changes and protein function. However, we should note that this environment differs from that experienced by a root growing in the field in terms of nutrient heterogeneity and structural properties. Split-root experiments and the introduction of nutrient patches have gone some way to address these problems. But if we are to truly understand what is happening in the field, then a move away from the sterile, nutrient-rich, Petri dish is required. Furthermore, the significant influence that bacterial and fungal associations exert in the rhizosphere in relation to N cycling, uptake and signaling cannot be ignored. These processes can now be addressed as new methodologies are developed. For example, there has been a great push in recent times to image the RSA of a growing plant so that complex mathematical models can be used to quantify important parameters of the RSA. While this information can be very useful in the quantification of root morphology, it often requires the use of less natural root growth environments, so exactly how useful these experimental systems are remains questionable. In attempting to bridge the gap between accurate quantification of root morphology and a more field-like growth environment, x-ray technology has been applied to crop root systems in soil, but the resolution remains insufficient for use with the genetically important *A. thaliana*.

Understanding the movement and uptake of NO_3^- within the soil and how plants sense and respond to the heterogeneous nature of soil will be key to the global challenges facing us over the coming century. Accepted estimates put the global population at nine billion people by 2050. Thus, housing and food will be needed for an extra three billion people from no more land than is currently available (Svirejeva-Hopkins and Schellnhuber 2008). Furthermore, the changing climate is likely to shift the traditional geographical ranges of crop production as a result of changing temperatures, precipitation patterns and severe weather episodes. NO_3^- can be lost from the soil readily via leaching and run-off. This can have significant economic costs for fertiliser application, with up to 60% of applied fertiliser lost to the environment (Miller and Cramer 2005). In turn, this can lead to detrimental effects to the surrounding ecosystems through pollution and eutrophication (Abit et al. 2008). Improving crop yield, fertiliser composition and application methods, and land management processes will remain central targets for agricultural and plant science as we try to achieve a jump in crop production similar to that seen in the green revolution (Borlaug 1992; Lynch 2007; Busov et al. 2008).

As external NO_3^- concentration regulates gene expression, RSA, and drives many of the processes discussed in this chapter, further work into linking the properties of the root growth environment with a given root morphology will help to understand what is happening in field systems. Quantification of the physical properties of many growth media would be useful when comparing RSA between experimental systems or indeed in attempting to extrapolate from the laboratory to the field.

In terms of cellular biology, there is obviously an overlap between NO_3^- signaling and hormones for several root proliferation responses. Further research into how, and at what point along the signaling cascades, these two networks interact will help us to fully understand why a plant produces a particular RSA. It is likely that intermediate signaling components, such as CIPK proteins and lipid messengers like PLDε, will be dominant in explaining NO_3^- signaling capabilities of other NRTs. As research is driven towards addressing global problems with climate change and food security, the study of the CLC family of transporters will become more important. This is due to their ability to regulate NO_3^- storage in aerial tissues, which has health and environmental benefits. Furthermore, manipulation of their regulation of Cl^- transport may be useful in developing salt-tolerant plants (Miller and Cramer 2005).

From a phylogeny view point, it would be interesting to identify whether NRT1.1 obtained regulatory power over NRT2.1 before a NO_3^- sensing/signaling role. NRT1.1 is generally accepted as having a signaling role in addition to its uptake ability. It is positioned upstream of NRT2.1 and regulates NRT2.1. Did it develop signaling capacity because it already regulated NRT2.1 or did it develop the ability to regulate NRT2.1 because it was already sensing the environmental conditions?

Although investigating crop RSA may be informative, the superior molecular kit available with *A. thaliana* is likely to be necessary to describe what is happening at a cellular level. We should not underestimate the value of gene disruption mutants in elucidating function, and to this end *A. thaliana* will continue to be the plant system of choice for many. Expression systems such as *X. laevis* oocytes and yeast will continue to be essential in determining protein transport function. But rather than simply characterising transport affinities of NRTs, it may prove more worthwhile to focus on designing a system to assess NO_3^- sensing capability. After all, it is the ability of the plant to sense NO_3^- availability that drives transporter function and root proliferation. Transcriptome approaches such as those adopted by Gifford et al. (2008), alongside systems approaches as undertaken by Vidal and Gutierrez (2008), Krouk et al. (2009) and Nero et al. (2009), are likely to elucidate further signaling mechanisms in response to NO_3^- supply. Indeed, identifying common signaling targets has proved useful in explaining interconnected nutrient-driven gene expression (Girin et al. 2007). A better understanding means that moving from *A. thaliana* to crop systems will be made easier through the identification of gene targets for improving crop yield or N use efficiency.

It should be clear to the reader that the sensing of NO_3^- availability, regulation of gene expression, proliferation of the root and the ultimate uptake function are interconnected by a network of complex interactions. We must continue to focus our research on the main organ for NO_3^- acquisition if we are to improve crop yield and N use efficiency. A holistic understanding of the complex interaction between NO_3^- availability and RSA will be required to address the global challenges facing us in the twenty-first century. Root architecture is very 'plastic' as roots are well adapted to the heterogeneous soil environment, but crop roots may have become lazy as they have been bred under luxury nutrient supplies. As lower more sustainable input agriculture is the requirement for the future, the characterisation of 'weed' or 'native' species may be advantageous as they are better adapted to low N supply.

References

Abit SM, Amoozegar A, Vepraskas MJ, Niewoehner CP (2008) Fate of nitrate in the capillary fringe and shallow groundwater in a drained sandy soil. Geoderma 146:209–215

Alboresi A, Gestin C, Leydecker MT, Bedu M, Meyer C, Truong HN (2005) Nitrate, a signal relieving seed dormancy in Arabidopsis. Plant Cell Environ 28:500–512

Almagro A, Lin SH, Tsay YF (2008) Characterization of the Arabidopsis nitrate transporter NRT1.6 reveals a role of nitrate in early embryo development. Plant Cell 20:3289–3299

Bao J, Chen FJ, Gu RL, Wang GY, Zhang FS, Mi GH (2007) Lateral root development of two Arabidopsis auxin transport mutants, auxl-7 and eirl-1, in response to nitrate supplies. Plant Sci 173:417–425

Bergsdorf EY, Zdebik AA, Jentsch TJ (2009) Residues important for nitrate/proton coupling in plant and mammalian CLC transporters. J Biol Chem 284:11184–11193

Bibikova T, Gilroy S (2002) Root hair development. J Plant Growth Regul 21:383–415

Borlaug NE (1992) World food security and the legacy of canadian wheat scientist Anderson, R., Glenn. Can J Plant Pathol 14:254–266

Busov VB, Brunner AM, Strauss SH (2008) Genes for control of plant stature and form. New Phytol 177:589–607

Casson SA, Lindsey K (2003) Genes and signaling in root development. New Phytol 158:11–38

Cerezo M, Tillard P, Filleur S, Munos S, Daniel-Vedele F, Gojon A (2001) Major alterations of the regulation of root NO3- uptake are associated with the mutation of Nrt2.1 and Nrt2.2 genes in arabidopsis. Plant Physiol 127:262–271

Chen YF, Wang Y, Wu WH (2008) Membrane transporters for nitrogen, phosphate and potassium uptake in plants. J Integr Plant Biol 50:835–848

Chiu CC, Lin CS, Hsia AP, Su RC, Lin HL, Tsay YF (2004) Mutation of a nitrate transporter, AtNRT1: 4, results in a reduced petiole nitrate content and altered leaf development. Plant Cell Physiol 45:1139–1148

Chopin F, Orsel M, Dorbe MF, Chardon F, Truong HN, Miller AJ, Krapp A, Daniel-Vedele F (2007a) The Arabidopsis ATNRT2.7 nitrate transporter controls nitrate content in seeds. Plant Cell 19:1590–1602

Chopin F, Wirth J, Dorbe MF, Lejay L, Krapp A, Gojon A, Daniel-Vedele F (2007b) The Arabidopsis nitrate transporter AtNRT2.1 is targeted to the root plasma membrane. Plant Physiol Biochem 45:630–635

Clark LJ, Whalley WR, Barraclough PB (2003) How do roots penetrate strong soil? Plant Soil 255:93–104

Crawford NM, Glass ADM (1998) Molecular and physiological aspects of nitrate uptake in plants. Trends Plant Sci 3:389–395

Daniel-Vedele F, Filleur S, Caboche M (1998) Nitrate transport: a key step in nitrate assimilation. Curr Opin Plant Biol 1:235–239

De Angeli A, Monachello D, Ephritikhine G, Frachisse JM, Thomine S, Gambale F, Barbier-Brygoo H (2006) The nitrate/proton antiporter AtCLCa mediates nitrate accumulation in plant vacuoles. Nature 442:939–942

De Angeli A, Monachello D, Ephritikhine G, Frachisse JM, Thomine S, Gambale F, Barbier-Brygoo H (2009) CLC-mediated anion transport in plant cells. Philos Trans R Soc Lond B Biol Sci 364:195–201

De Kroon H, Visser EJW, Huber H, Mommer L, Hutchings MJ (2009) A modular concept of plant foraging behaviour: the interplay between local responses and systemic control. Plant Cell Environ 32:704–712

Deak KI, Malamy J (2005) Osmotic regulation of root system architecture. Plant J 43:17–28

Desnos T (2008) Root branching responses to phosphate and nitrate. Curr Opin Plant Biol 11:82–87

Doddema H, Hofstra JJ, Feenstra WJ (1978) Uptake of nitrate by mutants of *Arabidopsis thaliana*, disturbed in uptake or reduction of nitrate. 1. Effect of nitrogen source during growth on uptake of nitrate and chlorate. Physiol Plant 43:343–350

Drew MC, Saker LR (1975) Nutrient supply and growth of seminal root system in barley. 2. Localized, compensatory increases in lateral root growth and rates of nitrate uptake when nitrate supply is restricted to only part of root system. J Exp Bot 26:79–90

Driouich A, Durand C, Vicre-Gibouin M (2007) Formation and separation of root border cells. Trends Plant Sci 12:14–19

Fan SC, Lin CS, Hsu PK, Lin SH, Tsay YF (2009) The Arabidopsis nitrate transporter NRT1.7, expressed in phloem, is responsible for source-to-sink remobilization of nitrate. Plant Cell 21:2750–2761

Feng CP, Mundy J (2006) Gene discovery and functional analyses in the model plant Arabidopsis. J Integr Plant Biol 48:5–14

Forde BG (2000) Nitrate transporters in plants: structure, function and regulation. Biochim Biophys Acta 1465:219–235
Fukaki H, Tasaka M (2009) Hormone interactions during lateral root formation. Plant Mol Biol 69:437–449
Gifford ML, Dean A, Gutierrez RA, Coruzzi GM, Birnbaum KD (2008) Cell-specific nitrogen responses mediate developmental plasticity. Proc Natl Acad Sci USA 105:803–808
Girin T, Lejay L, Wirth J, Widiez T, Palenchar PM, Nazoa P, Touraine B, Gojon A, Lepetit M (2007) Identification of a 150 bp cis-acting element of the AtNRT2.1 promoter involved in the regulation of gene expression by the N and C status of the plant. Plant Cell Environ 30:1366–1380
Gojon A, Nacry P, Davidian JC (2009) Root uptake regulation: a central process for NPS homeostasis in plants. Curr Opin Plant Biol 12:328–338
Gorska A, Ye Q, Holbrook NM, Zwieniecki MA (2008) Nitrate control of root hydraulic properties in plants: translating local information to whole plant response. Plant Physiol 148:1159–1167
Guo FQ, Wang RC, Crawford NM (2002) The Arabidopsis dual-affinity nitrate transporter gene AtNRT1.1 (CHL1) is regulated by auxin in both shoots and roots. J Exp Bot 53:835–844
Guo FO, Young J, Crawford NM (2003) The nitrate transporter AtNRT1.1 (CHL1) functions in stomatal opening and contributes to drought susceptibility in arabidopsis. Plant Cell 15:107–117
Gutierrez RA, Lejay LV, Dean A, Chiaromonte F, Shasha DE, Coruzzi GM (2007) Qualitative network models and genome-wide expression data define carbon/nitrogen-responsive molecular machines in Arabidopsis. Genome Biol 8:13
Ho CH, Lin SH, Hu HC, Tsay YF (2009) CHL1 functions as a nitrate sensor in plants. Cell 138:1184–1194
Hong YY, Devaiah SP, Bahn SC, Thamasandra BN, Li MY, Welti R, Wang XM (2009) Phospholipase D epsilon and phosphatidic acid enhance Arabidopsis nitrogen signaling and growth. Plant J 58:376–387
Hu HC, Wang YY, Tsay YF (2009) AtCIPK8, a CBL-interacting protein kinase, regulates the low-affinity phase of the primary nitrate response. Plant J 57(2):264–278
Huang NC, Liu KH, Lo HJ, Tsay YF (1999) Cloning and functional characterization of an Arabidopsis nitrate transporter gene that encodes a constitutive component of low-affinity uptake. Plant Cell 11:1381–1392
Iijima M, Morita S, Barlow PW (2008) Structure and function of the root cap. Plant Prod Sci 11:17–27
Jonassen EM, Lea US, Lillo C (2008) HY5 and HYH are positive regulators of nitrate reductase in seedlings and rosette stage plants. Planta 227:559–564
Jonassen EM, Sevin DC, Lillo C (2009) The bZIP transcription factors HY5 and HYH are positive regulators of the main nitrate reductase gene in Arabidopsis leaves, NIA2, but negative regulators of the nitrate uptake gene NRT1.1. J Plant Physiol 166:2071–2076
Komarova NY, Thor K, Gubler A, Meier S, Dietrich D, Weichert A, Grotemeyer MS, Tegeder M, Rentsch D (2008) AtPTR1 and AtPTR5 transport dipeptides in planta. Plant Physiol 148:856–869
Krapp A, Fraisier V, Scheible WR, Quesada A, Gojon A, Stitt M, Caboche M, Daniel-Vedele F (1998) Expression studies of Nrt2: 1Np, a putative high-affinity nitrate transporter: evidence for its role in nitrate uptake. Plant J 14:723–731
Krouk G, Tillard P, Gojon A (2006) Regulation of the high-affinity NO3- uptake system by NRT1.1-mediated NO3- demand signaling in Arabidopsis. Plant Physiol 142:1075–1086
Krouk G, Tranchina D, Lejay L, Cruikshank AA, Shasha D, Coruzzi GM, Gutierrez RA (2009) A systems approach uncovers restrictions for signal interactions regulating genome-wide responses to nutritional cues in Arabidopsis. PLoS Comput Biol 5:12
Krouk G, Crawford N, Coruzzi GM, Tsay YF (2010) Nitrate signaling: adaptation to fluctuating environments. Curr Opin Plant Biol 13:1–8

Lejay L, Tillard P, Lepetit M, Olive FD, Filleur S, Daniel-Vedele F, Gojon A (1999) Molecular and functional regulation of two NO3- uptake systems by N- and C-status of Arabidopsis plants. Plant J 18:509–519

Li WB, Wang Y, Okamoto M, Crawford NM, Siddiqi MY, Glass ADM (2007) Dissection of the AtNRT2.1: AtNRT2.2 inducible high-affinity nitrate transporter gene cluster. Plant Physiol 143:425–433

Lin CM, Koh S, Stacey G, Yu SM, Lin TY, Tsay YF (2000) Cloning and functional characterization of a constitutively expressed nitrate transporter gene, OsNRT1, from rice. Plant Physiol 122:379–388

Lin SH, Kuo HF, Canivenc G, Lin CS, Lepetit M, Hsu PK, Tillard P, Lin HL, Wang YY, Tsai CB, Gojon A, Tsay YF (2008) Mutation of the Arabidopsis NRT1.5 nitrate transporter causes defective root-to-shoot nitrate transport. Plant Cell 20:2514–2528

Linkohr BI, Williamson LC, Fitter AH, Leyser HMO (2002) Nitrate and phosphate availability and distribution have different effects on root system architecture of Arabidopsis. Plant J 29:751–760

Little DY, Rao HY, Oliva S, Daniel-Vedele F, Krapp A, Malamy JE (2005) The putative high-affinity nitrate transporter NRT2.1 represses lateral root initiation in response to nutritional cues. Proc Natl Acad Sci USA 102:13693–13698

Liu KH, Tsay YF (2003) Switching between the two action modes of the dual-affinity nitrate transporter CHL1 by phosphorylation. EMBO J 22:1005–1013

Liu KH, Huang CY, Tsay YF (1999) CHL1 is a dual-affinity nitrate transporter of arabidopsis involved in multiple phases of nitrate uptake. Plant Cell 11:865–874

Liu K, Kozono D, Kato Y, Agre P, Hazama A, Yasui M (2005) Conversion of aquaporin 6 from an anion channel to a water-selective channel by a single amino acid substitution. Proc Natl Acad Sci USA 102:2192–2197

Lopez-Bucio J, Cruz-Ramirez A, Herrera-Estrella L (2003) The role of nutrient availability in regulating root architecture. Curr Opin Plant Biol 6:280–287

Loque D, Tillard P, Gojon A, Lepetit M (2003) Gene expression of the NO3- transporter NRT1.1 and the nitrate reductase NIA1 is repressed in Arabidopsis roots by NO2-, the product of NO3-reduction. Plant Physiol 132:958–967

Loque D, Ludewig U, Yuan LX, von Wiren N (2005) Tonoplast intrinsic proteins AtTIP2;1 and AtTIP2;3 facilitate NH3 transport into the vacuole. Plant Physiol 137:671–680

Lynch JP (2007) Roots of the second green revolution. Aust J Bot 55:493–512

Martin Y, Navarro FJ, Siverio JM (2008) Functional characterization of the Arabidopsis thaliana nitrate transporter CHL1 in the yeast Hansenula polymorpha. Plant Mol Biol 68:215–224

Miller AJ, Cramer MD (2005) Root nitrogen acquisition and assimilation. Plant Soil 274:1–36

Miller AJ, Smith SJ (2008) Cytosolic nitrate ion homeostasis: could it have a role in sensing nitrogen status? Ann Bot 101:485–489

Miller AJ, Fan XR, Orsel M, Smith SJ and Wells DM (2007a) Nitrate transport and signaling. International Symposium on Nitrogen Nutrition in Plants, Lancaster, England

Miller AJ, Fan XR, Orsel M, Smith SJ, Wells DM (2007b) Nitrate transport and signaling. J Exp Bot 58:2297–2306

Miller AJ, Shen QR, Xu GH (2009) Freeways in the plant: transporters for N, P and S and their regulation. Curr Opin Plant Biol 12:284–290

Muller B, Touraine B (1992) Inhibition of NO_3^- uptake by various phloem-translocated amino-acids in soybean seedlings. J Exp Bot 43:617–623

Munos S, Cazettes C, Fizames C, Gaymard F, Tillard P, Lepetit M, Lejay L, Gojon A (2004) Transcript profiling in the chl1-5 mutant of Arabidopsis reveals a role of the nitrate transporter NRT1.1 in the regulation of another nitrate transporter, NRT2.1. Plant Cell 16:2433–2447

Nazoa P, Vidmar JJ, Tranbarger TJ, Mouline K, Damiani I, Tillard P, Zhuo DG, Glass ADM, Touraine B (2003) Regulation of the nitrate transporter gene AtNRT2.1 in Arabidopsis thaliana: responses to nitrate, amino acids and developmental stage. Plant Mol Biol 52:689–703

Nero D, Krouk G, Tranchina D, Coruzzi GM (2009) A system biology approach highlights a hormonal enhancer effect on regulation of genes in a nitrate responsive "biomodule". BMC Syst Biol 3:17

Okamoto M, Kumar A, Li WB, Wang Y, Siddiqi MY, Crawford NM, Glass ADM (2006) High-affinity nitrate transport in roots of Arabidopsis depends on expression of the NAR2-like gene AtNRT3.1. Plant Physiol 140:1036–1046

Orsel M, Filleur S, Fraisier V, Daniel-Vedele F (2002a) Nitrate transport in plants: which gene and which control? J Exp Bot 53:825–833

Orsel M, Krapp A, Daniel-Vedele F (2002b) Analysis of the NRT2 nitrate transporter family in Arabidopsis. Structure and gene expression. Plant Physiol 129:886–896

Orsel M, Eulenburg K, Krapp A, Daniel-Vedele F (2004) Disruption of the nitrate transporter genes AtNRT2.1 and AtNRT2.2 restricts growth at low external nitrate concentration. Planta 219:714–721

Orsel M, Chopin F, Leleu O, Smith SJ, Krapp A, Daniel-Vedele F, Miller AJ (2006) Characterization of a two-component high-affinity nitrate uptake system in Arabidopsis. Physiology and protein-protein interaction. Plant Physiol 142:1304–1317

Remans T, Nacry P, Pervent M, Filleur S, Diatloff E, Mounier E, Tillard P, Forde BG, Gojon A (2006a) The Arabidopsis NRT1.1 transporter participates in the signaling pathway triggering root colonization of nitrate-rich patches. Proc Natl Acad Sci USA 103:19206–19211

Remans T, Nacry P, Pervent M, Girin T, Tillard P, Lepetit M, Gojon A (2006b) A central role for the nitrate transporter NRT2.1 in the integrated morphological and physiological responses of the root system to nitrogen limitation in Arabidopsis. Plant Physiol 140:909–921

Rensink WA, Buell CR (2004) Arabidopsis to rice. Applying knowledge from a weed to enhance our understanding of a crop species. Plant Physiol 135:622–629

Rubio V, Bustos R, Irigoyen ML, Cardona-Lopez X, Rojas-Triana M, Paz-Ares J (2009) Plant hormones and nutrient signaling. Plant Mol Biol 69:361–373

Segal E, Kushnir T, Mualem Y, Shani U (2008) Water uptake and hydraulics of the root hair rhizosphere. Vadose Zone J 7:1027–1034

Segonzac C, Boyer JC, Ipotesi E, Szponarski W, Tillard P, Touraine B, Sommerer N, Rossignol M, Gibrat R (2007) Nitrate efflux at the root plasma membrane: Identification of an Arabidopsis excretion transporter. Plant Cell 19:3760–3777

Signora L, De Smet I, Foyer CH, Zhang HM (2001) ABA plays a central role in mediating them regulatory effects of nitrate on root branching in Arabidopsis. Plant J 28:655–662

Singh SK, Fischer U, Singh M, Grebe M, Marchant A (2008) Insight into the early steps of root hair formation revealed by the procuste1 cellulose synthase mutant of Arabidopsis thaliana. BMC Plant Biol 8:12

Somasundaram S, Fukuzono S, Iijima M (2008) Dynamics of root border cells in rhizosphere Soil of Zea mays L.: crushed cells during root penetration, survival in soil, and long term soil compaction effect. Plant Prod Sci 11:440–446

Svirejeva-Hopkins A, Schellnhuber HJ (2008) Urban expansion and its contribution to the regional carbon emissions: Using the model based on the population density distribution. Ecol Modell 216:208–216

Takei K, Takahashi T, Sugiyama T, Yamaya T and Sakakibara H (2001) Multiple routes communicating nitrogen availability from roots to shoots: a signal transduction pathway mediated by cytokinin. 6th International Symposium on Inorganic Nitrogen Assimilation, Reims, France

Takei K, Ueda N, Aoki K, Kuromori T, Hirayama T, Shinozaki K, Yamaya T, Sakakibara H (2004) AtIPT3 is a key determinant of nitrate-dependent cytokinin biosynthesis in Arabidopsis. Plant Cell Physiol 45:1053–1062

Teyker RH, Jackson WA, Volk RJ, Moll RH (1988) Exogenous (NO_3^-)-N-15 influx and endogenous (NO_3^-)-N-15 efflux by 2 maize (*Zea mays* L.) inbreds during nitrogen deprivation. Plant Physiol 86:778–781

Tian QY, Chen FJ, Liu JX, Zhang FS, Mi GH (2008) Inhibition of maize root growth by high nitrate supply is correlated with reduced IAA levels in roots. J Plant Physiol 165:942–951

Tian QY, Sun P, Zhang WH (2009) Ethylene is involved in nitrate- root growth and branching in Arabidopsis thaliana. New Phytol 184:918–931

Tillard P, Passama L, Gojon A (1998) Are phloem amino, acids involved in the shoot to root control of NO3- uptake in Ricinus communis plants? J Exp Bot 49:1371–1379

Tong Y, Zhou JJ, Li ZS, Miller AJ (2005) A two-component high-affinity nitrate uptake system in barley. Plant J 41:442–450

Tsay YF, Schroeder JI, Feldmann KA, Crawford NM (1993) The herbicide sensitivity gene CHL1 of *Arabidopsis* encodes a nitrate-inducible nitrate transporter. Cell 72:705–713

Tsay YF, Chiu CC, Tsai CB, Ho CH, Hsu PK (2007) Nitrate transporters and peptide transporters. FEBS Lett 581:2290–2300

van der Leij M, Smith SJ, Miller AJ (1998) Remobilisation of vacuolar stored nitrate in barley root cells. Planta 205:64–72

Vidal EA, Gutierrez RA (2008) A systems view of nitrogen nutrient and metabolite responses in Arabidopsis. Curr Opin Plant Biol 11:521–529

Visser EJW, Bogemann GM, Smeets M, de Bruin S, de Kroon H, Bouma TJ (2008) Evidence that ethylene signaling is not involved in selective root placement by tobacco plants in response to nutrient-rich soil patches. New Phytol 177:457–465

Walch-Liu P, Forde BG (2008) Nitrate signaling mediated by the NRT1.1 nitrate transporter antagonises L-glutamate-induced changes in root architecture. Plant J 54:820–828

Wang H, Inukai Y, Yamauchi A (2006) Root development and nutrient uptake. Crit Rev Plant Sci 25:279–301

Wang RC, Xing XJ, Wang Y, Tran A, Crawford NM (2009) A genetic screen for nitrate regulatory mutants captures the nitrate transporter gene NRT1.1. Plant Physiol 151:472–478

Wirth J, Chopin F, Santoni V, Viennois G, Tillard P, Krapp A, Lejay L, Daniel-Vedele F, Gojon A (2007) Regulation of root nitrate uptake at the NRT2.1 protein level in Arabidopsis thaliana. J Biol Chem 282:23541–23552

Xu LZ, Niu JF, Li CJ, Zhang FS (2009) Growth, nitrogen uptake and flow in maize plants affected by root growth restriction. J Integr Plant Biol 51:689–697

Yin LP, Li P, Wen B, Taylor D, Berry JO (2007) Characterization and expression of a high-affinity nitrate system transporter gene (TaNRT2.1) from wheat roots, and its evolutionary relationship to other NTR2 genes. Plant Sci 172:621–631

Zeuthen T, Meinild AK, Klaerke DA, Loo DDF, Wright EM, Belhage B, Litman T (1997) Water transport by the Na+/glucose cotransporter under isotonic conditions. Biol Cell 89:307–312

Zhang HM, Forde BG (1998) An Arabidopsis MADS box gene that controls nutrient-induced changes in root architecture. Science 279:407–409

Zhang HM, Jennings A, Barlow PW, Forde BG (1999) Dual pathways for regulation of root branching by nitrate. Proc Natl Acad Sci USA 96:6529–6534

Zhao XQ, Li YJ, Liu JZ, Li B, Liu QY, Tong YP, Li JY, Li ZS (2004) Isolation and expression analysis of a high-affinity nitrate transporter TaNRT2.3 from roots of wheat. Acta Bot Sin 46:347–354

Zhou JJ, Theodoulou FL, Muldin I, Ingemarsson B, Miller AJ (1998) Cloning and functional characterization of a Brassica napus transporter that is able to transport nitrate and histidine. J Biol Chem 273:12017–12023

Zhou JJ, Fernandez E, Galvan A, Miller AJ (2000) A high affinity nitrate transport system from Chlamydomonas requires two gene products. FEBS Lett 466:225–227

Sensing and Signaling of PO_4^{3-}

Lenin Sánchez-Calderón, Alejandra Chacón-López, Fulgencio Alatorre-Cobos, Marco Antonio Leyva-González, and Luis Herrera-Estrella

Abstract Phosphorus (P) is one of the most limiting factors for plant growth, development, and productivity in many natural and agricultural ecosystems. P is acquired by plants from the soil in the form of inorganic phosphate (Pi). Due to the high reactivity of Pi with cations present in the soil and its rapid conversion to organic forms by microbial activity, in most soils, Pi is not readily available for plant uptake. However, during millions of years of evolution, plants acquired complex physiological, biochemical, and molecular strategies to survive and reproduce in soils with low Pi availability. Thus, plants display a low Pi rescue system that includes a wide range of adaptive responses, which integrate external and internal signals to maintain Pi homeostasis at the cellular, organ, and whole plant levels. These adaptive responses allow plants to enhance Pi availability in the soil, increase its uptake, and improve the efficiency with which plants use Pi to support its metabolic activities to grow and reproduce. The development of a comprehensive understanding of how plants sense phosphate status and coordinate the local and systemic responses via network pathways has become a major area of research in plant biology. This review is aimed to succinctly present the current state of research covering most of process related with Pi uptake, transport, and in particular the plant responses to Pi-deprivation, making special emphasis on recent developments in the molecular mechanisms and the signaling pathways that regulate these responses.

L. Sánchez-Calderón
Laboratorio de Biología Molecular de Plantas, Unidad Académica de Biología Experimental, Universidad Autónoma de Zacatecas, Av. Revolución S/N Col. Tierra y Libertad, Guadalupe, Zacatecas 98600, Mexico

A. Chacón-López, F. Alatorre-Cobos, M.A. Leyva-González, and L. Herrera-Estrella (✉)
Laboratorio Nacional de Genómica para la Biodiversidad, Centro de Investigación y Estudios Avanzados, Campus Guanajuato, PO BOX 629, Irapuato, Guanajuato 36500, México
e-mail: lherrera@ira.cinvestav.mx

M. Geisler and K. Venema (eds.), *Transporters and Pumps in Plant Signaling*, Signaling and Communication in Plants 7,
DOI 10.1007/978-3-642-14369-4_7,

1 Introduction

1.1 Necessity and Low Availability

The "bearer of light" in Greek mythology, phosphorus (P), is required for practically all cellular metabolic processes. During development, reproduction, and environmental adaptation of all organisms, the fully oxidized form of P, inorganic phosphate (Pi), is essential. It takes part in physiological events such as photosynthesis, assimilation of carbon and nitrogen, oxidative respiration, energy conservation, cell cycle regulation, and lipid metabolism. Pi also plays a crucial role in the post-translational regulation of enzymes and the control of signal transduction cascades participating in the phosphorylation and dephosphorylation cycles of signaling proteins. Furthermore, phosphate homeostasis in the chloroplast regulates the transport of phosphorylated sugars across the membrane and in the synthesis of starch. Additionally, P is a structural component of several biomolecules such as nucleic acids, membrane phospholipids, and high-energy compounds such as ATP and NADPH. It is therefore obvious that Pi availability profoundly affects growth, development, fertility, and reproduction of all living beings (Ticconi and Abel 2004; Raghothama and Karthikeyan 2005), and can strongly limit primary production, structure, and diversity of species in all the ecosystems around the world (Tiessen 2008).

Almost all naturally occurring phosphorus (over 99%) is present as phosphates, either as inorganic phosphates (inorganic pool) or as organic phosphate esters (organic pool). The inorganic pool (Pi) is the form readily available for plant uptake and it is acquired from soil solution by the root system mainly in the form of $H_2PO_4^-$ and to a lesser extent as HPO_4^{-2}. Although the total amount of P in the soil may be high, it is one of the most unavailable plant nutrients in the rhizosphere. The reason behind low Pi availability in the soil is due to its physico-chemical properties that affect its rate of sorption/desorption and water solubility, as well as its rapid conversion into organic forms and mineralization by soil microorganisms (Holford 1997; Schachtman et al. 1998; White and Hammond 2008). Pi is negatively charged, making it highly reactive with positive charged cations on the surface of soil particles, restricting its mobility in the soil (Bieleski 1973; Tiessen 2008). Typically, in alkaline soils Pi is largely bound to calcium (Ca) and magnesium (Mg), and in acid soils to aluminum (Al) and iron (Fe), having, as a result, a strong reduction of its water solubility causing its fixation in the inorganic pool. Pi is also present in the soil forming part of organic compounds such as nucleic acids, phospholipids, and particularly as phytic acid (inositol hexaphosphate) that are not readily available for plant uptake (Schachtman et al. 1998; White and Hammond 2008).

As fixed Pi pools must be dissolved, desorbed, or degraded (mineralized) to release soluble Pi for plant nutrition, in most agricultural systems, the amount of Pi released in the soil is not sufficient to support the high growth rates of crop species. As a consequence, continuous Pi input in the form of fertilizers is used in arable soils. However, less than 20% of the applied Pi fertilizer is actually used by cultivated plants and the rest becomes fixed in the soil as part of the organic or inorganic pool (Holford 1997; Schachtman et al. 1998). To compensate the low fertilizer capture by plants, farmers frequently apply Pi fertilizers in excessive amounts to ensure crop production (Goldstein 1992). However, the excessive use of fertilizers has had severe environmental consequences in terrestrial freshwater and near-shore marine ecosystems due to leakage of Pi from cropping areas. Moreover, since the P cycle is very slow and the use of Pi fertilizers keeps increasing, the source of low-cost Pi-rock is being rapidly depleted.

2 Mechanisms of Acquisition and Transport of Phosphate

2.1 Going Through Membranes

It is widely accepted that the formation of a topologically closed membrane was an early event in cellular evolution. This membrane is a ubiquitous feature of all cellular life forms, which is essentially made out of phospholipid bilayers and proteins. The hydrophobic nature of membranes forces to maintain separate compartments with different chemical compositions between intra and extracellular sides, therefore strongly polar molecules such as ions have to traverse the bilayer assisted by membrane proteins (Peretó et al. 2004; Mulkidjanian et al. 2009). These protein systems also called transporter or carrier systems are highly specific. In plants, a dual uptake model for Pi is characterized by a high-affinity transporter operating at low (μM) concentration and a low-affinity transporter functioning at high concentration (mM) of Pi.

The concentrations of Pi available in cell root–soil interface is in the micro or even submicromolar range, while the Pi concentration within most plant cells is millimolar (Rausch and Bucher 2002). Thus, the uptake of Pi by cells must be carried out against a steep concentration gradient (Fig. 1). The energy required to transport Pi across membranes is supplied by a gradient of protons. The ATPases present in both plasma membrane and tonoplast use the energy derived from the hydrolysis of ATP to move protons (H^+) across the cell membrane. This leads to the formation of a large membrane potential with a negative potential of the cytoplasm. Consequently, the transport of Pi and other anions are usually coupled to protons in a secondary transport process (Hawkesford and Miller 2004; Poirier and Bucher 2009).

2.1.1 Phosphate Transporters H+/Pi Symporter

In plants, Pi transporters (PHT) are Pi/H+ symporters (Fig. 1). They are involved in the movement of Pi from the rhizosphere into the plant or in its distribution to

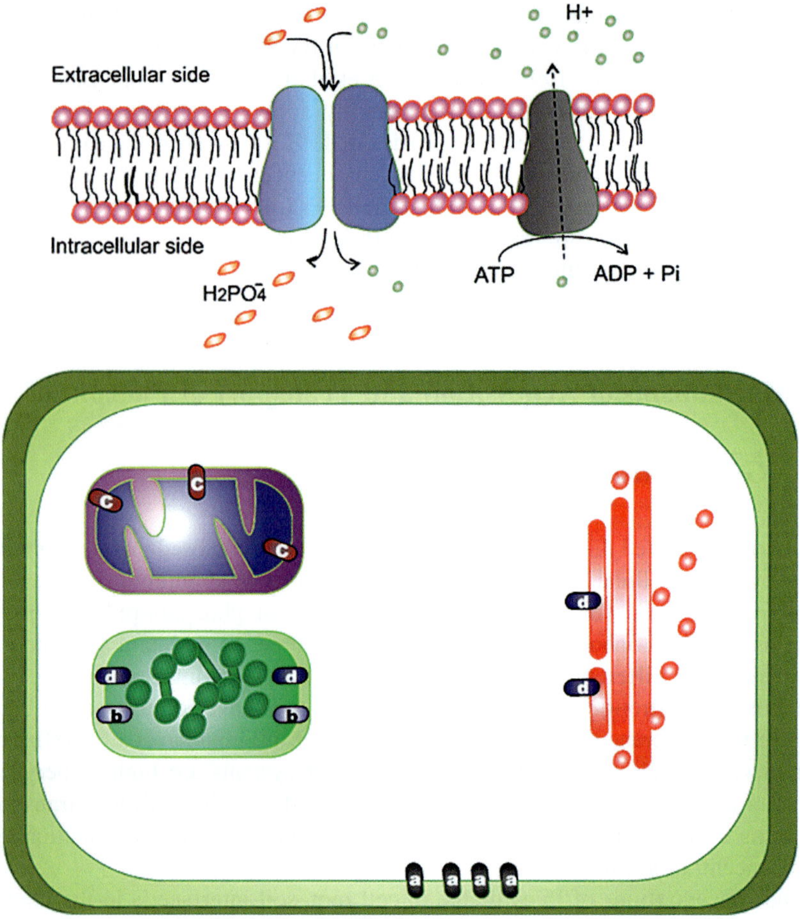

Fig. 1 PHT transporter family. Pi transport across cellular membranes is accomplished by transport of the anion coupled to the transport of protons (H+-symport; above). Concentrations of available Pi at the extracellular side are micromolar, whereas intracellular concentrations of Pi are in the milimolar range. Predicted membrane localization of PHTs (*bottom*). PHT1 (**a**) in plasma membrane, PHT2 (**b**) in photosynthetic plastid inner envelope, PHT3 (**c**) in mitochondrial inner membrane and PHT4 (**d**) in photosynthetic plastids and Golgi apparatus

different tissues and organs within the plant. The PHT are members of the Facilitator Superfamily (MFS;NCBI CDD cd06174), a large and diverse group of secondary transporters that facilitate the transport across cytoplasmic or internal membranes of a variety of substrates including phosphates. The majority of MFS proteins contain 12 transmembrane alpha helices connected by hydrophilic loops (Pao et al. 1998). PHTs are classified into four families based on their predicted membrane localization (Fig. 1 bottom): PHT1 in plasma membrane, PHT2 in plastid inner envelope, PHT3 in mitochondrial inner membrane and PHT4 in chloroplast, heterotrophic plastids and Golgi (Guo et al. 2008; Preuss et al. 2010).

Members of the specific plasma membrane phosphate transporter family called PHT1 are widely conserved in plants and they are considered as high-affinity Pi transporters. PHT1 transporters maintain their function even in heterologous systems such as animals and yeast (Muchhal et al. 1996; Daram et al. 1998; Guo et al. 2008; Preuss et al. 2010). In the *Arabidopsis* genome, the *PHT1* gene family is composed of nine members, *PHT1;1 to PHT1;9*, which exhibit sequence similarity to the yeast high-affinity Pi transporter *PHO84* (Bun-Ya et al. 1991; Okumura et al. 1998). Because four members of the PHT1 (*PHT1;1, PHT1;2, PHT1;3, and PHT1;4*) family are expressed in root epidermal cells, It is assumed that their function is to acquire Pi from the rhizosphere (Poirier and Bucher 2009). Members of *PHT1* gene family also could have an important role in Pi translocation within the plant, because their expression is detected also in flowers, hydathodes of cotyledons, pollen, leaf vasculature, and shoot buds (Karthikeyan et al. 2002; Mudge et al. 2002).

PHT transporters also catalyze a net import of Pi in plastids or vacuole and have an active role in the intracellular Pi dynamics. The PHT2 family of *Arabidopsis, Medicago truncatula*, and potato, harbors Pi low-affinity transporters. In such species, currently only one PHT2 member has been described: PHT2;1 is a chloroplast H+/Pi symporter with functions in Pi import and translocation. In *Arabidopsis*, *pht2;1* reveals a reduced Pi transport into the chloroplast and a decreased Pi allocation throughout the whole plant; however, the mutant is still viable, suggesting alternative mechanisms to Pi import into chloroplasts (Daram et al. 1999; Rausch and Bucher 2002; Versaw and Harrison 2002). PHT3 and PHT4 are PHT families recently reported. In *Arabidopsis*, three genes have been identified as members of the PHT3 family, which have been described as mitochondrial Pi transporters (Takabatake et al. 1999; Rausch et al. 2004). The detail function of PHT3 transporters in response to environmental stress and their molecular regulation is still poorly understood (Takabatake et al. 1999). Recently, Guo et al. (2008a, b) characterized all six members of PHT4 family in *Arabidopsis*. Subcellular localization of PHT4 proteins indicated that PHT4;1, PHT4;2, PHT4;4 and PHT4;5 are expressed in plastids; although the direction of Pi transport mediated by these transporters was not determined.

2.1.2 Pi Translocators

Once inside the cell, Pi taken from rhizosphere can join the metabolically active or inactive pool of Pi. It can become part of biomolecules (DNA, RNA, phospholipids), phosphorylated compounds, participate in energy transfer reactions and signaling cascades or stored in vacuole when plants are in Pi-optimal conditions (Schachtman et al. 1998; Kirkby and Johnston 2008; Cloutier et al. 2009). The interconversion of Pi between these different pools helps maintain Pi cellular homeostasis, in this process Pi translocators (PT) located in the root stele or organelle membranes play a critical role, since their activity regulates the Pi flow between organs (shoots and roots) and subcellular compartments.

Among the most studied PTs are those involved in sugar partitioning. PTs are proteins highly hydrophobic, with 6 to 8 transmembrane α-helix domains directly involved in energy metabolism trough carbon metabolism. These translocators act as dimers that catalyze an equimolar interchange of Pi and C3, C5 and C6 compounds between plastids and the cytosol (Fig. 2; Flügge 1999; Weber et al. 2005). Triose phosphate/phosphate translocator (TPT) was the first chloroplast PT discovered in plants and is considered the main Pi transport system into the chloroplast (Flügge and Heldt 1984; Flügge et al. 1989, 1991). TPT exports the fixed carbon during the Calvin cycle, at the expense of photosynthetically generated ATP and NADPH, in the form of glyceraldehydes-3-phosphate (GAP) and dihydroxyacetone-3-phosphate (DAP) from the plastid to the cytosol, where these metabolites are used for the biosynthesis of sucrose and amino acids (Flügge 1999; Weber 2004; Weber et al. 2005). Xylulose 5-phosphate/phosphate translocator (XPT)

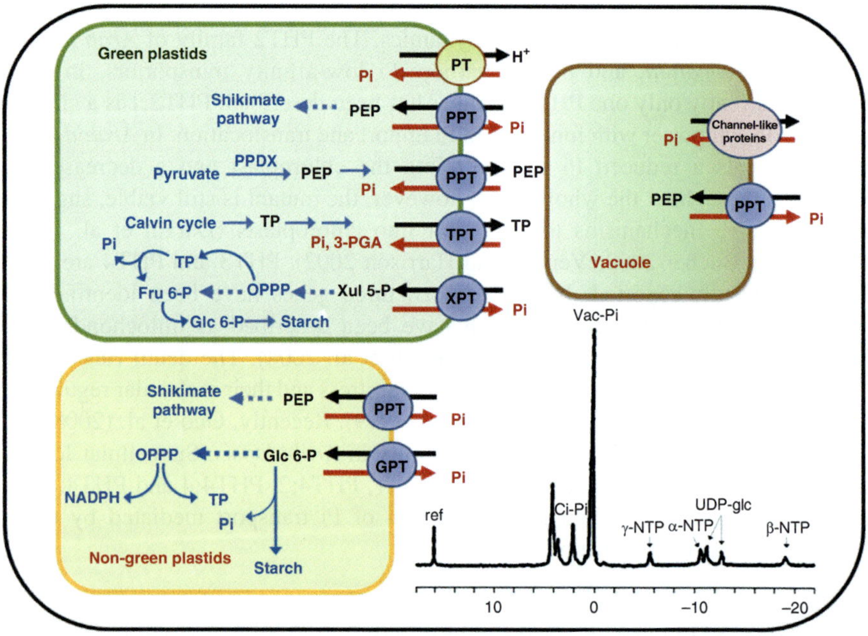

Fig. 2 Cellular Pi translocators. Intracellular Pi concentrations highly vary between organelles and cytosol as is shown by the in vivo ^{31}P-NMR spectrum from *Arabidopsis* inserted in the figure (Modified from Pratt et al. 2009). This Pi compartmentation together with intercelullar Pi changes play an important role in cellular Pi homeostasis and metabolic reactions. Several proteins anchored in organelle membranes are involved in this intracellular Pi dynamics. Phosphorylated sugar/Pi translocators (*blue circle*) have been found in green and non-green plastids and vacuoles, where they carry out a stoichiometric interchange of Pi and sugar molecules. Pi can also be interchanged between organelles and cytosol using others translocators; existence of vacuolar channel-like proteins and chloroplast low-affinity translocators, involved in Pi import, have been recently reported. Abbreviations: *ref* reference; *cyt-Pi* cytoplasmic Pi; *vac-Pi* vacuolar Pi; *NTP* nucleoside triphosphate; *UDP-glc* uridine-5′-diphosphate-α-D-Glc. See the text for the other abbreviations

and glucose 6-phosphate/phosphate translocator (GPT) are proteins with high amino acid similarity that interchange C-5 and C-6 compounds with Pi in the cytosol, respectively. XPT transports Xul 5-P and triose phosphates (TP) into the chloroplast from the cytosol (Eicks et al. 2002). It has been suggested that Xul 5-P allocated by XPT could feed the nonoxidative branch of the oxidative pentose phosphate pathway (OPPP), giving rise to four molecules of Fru 6-P and two molecules of TP, and then both metabolites are converted to Gluc 6-P (Eicks et al. 2002; Weber 2004; Weber et al. 2005). GPT is localized in heterotrophic tissues and imports Gluc 6-P into nongreen plastid, which can be a precursor for starch biosynthesis (Borchert et al. 1989; Kammerer et al. 1998). Disruption of Gluc 6-P import can seriously affect the oxidative phosphate cycle, which in turn affects fatty acid biosynthesis (Niewiadomski et al. 2005). Phosphoenolpyruvate/ phosphate translocator (PPT) is, meanwhile, a ubiquitously expressed PT. In green plastids of C4-plants, PPT is an abundant inner chloroplast membrane protein that exports PEP toward the cytosol, whereas in nongreen plastids and C3-metabolism chloroplast PPT imports PEP where this metabolite can be a precursor for the shikimate pathway (Fischer et al. 1997; Knappe et al. 2003a, b; Bräutigam et al. 2008). The roles of these Pi-dependent carriers (TPP, XP, GTP, and PPT) in the phosphorylated sugar partitioning between the plastid stroma and cytosol have been widely addressed in several studies on carbohydrate metabolism (for reviews see Emes and Neuhaus 1997; Weber 2004; Weber et al. 2005). However, their roles in shifting the phosphate balance of the cell have been less revised, especially when Pi availability is limited.

2.2 Pi Uptake and Translocation into the Plant

In the rhizosphere, Pi is generally found in fixed pools. Fixed Pi must be dissolved, desorbed, or degraded (mineralized) for plant uptake. Once present in the soil solution, Pi enters into the plant through symplastic pathway. First, Pi crosses the epidermal root cell membrane against a concentration gradient and is cotransported into the cell by PHTs (H+/Pi symporter). Pi moves from epidermis cells through the cortex into the endodermis and pericycle where it can be moved into the xylem for long distance transport (Schachtman et al. 1998).

The phosphate1 (PHO1) protein (which is not a Pi transporter) and its homologous *AtPHO1;H1 (Arabidopsis) and OsPHO1;2* (rice) are implicated in modulating Pi loading into the xylem. PHO1 is expressed in vascular root tissue mainly in stellar cells. Both *Ospho1;2, and Arabidopsis pho1* mutants have alteration in Pi partition, overaccumulating Pi in the root because of its increased incapacity to load phosphate into the xylem (Poirier et al. 1991; Hamburger et al. 2002; Secco et al. 2010). Interestingly, seven *AtPHO1* homologues have been found in a nonvascular plant, the moss *Physcomitrella patens*, although their function still remains to be determined (Wang et al. 2008). Pi transported to leaves exits the xylem and is distributed to the remaining tissues, where PHTs transport Pi into cells. Once inside cells, Pi would follow several pathways: (a) used in biosynthetic routes; (b) transported to plastids or mitochondria to be used in metabolic processes, (c) transported to vacuoles to be stored, or (d) transported to other cells (Rausch and Bucher 2002). The final or transient destination of Pi will depend on the kind of tissue and the metabolic requirements of cells. In senescence and Pi starvation conditions, Pi is remobilized from older to younger leaves and other sink tissues. In this process, PHO2 (phosphate2) and RNA mir399 are key regulators (Delhaize and Randall 1995; Fujii et al. 2005; Chiou 2007).

3 Adaptation of Plants to Pi Starvation

Plants as sessile organisms have to cope with highly variable environmental conditions. In the soil, despite a large amount of total Pi (Poirier and Bucher 2002), its availability is determinate by several factors, including the content of recalcitrant organic matter, the rhizosphere microbiota, soil pH and the presence of cations as calcium, iron, and aluminum. In this context, plants have evolved a series of adaptive responses that integrates external and internal signals to maintain a steady cellular Pi level (Fig. 4) (Schachtman et al. 1998; Abel et al. 2000; Vance et al. 2003; Ticconi and Abel 2004). In the follow paragraphs, we will describe these strategies to allow sustaining the plant metabolism during Pi stress.

3.1 Increasing the Pi Availability

3.1.1 Phosphorus Is Solubilized

To release Pi from organophosphates present in the rhizosphere (which account for approximately 30 to 80% of total soil phosphorus) and intracellular sources, the expression of genes encoding phosphatases (*APases*) and ribonucleases (*RNases*) is enhanced and these enzymes are released into the root vicinity and vacuoles (Köck et al. 1995, 1998; Zhang et al. 2003; Ticconi and Abel 2004; Zimmermann et al. 2004; Amtmann et al. 2006; Bozzo et al. 2006). Intracellular and secreted plant APases are up-regulated by low Pi availability and have been characterized as purple acid phosphatases (PAPs). In *Arabidopsis*, 29 PAP-like genes have been reported (Li et al. 2002). Among these PAPs, *AtACP5* (*AtPAP17*) (Pozo et al. 1999; Li et al. 2002), *PAP1* (*AtPAP12*) (Li et al. 2002) and *AtPAP26* (Veljanovski et al. 2006; Hurley et al. 2010) have been proposed to mediate phosphorus acquisition and redistribution based on their ability to hydrolyze Pi from organic sources. Two *Arabidopsis* mutants, *phosphatase underproducer 1* and *3* (*pup*) have been reported, which have reduced levels of PAPs in the root and accumulate less Pi in shoots when grown in media containing organic phosphorus forms (Trull and Deikman1998; Tomscha et al. 2004). In response to Pi starvation, APases are transcriptionally regulated by PHR1, WRKY75, and ZAT6 transcription factors (TFs) (Rubio et al. 2001; Devaiah et al. 2007a, b), and post-transcriptional mechanisms such as proteolysis appear to also be essential for the up-regulation of AtPAP26 during Pi stress (Bozzo et al. 2006; Hurley et al. 2010). In *pup3* Pi-responsive PAP12 activity is reduced with no change in its transcript level suggesting that PUP3 may be involved in the post-transcriptional modification of APase (Tomscha et al. 2004).

Nucleic acids in soil decaying organic matter are a source of extracellular Pi for deprived plants (Abel et al. 2000). *Arabidopsis* has two Pi-starvation induced RNAses, RNS1 and RNS2, but only RNS1 is secreted (Bariola et al. 1999). The transcriptional activation of these genes is a rapid and reversible process that is sensitive to changes in Pi concentration (Bariola et al. 1994; Köck et al. 1995, 1998; Chen et al. 2008), and is regulated at least by the TFs PHR1 and WRKY75 (Rubio et al. 2001; Devaiah et al. 2007a).

Roots of several plant species exude organic acids, such as malate and citrate, when the external Pi is low. These organic acids facilitate the dissolution of some insoluble compounds, such as $FePO_4$, $AlPO_4$, or $Ca_3(PO_4)_2$, by chelating the cations commonly associated with Pi in soil (Jones and Darrah 1994; Lan et al. 1995; Jones 1998; Vance et al. 2003). In a variety of plants including pigeon pea (*Cajanus cajan*), radish (*Raphanus sativus*), turnip (*Brassica napus*), barley (*Hordeum vulgare*), and rice (*Oryza sativa*), high rates of organic acids exudation are associated with a large capacity for Pi-mobilization (Otani et al. 1996; Zhang et al. 1997; Kirk et al. 1999; Gahoonia et al. 2000; López-Bucio et al. 2000). López-Bucio et al. (2000) showed that transgenic tobacco plants overexpressing a citrate synthase gene from *Pseudomonas aeruginosa*, increase citrate efflux from their roots and as a consequence the transgenic lines more effectively access Pi from $Ca_3(PO_4)_2$.

3.2 *Enhancing Pi Uptake,*

3.2.1 Phosphate Transport, and Translocation

In response to Pi deficiency, with the exception of PHT1-6, the expression of all members of the AtPHT1 family increases, particularly in the root system (Mudge et al. 2002; Rausch and Bucher 2002). In Pi stress PHT1-1 and PHT1-4 appear to have a major role in Pi acquisition: pth1-1 or pth1-4 single mutants display a 20 and 40% decrease in Pi uptake as compared to the WT, respectively, whereas the double mutant shows a 75% reduction (Misson et al. 2004). Proper plasma membrane targeting of at least some members of PHT1 family is dependent on PHF1 (phosphate transporter traffic facilitator1), an endoplasmic reticulum-located protein, structurally related to the SEC12 proteins of the early secretory pathway. *phf1* mutations lead to ER retention and reduced accumulation of PHT1;1 in the plasma membrane and consequently are impaired in Pi uptake and thus displays Pi starvation symptoms, including the constitutive expression of many Pi starvation–induced genes (Gonzalez et al. 2005).

The low-affinity Pi transporter PHT3;2 is induced in shoots and roots by Pi starvation (Misson et al. 2005; Morcuende et al. 2007; Müller et al. 2007). It is noteworthy that the expression of PHT2 and PHT4 members is not regulated by plant Pi status, suggesting that they are involved in the Pi intracellular dynamics but not with Pi uptake and remobilization.

Pi concentration in cytosol and plastids has been calculated in several studies but the values reported vary in several orders (from μM to mM). These large differences have been explained in function cell or tissue type analyzed (mature barley leaves, sycamore and *Arabidopsis* cells, root tips, isolated vacuoles) and the technique used to calculate Pi concentrations (spectrophometric assay, 31P-nuclear magnetic resonance) (Rébeillé et al. 1983; Bligny et al. 1990; Mimura et al. 1990; Ohnishi et al. 2007; Pratt et al. 2009). Nevertheless, many reports are consistent with the notion that the Pi concentration in the cytosol is 2–3-fold lower than that of plastids and mitochondria (Mimura et al. 1990; Versaw and Harrison 2002; Pratt et al. 2009) and under Pi starvation such concentration can drop, for example in leaves, up to 25-fold (Rausch and Bucher 2002; Pratt et al. 2009). To overcome this cellular Pi imbalance during Pi stress, a tight control of intracellular Pi flux and translocation into the plants must exist. In chloroplast, TPT and PPT regulate the endogenous Pi concentration (Shinano et al. 2005). TPT imports orthophosphate (Pi) or 3-phosphoglyceric acid (3-PGA), to replenish the ATP pool, to sustain the Calvin–Benson cycle and photosynthetic electron transport (Flügge et al. 1989; Weber 2004; Weber et al. 2005). PPT, meanwhile, can export or import Pi depending on the physiological status of the plastid where this transporter is located. PPT exports Pi in chloroplasts whereas in nongreen plastids such as amyloplastids imports Pi (Fischer et al. 1997; Knappe et al. 2003a, b; Bräutigam et al. 2008).

The expression of PPT and GTP is highly induced by prolonged Pi deficiency in *Arabidopsis*, bean, maize and rice, contrasting with the expression of TPT that remains constant in Pi-starved plants (Wasaki et al. 2003; Misson et al. 2005; Shinano et al. 2005; Hernández et al. 2007; Morcuende et al. 2007; Calderon-Vazquez et al. 2008). Although starch synthesis from pyruvate, imported by GTP into nongreen plastids, can release Pi (Kammerer et al. 1998), GTP appears to play a more important role in enhancing Pi use efficiency and carbon recycling under low Pi availability than in Pi storage.

Vacuoles can store significant amounts of Pi. Recent studies indicate that $H_2PO_4^-$ is the predominant form of vacuolar Pi, with concentration in the mM range. Under Pi deficiency, for short periods the cytoplasmic Pi concentration can be maintained through the efflux of Pi from the vacuole (Rébeillé et al. 1983). It has been shown that the cell responds rapidly to a decrease in Pi availability to maintain the concentration of cytoplasmic Pi by allowing Pi efflux from the vacuole to cytosol, and when this efflux is experimentally blocked, cellular metabolism and growth are completely abolished (Pratt et al. 2009), although the mechanisms of Pi transport across the tonoplast are little known so far. Analysis of Pi uptake in isolated vacuoles of *Catharanthus roseus* indicates that during this process the driving force is the electric potential generated by H+ pumps, for which the energy can be donated by ATP or pyrophosphate (Ohnishi et al. 2007; Pratt et al. 2009). Existence of Pi-specific channels was initially suggested in *Saccharomyces cerevisae* (Booth and Guidotti 1997; Kulakovskaya and Kulaev 1997). Dunlop and Phung (1998) characterized a $H_2PO_4^-$ permeable channel in *Beta vulgaris*. Similar channel-like proteins involved in Pi export and import have been also suggested to other plant species (Martinoia et al. 2007; Pratt et al. 2009).

3.2.2 Increasing in Root Absorption Area

The ability of plants to explore and exploit the spatially heterogeneous soil resources is determinate by the root system architecture (RSA), therefore changes in RSA can be crucial for plant adaptation to soils with limited water and/or nutrient availability. Since Pi availability is greater in the upper layers of the soil, plant species can enhance Pi acquisition by increasing root branching and root hair (RH) proliferation that increase the root absorption surface. (Lynch and van Beem 1993; Ma et al. 2001; Schmidt and Schikora 2001; Williamson et al. 2001; López-Bucio et al. 2002; Sánchez-Calderón et al. 2005). Several species undergo modifications in the root system architecture in response to Pi deficiency as a consequence of alterations in: (1) RH formation and elongation (2) cell growth and elongation, (3) root apical meristem activity, and (4) development of lateral, cluster, or proteoid roots.

In Pi-deprived plants, the root epidermis shows drastic changes. The rate and duration of RH growth increases, as well as the frequency of recruitment of atrichoblasts to form RH. The consequence of these responses to low Pi availability is an increase in root hair length and density (Bates and Lynch 2001). RHs contribute up to 80% of the surface contact area of the root, and are the main sites of Pi uptake in species that do not establish associations with mycorrhizae (Jungk 2001). Interestingly, the expression of high-affinity Pi transporters (PHT1;1 and ;4) is strongly enhanced in RH during Pi stress (Fig. 3), making the RHs an extremely efficient surface for Pi uptake. Due to this highly efficient Pi uptake system, available forms of Pi in soil are rapidly acquired by roots, generating in their surrounding "depletion zones" that will only slowly recharge through diffusion and mineralization (Hinsinger 2001).

In addition to changes in root hair formation and elongation, *Arabidopsis* shows an arrest of root tip growth and an increase in the initiation and elongation of lateral roots (LR) (Williamson et al. 2001; Linkohr et al. 2002; López-Bucio et al. 2002; Al-Ghazi et al. 2003; Nacry et al. 2005; Jain et al. 2007; Pérez-Torres et al. 2008). Common bean (*Phaseolus vulgaris*), maize and wheat display enhanced adventitious root formation and greater dispersion of LR (Lynch and van Beem 1993; Manske et al. 2000; Zhu and Lynch 2004). Lupinus (*Lupinus albus*) develops specialized root cluster structures which consist of massive numbers of secondary and tertiary roots covered by RH known as proteoid roots (Keerthisinghe et al. 1998; Vance et al. 2003; Dinkelaker et al. 2005; Lambers et al. 2006; Kirkby and

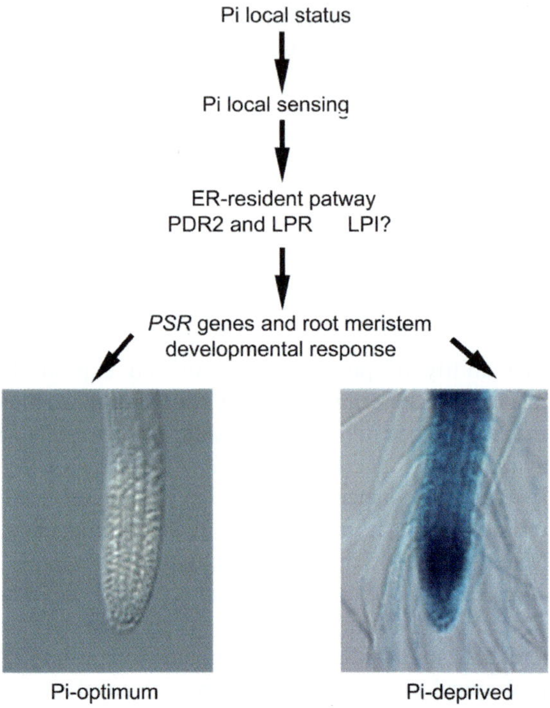

Fig. 3 Pi local sensing. Pi availability at the root tip is perceived by an unknown Pi starvation sensor (?) which triggers the endoplasmic reticulum (ER) resident pathway involved in changes in root architecture and in the activation of a subset of phosphate starvation responsive (PSR) genes. *Bottom* pictures: typical developmental changes in P transporter expression in primary roots of *Arabidopsis*. GUS staining for *pAtPT2:uidA* (pPHT1;4)

Johnston 2008). The formation of a highly branched root system, in response to Pi starvation produces a shallow root system capable of exploring large areas of the upper soil layer in search of nutrient-rich patches.

An example that shows the importance of the RSA came from the report of Yi et al. (2005) where the normally up-regulated transcription factor OsPTF1, a Pi starvation-induced transcription factor 1, was overexpressed. In low Pi condition, tiller number, shoot biomass, panicle weight, and P content of transgenic plants were higher in OsPTF1 overexpressing lines than in the WT, probably as a consequence of increased total root length and root surface area (20% compared to wild-type plants) (Yi et al. 2005).

In *Arabidopsis,* Pi limitation can induce an irreversible shift from an undeterminate to a determinate root developmental program in which cell division is arrested and cell differentiation is promoted. It starts with an inhibition of cell elongation followed by the progressive loss of meristematic cells in a response that requires the *LPR* multicopper oxidase gene. In this process, the quiescent center plays a central role (Sánchez-Calderón et al. 2005; Svistoonoff et al. 2007). As a consequence of the establishment of a determinate root growth developmental program under low P conditions, the primary root meristem differentiates with concomitant high expression levels of PAP and high-affinity Pi transporters (Sánchez-Calderón et al. 2005), suggesting its participation in Pi uptake.

3.3 Recovering Pi from the Cell Membrane

One of the largest internal Pi pools in plants is the cell membrane. During Pi deprivation, the relative phospholipids abundance is reduced and the abundance of nonphosphorus lipids such as sulphoquinovosyldiacylglycerol (SQDG), digalatosyldiacyglycerol (DGDG), and monogalatosyldiacyglycerol (MGDG) increases. (Essigmann et al. 1998; Härtel et al. 2000; Dörmann and Benning 2002; Andersson et al. 2003, 2005; Jouhet et al. 2004; Benning and Ohta 2005; Kobayashi et al. 2006; Li et al. 2006) (Fig. 4).

Membrane phospholipids are releasing Pi, to sustain other Pi-requiring processes, and diacylglycerol (DAG), which is subsequently converted into galactolipids or sulpholipids to compensate the reduction of phospholipids in the plasma membrane (Härtel et al. 2000; Calderon-Vazquez et al. 2008). There are two pathways to produce DAG from phosphatidylcholine (PC) and phosphatidylethanolamine (PE), either by phospholipase C (PLC) or phospholipases DZ1 and 2 (PLDZ1 and 2) in combination with phosphatidic acid phosphatases (PAPs; AtPAH1 and AtPAH2), these enzymes take essential part in adaptation mechanism to cope with phosphate starvation (Cruz-Ramírez et al. 2006; Li et al. 2006; Nakamura et al. 2009). The nonspecific phospholipase C (NPC5) is, at least partially, responsible for phospholipid degradation and DGDG accumulation in *Arabidopsis* leaves during Pi starvation (Gaude et al. 2008). A concluding evidence of the role of this gene is that whereas the *npc5-1* mutant attains normal growth, the double mutant *pho1* (leaves of *pho1* are permanently phosphate-deprived) *npc5-1* is severely retarded compared to *pho1*, indicating that NPC5 is required for normal growth under phosphate limitation (Gaude et al. 2008).

UDP-sulfoquinovose synthase1 (SQD1) and sulfoquinovosyl diacylglycerol synthase 2 (SQD2) (Yu et al. 2002), and the recently discovered UDP-glucose pyrophosphorylase3 (UGP3) (Okazaki et al. 2009) are three enzymes required for sulpholipid biosynthesis in *Arabidopsis* that are up-regulated under Pi deficient conditions (Misson et al. 2005; Morcuende et al. 2007; Okazaki et al. 2009). Galactolipids are synthesized by the transfer of galactose from UDP-galactose to diacylglycerol (DAG) by monogalactosyldiacylglycerol synthase (MGD) and digalactosyl diacylglycerol synthase (Dörmann and Benning 2002).

3.4 Metabolic Alterations

As an integral strategy to cope with Pi deficiency, plants modify their metabolic processes affecting respiration, glycolysis, and photosynthesis (Jeschke et al. 1997; Hammond et al. 2003; Plaxton 2004). During Pi stress, the electron transport chain is modified due to: (1) inhibition of the cytochrome pathway, (2) activation of alternative oxidase in respiration and (3) pyruvate accumulation in nonphosphorylating pathways that bypass energy requirement (Vance et al. 2003). Changes in photosynthetic rates are present, which is reflected in the decreased expression of genes involved in photosynthesis (Wu et al. 2003; Morcuende et al. 2007). The rate of photosynthetic carbon fixation as expected, declines in low Pi conditions (Wu et al. 2003). Metabolites further down in glycolysis such as glycerate-3-phosphate, glycerate-2-phosphate, phosphoenolpyruvate (PEP) also increase in the cytoplasm (Morcuende et al. 2007). There are also changes in the synthesis, translocation, and degradation of sucrose. The rate of transcription of genes encoding invertases, such as sucrose synthases, sucrose–phosphate synthases, and sucrose–phosphate phosphatases are highly affected (Hammond et al. 2003, 2005; Wu et al. 2003; Misson et al. 2005; Müller et al. 2007) (Fig. 4).

4 Regulation of the Low Pi Rescue System

All the changes that induce Pi stress as an adaptive response require highly coordinated programs of gene regulation. Through microarray technology, changes in the expression profiles of a broad number of genes during Pi stress have been determined (Hammond et al. 2003; Wu et al. 2003; Misson et al. 2005; Morcuende et al. 2007; Müller et al. 2007). The extent and complexity of the network of regulatory genes necessary to sense Pi status and regulate the Pi-low rescue system is now beginning to be revealed. In this context, the molecular mechanisms of how the plants sense Pi availability and how this signal evokes the adaptive responses to this deficient nutrition are important.

4.1 Pi Sensing

4.1.1 Pi Local Sensing

Responses such as root hair number and length, and primary root length are locally regulated. In *Arabidopsis* seedlings, when transferred to low Pi from high Pi conditions, newly formed root hairs are longer than those preexisting and vice versa, suggesting that those changes are regulated by local Pi-status (Bates and Lynch 2001).

Detailed analysis of root tips during Pi starvation suggest that Pi is locally sensed, since physical contact of the root tip with low Pi medium is necessary and sufficient to reduce primary root growth, suggesting that the meristem is the site where the external Pi concentration is sensed (Sánchez-Calderón et al. 2005; Svistoonoff et al. 2007). Mutants defective in this response such as *lpr* (low phosphate root), *pdr2* (phosphate deficiency response 2), and *lpi* (low phosphorus insensitive), have been proposed to function at a Pi-sensitive checkpoint in root development, which monitors environmental Pi status and adjust meristematic activity accordingly (Abel et al. 2000; Ticconi and Abel 2004; Sánchez-Cálderón

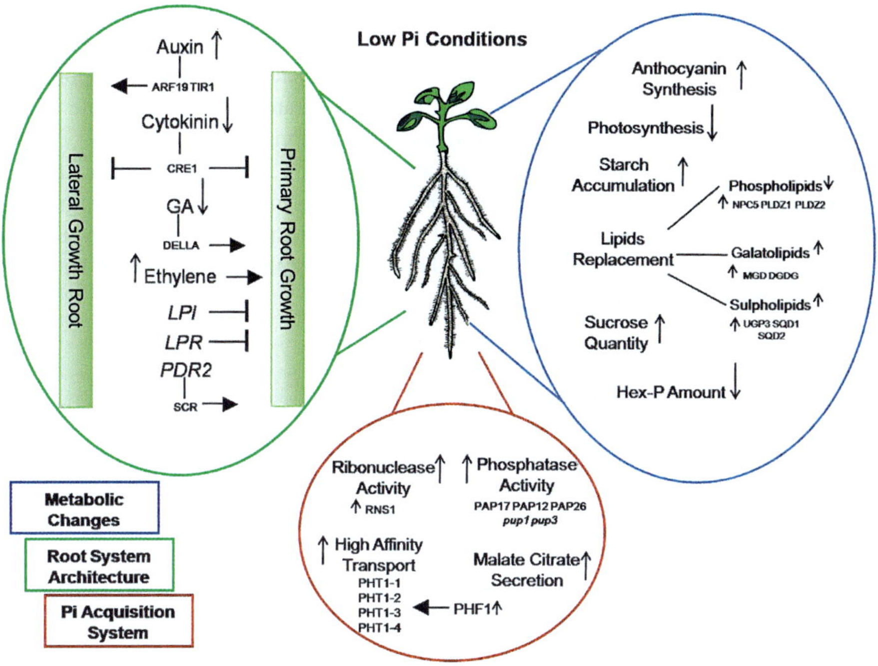

Fig. 4 Overview of the effects caused by Pi starvation conditions in three main aspects: Pi Acquisition System (*red circle*), Metabolic Changes (*blue circle*) and Root System Architecture (*green circle*) as indicated in the bottom *left* corner of the figure. Lines with *up arrows* symbolize upregulation (in case of genes) or increase (in case of metabolites or processes); lines with *down arrows* symbolize downregulation (in case of genes) or decrease (in case of metabolites or processes). Lines with *bold arrows* indicate positive effects and bars represent negative effects. Within Acquisition System are mentioned the four principal processes that account for enhanced Pi availability, the secretion of RNS1, the exudation of citrate and malate, the expression of four members of PHT family in de roots epidermal cells target to the plasma membrane by PHF1, and the secretion of ATPases into the vicinity of root tissues, mutants *pup*1 and 3 are also showed, since they underproduce the secreted ACP in root tissues. Metabolic changes like the increase of anthocyanin synthesis, starch accumulation and sucrose quantity, or the decrease of photosynthesis, hexoses-phosphate (Hex-P) amount, and the degradation of the major lipids of cellular membranes, phospholipids by NPC5, PLDZ1, PLDZ2, and their replacement by sulpholipids or galactolipids with the involvement of UGP3, SQD1, SQD2 and MGD DGDG, respectively, are also presented as a manner to enhance the Pi use efficiency. And finally, the effect on RSA is shown. First, the action of phytohormones like auxin, cytokinins, gibberellin (GA) and ethylene with some of their key participants (TIR1 ARF19; CRE1; DELLA; respectively) and second, some molecular components involved in Pi local signaling pathways, LPI, LPR and PDR2; triggering root growth responses to Pi starvation

et al. 2006; Svistoonoff et al. 2007). Analysis of these mutants reveals the mechanism of local Pi sensing. A multicopper oxidase encoded by *LPR1* is required for the root growth arrest caused by low-Pi availability (Svistoonoff et al. 2007) *PDR2* encodes the P5-type ATPase required for proper expression of SCARECROW (SCR), a key regulator of root patterning, and for stem-cell maintenance in

Pi-deprived roots, revealing a link between root patterning and the adjustment of meristem activity in response to Pi availability (Ticconi et al. 2009). LPR1 and PDR2 are thought to function together to adjust meristem activity in response to external Pi status (Ticconi et al. 2009). All these data attest the critical role of the root tip for local sensing and responding to Pi starvation.

Phosphite (Phi), a nonmetabolic Pi analog, is able to rescue the conditional short root phenotype of *pdr2* (Ticconi et al. 2001) and also interferes with Pi-signaling pathways attenuating developmental and molecular responses elicited by Pi-low (Carswell et al. 1996; Ticconi et al. 2001), supporting the notion that Pi by itself acts as a signaling molecule.

4.1.2 Long Distance Sensing

Split-root experiments demonstrated the existence of long-distance signals that report Pi status in the plant. In these experiments the root system of Pi-starved tomato plants was divided in two parts, one exposed to Pi-high and the other to Pi-low media. The root system exposed to Pi-deficient medium presented systemically repressed genes that are normally induced in response to Pi starvation (Liu et al. 1998), suggesting the existence of long-distance signals. Lateral root elongation is also affected by the shoot Pi status (Linkohr et al. 2002; Shane and Lambers 2006).

Sugars are good candidates to be systemical signals in the Pi-deprivation response. They are required for the activation of the low Pi rescue system including the induction of lateral root growth and the induction of Pi starvation-responsive genes (Karthikeyan et al. 2007). In plants grown in optimal Pi conditions, the addition of sucrose induces the accumulation of *Pht1;1* and *At4* transcripts, whereas in low Pi media the lack of sucrose represses transcript accumulation of these genes, suggesting that sucrose acts as one of the signals in the Pi starvation response pathway (Karthikeyan et al. 2007; Hammond and White 2008). Additionally, the mutant *pho3*, affected in *SUC2* gene encoding a sucrose–proton symporter, which is impaired in sucrose loading into the phloem, has several components of the Pi-low rescue system affected. In Pi sufficient conditions, the mutant exhibits a severely reduced growth, higher anthocyanin and starch content than wild-type plants (Zakhleniuk et al. 2001). This suggests that sugar could act as a systemic signal in the Pi starvation response. However, more research needs to be done to understand fully the role of sugars as components of the Pi starvation response pathway.

4.1.3 SPX Genes Domain Role in Pi Sensing

Recently, the role of proteins containing the SPX domain in the Pi starvation response has been examined in *Arabidopsis* and rice (Duan et al. 2008; Wang et al. 2009a, b). Most of the plant SPX domain genes are involved in responses to environmental cues or the internal regulation of nutrient homeostasis. IDS4 (iron-deficiency specific clone 4) from barley, contains part of the SPX domain and is expressed in

Fe-deficient roots (Nakanishi et al. 1993). The *Arabidopsis* genome encodes 20 genes with the SPX domain, four of them *AtSPX1*, *AtSPX2*, *AtSPX3* and *AtSPX4* are involved in Pi starvation response. Phosphate starvation strongly induced expression of *AtSPX1* and *AtSPX3*, whereas *AtSPX2* was only weakly induced and *AtSPX4* was repressed. Expression of *AtSPX1* is regulated through the SIZ1/PHR1 pathway. AtSPX1 is a nuclear located protein implicated in the transcriptional activation of genes involved in mobilization and scavenging of Pi, including *ACP5*, *RNS1* and *PAP2* (Duan et al. 2008). *PAP2* is involved in scavenging active oxygen species (Borevitz et al. 2000; Nagata et al. 2003). In rice, Wang et al. (2009a, b) showed that *OsPSX1*, an orthologue of *AtSPX1*, is specifically induced by Pi starvation and acts downstream of *OsPHR2* (an orthologue of *Arabidopsis* PHR1) and *OsPHO2*. SPX1 plays a regulatory role of several Pi starvation-induced genes: in Pi sufficient plants, silencing of OsSPX1 by RNAi induces the expression of the phosphate transporters OsPT2 and OsPT8. In contrast, in Pi starvation, *SPX1* overexpression suppressed the induction of several Pi starvation-induced genes including, *IPS1, IPS2, OsPAP10, OsSQD2, miR399d, miR399j, OsPT2, OsPT3, OsPT6* and *OsPT8*. This strongly suggests that OsSPX1 acts as a negative feedback factor of the PSI signaling pathway in rice (Wang et al. 2009a, b). Silencing of AtSPX3 by RNAi resulted in a hypersensitive response to Pi deprivation and leads to a decrease in total Pi allocation to roots and an increase of both total Pi and Pi in shoots. Thus AtSPX3 could have a positive role in plant adaptation to Pi deprivation improving Pi remobilization. Duan et al 2008 proposed that AtSPX3 might have a role in sensing internal Pi deficiency, and AtSPX1 in sensing external Pi deficiency. It has been proposed that through the evolutionary process some SPX domain transporters have acquired a function in signal reception, opening the possibility that some plant SPX domain proteins could also function as Pi sensors in plants (Lenburg and O'Shea 1996).

4.2 Regulation of Transcription

Pi availability regulates the expression of several hundreds or even thousands of genes (Hammond et al. 2003; Wu et al. 2003; Misson et al. 2005; Morcuende et al. 2007; Müller et al. 2007; Duan et al. 2008). Among the *Arabidopsis* genes regulated by Pi there are several TFs that seem to play a role as regulators of this response, including WRKY75 (which contains a conserved WRKYGQK sequence; (Devaiah et al. 2007a), WRKY6 (Chen et al. 2009), ZAT6 (zinc finger6) (Devaiah et al. 2007b), BHLH32 (basic helix_loop_helix 32) (Chen et al. 2007), MYB62 (Devaiah et al. 2009) and ARP6 (Smith et al. 2010). (Fig. 5)

4.2.1 Chromatin Structure Modifications

Smith et al. (2010) reported the role of ARP6-dependent H2A.Z deposition in controlling some Pi starvation responses in *Arabidopsis*. ARP6 is a nuclear

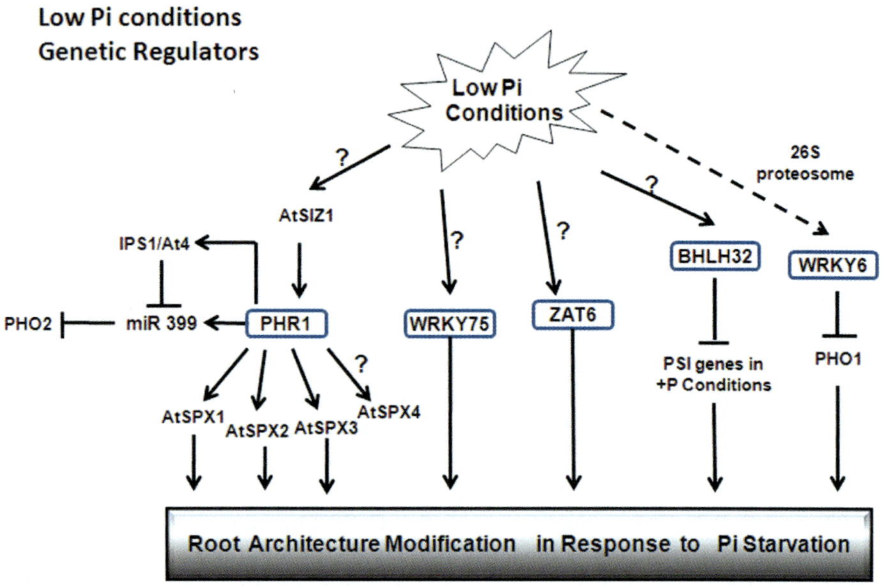

Fig. 5 Model for Pi starvation signaling pathways in *Arabidopsis thaliana*. Unknown signal molecule (?) which is activated in a Pi status dependent manner transduces the signal to each Transcription Factor (TF) involved in the Pi starvation response. The signaling pathway directed by each TF is represented as PHR1, WRKY75, ZAT6, BHLH32 and WRKY6. *Arrows* indicate positive regulation and bars indicate negative regulation, the *dashed arrow* indicates degradation of WRKY6 induced by low phosphate. PHO2 acts on the translocation of Pi from shoot to the roots through the xylem, and it also acts as inhibitor of some PSI genes and is downregulated by miRNA 399

actin-related protein conserved among all eukaryotes and is an essential component of the SWR1 chromatin-remodeling complex, which regulates transcription via deposition of the H2A.Z histone variant into the chromatin. *arp6* mutants exhibit short primary roots and increased number and length of root hairs, as well as increased phosphatase activity in shoots, suggesting that the deposition of this H2A.Z plays a role in the transcriptional regulation of genes related to this Pi-starvation phenotype.

4.2.2 Transcription Factors

The PHR1 transcription factor was the first signaling component involved in Pi starvation responses to be identified and characterized at the molecular level (Rubio et al. 2001). PHR1, encoding a member of the MYB family of DNA-binding proteins with a predicted coiled-coil domain involved in protein dimerization, regulates a large subset of Pi starvation-responsive genes through the binding to the GNATATNC P1BS DNA motif (PHR1 specific binding sequence) (Rubio et al. 2001). The *phr1* mutant displays a reduced concentration of Pi under both Pi-sufficient and Pi-deficient conditions and minimal induction of anthocyanin accumulation in response to Pi deprivation, whereas overexpression of PHR1 leads to an increased concentration of Pi in the shoots, together with an elevated expression of a range of Pi starvation-induced genes encoding Pi transporters, phosphatase and RNase (Nilsson et al. 2007). Therefore, PHR1 is a key transcriptional activator in controlling Pi uptake and allocation, anthocyanin accumulation, and carbon metabolism.

Recently, in rice, two homologous of *PHR1, OsPHR1* and *OsPHR2*, were identified (Zhou et al. 2008). Analysis of the transgenic rice lines indicated that both transcription factors are involved in the Pi starvation signaling pathway by regulating the expression of Pi starvation-induced genes, however only overexpression of *OsPHR2* results in increased Pi accumulation in the shoot. Under Pi sufficient conditions, plants overexpressing *OsPHR2* mimic Pi starvation stress with enhanced root elongation, enhanced root hair elongation and up-regulation of some Pi transporter genes which correlates with the excessive accumulation of Pi in shoots (Zhou et al. 2008).

PHR1 expression and its nuclear localization are independent of Pi status (Rubio et al. 2001). PHR1 has two predicted sumoylation sites, and in vitro assays showed that PHR1 sumoylation is mediated by the nuclear-located small ubiquitin-like modifier (SUMO) E3 ligase 1 (SIZ1) (Miura et al. 2005) (Fig. 5). Sumoylation can alter protein function in many different ways, for example, by controlling subcellular localization, activity or stability (Miura et al. 2007). *SIZ1* was first identified as a suppressor of *Arabidopsis sos3 (salt overly sensitive 3)* mutant, and then found to also be involved in Pi starvation responses (Miura et al. 2005). The *siz1* mutant has an exaggerated lateral root and root hair growth, and accumulates more anthocyanin in Pi deficient conditions. Expression of SIZ1 is up-regulated by Pi depletion and the intracellular Pi concentration is higher in shoots of *siz1* than in wild-type plants under Pi-sufficient conditions, suggesting that sumoylation could be a key control mechanism that acts on different Pi starvation-induced responses, at least partially via posttranscriptional regulation of PHR1.

The use of global expression analysis tools like microarrays, allowed the identification of three TFs, WRKY75, ZAT6, and BHLH32, that are transcriptionally induced by Pi deficiency (Misson et al. 2005; Chen et al. 2007; Devaiah et al. 2007a, b). Expression of WRKY75 is induced by Pi starvation and the gene product resides in the nucleus regardless of Pi status. RNAi suppression of WRKY75 impairs many aspects of Pi starvation responses as the early accumulation of anthocyanin, reduced expression of Pi responsive genes, and Pi uptake and alterations in RSA, producing more lateral roots and root hairs compared to wild-type plants under both Pi sufficient and deficient conditions (Devaiah et al. 2007a), suggesting a dual role for this transcription factor. WRKY75 contains a zinc finger motif that could bind to the W-box DNA motif (TTGACC/T), which is present in the promoters of at least 20 Pi starvation-induced genes (Devaiah et al. 2007a).

WRKY6 overexpression lines are more sensitive to low Pi stress and had lower Pi content in shoots compared to the WT, presenting a phenotype that resembles that of *pho1*. Interestingly, WRKY6 binds to the *PHO1* promoter and low Pi treatment reduces this binding, indicating that WRKY6 is a repressor of PHO1 expression. Moreover, it was proposed that low Pi–induced derepression of PHO1 may result from the 26S proteosome–mediated proteolysis of WRKY6 (Chen et al. 2007) (Fig. 5).

ZAT6 encodes a cysteine-2/histidine-2 (C2H2) zinc finger transcription factor that functions as a transcriptional repressor. Suppression of ZAT6 by RNAi is apparently lethal, whereas young Pi-deprived seedlings of *ZAT6* overexpressing lines showed low Pi content and uptake rate, increased anthocyanin accumulation and down-regulation of some Pi starvation-induced genes. ZAT6 overexpression also resulted in altered root architecture of older plants, with consequent changes in Pi acquisition. These results indicate that ZAT6 regulates the development of primary root growth and regulates Pi homeostasis independently of the Pi status (Devaiah et al. 2007b).

Chen et al. (2007) showed the participation of the *Arabidopsis* BHLH transcription factor 32 (BHLH32) in the response to Pi deprivation. In *bhlh*32 plants grown under Pi sufficient conditions, several components of the Pi rescue system such as the expression of several Pi-starvation responsive genes, anthocyanin accumulation, total Pi content and root hair formation are all significantly increased. BHLH32 thus seems to serve as a negative regulator of a variety of biochemical and morphological responses induced by Pi (Chen et al. 2007).

The importance of posttranscriptional mechanisms in the regulation of Pi starvation responses emerged again from the characterization of *PER1*, a gene encoding the ubiquitin-specific protease 14 (UBP14). *per1* seedlings show normal root hair development under optimal Pi but displayed an inhibited root hair elongation phenotype upon Pi deficiency. Transcriptional profiling of *per1* and wild-type plants subjected to short-term Pi starvation revealed genes that may be important for the signaling of Pi deficiency (Li et al. 2010).

4.2.3 Hormonal Regulation

Phytohormones are believed to play a crucial role in all the RSA alterations caused by Pi deficiency (Fig. 4). It has been shown by microarray data that Pi starvation can affect the expression of many genes involved in hormonal responses (Misson et al. 2005). Particularly auxins, play an important role in mediating the Pi starvation effects on RSA (López-Bucio et al. 2002; Al-Ghazi et al. 2003; Nacry et al. 2005; Jain et al. 2007). The roots of Pi deprived seedlings are more responsive, in terms of LR formation, to exogenous auxins than are roots of nondeprived seedlings (López-Bucio et al. 2002). The role of auxins in mediating root developmental responses to low Pi availability seems to be rather complex. López-Bucio et al. (2005) proposed that primary root growth inhibition in low Pi conditions was independent of polar auxin transport, whereas enhanced root branching was likely to be an auxin-dependent process. Recently, Pérez-Torres et al. (2008) showed that the increase in LR formation in Pi-deprived *Arabidopsis* seedlings is, at last partially, mediated by an increase in auxin sensitivity of root cells and that Pi availability modulates the expression of the TIR1 auxin receptor.

The involvement of ethylene signaling has been studied in the context of Pi-starvation changes in root morphology (López-Bucio et al. 2002; Ma et al. 2003; Zhang et al. 2003). The effects of ethylene on root development are opposite for Pi-sufficient and Pi-starved plants (Borch et al. 1999; Ma et al. 2003). Thus, ethylene decreases root elongation in Pi-sufficient plants and has an opposite effect on Pi-starved plants. Reciprocally, inhibitors of ethylene synthesis or action promote increased root elongation in Pi-sufficient plants and decreased elongation in Pi-starved plants. Likewise, inhibition of ethylene synthesis or activity results in root hair initiation closer to the root tip in Pi-starved plants and farther away from the root tip in Pi-sufficient plants (Ma et al. 2003). It is also noteworthy that the ethylene signaling that influences root hair formation in Pi-sufficient roots is not involved in the increase in root hair density and size that occurs in Pi-starved plants (Ma et al. 2001; Schmidt and Schikora 2001).

Early studies suggested the role of cytokinins in responses to nutrient deficiency. Cytokinin levels were found to decrease after phosphate or nitrate starvation (Salama and Wareing 1979; Horgan and Wareing 1980). Exogenous application of cytokinins represses in the root the induction of many Pi starvation-responsive genes, such as *ACP5* and *PHT1;1* (Martín et al. 2000), and this effect is impaired in *cre1* mutants, implicating cytokinin two component signaling in the negative regulation of the Pi starvation responses (Franco-Zorrilla et al. 2002). Cytokinin drastically inhibits primary root elongation in low Pi seedlings at concentrations of 10^{-7} M and has a gradual effect on primary root elongation at high P conditions. Lateral root formation is also inhibited by cytokinin at both low and high P conditions. However, there are no significant differences between cytokinin untreated plants grown at high P and low P grown plants treated with 10^{-6} M of this hormone. Cytokinin also inhibits galactolipid synthesis in response to Pi starvation (Kobayashi et al. 2006). Recently, it has been shown that cytokinin modulates the level of meristem cell cycle activity and that this, in turn, influences the expression of Pi starvation-responsive genes (Lai et al. 2007). Thus, cytokinin seems to play a negative role in Pi starvation responses.

Gibberellin (GA) plays an important role in regulating plant growth and development throughout the life cycle, including seed germination, root elongation, hypocotyl growth, leaf expansion, floral initiation, and floral development (Hooley 1994; Richards et al. 2001). Jiang et al. (2007) found that Pi starvation induced changes in the levels of transcripts encoding GA metabolism enzymes and decreases the level of bioactive GA. Root architecture was also altered by GA, since it increases primary root length of wild-type Pi-starved seedlings, but does not significantly alter the number of lateral roots formed by wild-type seedlings independently of Pi status. Pi starvation also causes the accumulation of DELLA proteins that are negative regulators of GA responses.

A more direct evidence of the role of GA in Pi homeostasis came from a recent report where the overexpression of the MYB transcription factor 62 (MYB62), which produces a characteristic GA-deficiency phenotype that can be partially reversed by exogenous application of GA, leads to a down-regulation of several phosphate starvation-induced genes, altered root architecture, lower Pi uptake and acid phosphatase activity, leading to decreased total Pi content in the shoots (Devaiah et al. 2009).

5 Pi Homeostasis

5.1 Phosphate Signaling Pathway Dependent on PHR1, PHO2 and MicroRNA399

One of the most important plant responses to cope with low Pi availability includes enhancement of external Pi acquisition by roots and the conservation of internal Pi resources, which is accomplished through a complex regulation of signaling by both local and systemic pathways. A step to start elucidating one of the signaling pathways of the Pi starvation response was the characterization of the *pho2 Arabidopsis* mutant, which overaccumulates Pi in the shoots in both Pi sufficient and Pi deficient conditions (Delhaize and Randall 1995). Map-based cloning of PHO2 revealed that it encodes an ubiquitin-conjugating E2 enzyme (UBC24; Aung et al. 2006; Bari et al. 2006). The mRNA level of PHO2 is regulated by miR399, a microRNA strongly induced upon Pi-starvation (Bari et al. 2006; Chiou et al. 2006). The expression of miR399 increases rapidly during Pi depletion and is partially inhibited in *phr1* mutants, suggesting that PHR1 is required for miR399 expression. When the expression of miR399 is induced by Pi starvation the transcript level of PHO2 is significantly reduced (Fujii et al. 2005). The overexpression of miR399 in WT *Arabidopsis* plants results in high levels of Pi in shoots. These results and the finding that miRNA399 and PHO2 are both expressed in vascular tissue support the notion that these genes play an important role in Pi translocation and remobilization (Aung et al. 2006; Bari et al. 2006; Chiou et al. 2006; Chiou 2007). Pant et al. (2008) found that miR399 is present in phloem sap of two plant species, rapeseed (*Brassica napus*) and pumpkin (*Cucurbita maxima*), and demonstrated that its expression is Pi-status-dependent. The movement of miR399 through the phloem sap strongly suggests this microRNA is implicated in long-distance signaling (Lin et al. 2008; Pant et al. 2008). The movement of miR399 establishes communication from shoots to roots through the modulation of the expression of PHO2, which is crucial to enhance Pi uptake and translocation during the Pi deficiency. In addition to the vascular system, miR399 is also expressed in root tips and mesophyll cells, suggesting the possibility of additional, unknown targets of miR399 in these tissues.

Other components of the pathway implicated in the redistribution of Pi in response to low Pi conditions is the Mt4/TPSI1 gene family (Shin et al. 2006). These genes produce noncoding mRNAS that rapidly increase upon Pi starvation and have been assigned a regulatory role in Pi starvation responses (Fig. 5). At4 is expressed in the vascular tissue and the *at4* mutant is defective in Pi redistribution to the roots in response to low Pi, hereby, At4 has been implicated in internal allocation of Pi between shoot and root during Pi stress. All members of the

At4 gene family present complementary sequences to a mir399 (Shin et al. 2006). Franco-Zorrilla et al. (2007) showed that IPS1 inhibits miR-399 activity, through a process denominated "target mimicry". In this process, IPS1 sequesters miR399 through sequence complementary interaction, thereby resulting in the accumulation of PHO2 mRNA which as mentioned above is a target of miR399. Overexpression of both IPS1 and At4 results in decreased shoot Pi accumulation, which suggests a redundant activity (Franco-Zorrilla et al. 2007). Several homologues of the *PHR1, PHO2 and MicroRNA399* systemic Pi signaling pathway have been identified in common bean and rice (Doerner 2008; Zhou et al. 2008), showing that systemic Pi homeostasis is an evolutionary conserved mechanism in vascular plant species.

5.2 Emerging Insights into Ca^{2+} Roles in Pi Homeostasis

It has been suggested that the Ca^{2+} could be involved in Pi homeostasis (Cheng et al. 2005). *CAX1* (Calcium Exchanger 1) a vacuolar H^+/Ca^{2+} transporter and its close homologue *CAX3* have been implicated in the Pi starvation response. *CAX1* is preferentially expressed in leaves, while *CAX3* is highly expressed in roots (Cheng et al. 2003, 2005). The *cax1/cax3* double mutant is impaired in growth and is also affected in the sensitivity to hormones and ion homeostasis. Cheng et al. (2005) proposed that *CAX1* and *CAX3* could be involved in Pi homeostasis because the *cax1/cax3* double mutant shows 66% more shoot Pi than the wild type.

Fluctuation of cytosolic Ca^{2+} concentrations triggers a signaling cascade where inositol 1,4,5-triphosphate [Ins(1,4,5)P3, InsP3] is generated by the activation of phospholipase C. InsP3 regulates the liberation of Ca^{2+} from the vacuole and endoplasmic reticulum through InsP3-gated Ca2+ channels (Allen et al. 1995; Sanders et al. 2002; Krinke et al. 2007). The characterization of the *ipk1* (inositol polyphosphate kinase 1) mutant revealed the first insight in the connection between inositol phosphate signals and Pi homeostasis (Stevenson-Paulik et al. 2005). The *ipk1* mutant present overaccumulation of Pi in leaves similar to *pho2* mutant. The *IPK1* gene encodes an IP4 and IP5 2-Kinase, which are involved in the biosynthesis of phytate or inositol hexakisphosphate (InsP6). In low Pi, *ipk1* shows root hairs longer than wild type and is less sensitive to high external Pi concentrations, that suggest that the involvement of InsP signals in the integration of Pi sensing starvation responses, is possible (Stevenson-Paulik et al. 2005). InsP4, InsP5 and InsP6 act downstream of InsP3, thus the involvement of calcium in Pi homeostasis could be through modulation of InsP3.

5.3 Soil Minerals Affecting Pi Response

Zinc deficiency can cause accumulation of Pi in plant tissues, which can produce toxicity when Pi is readily available. The high-affinity Pi transporters Hv-PT1 and

Hv-PT2 of barley are induced by zinc deficiency, independently of Pi availability. The regulation of the high-affinity Pi transporter cannot be mimicked by any other divalent cation, suggesting that zinc may have a specific role in the signal transduction pathway regulating the expression of high-affinity Pi transporter genes (Huang et al. 2000).

In *Arabidopsis* iron is necessary to trigger the inhibition of primary root growth upon Pi starvation (Ward et al. 2008). This response has been liked to iron toxicity, however since low concentrations of Fe are sufficient for the inhibition of primary root growth by Pi-deficiency, how Fe affect this response it are still poorly understood.

6 Conclusions

Modern agriculture faces new challenges to feed a global growing population: higher crop yields, lower production costs and minimum negative impacts on the environment seem to be mutually exclusive features but are indispensable traits for a sustainable agriculture. Enhancing crop yields through excessive fertilizer use is not an adequate solution and, therefore developing crops with a better nutrient use efficiency are an important component of an integral strategy to solve this problem. To achieve this goal it is necessary to know how plants acquire, transport and translocate nutrients from the soil and how they cope with nutrient deficiency.

Pi is an essential nutrient for plant development and reproduction, which is largely unavailable in most of the soils. Experimental biology together with molecular and genetic strategies, supported by high-throughput techniques for gene expression analysis, large-scale forward genetic screening, and complete genome sequencing of model plants, have allowed the identification of important components of the Pi sensing and signaling pathway in plants. The study of model species such as *Arabidopsis*, rice, and *Lupinus* has shown that angiosperms present general and specific mechanisms to cope with Pi deprivation. Several transcriptional modulators have also been described and integrated in a Pi regulatory network where PHR1 plays a central role. However, despite all the identified components so far, the Pi sensing mechanism through which physiological and biochemical processes are triggered remains to be completely elucidated. Nevertheless, these recent advances show that it is feasible to produce crop varieties with a better Pi use efficiency either by exploiting natural genetic variation or by transgenic strategies. For example, overexpression of PHR1 leads to an increased concentration of Pi in the shoots, (Nilsson et al. 2007), transgenic rice plants overexpressing OsPTF1 have a better yield in low Pi conditions (Yi et al. 2005). Therefore, the application of the increasing knowledge of the plant responses to Pi availability will be enormously useful for designing new plant varieties with an increased nutrient use efficiency to extend agricultural frontiers to poor soils and cope with the increased demand for food and feed, as well as biomass for industrial purposes.

References

Abel S, Nurnberger T, Ahnert V, Krauss G, Glund K (2000) Induction of an extracellular cyclic nucleotide phosphodiesterase as an accessory ribonucleolytic activity during phosphate starvation of cultured tomato cells. Plant Physiol 122:543–552

Al-Ghazi Y, Muller B, Pinloche S, Tranbarger TG, Nacry P, Rossignol M, Tardieu F, Doumas P (2003) Temporal responses of Arabidopsis root architecture to phosphate starvation: evidence for the involvement of auxin signaling. Plant Cell Environ 26:1053–1066

Allen GJ, Muir SR, Sanders D (1995) Release of Ca2+ from individual plant vacuoles by both InsP3 and cyclic ADP-ribose. Science 268:735–737

Amtmann A, Hammond JP, Armengaud P, White PJ (2006) Nutrient sensing and signalling in plants: potassium and phosphorus. Adv Bot Res 43:209–257

Andersson MX, Stridh MH, Larsson KE, Liljenberg C, Sandelius AS (2003) Phosphate-deficient oat replaces a major portion of the plasma membrane phospholipids with the galactolipid digalactosyldiacylglycerol. FEBS Lett 537:128–132

Andersson MX, Larsson KE, Tjellstrom H, Liljenberg C, Sandelius AS (2005) The plasma membrane and the tonoplast as major targets for phospholipid-to-glycolipid replacement and stimulation of phospholipases in the plasma membrane. J Biol Chem 280:27578–27586

Aung K, Lin S-I, Wu C-C, Huang Y-T, Su C-I, Chiou T-J (2006) pho2, a phosphate over-accumulator, is caused by a nonsense mutation in a microRNA399 target gene. Plant Physiol 141:1000–1011

Bari R, Datt Pant B, Stitt M, Scheible W-R (2006) PHO2, microRNA399, and phr1 define a phosphate-signaling pathway in plants. Plant Physiol 141:988–999

Bariola PA, Howard CJ, Taylor CB, Verburg MT, Jaglan VD, Green PJ (1994) The *Arabidopsis* ribonuclease gene *RNS1* is tightly controlled in response to phosphate limitation. Plant J 6:673–685

Bariola PA, MacIntosh GC, Green PJ (1999) Regulation of S-Like ribonuclease levels in *Arabidopsis*. Antisense inhibition of RNS1 or RNS2 elevates anthocyanin accumulation. Plant Physiol 119(1):331–342

Bates TR, Lynch JP (2001) Root hairs confer a competitive advantage under low phosphorus availability. Plant Soil 236:243–250

Benning C, Ohta H (2005) Three enzyme systems for galactoglycerolipid biosynthesis are coordinately regulated in plants. J Biol Chem 280:2397–2400

Bieleski RL (1973) Phosphate pools, phosphate transport, and phosphate availability. Annu Rev Plant Physiol 24:225–252

Bligny R, Gardestrom P, Roby C, Douce R (1990) ^{31}P NMR studies of spinach leaves and their chloroplasts. J Biol Chem 265:1319–1326

Booth J, Guidotti G (1997) Phosphate transport in yeast vacuoles. J Biol Chem 272:20408–20413

Borch K, Bouma TJ, Lynch JP, Brown KM (1999) Ethylene: a regulator of root architectural responses to soil phosphorus availability. Plant Cell Environ 22:425–431

Borchert S, Grosse H, Heldt H (1989) Specific transport of inorganic phosphate, glucose 6-phosphate, dihyroxyacetone phosphate and 3-phosphoglycerate into amyloplasts from pea roots. FEBS Lett 253:183–186

Borevitz JO, Xia Y, Blount J, Dixon RA, Lamb C (2000) Activation tagging identifies a conserved MYB regulator of phenylpropanoid biosynthesis. Plant Cell 12:2383–2394

Bozzo GG, Dunn EL, Plaxton WC (2006) Differential synthesis of phosphate-starvation inducible purple acid phosphatase isozymes in tomato (Lycopersicon esculentum) suspension cells and seedlings. Plant Cell Environ 29:303–313

Bräutigam A, Hoffmann-Benning S, Weber APM (2008) Comparative proteomics of chloroplast envelopes from C3 and C4 plants reveals specific adaptations of the plastid envelope to C4 photosynthesis and candidate proteins required for maintaining C4 metabolite fluxes. Plant Physiol 148:568–579

Bun-Ya M, Nishimura S, Oshima Y (1991) The PHO84 gene of *Saccharomyces cerevisiae* encodes an inorganic phosphate transporter. Mol Cell Biol 11:3229–3238

Calderon-Vazquez C, Ibarra-Laclette E, Caballero-Perez J, Herrera-Estrella L (2008) Transcript profiling of Zea mays roots reveals gene responses to phosphate deficiency at the plant- and species-specific levels. J Exp Bot 59:2479–2497

Carswell C, Grant BR, Theodorou ME, Harris J, Niere JO, Plaxton WC (1996) The fungicide phosphonate disrupts the phosphate-starvation response in Brassica nigra seedlings. Plant Physiol 110:105–110

Chen YF, Li LQ, Xu Q, Kong Y H, Wang H, Wu WH (2009) The WRKY6 transcription factor modulates PHOSPHATE1 expression in response to low Pi stress in *Arabidopsis*. Plant Cell 21:3554–3566

Chen Z-H, Nimmo GA, Jenkins GI, Nimmo HG (2007) BHLH32 modulates several biochemical and morphological processes that respond to Pi starvation in Arabidopsis. Biochem J 405:191–198

Chen Y, Wang Y, Wu W (2008) Membrane transporters for nitrogen, phosphate and potassium uptake in plants. J Integr Plant Biol 50:835–848

Cheng N-H, Pittman JK, Barkla BJ, Shigaki T, Hirschi KD (2003) The Arabidopsis cax1 mutant exhibits impaired ion homeostasis, development, and hormonal responses and reveals interplay among vacuolar transporters. Plant Cell 15:347–364

Cheng N-H, Pittman JK, Shigaki T, Lachmansingh J, LeClere S, Lahner B, Salt DE, Hirschi KD (2005) Functional association of Arabidopsis CAX1 and CAX3 is required for normal growth and ion homeostasis. Plant Physiol 138:2048–2060

Chiou TJ (2007) The role of microRNAs in sensing nutrient stress. Plant Cell Environ 30:323–332

Chiou T-J, Aung K, Lin S-I, Wu C-C, Chiang S-F, Su C-I (2006) Regulation of phosphate homeostasis by microRNA in Arabidopsis. Plant Cell 18:412–421

Cloutier M, Chen J, Tatge F, McMurray-Beaulieu V, Perrier M, Jolicoeur M (2009) Kinetic metabolic modelling for the control of plant cells cytoplasmic phosphate. J Theor Biol 259:118–131

Cruz-Ramírez A, Oropeza-Aburto A, Razo-Hernández F, Ramírez-Chávez E, Herrera-Estrella L (2006) Phospholipase DZ2 plays an important role in extraplastidic galactolipid biosynthesis and phosphate recycling in Arabidopsis roots. Proc Natl Acad Sci USA 103:6765–6770

Daram P, Brunner S, Persson BL, Amrhein N, Bucher M (1998) Functional analysis and cell-specific expression of a phosphate transporter from tomato. Planta 206:225–233

Daram P, Brunner S, Rausch C, Steiner C, Amrhein N, Bucher M (1999) Pht2;1 encodes a low-affinity phosphate transporter from Arabidopsis. Plant Cell 11:2153–2166

Delhaize E, Randall PJ (1995) Characterization of a phosphate accumulator mutant of Arabidopsis thaliana. Plant Physiol 107:207–213

Devaiah BN, Karthikeyan AS, Raghothama KG (2007a) WRKY75 transcription factor is a modulator of phosphate acquisition and root development in Arabidopsis. Plant Physiol 143:1789–1801

Devaiah BN, Nagarajan VK, Raghothama KG (2007b) Phosphate homeostasis and root development in Arabidopsis are synchronized by the zinc finger transcription factor ZAT6. Plant Physiol 145:147–159

Devaiah BN, Madhuvanthi R, Karthikeyan AS, Raghothama KG (2009) Phosphate starvation responses and gibberellic acid biosynthesis are regulated by the MYB62 transcription factor in Arabidopsis. Mol Plant 2:43–58

Dinkelaker B, Hengeler C, Marschner H (2005) Distribution and function of proteoid roots and other root clusters. Bot Acta 108:183–200

Doerner P (2008) Phosphate starvation signaling: a threesome controls systemic Pi homeostasis. Curr Opin Plant Biol 11:536–540

Dörmann P, Benning C (2002) Galactolipids rule in seed plants. Trends Plant Sci 7:112–118

Duan K, Yi K, Dang L, Huang H, Wu W, Wu P (2008) Characterization of a sub-family of Arabidopsis genes with the SPX domain reveals their diverse functions in plant tolerance to phosphorus starvation. Plant J 54:965–975

Dunlop J, Phung T (1998) Phosphate and slow vacuolar channels in *Beta vulgaris*. Aust J Plant Physiol 25:709–718

Eicks M, Maurino V, Knappe S, Flügge U-I, Fischer K (2002) The plastidic pentose phosphate translocator represents a link between the cytosolic and the plastidic pentose phosphate pathways in plants. Plant Physiol 128:512–522

Emes MJ, Neuhaus H (1997) Metabolism and transport in non-photosynthetic plastids. J Exp Bot 48:1995–2005

Essigmann B, Güler S, Narang RA, Linke D, Benning C (1998) Phosphate availability affects the thylakoid lipid composition and the expression of SQD1, a gene required for sulfolipid biosynthesis in Arabidopsis thaliana. Proc Natl Acad Sci USA 95:1950–1955

Fischer K, Kammerer B, Gutensohn M, Arbinger B, Weber A, Hausler RE, Flugge UI (1997) A new class of plastidic phosphate translocators: a putative link between primary and secondary metabolism by the phosphoenolpyruvate/phosphate antiporter. Plant Cell 9:453–462

Flügge U-I (1999) Phosphate translocators in plastids. Annu Rev Plant Physiol Plant Mol Biol 50:27–45

Flügge U-I, Heldt HW (1984) The phosphate-triose phosphate-phosphoglycerate translocator of the chloroplast. Trends Biochem Sci 9:530–533

Flügge U-I, Fischer K, Gross A, Sebald W, Lottspeich F, Eckerskorn C (1989) The triose phosphate-3-phosphoglycerate-phosphate translocator from spinach chloroplasts: nucleotide sequence of a full-length cDNA clone and import of the in vitro synthesized precursor protein into chloroplasts. EMBO J 8:39–46

Flügge U-I, Weber A, Fischer K, Lottspeich F, Eckerskorn C, Waegemann K, Soll J (1991) The major chloroplast envelope polypeptide is the phosphate translocator and not the protein import receptor. Nature 353:364–367

Franco-Zorrilla JM, Martin AC, Solano R, Rubio V, Leyva A, Paz-Ares J (2002) Mutations at *CRE1* impair cytokinin-induced repression of phosphate starvation responses in *Arabidopsis*.. Plant J 32:353–360

Franco-Zorrilla JM, Valli A, Todesco M, Mateos I, Puga MI, Rubio-Somoza I, Leyva A, Weigel D, Garcia JA, Paz-Ares J (2007) Target mimicry provides a new mechanism for regulation of microRNA activity. Nat Genet 39:1033–1037

Fujii H, Chiou T-J, Lin S-I, Aung K, Zhu J-K (2005) A miRNA Involved in phosphate-starvation response in Arabidopsis. Curr Biol 15:2038–2043

Gahoonia TS, Asmar F, Giese H, Nielsen GG, Nielsen NE (2000) Root released organic acids and phosphorus uptake of two barley cultivars in laboratory and field experiments. Eur J Agron 12:281–289

Gaude N, Nakamura Y, Scheible WR, Ohta H, Dörmann P (2008) Phospholipase C5 (NPC5) is involved in galactolipid accumulation during phosphate limitation in leaves of *Arabidopsis*. The Plant Journal 56(1):28–39

Goldstein AH (1992) Phosphate starvation indicuble enzymes and proteins in higher plants. In: Wray JL (ed) Inducible plant proteins, their biochemistry and molecular biology. Cambridge University Press, Cambridge, UK, pp 25–44

Gonzalez E, Solano R, Rubio V, Leyva A, Paz-Ares J (2005) Phosphate transporter traffic facilitator1 is a plant-specific SEC12-related protein that enables the endoplasmic reticulum exit of a high-affinity phosphate transporter in Arabidopsis. Plant Cell 17(12):3500–3512

Guo B, Jin Y, Wussler C, Blancaflor EB, Motes CM, Versaw WK (2008) Functional analysis of the Arabidopsis PHT4 family of intracellular phosphate transporters. New Phytol 177:889–898

Hamburger D, Rezzonico E, MacDonald-Comber P, Etétot J, Somerville C, Poirier Y (2002) Identification and characterization of Arabidopsis PHO1 gene involved in phosphate loading to the xylem. Plant Cell 14:889–902

Hammond JP, White PJ (2008) Sucrose transport in the phloem: integrating root responses to phosphorus starvation.s. J Exp Bot 59:93–109

Hammond JP, Bennett MJ, Bowen HC, Broadley MR, Eastwood DC, May ST, Rahn C, Swarup R, Woolaway KE, White PJ (2003) Changes in gene expression in Arabidopsis shoots during phosphate starvation and the potential for developing smart plants. Plant Physiol 132:578–596

Hammond JP, Broadley MR, Craigon DJ, Higgins J, Emmerson Z, Townsend H, White PJ, May ST (2005) Using genomic DNA-based probe-selection to improve the sensitivity of high-density oligonucleotide arrays when applied to heterologous species. Plant Methods 1:10

Härtel H, Dörmann P, Benning C (2000) DGD1-independent biosynthesis of extraplastidic galactolipids after phosphate deprivation in Arabidopsis. Proc Natl Acad Sci USA 97:10649–10654

Hawkesford MJ, Miller AJ (2004) Ion-coupled transport of inorganic solutes. In: Blatt M (ed) Membrane transport in plants annual plant reviews., vol 15. Blackwell CRC press, USA

Hernández G, Ramírez M, Valdés-López O, Tesfaye M, Graham MA, Czechowski T, Schlereth A, Wandrey M, Erban A, Cheung F, Wu HC, Lara M, Town CD, Kopka J, Udvardi MK, Vance CP (2007) Phosphorus stress in common bean: root transcript and metabolic responses. Plant Physiol 144:752–767

Hinsinger P (2001) Bioavailability of soil inorganic P in the rhizosphere as affected by root-induced chemical changes: a review Plant and Soil 237:173–195

Holford ICR (1997) Soil phosphorus: its measurement and its uptake by plants. Aust J Soil Res 35:227–239

Hooley R (1994) Gibberellins: perception, transduction and responses. Plant Mol Biol 26:1529–1555

Horgan JM, Wareing PF (1980) Cytokinins and the growth responses of seedlings of Betula pendula Roth. and Acer pseudoplatanus L. to nitrogen and phosphorus deficiency. J Exp Bot 31:525–532

Huang C, Barker SJ, Langridge P, Smith FW, Graham RD (2000) Zinc deficiency up-regulates expression of high-affinity phosphate transporter genes in both phosphate-sufficient and -deficient barley roots. Plant Physiol 124:415–422

Hurley BA, Tran HT, Marty NJ, Park J, Snedden WA, Mullen RT, Plaxton WC (2010) The dual-targeted purple acid phosphatase isozyme AtPAP26 is essential for efficient acclimation of arabidopsis thaliana to nutritional phosphate deprivation. Plant Physiol 110.153270

Jain A, Poling MD, Karthikeyan AS, Blakeslee JJ, Peer WA, Titapiwatanakun B, Murphy AS, Raghothama KG (2007) Differential effects of sucrose and auxin on localized phosphate deficiency-induced modulation of different traits of root system architecture in *Arabidopsis*. Plant Physiol 144:232–247

Jeschke WD, Peuke AD, Pate JS, Hartung W (1997) Transport, synthesis and catabolism of abscisic acid (ABA) in intact plants of castor bean (*Ricinus communis* L.) under phosphate deficiency and moderate salinity. J Exp Bot 48:1737–1747

Jones DL (1998) Organic acids in the rhizosphere – a critical review. Plant Soil 205:25–44

Jones DL, Darrah PR (1994) Role of root derived organic-acids in the mobilization of nutrients from the rhizosphere. Plant Soil 166:247–257

Jouhet J, Maréchal E, Baldan B, Bligny R, Joyard J, Block MA (2004) Phosphate deprivation induces transfer of DGDG galactolipid from chloroplast to mitochondria. J Cell Biol 167:863–874

Jungk A (2001) Root hairs and the acquisition of plant nutrients from soil. J Plant Nutr Soil Sci 164:121–129

Kammerer B, Fischer K, Hilperta B, Schuberta S, Gutensohna M, Weber A, Flügge U-I (1998) Molecular characterization of a carbon transporter in plastids from heterotrophic tissues: the glucose 6-phosphate/phosphate antiporter. Plant Physiol 10(1):105–118

Karthikeyan AS, Varadarajan DK, Mukatira UT, D'Urzo MP, Damsz B, Raghothama KG (2002) Regulated expression of arabidopsis phosphate transporters. Plant Physiol 130:221–233

Karthikeyan A, Varadarajan D, Jain A, Held M, Carpita N, Raghothama K (2007) Phosphate starvation responses are mediated by sugar signaling in Arabidopsis. Planta 225:907–918

Keerthisinghe G, Hocking PJ, Ryan PR, Delhaize E (1998) Effect of phosphorus supply on the formation and function of proteoid roots of white lupin .(*Lupinus albus* L.). Plant Cell Environ 21:467–478

Kirk GJ, Santos E, Findenegg GR (1999) Phosphate solubilization by organic anion secretion from rice (Oryza sativa L.) growing in aerobic soil. Plant Soil 211:11–18

Kirkby EA, Johnston AE (2008) Soil and fertilizer phosphorus in relation to crop nutrition. In: White PJ, Hammond JP (eds) The ecophysiology of plant-phosphorus interactions. Springer, Dordrecht, The Netherlands, pp 177–223

Knappe S, Flügge U-I, Fischer K (2003a) Analysis of the plastidic phosphate translocator gene family in Arabidopsis and identification of new phosphate translocator-homologous transporters, classified by their putative substrate-binding site. Plant Physiol 131:1178–1190

Knappe S, Löttgert T, Schneider A, Voll L, Flügge U-I, Fischer K (2003b) Characterization of two functional phosphoenolpyruvate/phosphate translocator (PPT) genes in Arabidopsis–AtPPT1 may be involved in the provision of signals for correct mesophyll development. Plant J 36:411–420

Kobayashi K, Masuda T, Takamiya K, Ohta H (2006) Membrane lipid alteration during phosphate starvation is regulated by phosphate signaling and auxin/cytokinins cross-talk. Plant J 47:238–248

Köck M, Loffler A, Abel S, Glund K (1995) cDNA structure and regulatory properties of a family of starvation-induced ribonucleases from tomato. Plant Mol Biol 27:477–485

Köck M, Theierl K, Stenzel I, Glund K (1998) Extracellular administration of phosphate sequestering metabolites induces ribonucleases in cultured tomato cells. Planta 204:404–407

Krinke O, Novotná Z, Valentová O, Martinec J (2007) Inositol trisphosphate receptor in higher plants: is it real? J Exp Bot 58:361–376

Kulakovskaya T, Kulaev I (1997) Transport of phosphate into vacuoles of Saccharomyces cerevisiae. Microbiologia 13:71–74

Lai F, Thacker J, Li Y, Doerner P (2007) Cell division activity determines the magnitude of phosphate starvation responses in Arabidopsis. Plant J 50:545–556

Lambers HY, Shane MW, Cramer MD, Pearse SJ, Veneklass EJ (2006) Root structure and functioning for efficient acquisition of phosphorus: matching morphological and physiological traits. Ann Bot 98:693–713

Lan M, Comerford NB, Fox TR (1995) Organic anions effect on phosphorus release from spodic horizons. Soil Sci Soc Am J 59:1745–1749

Lenburg ME, O'Shea EK (1996) Signalling phosphate starvation. Trends in Biochemical Sciences. 21:383–387

Li D, Zhu H, Liu K, Liu X, Leggewie G, Udvardi M, Wang D (2002) Purple acid phosphatases of Arabidopsis thaliana. J Biol Chem 277:27772–27781

Li MY, Welti R, Wang XM (2006) Quantitative profiling of Arabidopsis polar glycerolipids in response to phosphorus starvation. Roles of phospholipases Df1 and Df2 in phosphatidylcholine hydrolysis and digalactosyldiacylglycerol accumulation in phosphorus-starved plants. Plant Phys 142:750–761

Li W-F, Perry PJ, Prafulla NN, Schmidt W (2010) Ubiquitin-specific protease 14 (UBP14) is involved in root responses to phosphate deficiency in Arabidopsis. Mol Plant 3:212–223

Lin S-I, Chiang S-F, Lin W-Y, Chen J-W, Tseng C-Y, Wu P-C, Chiou T-J (2008) Regulatory network of microRNA399 and PHO2 by systemic signaling. Plant Physiol 147:732–746

Linkohr BI, Williamson LC, Fitter AH, Leyser O (2002) Nitrate and phosphate availability and distribution have different effects on root system architecture of Arabidopsis. Plant J 29:751–760

Liu C, Muchhal US, Uthappa M, Kononowicz AK, Raghothama KG (1998) Tomato phosphate transporter genes are differentially regulated in plant tissues by phosphorus. Plant Physiol 116:91–99

López-Bucio J, de la Vega O, Guevara-Garcia A, Herrera-Estrella L (2000) Enhanced phosphorus uptake in transgenic tobacco plants that overproduce citrate. Nat Biotechnol 18:450–453

López-Bucio J, Hernández-Abreu E, Sánchez-Calderón L, Nieto-Jacobo MF, Simpson J, Herrera-Estrella L (2002) Phosphate availability alters architecture and causes changes in hormone sensitivity in the Arabidopsis root system. Plant Physiol 129:244–256

López-Bucio J, Hernández-Abreu E, Sánchez-Calderón L, Pérez-Torres A, Rampey RA, Bonnie B, Herrera-Estrella L (2005) An auxin transport independent pathway is involved in phosphate stress-induced root architectural alterations in *Arabidopsis*. Identification of BIG as a mediator of auxin in pericycle cell activation. Plant Physiol 137:681–691

Lynch J, van Beem J (1993) Growth and architecture of seedling roots of common bean genotypes. Crop Sci 33:1253–1257

Ma Z, Bielenberg DG, Brown KM, Lynch JP (2001) Regulation of root hair density by phosphorus availability in Arabidopsis thaliana. Plant Cell Environ 24:459–467

Ma Z, Baskin TI, Brown KM, Lynch JP (2003) Regulation of root elongation under phosphorus stress involves changes in ethylene responsiveness. Plant Physiol 131:1381–1390

Manske G, Ortiz-Monasterio J, Van Grinkel M, González R, Rajaram S, Molina E, Vlek P (2000) Traits associated with improved P-uptake efficiency in CIMMYT's semidwarf spring bread wheat grown on an acid Andisol in Mexico. Plant Soil 221:189–204

Martín AC, del Pozo JC, Iglesias J, Rubio V, Solano R, de la Peña A, Leyva A, Paz-Ares J (2000) Influence of cytokinins on the expression of phosphate starvation responsive genes in Arabidopsis. Plant J 24:559–567

Martinoia E, Maeshima M, Neuhaus HE (2007) Vacuolar transporters and their essential role in plant metabolism. J Exp Bot 58:83–102

Mimura T, Dietz K-J, Kaiser W, Schramm MJ, Kaiser G, Heber U (1990) Phosphate transport across biomembranes and cytosolic phosphate homeostasis in barley leaves. Planta 180:139–146

Misson J, Raghothama KG, Jain A, Jouhet J, Block MA, Bligny R, Ortet P, Creff A, Somerville S, Rolland N, Doumas P, Nacry P, Herrerra-Estrella L, Nussaume L, Thibaud M-C (2005) A genome-wide transcriptional analysis using Arabidopsis thaliana Affymetrix gene chips determined plant responses to phosphate deprivation. Proc Natl Acad Sci USA 102: 11934–11939

Misson J, Thibaud MC, Bechtold N, Raghothama K, Nussaume L (2004) Transcriptional regulation and functional properties of *Arabidopsis* Pht1;4, a high affinity transporter contributing greatly to phosphate uptake in phosphate deprived plants. Plant Molecular Biology 55:727–741

Miura K, Rus A, Sharkhuu A, Yokoi S, Karthikeyan AS, Raghothama KG, Baek D, Koo YD, Jin JB, Bressan RA, Yun D-J, Hasegawa PM (2005) The Arabidopsis SUMO E3 ligase SIZ1 controls phosphate deficiency responses. Proc Natl Acad Sci USA 102:7760–7765

Miura K, Jin JB, Lee J, Yoo CY, Stirm V, Miura T, Ashworth EN, Bressan RA, Yun D-J, Hasegawa PM (2007) SIZ1-mediated sumoylation of ICE1 controls CBF3/DREB1A expression and freezing tolerance in Arabidopsis. Plant Cell 19:1403–1414

Morcuende R, Bari R, Gibon Y, Zheng W, Pant BD, Bläsing O, Usadel B, Czechowski T, Udvardi MK, Stitt M, Scheible W-R (2007) Genome-wide reprogramming of metabolism and regulatory networks of *Arabidopsis* in response to phosphorus. Plant Cell Environ 30:85–112

Muchhal US, Pardo JM, Raghothama KG (1996) Phosphate transporters from the higher plant Arabidopsis thaliana. Proc Natl Acad Sci USA 93:10519–10523

Mudge SR, Rae AL, Diatloff E, Smith FW (2002) Expression analysis suggests novel roles for members of Pht1 family of phosphate transporters in Arabidopsis. Plant J 31:341–353

Mulkidjanian AY, Galperin MY, Koonin EV (2009) Co-evolution of primordial membranes and membrane proteins. Trends Biochem Sci 34:206–215

Müller R, Morant M, Jarmer H, Nilsson L, Nilsson TH (2007) Genome-wide analysis of the Arabidopsis leaf transcriptome reveals interaction of phosphate and sugar metabolism. Plant Physiol 143:156–171

Nacry P, Canivenc G, Muller B, Azmi A, Van Onckelen H, Rossignol M, Doumas P (2005) A role for auxin redistribution in the responses of the root system architecture to phosphate starvation in Arabidopsis. Plant Physiol 138:2061–2074

Nagata T, Todoriki S, Masumizu T, Suda I, Furuta S, Du Z, Kikuchi S (2003) Levels of active oxygen species are controlled by ascorbic acid and anthocyanin in Arabidopsis. J Agric Food Chem 51:2992–2999

Nakamura Y, Koizumi R, Shui G, Shimojima M, Wenk MR, Ito T, Ohta H (2009) Arabidopsis lipins mediate eukaryotic pathway of lipid metabolism and cope critically with phosphate starvation. Proc Natl Acad Sci USA 106:20978–20983

Nakanishi H, Okumura N, Umehara Y, Nishizawa N-K, Chino M, Mori S (1993) Expression of a gene specific for iron deficiency (Ids3) in the roots of Hordeum vulgare. Plant Cell Physiol 34:401–410

Niewiadomski P, Knappe S, Geimer S, Karsten F, Schulz B, Unte US, Rosso MG, Ache P, Flügge U-I, Schneidera A (2005) The Arabidopsis plastidic glucose 6-phosphate/phosphate translocator GPT1 is essential for pollen maturation and embryo sac development. Plant Cell 17:760–775

Nilsson L, Müller R, Nielsen TH (2007) Increased expression of the MYB-related transcription factor, PHR1, leads to enhanced phosphate uptake in Arabidopsis thaliana. Plant Cell Environ 30:1499–1512

Ohnishi M, Mimura T, Tsujimura T, Mitsuhashi N, Washitani-Nemoto S, Maeshima M, Martinoia E (2007) Inorganic phosphate uptake in intact vacuoles isolated from suspension-cultured cells of *Catharanthus roseus* (L.) G. Don under varying Pi status. Planta 225:711–718

Okazaki Y, Shimojima M, Sawada Y, Toyooka K, Narisawa T, Mochida K, Tanaka H, Matsuda F, Hirai A, Hirai MY, Ohta H, Saito K (2009) A chloroplastic UDP-glucose pyrophosphorylase from Arabidopsis is the committed enzyme for the first step of sulfolipid biosynthesis. Plant Cell 21:892–909

Okumura S, Mitsukawa N, Shirano Y, Shibata D (1998) Phosphate transporter gene family of Arabidopsis thaliana. DNA Res 5:261–269

Otani T, Ae N, Tanaka H (1996) Uptake mechanisms of crops grown in soils with low P status. II. Significance of organic acids in root exudates of pigeon pea. Soil Sci Plant Nutr 42:553–560

Pant BD, Buhtz A, Kehr J, Scheible W-R (2008) MicroRNA399 is a long-distance signal for the regulation of plant phosphate homeostasis. Plant J 53:731–738

Pao SS, Paulsen IT, Saire MH (1998) Major facilitator superfamily. Microbiol Mol Biol Rev 62:1–34

Peretó J, López-García P, Moreira D (2004) Ancestral lipid biosynthesis and early membrane evolution. Trends Biochem Sci 29:469–477

Pérez-Torres C, López-Bucio J, Cruz-Ramírez A, Ibarra-Laclette E, Dharmasiri S, Estelle M, Herrera-Estrella L (2008) Phosphate availability alters lateral root development in Arabidopsis by modulating auxin sensitivity via a mechanism involving the TIR1 auxin receptor. Plant Cell 20:3258–3272

Plaxton WC (2004) Plant response to stress: biochemical adaptations to phosphate deficiency. In: Goodman RM (ed) Encyclopedia of plant and crop science. Taylor and Francis, London, pp 976–980

Poirier Y, Bucher M (2002) Phosphate transport and homeostasis in Arabidopsis. In: Sommerville CR, Meyerowitz EM (eds) The arabidopsis book. The American society of plant biologists, USA, MD, pp 1–35

Poirier Y, Bucher M (2009) Phosphate transport and homeostasis in Arabidopsis. In: Sommerville CR, Meyerowitz EM (eds) The Arabidopsis book.. The American society of plant biologists, Rockville, MD, pp 1–35

Poirier Y, Thoma S, Somerville C, Schiefelbein J (1991) A mutant of Arabidopsis deficient in xylem loading of phosphate. Plant Physiol 97:1087–1093

Pozo JCD, Allona I, Rubio V, Leyva A, de la Peña A, Aragoncillo C, Paz-Ares J (1999) A type 5 acid phosphatase gene from Arabidopsis thaliana is induced by phosphate starvation and by some other types of phosphate mobilising/oxidative stress conditions. Plant J 19: 579–589

Pratt J, Boisson A-M, Gout E, Bligny R, Douce R, Aubert S (2009) Phosphate (Pi) starvation effect on the cytosolic Pi concentration and Pi exchanges across the tonoplast in plant cells: an in vivo 31P-nuclear magnetic resonance study using methylphosphonate as a Pi analog. Plant Physiol 151:1646–1657

Preuss CP, Huang CY, Gilliham M, Tyerman SD (2010) Channel-like characteristics of the low-affinity barley phosphate transporter PHT1;6 when expressed in Xenopus oocytes. Plant Physiol 152:1431–1441

Raghothama KG, Karthikeyan AS (2005) Phosphate acquisition. Plant Soil 274:37–49

Rausch C, Bucher M (2002) Molecular mechanisms of phosphate transport in plants. Planta 216:23–37

Rausch C, Zimmermann P, Amrhein N, Bucher M (2004) Expression analysis suggests novel roles for the plastidic phosphate transporter Pht2;1 in auto- and heterotrophic tissues in potato and *Arabidopsis*.. Plant J 39:13–28

Rébeillé F, Bligny R, Martin JB, Douce R (1983) Relationship between the cytoplasm and the vacuole phosphate pool in Acer pseudoplatanus cells. Arch Biochem Biophys 225:143–148

Richards DE, King KE, Ait-ali T, Harberd NP (2001) How gibberellin regulates plant growth and development: a molecular genetic analysis of gibberellin signaling. Annu Rev Plant Physiol Plant Mol Biol 52:67–88

Rubio V, Linhares F, Solano R, Martín AC, Iglesias J, Leyva A, Paz-Ares J (2001) A conserved MYB transcription factor involved in phosphate starvation signaling both in vascular plants and in unicellular algae. Genes Dev 15:2122–2133

Salama A, Wareing PF (1979) Effects of mineral nutrition on endogenous cytokinins in plants of sunflower (*Helianthus annuus* L.). J Exp Bot 30:971–981

Sanchez-Calderon L, Lopez-Bucio J, Chacon-Lopez A, Gutierrez-Ortega A, Hernandez-Abreu E, Herrera-Estrella L (2006) Characterization of low phosphorus insensitive mutants reveals a crosstalk between low phosphorus-induced determinate root development and the activation of genes involved in the adaptation of Arabidopsis to phosphorus deficiency. Plant Physiol 140:879–889

Sánchez-Calderón L, López-Bucio J, Chacón-López A, Cruz-Ramírez A, Nieto-Jacobo F, Dubrovsky JG, Herrera-Estrella L (2005) Phosphate starvation induces a determinate developmental program in the roots of Arabidopsis thaliana. Plant Cell Physiol 46:174–184

Sanders D, Pelloux J, Brownlee C, Harper JF (2002) Calcium at the crossroads of signaling. Plant Cell 14:S401–S417

Schachtman DP, Reid RJ, Ayling SM (1998) Phosphorus uptake by plants: from soil to cell. Plant Physiol 116:447–453

Schmidt W, Schikora A (2001) Different pathways are involved in phosphate and iron stress-induced alterations of root epidermal cell development. Plant Physiol 125:2078–2084

Secco D, Baumann A, Poirier Y (2010) Characterization of the rice PHO1 gene family reveals a key role for OsPHO1;2 in phosphate homeostasis and the evolution of a distinct clade in dicotyledons. Plant Physiol 152:1693–1704

Shane MW, Lambers H (2006) Systemic suppression of cluster-root formation and net P-uptake rates in Grevillea crithmifolia at elevated P supply: a proteacean with resistance for developing symptoms of 'P toxicity'. J Exp Bot 57:413–423

Shin H, Shin H-S, Chen R, Harrison MJ (2006) Loss of *At4* function impacts phosphate distribution between the roots and the shoots during phosphate starvation. Plant J 45:712–726

Shinano T, Nanamori M, Dohi M, Wasaki J, Osaki M (2005) Evaluation of phosphorus starvation inducible genes relating to efficient phosphorus utilization in rice. Plant Soil 269:81–87

Smith AP, Jain A, Deal RB, Nagarajan VK, Poling MD, Raghothama KG, Meagher RB (2010) Histone H2A.Z regulates the expression of several classes of phosphate starvation response genes but not as a transcriptional activator. Plant Physiol 152:217–225

Stevenson-Paulik J, Bastidas RJ, Chiou S-T, Frye RA, York JD (2005) Generation of phytate-free seeds in Arabidopsis through disruption of inositol polyphosphate kinases. Proc Natl Acad Sci USA 102:12612–12617

Svistoonoff S, Creff A, Reymond M, Sigoillot-Claude C, Ricaud L, Blanchet A, Nussaume L, Desnos T (2007) Root tip contact with low-phosphate media reprograms plant root architecture. Nat Genet 39:792–796

Takabatake R, Hata S, Taniguchi M, Kouchi H, Sugiyama T, Izui K (1999) Isolation and characterization of cDNAs encoding mitochondrial phosphate transporters in soybean, maize, rice, and *Arabidopsis*.. Plant Mol Biol 40:479–486

Ticconi CA, Abel S (2004) Short on phosphate: plant surveillance and countermeasures. Trends Plant Sci 9:548–555

Ticconi CA, Delatorre CA, Abel S (2001) Attenuation of phosphate starvation responses by phosphite in Arabidopsis. Plant Physiol 127:963–972

Ticconi CA, Lucero RD, Sakhonwasee S, Adamson AW, Creff A, Nussaume L, Desnos T, Abel S (2009) ER-resident proteins PDR2 and LPR1 mediate the developmental response of root meristems to phosphate availability. Proc Natl Acad Sci USA 106:14174–14179

Tiessen H (2008) Phosphorus in the global environment. In: White PJ, Hammond JP (eds) The ecophysiology of plant-phosphorus interactions., vol 7. Springer, Netherlands, pp 1–7

Tomscha JL, Trull MC, Deikman J, Lynch JP, Guiltinan MJ (2004) Phosphatase under-producer mutants have altered phosphorus relations. Plant Physiol 135:334–345

Trull MC, Deikman J (1998) An Arabidopsis mutant missing one acid phosphatase isoform. Planta **206**:544–550

Vance CP, Uhde-Stone C, Allan DL (2003) Phosphorus acquisition and use: critical adaptations by plants for securing a nonrenewable resource. New Phytol 157:423–447

Veljanovski V, Vanderbeld B, Knowles VL, Snedden WA, Plaxton WC (2006) Biochemical and molecular characterization of AtPAP26, a vacuolar purple acid phosphatase up-regulated in phosphate-deprived Arabidopsis suspension cells and seedlings. Plant Physiol 142: 1282–1293

Versaw WK, Harrison MJ (2002) A chloroplast phosphate transporter, PHT2;1 influences allocation of phosphate within the plant and phosphate-starvation responses. Plant Cell 14:1751–1766

Wang Y, Secco D, Poirier Y (2008) Characterization of the PHO1 gene family and the responses to phosphate deficiency of Physcomitrella patens. Plant Physiol 146:646–656

Wang C, Ying S, Huang H, Li K, Wu P, Shou H (2009a) Involvement of *OsSPX1* in phosphate homeostasis in rice. Plant J 57:895–904

Wang Z, Hu H, Huang H, Duan K, Wu Z, Wu P (2009b) Regulation of OsSPX1 and OsSPX3 on expression of OsSPX domain genes and Pi-starvation signaling in rice. J Integr Plant Biol 51:663–674

Ward JT, Lahner B, Yakubova E, Salt DE, Raghothama KG (2008) The effect of Iron on the primary root elongation of Arabidopsis during phosphate deficiency. Plant Physiol 147:1181–1191

Wasaki J, Yonetani R, Kuroda S, Shinano Y, Yazaki J, Fujii F, Shimbo K, Yamamoto K, Sakata K, Sasaki T, Kishimoto N, Kikuchi S, Yamagishi M, Osaki M (2003) Transcriptomic analysis of metabolic changes by phosphorus stress in rice plant roots. Plant Cell Environ 26:1515–1523

Weber APM (2004) Solute transporters as connecting elements between cytosol and plastid stroma. Curr Opin Plant Biol 7:247–253

Weber APM, Schwacke R, Flügge U-I (2005) Solute transporter of the plastid envelope membrane. Annu Rev Plant Biol 56:133–164

White PJ, Hammond JP (2008) Phosphorous nutrition of terrestrial plants. In: White PJ, Hammond JP (eds) The ecophysiology of plant-phosphorus interactions. Springer, Netherlands, pp 51–81

Williamson LC, Ribrioux SP, Fitter AH, Leyser O (2001) Phosphate availability regulates root system architecture in Arabidopsis. Plant Physiol 126:875–882

Wu P, Ma L, Hou X, Wang M, Wu Y, Liu F, Deng XW (2003) Phosphate starvation triggers distinct alterations of genome expression in Arabidopsis roots and leaves. Plant Physiol 132:1260–1271

Yi K, Wu Z, Zhou J, Du L, Guo L, Wu Y, Wu P (2005) OsPTF1, a novel transcription factor involved in tolerance to phosphate starvation in rice. Plant Physiol 138:2087–2096

Yu B, Xu C, Benning C (2002) Arabidopsis disrupted in SQD2 encoding sulfolipid synthase is impaired in phosphate-limited growth. Proc Natl Acad Sci USA 99:5732–5737

Zakhleniuk OV, Raines CA, Lloyd JC (2001) pho3: a phosphorus-deficient mutant of Arabidopsis thaliana (L.) Heynh. Planta 212:529–534

Zhang F, Ma J, Cao YP (1997) Phosphorus deficiency enhances root exudation of low molecular weight organic acids and utilization of sparingly soluble inorganic phosphates by radish (Raphanus sativus L.) and rape (Brasicca napus L.) plants. Plant Soil 196:261–264

Zhang YJ, Lynch JP, Brown KM (2003) Ethylene and phosphorus availability have interacting yet distinct effects on root hair development. J Exp Bot 54:2351–2361

Zhou J, Jiao F, Wu Z, Li Y, Wang X, He X, Zhong W, Wu P (2008) OsPHR2 is involved in phosphate-starvation signaling and excessive phosphate accumulation in shoots of plants. Plant Physiol 146:1673–1686

Zhu J, Lynch JP (2004) The contribution of lateral rooting to phosphorus acquisition efficiency in maize (Zea mays) seedlings. Funct Plant Biol 31:949–958

Zimmermann P, Regierer B, Kossmann J, Frossard E, Amrhein N, Bucher M (2004) Differential synthesis of three purple acid phosphatases from potato. Plant Biol 6:519–528

Sucrose Transporters and Plant Development

Christina Kühn

Abstract Sucrose transporters are essential proteins for the allocation of assimilates from source to sink. With increasing information about their function and the role of the sucrose molecule, acting as an informational and signaling molecule, new questions arise on how sucrose transporters could interact or coordinate metabolic pathways and developmental processes and to integrate whole plant communication. New information about the phloem mobility of sucrose transporter mRNAs and other phloem mobile signals is also available, shedding light on the long-distance transport of leaf-derived information to terminal sink organs. Recent advances on the subcellular localisation and function of sucrose transporters, sucrose facilitators and sucrose transporter-like proteins open new questions about the role of membrane compartmentation on the dimerization, endocytosis, degradation and signaling of plant sucrose transporters.

1 Plants Contain More than One Sucrose Transporter

The increasing availability of full genome sequence information provides new opportunities for phylogenetic analysis of the sucrose transporter gene family. There are nine sucrose transporter genes (*SUTs or SUCs*) described in Arabidopsis (initiative 2000), whereas the rice genome contains five *SUT* genes (Aoki et al. 2003). An analysis of the genomes of the monocots sorghum, maize and *Brachypodium* has led to a proposed separation of sucrose transporters into five groups, where the fourth and fifth groups are made up exclusively of monocot transporters (Braun and Slewinski 2009). Sucrose transporters belong to the major facilitator superfamily (MFS) and show similarities to bacterial sugar transporters like the lactose permease

C. Kühn
Institute of Biology, Plant Physiology, Humboldt University of Berlin, Philippstrasse 13, Building 12, Berlin 10115, Germany
e-mail: kuehnchr@cms.hu-berlin.de

M. Geisler and K. Venema (eds.), *Transporters and Pumps in Plant Signaling*, Signaling and Communication in Plants 7,
DOI 10.1007/978-3-642-14369-4_8,

LacS from *Escherichia coli* or LacY from *Streptococcus*. Sucrose transporters have been identified in primitive multicellular lycophytes like *Selaginalla lepidophylla* and mosses like *Physcomitrella patens*; however, no ESTs of sucrose transporters could be identified in the unicellular green alga *Chlamydomonas rheinhardtii*. Even in fungi, like *Schizosaccharomyces pombe*, a functional sucrose transporter has been characterised (Reinders and Ward 2001), arguing for a very early evolutionary origin of sucrose transport.

Interestingly, three out of the five phylogenetic clades seem to be specific for either monocotyledonous or dicotyledonous species (Fig. 1). Only two out of the five clades contain members of both groups: namely the SUT2 and the SUT4 subfamily. The SUT2 and SUT4 transporters comprising both monocot and dicot members have a lower affinity for sucrose than mixed clades with reported K_m values of 4–20 mM. It remains an open question, if members of these two subfamilies fulfil similar functions in the different plants species.

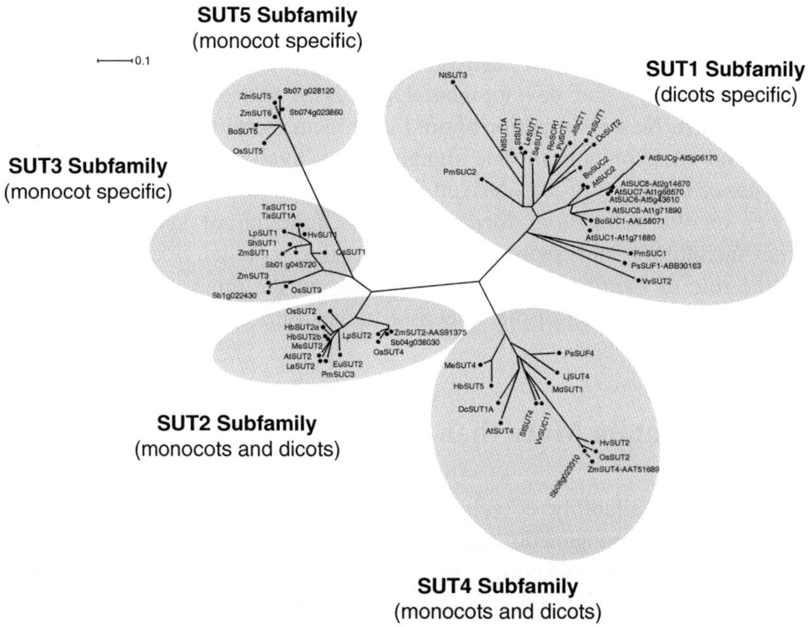

Fig. 1 Phylogenetic tree of sucrose transporters from monocotyledonous and dicotyledonous species displayed using Dendroscope (Huson et al. 2007). *Arabidopsis thaliana*: At SUC1, At1g71880; AtSUC2, At1g22710; AtSUT2, At2g02860; AtSUT4, At1g09960; AtSUC5, At1g71890; AtSUC6, At5g43610; AtSUC7, At1g66570; AtSUC8, At2g14670; AtSUC9, At5g06170. *Brassica oleracea*: BoSUC1, AAL58071; BoSUC2, AAL58072. *Bambusa oldhami*: BoSUT5, AAY43226. *Citrus sinensis*: CsSUT2, AAM29153. *Daucus carota*: DcSUT1A, CAA76367; DcSUT2, CAA76369. *Eucommia ulmoides*: AAX49396. *Hevea brasiliensis*: HbSUT2a, ABJ51934; HbSUT2b, ABJ51932; HbSUT5, ABK60189. *Hordeum vulgare*: HvSUT1, CAB75882; HvSUT2, CAB75881. *Juglans regia*: JrSUT1, AAU11810. *Lycopersicum esculentum* renamed *Solanum lycopersicum*: LeSUT2, AAG12987; LeSUT4, AAG09270.

2 Sucrose Transporters from Monocotyledonous Plants

Unlike dicot SUTs, monocotyledonous SUT1 proteins like ShSUT1 and HvSUT1 are highly selective for sucrose (Sivitz et al. 2007). ShSUT1 from sugarcane is expressed in maturing stems and is assumed to play an important role in the accumulation of sucrose in sugarcane stalks. Its K_m is close to 8 mM (Reinders et al. 2006). So far, only few of the monocot sucrose transporters have been functionally characterised and apparent K_m values between 2 and 8 mM have been measured (Rae et al. 2005; Reinders et al. 2008; Sun et al. 2010).

Since monocots are agriculturally important, SUTs from cereals have been the focus of increasing research interest. Antisense suppression of OsSUT1 affected grain filling, grain weight, as well as germination and growth of the seed, suggesting a reduction of the starch mobilisation in developing seedlings (Furbank et al. 2001; Ishimaru et al. 2001). OsSUT1 promoter::reporter gene fusion together with immunolocalisation of the SUT1 protein revealed OsSUT1 presence in the mature transport phloem during grain filling. Whereas the promoter activity seemed to be restricted to the phloem companion cells, the protein localised to both sieve elements and companion cells (Scofield et al. 2007a). OsSUT1 is assumed to be involved in the retrieval of sucrose from the apoplasm into the vasculature (Scofield et al. 2007b).

The orthologous transporter of wheat, TaSUT1, is assumed to be directly involved in sugar transfer across the scutellar epithelium and to have a transport

Fig. 1 (Continued) Lotus japonicus: LjSUT4, CAD61275. *Lolium perenne*: LpSUT1, EU255258; LpSUT2, ACU87542. *Malus x domestica*: MdSUT1, AAR17700. *Manihot esculenta*: MeSUT2, ABA08445; MeSUT4, ABA08443. *Nicotiana tabacum*: NtSUT1A, CAA57727; NtSUT3, AAD34610. *Oryza sativa*: OsSUT1, AAF90181; OsSUT2, BAC67163; OsSUT3, BAB68368; OsSUT4, BAC67164; OsSUT5, BAC67165. *Plantago major*: PmSUC1, CAI59556; PmSUC2, X75764; PmSUC3, CAD58887. *Pisum sativum*: PsSUT1, AAD41024; PsSUF1, ABB30163; PsSUF4, A3DSX1. *Populus tremula x Populus tremuloides*: PtSUT1-1, CAJ33718. *Ricinus communis*: RsSCR1, CAA83436. *Saccharum* hybrid: ShSUT1, AAV41028. *Spinacea oleracea*: SoSUT1, Q03411. *Solanum tuberosum*: StSUT1, CAA48915; StSUT4, AAG25923. *Sorghum bicolor*[a]: SbSUT1, Sb01g045720; SbSUT2, Sb04g038030; SbSUT3, Sb01g022430; SbSUT4, Sb08g023310; SbSUT5, Sb04g023860; SbSUT6, Sb07g028120. *Triticum aestivum*: TaSUT1A, AAM13408; TaSUT1B, AAM13409; TaSUT1D, AAM13410. *Vitus vinifera*: VvSUC11, AAF08329; VvSUC12, AAF08330; VvSUC27, AAF08331; VvSUT2, AAL32020. *Zea mays*[b]: ZmSUT1, BAA83501; ZmSUT2, AAS91375; ZmSUT3, ACF86653; ZmSUT4, AAT51689; ZmSUT5, ACF85284; ZmSUT6, ACF85673. Accession numbers are also used as additional descriptors in the tree in those instances where confusion may arise due to variations in nomenclature

[a]Sorghum Genome Project: www.phytozome.net.

[b]Maize Genome Project: http://www.maizesequence.org/index.html. The Phylogeny website (http://www.phylogeny.fr/version2_cgi/index.cgi) was used extensively for sequence alignment and analysis (Dereeper et al. 2008). The data were converted into Newick format prior to transfer to Dendroscope (http://www-ab.informatik.uni-tuebingen.de/software/dendroscope). Phylogenetic analysis was performed and kindly provided by Christopher Grof (University of Newcastle, Australia).

function in enucleate sieve elements (Aoki et al. 2004). The function of the SUT1 orthologue from maize was analysed with the help of a SUT1 defective mutant and accumulated high amount of carbohydrates in leaves which led to leaf chlorosis and early senescence. Its expression pattern as well as its activity is consistent with an important role in phloem loading (Slewinski et al. 2009).

3 Sucrose Efflux

Sucrose efflux, although measured in leaf discs from sugar beat source leaves (Secor 1987), is not well understood, and so far no sucrose efflux carrier has been identified that catalyses the efflux of sucrose from mesophyll cells in leaves or from sink parenchyma cells during phloem unloading. However, sucrose transporters are also found to be expressed in sink organs such as potato tubers, pollen grains, seeds and flowers, and sucrose transporters have traditionally assumed to be involved in phloem unloading (Aloni et al. 1986). Tuber-specific inhibition of the SUT1 expression, which is localised in sieve elements, leads to changes in tuber development suggesting a function of StSUT1 in phloem unloading (Kühn et al. 2003). Also in developing seeds of leguminous plants, where phloem unloading includes necessarily an apoplasmic step, SUT1 has been localised to the phloem (Zhang et al. 2007). Since the sucrose transporter ZmSUT1 from maize is able to catalyse the in vitro transport of the sucrose molecule in both directions depending on the driving force, the ZmSUT1-mediated sucrose-coupled proton current measured in Xenopus oocytes was reversible and depended on the direction of the sucrose and pH gradient as well as the membrane potential (Carpaneto et al. 2005). Therefore, it is possible that the sucrose proton co-transporters identified to date might be responsible for both sucrose uptake into the phloem cells in source leaves as well as unloading of sucrose from the phloem sieve elements into the cells of sink organs.

4 Phloem Mobility of Sucrose Transporter mRNAs

Whereas the sucrose transporter AtSUC2 form Arabidopsis as well as PmSUC2 from *Plantago major* have been localised in phloem companion cells (Stadler et al. 1995; Stadler and Sauer 1996), the sucrose transporters of the Solanaceae are found in phloem sieve elements and their mRNA in both cell types (Kühn et al. 1997). Thus, the mRNA of the StSUT1 transporter was localised by electron microscopy in the companion cells and in the neighbouring sieve elements, and the signal density was highest close to the plasmodesmal connections between these two cells (Kühn et al. 1997).

SUT1 transcription takes place in companion cells since the use of the companion cell-specific rolC promoter to drive an antisense construct successfully inhibited its expression (Kühn et al. 1996). It is assumed that the sucrose transporter mRNA moves from the companion cells into the sieve elements via plasmodesmata, and in several plants species, SUT1 mRNA was detected in the phloem sap (Ruiz-Medrano et al. 1999; Knop et al. 2001, 2004; Doering-Saad et al. 2006). The ability of the StSUT1 mRNA to move through plasmodesmata was shown by microinjection experiments (Xoconostle-Cázares et al. 1999).

In Solanaceae and Arabidopsis, the phloem mobility of sucrose transporter transcripts has been shown by various independent techniques, and the presence of sucrose transporter mRNAs in the phloem sap is also postulated.

The phloem mobility of sucrose transporter mRNAs has been shown by heterograft and parasitic experiments for solanaceous NtSUT1, StSUT1, SlSUT1, as well as SoSUT1 (He et al. 2008). The full-length mRNA of StSUT1 from potato and NtSUT1 from tobacco was detected by RT-PCR using reversibly transcribed mRNA from *Cuscuta reflexa* after 2 weeks of parasitism on the respective host plant (He et al. 2008). Heterograft experiments revealed phloem mobility of SoSUT1 mRNA, which was over-expressed in potato plants, as well as SlSUT1 mRNA if expressed under its own promoter in transgenic plants (He et al. 2008).

The phloem mobility of AtSUC2 and AtSUT4 mRNA from Arabidopsis was investigated by laser micro-dissection coupled to laser pressure catapulting (LMPC) and by microarray analysis (Deeken et al. 2008).

5 Sucrose Import into the Sieve Element–Companion Cell Complex

Members of the high-affinity dicotyledonous SUT1 subfamily have been localised on the plasma membranes of sieve elements (Kühn et al. 1997; Barker et al. 2000; Weise et al. 2000), companion cells (Stadler et al. 1995; Stadler and Sauer 1996) or in both cell types (Knop et al. 2004). Immunlolocalisation of StSUT1 in the sieve elements of the internal and external phloem and in guard cells is consistent with the SUT1 promoter reporter gene studies showing promoter activity not only in the phloem but also in guard cells and trichomes (Weise et al. 2008). A SUT1-GFP fusion protein was localized to the plasma membrane and the endoplasmic reticulum (ER) if co-infiltrated with a fluorochrome-coupled ER marker in tobacco leaves (Krügel et al. 2008).

Fusion of SUT1 to GFP or GUS in stably transformed plants blocked SUT1 localisation in sieve elements, and the reporter genes were detected only in companion cells (Lalonde et al. 2003; Weise et al. 2008). The use of the companion cell-specific AtSUC2 promoter for the expression of soluble proteins of sizes up to 67 kDa led to non-specific trafficking of proteins into the sieve elements (Stadler et al. 2005). Companion cell-specific expression of a membrane-anchored fluorescent protein of 6 kDa containing two transmembrane domains under control of the AtSUC2 promoter led to localisation of the membrane protein in the plasma membrane of neighbouring sieve elements in Arabidopsis and tobacco and in vesicle-like structures in the lumen of sieve elements suggesting the existence of vesicle trafficking machinery in mature sieve elements (Thompson and Wolniak 2008). Sieve elements and companion cells are coupled via the desmotubulus, which is derived from ER cisternae from both cell types. ER coupling via the desmotubulus has been shown using ER-specific fluorochromes and the fluorescence recovery after photobleaching (FRAP) (Martens et al. 2006). The continuity of the ER coupling in these two cell types allows the exchange of membrane proteins between them. Thus, we assume that both the SUT1 mRNA and the protein move independently of each other from companion cells into the sieve elements. Translation of the protein can take place in the companion cells and the translated protein can move via the desmotubulus without leaving the ER during cell-to-cell transport. This hypothesis is summarised in the model in Fig. 2.

From these findings, it was concluded that phloem loading occurs directly at the plasma membrane of sieve elements. Nevertheless, a recent report using antiserum raised against a NtSUT1–MBP fusion protein demonstrated localisation of SUT1 to companion cells and not sieve elements of *Solanaceae* (Schmitt et al. 2008). These findings are reproducible with several SUT1-specific antibodies, using dot blots of nitrocellulose for antibody purification (Fig. 3), provided that the purification protocol described by Schmitt et al. (2008) was strictly followed. However, if the

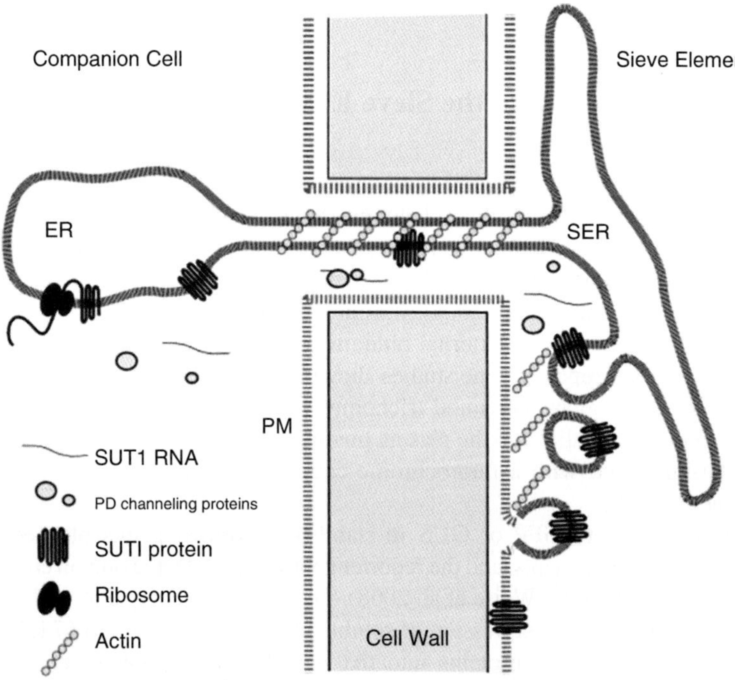

Fig. 2 Hypothetical model of SUT1 mRNA and protein targeting through plasmodesmata via the desmotubulus. Co-localisation of a SUT1-GFP fusion with an ER tracker (Krügel et al. 2008), together with the quantification of plasmodesmal ER coupling between sieve elements and companion cells (Martens et al. 2006) opens the possibility for exchange of membrane proteins between SE and CC via the ER . The picture was drawn and kindly provided by Johannes Liesche (University of Kopenhagen)

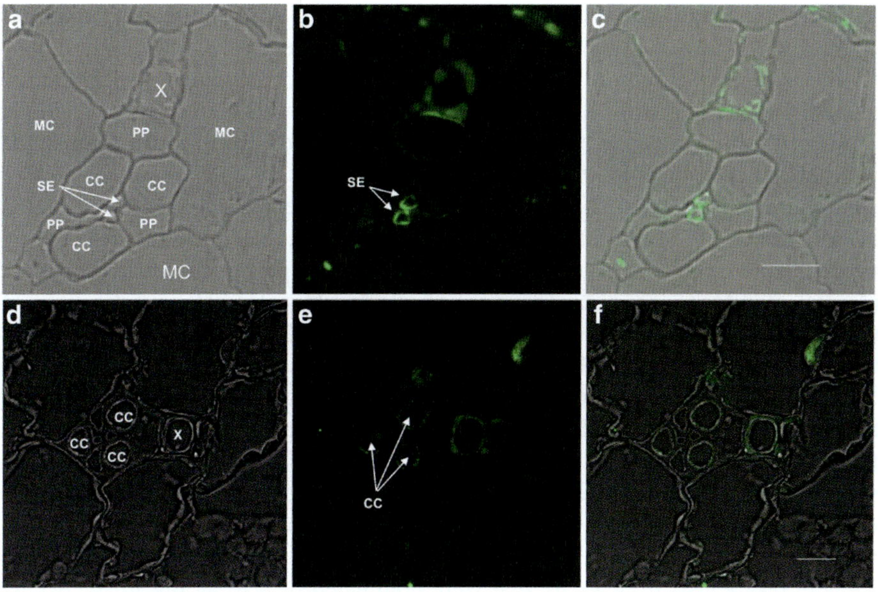

Fig. 3 Immunolocalisation of StSUT1 protein in source leaf minor veins. (**a–c**) Immunolocalisation of StSUT1 protein with sepharose column affinity-purified peptide antibody in potato source leaf material after formaldehyde fixation and embedding in LR White. (**d–f**) Immunolocalisation of StSUT1 protein with the same peptide antiserum used in **b**, but purified via pieces of nitrocellulose. Immunodetection was performed in potato source leaf material after acetic acid: ethanol fixation and embedding in methacrylate (**a**, **d**). Transmission picture (**b**, **e**). Immunodetection with FITC-coupled secondary antibody (**c**, **f**) overlay. Scale bar represents 10 μm. Experiments were performed and pictures were taken by Johannes Liesche (University of Kopenhagen, Denmark)

antiserum, regardless of whether it was raised against a synthetic peptide of SUT1 or a MPB fusion protein, is purified via sepharose column chromatography, all SUT1 immune sera label the sieve elements of the phloem. The titre, quality and specificity of sepharose-purified antibodies are high enough to allow further applications such as immunolabelling at the EM level (Liesche et al. 2008), western blot analysis with plant extracts (Kühn et al. 1996) or immunoprecipitation of the protein (Krügel et al. 2008, 2010) to be undertaken, whereas nitrocellulose-purified antibodies are not sufficiently concentrated for these applications. Therefore, the manner of antibody purification seems crucial for the specificity of the purified serum. Unspecific binding of proteins from immune sera to the empty sepharose column was also tested and no unspecific labelling of either companion cells or sieve elements in plant tissue sections was observed. Obviously, the technique of immunolocalisation is sometimes an error-prone method and in case of AtSUC3, the immunolocalisation performed with a nitrocellulose-purified antibody raised against a sucrose transporter–MBP fusion protein did not concur with the promoter::reporter gene expression (Meyer et al. 2000). A few years later, the localisation of AtSUC3 was detected in phloem sieve elements using specific peptide antibodies against AtSUC3 (Meyer et al. 2004). In case of NtSUT1, localisation using MBP fusion protein antibodies is not fully in accordance with promoter::reporter gene studies in tobacco; promoter activity was detected in the internal and external phloem cells and in guard cells (Weise et al. 2008). Schmitt et al. localised the NtSUT1 protein only in the external phloem and not in guard cells (Schmitt et al. 2008). Peptide antibodies purified via sepharose columns coupled to the antigene, however, show SUT1 localisation in the internal and external phloem, which is in a good agreement with promoter–reporter gene studies (Weise et al. 2008). In summary, the two reports on SUT1 localisation in Solanaceae are still contradictory and additional further technologies will be needed to clarify the question if sucrose transporters are present in the sieve element plasma membrane or not. The use of electrophysiological investigation of uncoupled and isolated sieve element protoplasts may be one possibility to clarify this problem (Hafke et al. 2007).

Sucrose loading into phloem cells via a proton symport mechanism is linked to the activity of phloem-localised, inward-rectifying potassium channels, which are thought to play an important role in the repolarisation of the plasma membrane after sucrose import. The sucrose content in the phloem sap of Arabidopsis potassium channel AtAKT2/3 mutants was only one-half of that measured in wild-type plants (Deeken et al. 2002). Interestingly, the current–voltage relations of isolated sieve element protoplasts from *Vicia faba* were dominated by a weak inward-rectifying potassium channel with electrical properties that are reminiscent of those of the AKT2 channel family (Hafke et al. 2007). The presence of AKT2/3-like channels at the sieve element plasma membrane argues for sucrose retrieval directly at the sieve element plasma membrane. Electrophysiological measurements with electrically uncoupled sieve element protoplasts would be an appropriate tool to answer the question of whether sucrose-induced currents can be detected at the plasma membrane of sieve elements.

It is an old dogma that enucleate mature sieve elements are devoid of ribosomes and unable to translate proteins. A large-scale proteomics approach to analyse pumpkin phloem exudates challenges this assertion as sets of phloem proteins that function in RNA binding, mRNA translation, ubiquitin-mediated proteolysis as well as macromolecular and vesicle trafficking were reported (Lin et al. 2009). Functional sieve elements may retain Golgi, endosomes and also small vacuoles (Lin et al. 2009). A recent report of small non-coding RNAs in the phloem sap of pumpkin revealed the presence of rRNAs, including 5S, 18S and 26S rRNA, and most of the tRNAs tested indicated the presence of a translational machinery in mature phloem sieve elements (Zhang et al. 2009). Phloem proteomic and transcriptomic approaches have also revealed the significance of redox active systems in the phloem (Walz et al. 2002, 2004; Lin et al. 2009).

6 Vacuolar Sucrose Transporters

The vacuole stores large concentrations of the disaccharide sucrose, but hexoses like glucose and fructose are also present at high levels (Neuhaus 2007). Sucrose accumulating species like sugar beet (*Beta vulgaris*) and sugar cane (*Saccharum officinarum*) use the vacuole as a sucrose storage compartment. The transport of sucrose against a concentration gradient seems to be partially dependent on Mg-ATP, and a sucrose proton antiport mechanism has been postulated (Willenbrink 1987).

The process of sucrose transport was investigated in tonoplast vesicles isolated from sugar beet taproot. The presence of an ATP-dependent pH gradient allowed transport of labelled sucrose into vesicles to occur at a rate tenfold higher than the rate observed in the absence of an imposed pH gradient Thus, an electrogenic H^+-sucrose antiport mechanism was assumed to be responsible for sucrose import at the tonoplast vesicle membrane (Briskin et al. 1985). The same system was used to demonstrate a sucrose-inducible medium acidification, which was sensitive to the same inhibitors that were efficient in inhibition of sucrose transport. Apparent K_m for H^+ export corresponded to those obtained for sucrose uptake, and a stoichiometry of one proton per transported sucrose molecule was assumed (Getz and Klein 1995). In contrast, in *Ananas comosus* L., which accumulates high amounts of carbohydrates in the vacuole, sucrose uptake into tonoplast-enriched microsomal membranes was not dependent on Mg-ATP, suggesting that the sucrose transport into the vacuole in this system does not involve a H^+-coupled co-transport (McRae et al. 2002).

Members of the dicotyledonous SUT4 subfamily have recently been assigned to the vacuolar membrane. The SUT4 homolog in Arabidopsis (AtSUT4, Endler et al. 2006), barley (HvSUT2, Endler et al. 2006) and *Lotus Japonica* (LjSUT4, Reinders et al. 2008) localise to the tonoplast when transiently expressed in plants as GFP fusion proteins. As all three sucrose transporters function as sucrose proton symporters rather than antiporters, it was postulated that they are most likely involved in the sucrose efflux from the vacuole (Neuhaus 2007), rather than catalysing the influx of sucrose into the vacuole. To date, vacuolar sucrose importer systems have not been identified.

7 Substrate Specificity of Sucrose Transporters

Sucrose transporters of dicotyledonous species have been shown to transport not only sucrose but also other substrates. Members of the SUT1 clade are also able to transport maltose with an affinity lower than sucrose (K_m 10 mM) (Riesmeier et al. 1992; Schulze et al. 2000).

Arabidopsis AtSUC5, another member of the SUT1 subfamily, mediates the energy-dependent transport of biotin in addition to H^+-dependent sucrose co-transport. AtSUC5 is able to complement a biotin transport-defective yeast mutant. Although the AtSUC5-mediated biotin uptake was concentration dependent, it was not saturable (Ludwig et al. 2000).

AtSUC2, another Arabidopsis member of the SUT1 clade, can transport sucrose, maltose and other glucosides such as arbutin, salicin, phenyl-α-glucoside, phenyl-β-glucoside, p-nitrophenyl-α-glucoside, p-nitrophenyl-β-glucoside, p-nitrophenyl-β-thioglucoside, turanose and α-methylglucoside with low affinity (Chandran et al. 2003). AtSUC9 transports glucosides such as helicin, salicin, arbutin, maltose, fraxin, esculin, turanose and α-methyl-D-glucose. In comparison, monocotyledonous SUT1 members like HvSUT1 and ShSUT1 revealed a much higher substrate specificity than their dicotyledonous homologs (Sivitz et al. 2007).

8 Sucrose Facilitators (SUFs)

All the sucrose transporter proteins identified so far were shown to function in a H^+-dependent manner. Recently, a new group of facilitators have been described which due to their high sequence similarities belong to the phylogenetic family of sucrose transporters, but act as sucrose facilitators (SUFs) in a pH- and energy-independent manner. SUFs from pea (*Pisum sativum)* and bean *(Phaseolus vulgaris)* can mediate transport of sucrose in a bi-directional manner (Zhou et al. 2007). Whereas sucrose influx was positively correlated with the expression of the sucrose transporter PsSUT1 from *P. sativum*, a negative correlation between sucrose influx and the two sucrose facilitators, PsSUF1 and PsSUF4, was observed. The effect of the intracellular sugar concentration on transporter expression is also different. Whereas PsSUT1 expression was inhibited, PsSUF1 expression was unaffected and PsSUF4 expression was enhanced by intracellular sucrose and hexoses (Zhou et al. 2009).

9 Regulation of Sucrose Transporters

9.1 Regulation at the Transcriptional Level

The sucrose substrate has a negative effect on sucrose transporter expression in most cases studied (BvSUT1 from *Beta vulgaris*, Vaughn et al. 2002; Ransom-Hodgkins et al. 2003) and VfSUT1 from *Vicia faba*, Weber et al. 1997). Only in the case of SUT2 from tomato plants was a positive effect on sucrose transporter expression detected (Barker et al. 2000).

The accumulation of the SUT1 mRNA from potato is inducible by auxin and cytokinin (Harms et al. 1994), and this up-regulation is also reflected at the protein level (He et al. 2008). Analysis of the genomic StSUT1 sequence revealed the presence of a putative-binding domain of the auxin response factor, ARF (auxin responsive element AuxRE), in the third intron. Also the mRNAs of SUT2 and SUT4 genes from Solanaceae show auxin responsive elements. The genomic sequence of the SlSUT2 gene contains one AuxRE in the promoter and an additional one in the tenth intron region (He et al. 2008).

The expression patterns of the three known sucrose transporters from potato show circadian oscillation (Chincinska et al. 2008). SUT1 and SUT4 genes show highest expression in source leaves at the end of the light period, whereas transcript accumulation decreases during the night to a minimum in the early morning. Replacement of the endogenous SUT1 promoter by the CaMV 35S promoter abolishes the circadian oscillation of SUT1 expression suggesting a transcriptional control of the circadian rhythm via *cis*-acting factors (He et al. 2008).

The promoter of circadian genes that are expressed predominantly in the evening contain a so-called evening element (EE), and genes preferentially expressed in the morning contain morning elements (ME) (Harmer and Kay 2005). The promoter of the SlSUT1 gene, which shows maximal transcript levels at the end of the light period, contains an imperfect evening element (AAAATATGT instead of AAAATATCT), arguing for transcriptional control of circadian oscillation. Many other circadian genes have been found to be affected at the post-transcriptional level by a sequence-specific mRNA decay mechanism, i.e. in mouse (Kwak et al. 2006; Woo et al. 2009), Xenopus (Baggs and Green 2003) or Arabidopsis (Lidder et al. 2005).

9.2 Post-transcriptional Regulation of Sucrose Transporters

Post-transcriptional control seems to play an important role in the regulation of sucrose transporters belonging to the SUT2 and SUT4 clade. Members of this clade are found in both mono- and dicotyledonous species.

The half-life for sucrose transporter mRNAs was determined to be in the range of 1–2 h (Vaughn et al. 2002; He et al. 2008). Detailed inhibitor studies helped to analyse the factors affecting the stability of solanaceous sucrose transporter transcripts.

Whereas tomato SlSUT1 and potato StSUT1 transcript accumulation decreases upon inhibition of translation with Cycloheximide (CHX), indicating that de novo protein synthesis is needed to guarantee a high level of SUT1 mRNA, the amount of SUT2 and SUT4 transcripts increases more than fourfold within only 2 h after CHX application (He et al. 2008).

This accumulation of SUT2 and SUT4 mRNA in the absence of de novo protein synthesis can either be explained by the lack of short-lived negative transcription factors, which under normal conditions inhibit efficient SUT2 and SUT4 transcription, or by short-lived RNA-binding proteins, which under normal conditions are responsible for a sequence-specific mRNA decay via ribonucleases. The simultaneous incubation with actinomycin D to prevent transcription and CHX did not affect SUT2 and SUT4 mRNA accumulation. Thus, a post-transcriptional control seems likely since transcripts accumulate even in the absence of transcription, and the stability of the SUT2 and SUT4 mRNA is most likely affected by short-lived proteins (He et al. 2008).

A similar phenomenon has been observed for the Arabidopsis SUT2 orthologue. The transcript abundance of *AtSUT2/AtSUC3* (accession no: At2g02860) increases dramatically upon 3-h treatment with 10 μM CHX as shown by microarray analysis (http://csbdb.mpimp-golm.mpg.de/csbdb/dbcor/ath/ath_txp.html), suggesting a similar regulatory effect on AtSUT2 mRNA stability at the post-transcriptional level.

Further confirmation of the post-transcriptional control of SUT4 mRNA stability came from the analysis of transgenic potato plants with reduced SUT4 expression. The sucrose transporter StSUT4 seems to play a crucial role in the development of the shade avoidance syndrome (SAS) when plants are shaded or exposed to increased intensities of far-red light (Chincinska et al. 2008). Under conditions where the red:far-red ratio is decreased, the StSUT4 mRNA amount increases. In addition, inhibitor studies revealed a prolonged half-life of the mRNA if far-red light is enriched, arguing for a post-transcriptional increase of the StSUT4 mRNA stability (I. Chincinska, H. He, C. Kühn, unpublished data).

Identification of a non-redundant set of more than 75,000 microRNAs from Arabidopsis revealed miRNAs targeted against seven out of the nine known Arabidopsis sucrose transporter genes (Lu et al. 2005). It is worthwhile mentioning that only AtSUT2 and AtSUT4 are not targets of miRNAs, suggesting a different regulatory mechanism for the members of the phylogenetic SUT1 clade.

9.3 *Translational Control of Sucrose Transporter Expression*

Translational control might be achieved by the efficiency of translation. So far only for SlSUT2 and AtSUT2, a low codon bias together with a low homology with the Kozak consensus sequence close to the start codon (Kozak 1996) has been discussed to be responsible for a low translational efficiency and a low protein abundance (Barker et al. 2000).

9.4 *Post-translational Control of Sucrose Transporter Expression*

9.4.1 Phosphorylation/Dephosphorylation

The impact of phosphorylation events on the activity of sucrose transporters has first been investigated in *B. vulgaris*. Inhibition of protein phosphatases by ocadaic acid affected not only the activity of BvSUT1 but also the BvSUT1 mRNA amount and its transcriptional rate (Roblin et al. 1998). Protein kinase inhibitors affected neither BvSUT1 transcription nor activity (Ransom-Hodgkins et al. 2003). Phosphoproteomics of the Arabidopsis plasma membrane provided evidence for the phosphorylation of the AtSUC5 N-terminus (Nühse et al. 2004). The exact position of AtSUC1 phopshorylation was determined by mass spectrometry (Ser_{20}, Thr_{393}; Niittyla et al. 2007). Serine 20 represents a highly conserved potential phosphorylation site which is present in all Arabidopsis sucrose transporters, except AtSUC2.

9.4.2 Redox-dependent Targeting and Dimerization

The quaternary structure of sucrose transporters might also be of functional relevance, since StSUT1 from potato as well as SlSUT1 from tomato were shown to dimerize in a redox-dependent manner, and GFP fusion constructs showed increased plasma membrane targeting in an oxidative environment if expressed in yeast (Krügel et al. 2008). Blue native PAGE, as well as SDS-PAGE under non-reducing conditions and immunoprecipitation, revealed the ability of the SUT1 proteins from potato and tomato to form dimers in yeast and plants (Krügel et al. 2008). Biochemical studies confirmed the capacity of SUT1 to form a dimer in plants, yeast cells, *Lactococcus lactis* and *Xenopus* oocytes in a redox-dependent manner. The spinach sucrose transporter SoSUT1 also migrates as a homodimeric complex under native-like conditions if over-expressed in transgenic potato plants (Liesche et al. 2008) leading to increased sucrose uptake capacity (Leggewie et al. 2003). It is still unclear if redox-dependent dimerization of the SUT1 protein has an impact on its plasma membrane targeting *in planta*.

9.4.3 Protein–Protein Interactions of Sucrose Transporters

With the help of the yeast two-hybrid split-ubiquitin system (SUS), sucrose transporter-interacting proteins were identified. Using a combination of SUS, immunoprecipitation and Bimolecular Fluorescence Complementation (BiFC), the interaction between the apple sucrose transporter MdSUT1 with the ER-anchored plant cytochrome b5, MdCYB5, was demonstrated. Interaction of this plant MdCYB5 increases the affinity of the sucrose transporter in yeast (Fan et al. 2009) and revealed the ability of plasma membrane-localised sucrose transporters to interact with endomembrane proteins.

A more recent work revealed the interaction between sucrose transporters and a ER-associated protein disulfide isomerase (PDI) lacking the characteristic C-terminal ER-retention signal, which is present in most PDI-like proteins (Krügel et al. 2010). The subcellular localisation of this protein disulfide isomerase is in the ER, the plasma membrane in raft-like microdomains and in vesicle-like structures when expressed as YFP fusion protein in infiltrated tobacco leaves. Therefore, a function of this PDI-like protein as escort protein or secretion factor is discussed.

9.4.4 Subcellular Localisation of Sucrose Transporters

The concept of membrane lipid rafts provides an explanation for the temporal and spatial organisation of membranes and allows the specific clustering of membrane proteins in a dynamic manner (see also Chapter “Regulation of Plant Transporters by Lipids and Microdomains”). The lipid composition of the membrane plays an important role in this compartmentalisation, and raft-like microdomains are enriched in sphingolipids and sterols. Membrane rafts are defined as ‘small (10–200 nm), heterogeneous, highly dynamic, sterol- and sphingolipid-enriched domains that compartmentalise cellular processes. Small rafts can sometimes be stabilised to form larger platforms through protein–protein and protein–lipid interactions' (Pike 2006). Clustering of membrane proteins in lipid rafts is assumed to be involved in signaling, function, oligomerization, endocytosis, degradation and lipid or protein transport (Opekarova et al. 2005; Grossmann et al. 2008). The molecular mechanism behind how raft organisation allows separation of liquid-ordered and liquid-disordered phases is still unclear.

StSUT1–GFP fusion proteins appear in lipid raft-like microdomains if expressed in yeast cells. After two-phase partitioning, solubilisation for 30 min in 1% Triton X 100 and separation by sucrose density centrifugation, the sucrose transporter StSUT1 was found in the detergent-resistant membrane fraction (DRM) of the potato leaf plasma membranes (Krügel et al. 2008). The StSUT1–GFP fusion protein is internalised by endocytic process in response to brefeldin A and CHX treatments in yeast as well as in plant cells. The oligomeric structure of the StSUT1 protein is affected by the redox environment, and a homodimeric form is detectable in the absence of reducing agents (Krügel et al. 2008). A homodimeric protein is also detectable by Blue native PAGE if the tagged SoSUT1 protein is overexpressed in transgenic plants under control of the CaMV 35S promoter (Liesche et al. 2008). Since no monomeric form of the SoSUT1 protein could be detected and since these transgenic potato plants show significantly higher sucrose uptake rates in plasma membrane vesicles (Leggewie et al. 2003), it is assumed that the dimeric form of the protein is functional in sucrose transport. It is still under investigation whether the raft association of the SUT1 protein is related to its dimerization, internalisation or degradation.

10 Potential Function of Sucrose Transporters in Sink Organs

10.1 Members of the SUT2 Subfamily of Sucrose Transporters

Sucrose transporters belonging to the SUT2 clade have been localised to the plasma membrane of sieve elements in tomato (Barker et al. 2000), Plantago (PmSUC3; Barth et al. 2003) and Arabidopsis (AtSUC3; Meyer et al. 2004). SUT2 proteins are localised not only to the phloem in various plant species but also in sink organs such as pollen, pollen tubes (Meyer et al. 2004; Hackel et al. 2006) and seeds (Zhang et al. 2007).

No ESTs from SUT2 orthologous sucrose transporters were identified in the tobacco species *Nicotiana tabaccum*, *N. benthamiana* and *N. sylvestris*. A member of the SUT1 clade, which is not present in the phloem or other tissues, is exclusively expressed in the anthers of mature tobacco flowers (Lemoine et al. 1999). NtSUT3 seems to be responsible in pollen loading and pollen tube growth, likely replacing the SUT2 function in tobacco. In vitro germinated pollen tubes are able of taking up not only hexoses but also sucrose, as has been demonstrated by uptake experiments using radioactively labelled sucrose (Lemoine et al. 1999).

SUT2-defective tomato plants do not show obvious modifications of leaf morphology or development. However, the fruit size and total fruit yield are strongly reduced in the absence of SUT2 expression, which may be the consequence of either reduced phloem unloading at the fruit level and/or a reduction in pollen development and pollen tube growth, thereby preventing efficient pollination (Hackel et al. 2006). The pollen morphology of SUT2-defective tomato plants is dramatically affected as analysed by raster electron microscopy (Fig. 4). Fruits of SUT2-inhibited plants are occasionally sterile and since the tomato fruit size is correlated with the number of seeds per fruit, a reduction in tuber yield can therefore be explained simply by an inefficient pollination of autogamous tomato flowers.

Interestingly, transgenic tomato plants with inhibited expression of invertases display a very similar phenotype to SlSUT2-inhibited tomato plants with regard to pollen viability, pollen morphology, reduction of seed number and fruit development (Goetz et al. 2001; Zanor et al. 2009), suggesting a close link between sucrose transport and cleavage. Gene silencing of the invertase LIN5, which is expressed in stamen, petals, ovaries and small fruits but not in pollen, affects pollen development in tomato. This is most likely a consequence of altered sucrose supply to the stamen (Zanor et al. 2009). However, the transcript profile of these plants also revealed significant changes in expression of sugar-responsive genes involved in hormonal metabolism which could also account for these defects.

The Arabidopsis SUT2 orthologue is localised not only to phloem sieve elements but also to guard cells, trichomes, germinating pollen, roots and seeds. Its expression is inducible upon wounding as revealed by promoter::reporter gene studies (Meyer et al. 2004). As the promoter of the SlSUT2 gene also contains wound-inducible *cis*-regulatory elements (He et al. 2008), a similar function can be assumed for these two members of the SUT2 clade.

Due to its low expression, its characteristic structure with extended cytoplasmic domains at the central loop region and the N terminus, members of the SUT2 subfamily have been discussed to play a role as sucrose sensor or regulatory protein (Barker et al. 2000).

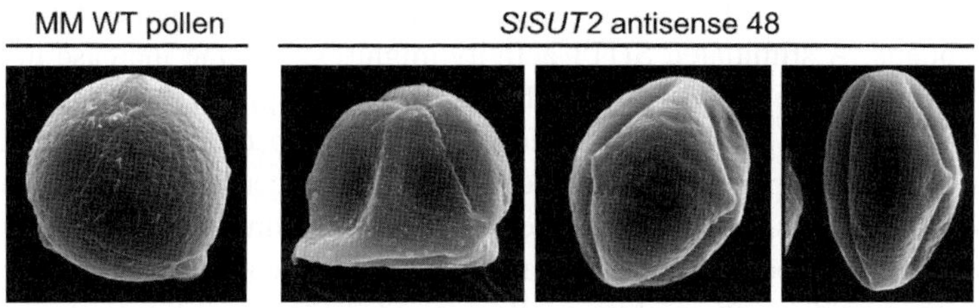

Fig. 4 Pollen morphology of wild-type (WT) and SlSUT2 antisense tomato plants (Hackel et al. 2006). Raster electron microscopy images of wild-type (*left*) and SlSUT2 transgenic pollen (*right*) were taken and kindly provided by Wilfried Bleiss (Humboldt University of Berlin). Representative examples were chosen

10.2 Members of the SUT4 Subfamily of Sucrose Transporters

Members of the SUT4 subfamily have been assigned to the sieve element plasma membrane in solanaceous plants (Weise et al. 2000) and to the tonoplast membrane in species like Arabidopsis, barley (Endler et al. 2006) or *L. japonicus* (Reinders et al. 2008). SUT4 members seem to be mainly expressed in sink organs (Weise et al. 2000; Weschke et al. 2000). LjSUT4 is highly expressed in roots and nodules of *L. japonicus* and its expression is induced during nodulation (Flemetakis et al. 2003).

Selective over-expression of sucrose transporters in transgenic plants is able to increase the sucrose transport capacity (Rosche et al. 2002; Leggewie et al. 2003). The functional characterisation of SUT4 was performed by an RNAi approach inhibiting SUT4 expression in transgenic potato plants. Surprisingly, as previously seen in SoSUT1 over-expressing tobacco plants (Riesmeier and Frommer 1994), the StSUT4 RNAi plants flowered earlier than the wild-type (Chincinska et al. 2008). StSUT4 inhibition affects not only the flowering behaviour but also tuberization of transgenic potato plants; inhibition of StSUT4 expression in *Solanum tuberosum andigena* plants, which tuberize only under short-day conditions, leads to tuber production even under long-day conditions. This effect is graft transmissible in transgenic andigena scions that were grafted on WT stocks. These findings argue for a similar inducing signal affecting both flowering and tuberization. Since decades, it is discussed whether the so-called florigen is identical to the tuberigen (Rodriguez-Falcon et al. 2006) and whether CONSTANS-like and FT-like proteins are involved.

Interestingly, the StSUT4-inhibited plants do not respond to a reduced red:far-red light ratio that is similar to canopy shade (Chincinska et al. 2008). This may indicate a photoreceptor-dependent regulation of the SUT4 expression. In wild-type potato plants, the StSUT4 expression is induced under shade conditions, and indeed no difference in SUT4 expression is observed in phytochrome B antisense plants (unpublished data), supporting the hypothesis that SUT4 expression is phyB dependent.

Several arguments underlie the assumption that SUT4 might act as regulatory protein: (1) the sucrose efflux from source leaves of SUT4-inhibited plants is increased (Chincinska et al. 2008), and sucrose as well as starch accumulation is higher in sink organs such as shoot apical meristems or in vitro induced microtubers of these plants. A stimulated SUT1 activity might explain the increase in sucrose transport from source to sink. (2) SUT1 over-expressing plants show a similar early flowering phenotype as SUT4 inhibited plants with respect to early flowering and shade avoidance response (unpublished). (3) Starch accumulation is affected in StSUT4 RNAi plants, and the redox-regulated key enzyme of starch biosynthesis, AGPase, seems to be activated in day time-independent manner. One possible explanation would be an inhibitory effect of SUT4 on SUT1 activity, which in StSUT4 RNAi plants would be missing. Protein–protein interaction between SUT4 and SUT1 was shown in yeast using the split-ubiquitin system (Reinders et al. 2002), as well as in plants using the BiFC.

11 Phloem Mobile Signals

11.1 *Phloem RNAs*

Not only the transcripts of sucrose transporters show phloem mobility (see above), but other phloem RNAs or small RNA populations were also graft transmissible and may potentially play a role in long-distance signaling. For phloem mobile RNA molecules such as CmPP16 (Xoconostle-Cázares et al. 1999), CmNCAP (Ruiz-Medrano et al. 1999), StBEL5 (Banerjee et al. 2006; Hannapel 2010) and GA-insensitive (GAI) (Haywood et al. 2005; Huang and Yu 2009), a regulatory function has been assumed. GAI belongs to the GRAS family and acts as a negative regulator of GA responses. BEL proteins bind specific DNA sequences in the promoter of *ga20oxidase1*, a key enzyme of GA biosynthesis, and repress its activity (Chen et al. 2004).

Many of these mRNAs are involved in the gibberellin biosynthesis and perception. In a recent phloem proteomic approach, phloem RNAs and proteins were analysed from pumpkin by LC-MS/MS. The identified proteins are mainly involved in GA biosynthesis, antioxidation and defence. Seven new GA biosynthetic enzymes have been identified, and gibberellins are discussed to use the phloem for a whole plant developmental control (Cho et al. 2010).

11.2 *MicroRNAs*

MicroRNAs are also suspected to be involved in the long-distance regulation of their target genes via the phloem path. A detailed analysis of the vascular exudates form oilseed rape (*Brassica napus*) revealed increased amounts of miR395, miR398 and miR399, and miRNAs which are known to respond to nutrient starvation in non-vascular tissues (sulphate, copper and sulphate, respectively) in response to starvation conditions (Buhtz et al. 2008). The Arabidopsis microRNA miR399 is targeted to PHO2, involved in phosphate homeostasis and shows phloem mobility in micrograft experiments (Pant et al. 2008).

The microRNA miR172 plays an important role in the control of the flowering time and floral patterning of several plant species (Mlotshwa et al. 2006; Jung et al. 2007; Glazinska et al. 2009; Mathieu et al. 2009). The target genes of miR172 belong to the APETALA2 family of transcription factors, and in Arabidopsis, over-expression of miR172 leads to early flowering, since miR172 is thought to induce FLOWERING LOCUS T (FT) expression by repression of its inhibitor in a GIGANTEA-dependent manner (Jung et al. 2007).

In potato plants, miR172 plays a role in the induction of tuberization. Over-expression of miR172 in potato plants promotes flowering, accelerates tuberization and induces tuber formation under long days (Martin et al. 2009). These effects are graft transmissible. Plants over-expressing miR172 show up-regulation of the phloem mobile StBEL5 mRNA, which is known to promote tuberization. It is assumed that the AP2-like target gene of miR172 normally inhibits StBEL5 expression. Interestingly, the level of mR172 (and BEL5) mRNA is reduced in phyB antisense plants indicating a phyB-dependent regulation of miR172. All in all, the phenotype of miR172 over-expressing potato plants is strikingly similar to the phenotype of potato plants with reduced expression of the sucrose transporter StSUT4 (Chincinska et al. 2008), which is also assumed to be controlled by phytochrome B. It is therefore very likely that both genes are members of the same photoperiod-dependent signaling pathway.

11.3 Metabolites (Sucrose)

The sucrose molecule is also discussed to represent a phloem mobile signal since it is known to be a potent florigen. Shortly before floral initiation, the concentration of sucrose as well as gibberellin (GA) increases strongly in the shoot apex of Arabidopsis (Eriksson et al. 2006). Sucrose is required for the GA-dependent up-regulation of the flower meristem identity gene, LEAFY (Blazquez et al. 1998). Sucrose supply is able to complement the phenotype of several flowering mutants like *phyA*, *gi* and *co* in the dark, but is unable to rescue the late-flowering phenotype of *ft* mutants (Roldan et al. 1999; Bagnall and King 2001).

Sucrose is assumed to be an important component of the phloem mobile florigenic signal (Bernier and Perilleux 2005). A side effect of the sucrose molecule might be an effect on the efficiency of phloem transport of phloem mobile signals by driving the mass flow as the major component of the phloem translocation stream.

11.4 Ca^{2+}, ROS, Electric Potential Waves

Translocation of phloem mobile metabolites or macromolecules like mRNAs or proteins depends on the velocity of the phloem translocation stream which is driven by mass flow and therefore much faster than simple diffusion. Nevertheless, the signal transduction velocity is limited. A much higher translocation velocity can be achieved in the phloem by electric signals. The presence of electrical signals, such as action potentials (AP), in higher plants is assumed to make use of ion channels to transmit information over long distances (Fromm and Lautner 2007).

Wounding or other stimuli from one leaf to a distant target leaf induce propagation of electric signals with a velocity of 5–10 cm min^{-1} in both monocot and dicot plants (Zimmermann et al. 2009). It is assumed that the response is due to stimulation of the plasma membrane proton-ATPase. Furthermore, Ca^{2+}-, K^+-, as well as Cl^--channels might be involved in the transduction of electric signals.

Redox signals can also be mediated via the phloem sap over long distances, and one of the most abundant phloem proteins in rice is thioredoxin h (Ishiwatari et al. 1995). Light information in leaves is suggested to be recognised as a thiol signal in leaf chloroplasts that is transformed into a sugar signal for long-distance transport to the sink, where it is re-converted into a thiol signal to trigger starch metabolism in amyloplasts (Balmer et al. 2006).

A rapid (8.4 cm min^{-1}), long-distance and cell-to-cell signal was observed in Arabidopsis plants associated with reactive oxygen species (ROS) in response to wounding, heat, high light and salinity stress. Propagation of this rapid signal was accompanied by ROS accumulation in the extracellular space and was inhibited by suppression of ROS accumulation (Miller et al. 2009).

The occlusion of phloem sieve plates can reversibly be induced by distant stimuli involving electropotential waves (EPWs). In V. faba sieve plates, occlusion involves a giant protein body, called the forisom, which can specifically be found in sieve elements of legumes. Electrophysiological measurements revealed a sieve element occlusion mechanism depending on the longitudinal propagation of an EPW releasing Ca^{2+} into the sieve element lumen (Furch et al. 2007). An increased density of Ca^{2+} channels was observed in the ER at sieve plates and pore-plasmodesma units allowing transiently high levels of parietal Ca^{2+} in sieve elements, and allowing forisom dispersion at a threshold level of >100 nM intracellular Ca^{2+} (Furch et al. 2009).

12 Sucrose Transport and Plant Signaling

Sucrose transporters affect many developmental processes in higher plants, and interplay with phytohormonal signaling pathways is postulated. Many phytohormones such as gibberellin, cytokinin, jasmonic, tuberonic and abscisic acids play an important role in tuberization. Gibberellins are known to inhibit tuber formation, whereas abscisic acid affects tuberization positively by antagonising GA. Sucrose plays a role as a tuber-inducing molecule at high concentrations, which is thought to occur by regulating the level of gibberellins. Low sucrose concentrations or high sucrose plus GA are not able to induce tuber formation in vitro. And sucrose levels in the medium were negatively correlated with the amount of GA_1 levels, the active GA during tuber formation (Xu et al. 1998).

Regarding internode elongation, stem length and tuber induction, the StSUT4-RNAi plants resemble transgenic plants with reduced GA biosynthesis (Carrera et al. 2000). Reduction of the GA biosynthetic key enzyme GA20ox1 in StSUT4-RNAi plants indicates that GA biosynthesis is affected in these plants. The phenotype of StSUT4-inhibited plants could not be rescued by external supply of gibberellins. However, paclobutrazol treatment, an inhibitor of GA biosynthesis, mimicked the phenotype of StSUT4-RNAi plants in wild-type potatoes, indicating that GA biosynthesis was inhibited in SUT4-deficient plants (Chincinska et al. 2008). Interconnection of the ethylene and the GA responsive pathway in the control of the phytochrome-dependent response to shading has been described for tobacco plants, and it is assumed that ethylene modulated the GA action (Pierik et al. 2004).

The expression of the StSUT4 gene from potato is inducible by ethephon, an ethylene precursor and by GA_3 treatment, suggesting a reciprocal regulation of StSUT4 by ethylene and GA. Indeed, the key enzymes in ethylene and gibberellin biosynthesis, the ACC oxidase as well as the GA20oxidase1, respectively, are expressed at lower levels in StSUT4-inhibited plants (Chincinska et al. 2008). The hypothesis that StSUT4 might be involved in the ethylene-dependent signal transduction pathway is supported by protein–protein interactions. A split-ubiquitin screen for SUT4-interacting proteins helped to identify several ethylene responsive proteins as SUT4-interacting proteins (J. Reins, C. Kühn, unpublished data).

Subcellular localisation of sucrose transporters revealed that they are not exclusively localised to the plasma membrane (Chincinska et al. 2008; Krügel et al. 2008). StSUT1 is associated to lipid raft-like microdomains and is internalised in response to brefeldin A treatment (Krügel et al. 2008, 2010). A similar phenomenon can be observed if brefeldin A-treated plant material is embedded after formaldehyde fixation, and immunodetection is performed with StSUT4-specific antibodies (Fig. 5). StSUT4 is probably also associated with the detergent-resistant membrane fraction in plants.

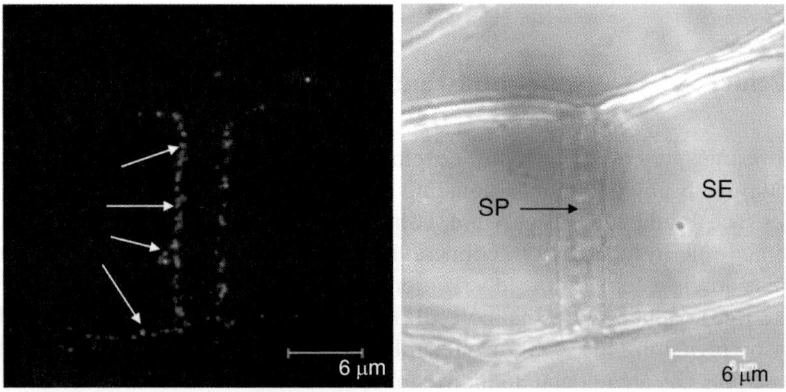

Fig. 5 Immunolocalisation of StSUT4 in longitudinal stem sections of potato plants pre-treated with brefeldin A. As previously shown for StSUT1 by immunolocalisation as well as for transiently over-expressed StSUT1-GFP fusion proteins (Krügel et al. 2008), internalisation of StSUT4 is detectable in response to brefeldin A treatment. Stem tissue was incubated in 50 μM brefeldin A for 1 h, fixed with formaldehyde and embedded in LR White. Immunodetection was performed on semi-thin sections with affinity-purified peptide antibodies against the StSUT4 protein. Localization was visualised by the help of FITC-coupled secondary antibodies (*left* panel). A transmission picture of the same region (*right* panel) reveals the presence of a sieve plate in the immunodecorated cell. StSUT4 association with brefeldin A compartments is detectable in mature sieve elements. *SE* sieve element, *SP* sieve plates

The question is whether or not the concentration of sucrose transporters in liquid-ordered membrane platforms enables the transporters to form oligomers with themselves, or heteromers with other putative signaling molecules or if raft-dependent endocytosis allows interaction of sucrose transporters with intracellularly localised interaction partners like the ER-anchored cytochrome b5, the protein disulfide isomerase or ethylene receptor proteins.

13 Conclusion

Sucrose transporters proteins are obviously able to fulfil different functions in sink and source tissues. Recent discovery of sucrose transporter localisation to brefeldin A-induced compartments within mature sieve elements opens interesting new possibilities for the regulation of the sucrose transport capacity of a given membrane. Inter- and intracellular trafficking of sucrose transporter proteins might play an important role in plant signaling and development.

Acknowledgements I gratefully acknowledge critical reading by Tom Buckhout and experimental work by Johannes Liesche, Izabela Chincinska, Aleksandra Hackel, Hongxia He and Undine Krügel. I thank Wilfried Bleiss for the raster electronic analysis of tomato pollen and Christopher Grof for the Phylogenetic analysis of the sucrose transporter gene family. Many thanks to Salome Prat for providing phytochrome B antisense potato plants. Financial support came from DFG.

References

Aloni B, Wyse RE, Griffith S (1986) Sucrose transport and phloem unloading in stem of Vicia faba: possible involvement of a sucrose carrier and osmotic regulation. Plant Physiol 81:482–486

Aoki N, Hirose T, Scofield GN, Whitfeld PR, Furbank RT (2003) The sucrose transporter gene family in rice. Plant Cell Physiol 44:223–232

Aoki N, Scofield GN, Wang XD, Patrick JW, Offler CE, Furbank RT (2004) Expression and localisation analysis of the wheat sucrose transporter TaSUT1 in vegetative tissues. Planta 219:176–184

Baggs JE, Green CB (2003) Nocturnin, a deadenylase in Xenopus laevis retina: a mechanism for posttranscriptional control of circadian-related mRNA. Curr Biol 13:189–198

Bagnall DJ, King RW (2001) Phytochrome, photosynthesis and flowering in *Arabidopsis thaliana*: photophysiological studies using mutants and transgenic lines. Aust J Plant Physiol 28:401–408

Balmer Y, Vensel WH, Cai N, Manieri W, Schurmann P, Hurkman WJ, Buchanan BB (2006) A complete ferredoxin/thioredoxin system regulates fundamental processes in amyloplasts. Proc Natl Acad Sci USA 103:2988–2993

Banerjee AK, Chatterjee M, Yu Y, Suh SG, Miller WA, Hannapel DJ (2006) Dynamics of a mobile RNA of potato involved in a long-distance signaling pathway. Plant Cell 18:3443–3457

Barker L, Kühn C, Weise A, Schulz A, Gebhardt C, Hirner B, Hellmann H, Schulze W, Ward JM, Frommer WB (2000) SUT2, a putative sucrose sensor in sieve elements. Plant Cell 12:1153–1164

Barth I, Meyer S, Sauer N (2003) PmSUC3: characterization of a SUT2/SUC3-type sucrose transporter from Plantago major. Plant Cell 15:1375–1385
Bernier G, Perilleux C (2005) A physiological overview of the genetics of flowering time control. Plant Biotechnol J 3:3–16
Blazquez MA, Green R, Nilsson O, Sussman MR, Weigel D (1998) Gibberellins promote flowering of arabidopsis by activating the LEAFY promoter. Plant Cell 10:791–800
Braun DM, Slewinski TL (2009) Genetic control of carbon partitioning in grasses: roles of sucrose transporters and tie-dyed loci in phloem loading. Plant Physiol 149:71–81
Briskin DP, Thornley WR, Wyse RE (1985) Membrane transport in isolated vesicles from sugarbeet taproot. Plant Physiol 78:871–875
Buhtz A, Springer F, Chappell L, Baulcombe DC, Kehr J (2008) Identification and characterization of small RNAs from the phloem of Brassica napus. Plant J 53:739–749
Carpaneto A, Geiger D, Bamberg E, Sauer N, Fromm J, Hedrich R (2005) Phloem-localized, proton-coupled sucrose carrier ZmSUT1 mediates sucrose efflux under the control of the sucrose gradient and the proton motive force. J Biol Chem 280:21437–21443
Carrera E, Bou J, Garcia-Martinez JL, Prat S (2000) Changes in GA 20-oxidase gene expression strongly affect stem length, tuber induction and tuber yield of potato plants. Plant J 22:247–256
Chandran D, Reinders A, Ward JM (2003) Substrate specificity of the Arabidopsis thaliana sucrose transporter AtSUC2. J Biol Chem 278:44320–44325
Chen H, Banerjee AK, Hannapel DJ (2004) The tandem complex of BEL and KNOX partners is required for transcriptional repression of ga20ox1. Plant J 38:276–284
Chincinska IA, Liesche J, Krügel U, Michalska J, Geigenberger P, Grimm B, Kühn C (2008) Sucrose transporter StSUT4 from potato affects flowering, tuberization, and shade avoidance response. Plant Physiol 146:515–528
Cho WK, Chen XY, Rim Y, Chu H, Kim S, Kim SW, Park ZY, Kim JY (2010) Proteome study of the phloem sap of pumpkin using multidimensional protein identification technology. J Plant Physiol 167(10):771–778
Deeken R, Geiger D, Fromm J, Koroleva O, Ache P, Langenfeld-Heyser R, Sauer N, May ST, Hedrich R (2002) Loss of the AKT2/3 potassium channel affects sugar loading into the phloem of Arabidopsis. Planta 216:334–344
Deeken R, Ache P, Kajahn I, Klinkenberg J, Bringmann G, Hedrich R (2008) Identification of Arabidopsis thaliana phloem RNAs provides a search criterion for phloem-based transcripts hidden in complex datasets of microarray experiments. Plant J 55:746–759
Dereeper A, Guignon V, Blanc G, Audic S, Buffet S, Chevenet F, Dufayard JF, Guindon S, Lefort V, Lescot M, Claverie JM, Gascuel O (2008) Phylogeny.fr: robust phylogenetic analysis for the non-specialist. Nucleic Acids Res 36:W465–W469
Doering-Saad C, Newbury HJ, Couldridge CE, Bale JS, Pritchard J (2006) A phloem-enriched cDNA library from Ricinus: insights into phloem function. J Exp Bot 57:3183–3193
Endler A, Meyer S, Schelbert S, Schneider T, Weschke W, Peters SW, Keller F, Baginsky S, Martinoia E, Schmidt UG (2006) Identification of a vacuolar sucrose transporter in barley and Arabidopsis mesophyll cells by a tonoplast proteomic approach. Plant Physiol 141:196–207
Eriksson S, Bohlenius H, Moritz T, Nilsson O (2006) GA4 Is the active gibberellin in the regulation of LEAFY transcription and Arabidopsis floral initiation. Plant Cell 18 (9):2172–2181
Fan RC, Peng CC, Xu YH, Wang XF, Li Y, Shang Y, Du SY, Zhao R, Zhang XY, Zhang LY, Zhang DP (2009) Apple sucrose transporter SUT1 and sorbitol transporter SOT6 interact with cytochrome b5 to regulate their affinity for substrate sugars. Plant Physiol 150:1880–1901
Flemetakis E, Dimou M, Cotzur D, Efrose RC, Aivalakis G, Colebatch G, Udvardi M, Katinakis P (2003) A sucrose transporter, LjSUT4, is up-regulated during Lotus japonicus nodule development. J Exp Bot 54:1789–1791
Fromm J, Lautner S (2007) Electrical signals and their physiological significance in plants. Plant Cell Environ 30:249–257
Furbank RT, Scofield GN, Hirose T, Wang X-D, Patrick JW, Offler CE (2001) Cellular localisation and function of a sucrose transporter OsSUT1 in developing rice grains. Aust J Plant Physiol 28:1187–1196

Furch AC, Hafke JB, Schulz A, van Bel AJ (2007) Ca2+-mediated remote control of reversible sieve tube occlusion in Vicia faba. J Exp Bot 58:2827–2838
Furch AC, van Bel AJ, Fricker MD, Felle HH, Fuchs M, Hafke JB (2009) Sieve element Ca2+ channels as relay stations between remote stimuli and sieve tube occlusion in Vicia faba. Plant Cell 21:2118–2132
Getz HP, Klein M (1995) Characteristics of sucrose transport and sucrose-induced H+ transport on the tonoplast of red beet (Beta vulgaris L.) storage tissue. Plant Physiol 107:459–467
Glazinska P, Zienkiewicz A, Wojciechowski W, Kopcewicz J (2009) The putative miR172 target gene InAPETALA2-like is involved in the photoperiodic flower induction of Ipomoea nil. J Plant Physiol 166:1801–1813
Goetz M, Godt DE, Guivarc'h A, Kahmann U, Chriqui D, Roitsch T (2001) Induction of male sterility in plants by metabolic engineering of the carbohydrate supply. Proc Natl Acad Sci USA 98:6522–6527
Grossmann G, Malinsky J, Stahlschmidt W, Loibl M, Weig-Meckl I, Frommer WB, Opekarova M, Tanner W (2008) Plasma membrane microdomains regulate turnover of transport proteins in yeast. J Cell Biol 183:1075–1088
Hackel A, Schauer N, Carrari F, Fernie AR, Grimm B, Kühn C (2006) Sucrose transporter LeSUT1 and LeSUT2 inhibition affects tomato fruit development in different ways. Plant J 45:180–192
Hafke JB, Furch AC, Reitz MU, van Bel AJ (2007) Functional sieve element protoplasts. Plant Physiol 145:703–711
Hannapel DJ (2010) A model system of development regulated by the long-distance transport of mRNA. J Integr Plant Biol 52:40–52
Harmer SL, Kay SA (2005) Positive and negative factors confer phase-specific circadian regulation of transcription in Arabidopsis. Plant Cell 17:1926–1940
Harms K, Wohner RV, Schulz B, Frommer WB (1994) Isolation and characterization of P-type H(+)-ATPase genes from potato. Plant Mol Biol 26:979–988
Haywood V, Yu TS, Huang NC, Lucas WJ (2005) Phloem long-distance trafficking of GIBBERELLIC ACID-INSENSITIVE RNA regulates leaf development. Plant J 42:49–68
He H, Chincinska I, Hackel A, Grimm B, Kühn C (2008) Phloem mobility and stability of sucrose transporter transcripts. Open Plant Sci J 2:15–26
Huang NC, Yu TS (2009) The sequences of Arabidopsis GA-INSENSITIVE RNA constitute the motifs that are necessary and sufficient for RNA long-distance trafficking. Plant J 59:921–929
Huson DH, Richter DC, Rausch C, Dezulian T, Franz M, Rupp R (2007) Dendroscope: an interactive viewer for large phylogenetic trees. BMC Bioinform 8:460
Ishimaru K, Hirose T, Aoki N, Takahashi S, Ono K, Yamamoto S, Wu J, Saji S, Baba T, Ugaki M, Matsumoto T, Ohsugi R (2001) Antisense expression of a rice sucrose transporter OsSUT1 in rice (Oryza sativa L.). Plant Cell Physiol 42:1181–1185
Ishiwatari Y, Honda C, Kawashima I, Nakamura S, Hirano H, Mori S, Fujiwara T, Hayashi H, Chino M (1995) Thioredoxin h is one of the major proteins in rice phloem sap. Planta 195:456–463
Jung JH, Seo YH, Seo PJ, Reyes JL, Yun J, Chua NH, Park CM (2007) The GIGANTEA-regulated microRNA172 mediates photoperiodic flowering independent of CONSTANS in Arabidopsis. Plant Cell 19:2736–2748
Knop C, Voitsekhovskaja O, Lohaus G (2001) Sucrose transporters in two members of the Scrophulariaceae with different types of transport sugar. Planta 213:80–91
Knop C, Stadler R, Sauer N, Lohaus G (2004) AmSUT1, a sucrose transporter in collection and transport phloem of the putative symplastic phloem loader Alonsoa meridionalis. Plant Physiol 134:204–214
Kozak M (1996) Interpreting cDNA sequences: some insights from studies on translation. Mamm Genome 7:563–574
Krügel U, Veenhoff LM, Langbein J, Wiederhold E, Liesche J, Friedrich T, Grimm B, Martinoia E, Poolman B, Kühn C (2008) Transport and sorting of the *Solanum tuberosum* sucrose transporter SUT1 is affected by posttranslational modification. Plant Cell 20:1–17

Krügel U, He H, Reins J, Liesche J, Chincinska I, Grimm B, Schulze WX, Kühn C (2010) Identification of sucrose transporter StSUT1 interacting proteins by split ubiquitin and mass spectrometry revealed SUT1 association to the detergent resistant membrane fraction in plants. (Submitted)

Kühn C, Quick WP, Schulz A, Riesmeier J, Sonnewald U, Frommer WB (1996) Companion cell-specific inhibition of the potato sucrose transporter SUT1. Plant Cell Environ 19: 1115–1123

Kühn C, Franceschi VR, Schulz A, Lemoine R, Frommer WB (1997) Macromolecular trafficking indicated by localization and turnover of sucrose transporters in enucleate sieve elements. Science 275:1298–1300

Kühn C, Hajirezaei MR, Fernie AR, Roessner-Tunali U, Czechowski T, Hirner B, Frommer WB (2003) The sucrose transporter StSUT1 localizes to sieve elements in potato tuber phloem and influences tuber physiology and development. Plant Physiol 131:102–113

Kwak E, Kim TD, Kim KT (2006) Essential role of 3'-untranslated region-mediated mRNA decay in circadian oscillations of mouse Period3 mRNA. J Biol Chem 281:19100–19106

Lalonde S, Weise A, Walsh RP, Ward JM, Frommer WB (2003) Fusion to GFP blocks intercellular trafficking of the sucrose transporter SUT1 leading to accumulation in companion cells. BMC Plant Biol 3:8

Leggewie G, Kolbe A, Lemoine R, Roessner U, Lytovchenko A, Zuther E, Kehr J, Frommer WB, Riesmeier JW, Willmitzer L, Fernie AR (2003) Overexpression of the sucrose transporter SoSUT1 in potato results in alterations in leaf carbon partitioning and in tuber metabolism but has little impact on tuber morphology. Planta 217:158–167

Lemoine R, Burkle L, Barker L, Sakr S, Kuhn C, Regnacq M, Gaillard C, Delrot S, Frommer WB (1999) Identification of a pollen-specific sucrose transporter-like protein NtSUT3 from tobacco. FEBS Lett 454:325–330

Lidder P, Gutierrez RA, Salome PA, McClung CR, Green PJ (2005) Circadian control of messenger RNA stability. Association with a sequence-specific messenger RNA decay pathway. Plant Physiol 138:2374–2385

Liesche J, Schulz A, Krügel U, Grimm B, Kühn C (2008) Dimerization and endocytosis of the sucrose transporter SUT1 in mature sieve elements. Plant Signal Behav 3:1136–1137

Lin MK, Lee YJ, Lough TJ, Phinney BS, Lucas WJ (2009) Analysis of the pumpkin phloem proteome provides insights into angiosperm sieve tube function. Mol Cell Proteomics 8:343–356

Lu C, Tej SS, Luo S, Haudenschild CD, Meyers BC, Green PJ (2005) Elucidation of the small RNA component of the transcriptome. Science 309:1567–1569

Ludwig A, Stolz J, Sauer N (2000) Plant sucrose-H+ symporters mediate the transport of vitamin H. Plant J 24:503–509

Martens HJ, Roberts AG, Oparka KJ, Schulz A (2006) Quantification of plasmodesmatal endoplasmic reticulum coupling between sieve elements and companion cells using fluorescence redistribution after photobleaching. Plant Physiol 142:471–480

Martin A, Adam H, Diaz-Mendoza M, Zurczak M, Gonzalez-Schain ND, Suarez-Lopez P (2009) Graft-transmissible induction of potato tuberization by the microRNA miR172. Development 136:2873–2881

Mathieu J, Yant LJ, Murdter F, Kuttner F, Schmid M (2009) Repression of flowering by the miR172 target SMZ. PLoS Biol 7:e1000148

McRae SR, Christopher JT, Smith JAC, Holtum JAM (2002) Sucrose transport across the vacuolar membrane of *Ananas comosus*. Funct Plant Biol 29:717–724

Meyer S, Melzer M, Truernit E, Hummer C, Besenbeck R, Stadler R, Sauer N (2000) AtSUC3, a gene encoding a new Arabidopsis sucrose transporter, is expressed in cells adjacent to the vascular tissue and in a carpel cell layer. Plant J 24:869–882

Meyer S, Lauterbach C, Niedermeier M, Barth I, Sjolund RD, Sauer N (2004) Wounding enhances expression of AtSUC3, a sucrose transporter from Arabidopsis sieve elements and sink tissues. Plant Physiol 134:684–693

Miller G, Schlauch K, Tam R, Cortes D, Torres MA, Shulaev V, Dangl JL, Mittler R (2009) The plant NADPH oxidase RBOHD mediates rapid systemic signaling in response to diverse stimuli. Sci Signal 2:ra45

Mlotshwa S, Yang Z, Kim Y, Chen X (2006) Floral patterning defects induced by Arabidopsis APETALA2 and microRNA172 expression in Nicotiana benthamiana. Plant Mol Biol 61:781–793

Neuhaus HE (2007) Transport of primary metabolites across the plant vacuolar membrane. FEBS Lett 581:2223–2226

Niittyla T, Fuglsang AT, Palmgren MG, Frommer WB, Schulze WX (2007) Temporal analysis of sucrose-induced phosphorylation changes in plasma membrane proteins of Arabidopsis. Mol Cell Proteomics 6:1711–1726

Nühse TS, Stensballe A, Jensen ON, Peck SC (2004) Phosphoproteomics of the Arabidopsis plasma membrane and a new phosphorylation site database. Plant Cell 16:2394–2405

Opekarova M, Malinska K, Novakova L, Tanner W (2005) Differential effect of phosphatidylethanolamine depletion on raft proteins: further evidence for diversity of rafts in Saccharomyces cerevisiae. Biochim Biophys Acta 1711:87–95

Pant BD, Buhtz A, Kehr J, Scheible WR (2008) MicroRNA399 is a long-distance signal for the regulation of plant phosphate homeostasis. Plant J 53:731–738

Pierik R, Cuppens ML, Voesenek LA, Visser EJ (2004) Interactions between ethylene and gibberellins in phytochrome-mediated shade avoidance responses in tobacco. Plant Physiol 136:2928–2936

Pike LJ (2006) Rafts defined: a report on the Keystone Symposium on lipid rafts and cell function. J Lipid Res 47:1597–1598

Rae AL, Perroux JM, Grof CP (2005) Sucrose partitioning between vascular bundles and storage parenchyma in the sugarcane stem: a potential role for the ShSUT1 sucrose transporter. Planta 220:817–825

Ransom-Hodgkins WD, Vaughn MW, Bush DR (2003) Protein phosphorylation plays a key role in sucrose-mediated transcriptional regulation of a phloem-specific proton-sucrose symporter. Planta 217:483–489

Reinders A, Ward JM (2001) Functional characterization of the alpha-glucoside transporter Sut1p from Schizosaccharomyces pombe, the first fungal homologue of plant sucrose transporters. Mol Microbiol 39:445–454

Reinders A, Schulze W, Kühn C, Barker L, Schulz A, Ward JM, Frommer WB (2002) Protein-protein interactions between sucrose transporters of different affinities colocalized in the same enucleate sieve element. Plant Cell 14:1567–1577

Reinders A, Sivitz AB, Hsi A, Grof CP, Perroux JM, Ward JM (2006) Sugarcane ShSUT1: analysis of sucrose transport activity and inhibition by sucralose. Plant Cell Environ 29:1871–1880

Reinders A, Sivitz AB, Starker CG, Gantt JS, Ward JM (2008) Functional analysis of LjSUT4, a vacuolar sucrose transporter from Lotus japonicus. Plant Mol Biol 68:289–299

Riesmeier JW, Frommer WB (1994) Einfluß der Überexpression von Saccharosetransportern auf das Blühverhalten von Pflanzen (Germany). German Patent P 44 39 748.8

Riesmeier JW, Willmitzer L, Frommer WB (1992) Isolation and characterization of a sucrose carrier cDNA from spinach by functional expression in yeast. EMBO J 11:4705–4713

Roblin G, Sakr S, Bonmort J, Delrot S (1998) Regulation of a plant plasma membrane sucrose transporter by phosphorylation. FEBS Lett 424:165–168

Rodriguez-Falcon M, Bou J, Prat S (2006) Seasonal control of tuberization in potato: conserved elements with the flowering response. Annu Rev Plant Biol 57:151–180

Roldan M, Gomez-Mena C, Ruiz-Garcia L, Salinas J, Martinez-Zapater JM (1999) Sucrose availability on the aerial part of the plant promotes morphogenesis and flowering of Arabidopsis in the dark. Plant J 20:581–590

Rosche E, Blackmore D, Tegeder M, Richardson T, Schroeder H, Higgins TJ, Frommer WB, Offler CE, Patrick JW (2002) Seed-specific overexpression of a potato sucrose transporter increases sucrose uptake and growth rates of developing pea cotyledons. Plant J 30:165–175

Ruiz-Medrano R, Xoconostle-Cazares B, Lucas WJ (1999) Phloem long-distance transport of CmNACP mRNA: implications for supracellular regulation in plants. Development 126:4405–4419

Schmitt B, Stadler R, Sauer N (2008) Immunolocalization of solanaceous SUT1 proteins in companion cells and xylem parenchyma: new perspectives for phloem loading and transport. Plant Physiol 148:187–199

Schulze W, Weise A, Frommer WB, Ward JM (2000) Function of the cytosolic N-terminus of sucrose transporter AtSUT2 in substrate affinity. FEBS Lett 485:189–194

Scofield GN, Hirose T, Aoki N, Furbank RT (2007a) Involvement of the sucrose transporter, OsSUT1, in the long-distance pathway for assimilate transport in rice. J Exp Bot 58:3155–3169

Scofield GN, Aoki N, Hirose T, Takano M, Jenkins CL, Furbank RT (2007b) The role of the sucrose transporter, OsSUT1, in germination and early seedling growth and development of rice plants. J Exp Bot 58:483–495

Secor J (1987) Regulation of sucrose efflux from soybean leaf discs. Plant Physiol 83:143–148

Sivitz AB, Reinders A, Johnson ME, Krentz AD, Grof CP, Perroux JM, Ward JM (2007) Arabidopsis sucrose transporter AtSUC9. High-affinity transport activity, intragenic control of expression, and early flowering mutant phenotype. Plant Physiol 143:188–198

Slewinski TL, Meeley R, Braun DM (2009) Sucrose transporter1 functions in phloem loading in maize leaves. J Exp Bot 60:881–892

Stadler R, Sauer N (1996) The Arabidopsis thaliana AtSUC2 gene is specifically expressed in companion cells. Bot Acta 109:299–306

Stadler R, Brandner J, Schulz A, Gahrtz M, Sauer N (1995) Phloem loading by the PmSUC2 sucrose carrier from Plantago major occurs into companion cells. Plant Cell 7:1545–1554

Stadler R, Wright KM, Lauterbach C, Amon G, Gahrtz M, Feuerstein A, Oparka KJ, Sauer N (2005) Expression of GFP-fusions in Arabidopsis companion cells reveals non-specific protein trafficking into sieve elements and identifies a novel post-phloem domain in roots. Plant J 41:319–331

Sun Y, Reinders A, Lafleur KR, Mori T, Ward JM (2010) Transport activity of rice sucrose transporters OsSUT1 and OsSUT5. Plant Cell Physiol 51(1):114–122

The Arabidopsis genome initiative (2000) Analysis of the genome sequence of the flowering plant Arabidopsis thaliana. Nature 408:796–815

Thompson MV, Wolniak SM (2008) A plasma membrane-anchored fluorescent protein fusion illuminates sieve element plasma membranes in Arabidopsis and tobacco. Plant Physiol 146:1599–1610

Vaughn MW, Harrington GN, Bush DR (2002) Sucrose-mediated transcriptional regulation of sucrose symporter activity in the phloem. Proc Natl Acad Sci USA 99:10876–10880

Walz C, Juenger M, Schad M, Kehr J (2002) Evidence for the presence and activity of a complete antioxidant defence system in mature sieve tubes. Plant J 31:189–197

Walz C, Giavalisco P, Schad M, Juenger M, Klose J, Kehr J (2004) Proteomics of curcurbit phloem exudate reveals a network of defence proteins. Phytochemistry 65:1795–1804

Weber H, Borisjuk L, Heim U, Sauer N, Wobus U (1997) A role for sugar transporters during seed development: molecular characterization of a hexose and a sucrose carrier in fava bean seeds. Plant Cell 9:895–908

Weise A, Barker L, Kühn C, Lalonde S, Buschmann H, Frommer WB, Ward JM (2000) A new subfamily of sucrose transporters, SUT4, with low affinity/high capacity localized in enucleate sieve elements of plants. Plant Cell 12:1345–1355

Weise A, Lalonde S, Kühn C, Frommer WB, Ward JM (2008) Introns control expression of sucrose transporter LeSUT1 in trichomes, companion cells and in guard cells. Plant Mol Biol 68:251–262

Weschke W, Panitz R, Sauer N, Wang Q, Neubohn B, Weber H, Wobus U (2000) Sucrose transport into barley seeds: molecular characterization of two transporters and implications for seed development and starch accumulation. Plant J 21:455–467

Willenbrink J (1987) Die pflanzliche Vakuole als Speicher. Naturwissenschaften 74:22–29

Woo KC, Kim TD, Lee KH, Kim DY, Kim W, Lee KY, Kim KT (2009) Mouse period 2 mRNA circadian oscillation is modulated by PTB-mediated rhythmic mRNA degradation. Nucleic Acids Res 37:26–37

Xoconostle-Cázares B, Xiang Y, Ruiz-Medrano R, Wang HL, Monzer J, Yoo BC, McFarland KC, Franceschi VR, Lucas WJ (1999) Plant paralog to viral movement protein that potentiates transport of mRNA into the phloem. Science 283:94–98

Xu X, van Lammeren AA, Vermeer E, Vreugdenhil D (1998) The role of gibberellin, abscisic acid, and sucrose in the regulation of potato tuber formation in vitro. Plant Physiol 117:575–584

Zanor MI, Osorio S, Nunes-Nesi A, Carrari F, Lohse M, Usadel B, Kuhn C, Bleiss W, Giavalisco P, Willmitzer L, Sulpice R, Zhou YH, Fernie AR (2009) RNA interference of LIN5 in Solanum lycopersicum confirms its role in controlling Brix content, uncovers the influence of sugars on the levels of fruit hormones and demonstrates the importance of sucrose cleavage for normal fruit development and fertility. Plant Physiol 150(3):1204–1218

Zhang W-H, Zhou Y, Dibley K, Tyerman S, Furbank R, Patrick J (2007) Nutrient loading of developing seeds. Funct Plant Biol 34:314–331

Zhang S, Sun L, Kragler F (2009) The phloem-delivered RNA pool contains small noncoding RNAs and interferes with translation. Plant Physiol 150:378–387

Zhou Y, Qu H, Dibley KE, Offler CE, Patrick JW (2007) A suite of sucrose transporters expressed in coats of developing legume seeds includes novel pH-independent facilitators. Plant J 49:750–764

Zhou Y, Chan K, Wang TL, Hedley CL, Offler CE, Patrick JW (2009) Intracellular sucrose communicates metabolic demand to sucrose transporters in developing pea cotyledons. J Exp Bot 60:71–85

Zimmermann MR, Maischak H, Mithofer A, Boland W, Felle HH (2009) System potentials, a novel electrical long-distance apoplastic signal in plants, induced by wounding. Plant Physiol 149:1593–1600

Part IV
Signaling Molecules

Auxin Transporters Controlling Plant Development

J. Petrášek, K. Malínská, and E. Zažímalová

Abstract The plant hormone auxin regulates many important aspects of plant development from the early embryogenesis to the seed production. On genomic and nongenomic levels, finely tuned auxin gradients form an important morphoregulatory trigger. *In planta*, auxin is transported to long distances using symplastic transport in the phloem. Besides, it is also a subject to the cell-to-cell transport, being able to move across the plasma membrane (PM) by diffusion and/or utilizing several types of auxin transporters. The role of the PM-localized auxin influx and efflux carriers lies mainly in the fast directional transport of auxin across the PM forming the basis for the establishment of developmentally significant auxin gradients. In addition to PM-localized auxin transporters, there is also a population of intracellular auxin transporters, which are probably involved in the regulation of auxin homeostasis and thus in the control of availability of free auxin molecules. This chapter summarizes recent knowledge on the types of auxin transporters, their functional characterization, and involvement in developmental processes in plants.

1 Introduction: Auxin as the Signaling Molecule

1.1 Auxin: The "Slavey" in Plant Growth and Development

Auxin is the plant growth regulatory compound that is involved in control of almost all developmentally preprogrammed processes in plants, as well as in the responses of plant organs to various environmental stimuli. The typical representative of native auxins is indole-3-acetic acid (IAA) but in plant tissues, also less abundant

J. Petrášek, K. Malínská, and E. Zažímalová (✉)
Institute of Experimental Botany, The Academy of Sciences of the Czech Republic, Rozvojová 263, 165 02 Prague 6, Czech Republic
e-mail: petrasek@ueb.cas.cz, malinska@ueb.cas.cz, eva.zazim@ueb.cas.cz

M. Geisler and K. Venema (eds.), *Transporters and Pumps in Plant Signaling*, Signaling and Communication in Plants 7,
DOI 10.1007/978-3-642-14369-4_9,

auxins and IAA-related compounds are present, such as phenyl acetic, indole-3-butyric, and 5-chloro-indole-3-acetic acids. The spectrum of auxin-dependent physiological effects is very broad, from regulation of cell cycle, establishment and maintenance of cell polarity and cell differentiation to control of embryogenesis, pattern formation, and morphogenesis, including the whole-plant coordinative phenomena as, for example, apical dominance. Auxin acts also in responses of plants to external factors, for example, toward light and gravity. The extraordinary wide spectrum of morphoregulatory actions of auxin is underlain by a common principle – that is, formation of so-called auxin maxima (auxin gradients). This unequal auxin distribution within plant tissue is formed with the assistance of both metabolic and transport processes and provides cells with distinct auxin amounts. Thus, auxin molecules are distributed in the tissue differentially, and so they are available unevenly to particular auxin signaling pathways (Badescu and Napier 2006; Santner and Estelle 2009). Therefore, differential distribution of auxin in tissue(s) together with the variety of and cross-talk between auxin signaling pathways, and especially various coactions of multiple Aux/IAA repressors and auxin response factor (ARF)-type transcription regulators (Guilfoyle and Hagen 2007) contribute to the pleiotropic nature of auxin effects, thus making auxin a real "slavey" in plant development.

1.2 Physical–Chemical Properties of Auxin Molecule: The (Molecular) Form Does Matter!

Generally, all active auxins (both native compounds and their synthetic analogs) known so far are weak acids, possessing carboxyl group at the end of side chain, which is attached to aromatic or heterocyclic ring. As with all weak acids, the carboxyl-proton can dissociate from the rest of auxin molecule, and the dynamic equilibrium establishes between proton-dissociated and nondissociated forms of auxin molecules. The amounts of both these molecular forms depend directly and extensively on pH. In plant tissues, the pH differs significantly between apoplast and cell walls (ca. 5.5) and cytoplasm (ca. 7.0) – see also Chapter "Plant Proton Pumps: Regulatory Circuits Involving H^+ PATPase and H^+-PPase". The amounts of dissociated and nondissociated molecular forms can be calculated from pK_a of particular auxins. For the native auxin IAA, pK_a varies around 4.8 (depending on temperature, as well as on the method used for its determination). For this value it can be calculated that at pH 5.5, approximately 4/5 of the total amount of auxin molecules are dissociated (and thus present in the form of anion IAA^-), while 1/5 of auxin molecules are nondissociated (i.e., proton-associated). In contrast, at pH 7, almost all auxin molecules are dissociated.

It is known that only some nonpolar and/or slightly polar molecules can penetrate through membranes (including PM) by passive diffusion. This is the case also for nondissociated forms of some auxins and, therefore, some portion of auxin molecules can be taken up passively from the apoplast into cells. However, in the cytoplasm all auxin molecules are dissociated and thus negatively charged. As such, they cannot penetrate across the PM out of cells passively, and they need the assistance of a carrier protein. The cell itself then functions as a weak acid trap (anion trap). So, the physical–chemical properties of auxin molecules together with the pH values inside and outside plant cells imply the need for transporter proteins assisting at least the efflux of auxin molecules from cells. Since the capacity of these carrier proteins is always limited, it is the efflux of auxin molecules from cells that creates the bottleneck in the intercellular auxin flow.

It is obvious from the above description that the auxin efflux carriers have a rate-limiting role in cell-to-cell auxin transport. Moreover, provided the auxin efflux transporters are localized asymmetrically but always in the analogous position in the file of adjacent cells, their localization would determine the direction of auxin flow (chemiosmotic polar diffusion model or chemiosmotic hypothesis; Rubery and Sheldrake 1974; Raven 1975; Goldsmith 1977).

2 Auxin Transporters: Inward, Outward, and Within the Cell

2.1 Auxin Carriers: Source of Energy and Topology

The chemiosmotically driven auxin concentration gradients are regulated at the point of cellular uptake and efflux by the complex system of primary and secondary membrane transporters. There are at least three families of transport proteins coordinately regulating auxin flow. Besides diffusion, the auxin influx is actively mediated by the AUX1/LIKE AUX1 (AUX1/LAX) family of transmembrane proteins (TC 2.A.18). In contrast, PIN-FORMED (PIN) auxin efflux carrier proteins (PINs; TC 2.A.69) and members of the subgroup B family of ATP-binding cassette (ABC) proteins/P-glycoproteins (ABCB/PGPs; TC 3.A.1) are involved in auxin efflux (reviews by Křeček et al. 2009; Geisler and Murphy 2006, respectively). Moreover, a reversible transport mechanism described for ABCB4/PGP4 (Yang and Murphy 2009; Cho et al. 2007; Santelia et al. 2005; Terasaka et al. 2005) may contribute to both auxin uptake and efflux in some cells (see Sect. 2.2). Continually, so far unknown carriers modulating auxin transport machinery are being identified. The newest members of the auxin transporter group are the nitrate permease, NRT1.1/Chl1 (TC #2.A.17.3.1; Krouk et al. 2010) and ABCG37/PIS1/PDR9 (Růžička et al. 2010).

The secondary structures of the main classes of auxin carriers are depicted in Fig. 1. Schemes are based on the primary sequence and reflect the size/mass differences among individual proteins AUX1, PIN1, PIN5, and ABCB19, in ratio of 138:177:100:356 percents, respectively, of number of amino acids. This quantification implies, perhaps a bit surprisingly, that the size of PIN5 is about one half of PIN1, and that of PIN1 is about one half of ABCB19.

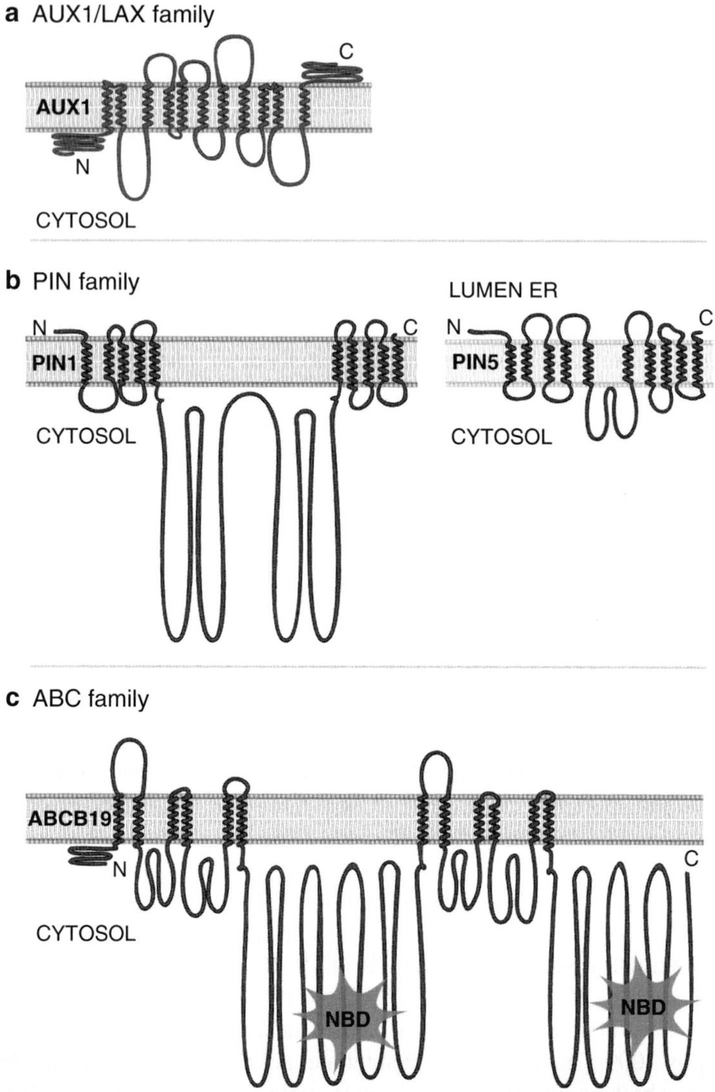

Fig. 1 Predicted topology of auxin carriers. Auxin transporters of AUX1/LAX family represented by the auxin influx carrier AtAUX1 (**a**), auxin transporters of PIN family represented by the auxin efflux carrier AtPIN1 with long cytosolic loop and the endoplasmic reticulum (ER)-localized auxin carrier AtPIN5 with short cytosolic loop (**b**), and auxin transporters of ABCB family represented by the auxin efflux carrier AtABCB19 (**c**). Predicted structure based on the protein sequence was created by TMRPres2D software (http://bioinformatics.biol.uoa.gr/TMRPres2D) and reflects the size of individual transporters (AUX1 – 485 amino acids (AA), PIN1 – 622 AA, PIN5 – 351 AA, ABCB19 – 1252 AA). Transmembrane domain (TMD) predictions have been performed using servers HMMTOP (http://www.enzim.hu/hmmtop/index.html; Tusnady and Simon 1998) and PRED-TMR (http://athina.biol.uoa.gr/PRED-TMR; Pasquier et al. 1999)

AUX1/LAX proteins (Bennett et al. 1996; Swarup et al. 2008) belong to the Amino Acid/Auxin Permease (AAAP) family that includes hundreds of proteins from plants, animals, yeast, and fungi (Young et al. 1999); members of this family transport not only auxin but also a single or multiple amino acid(s). Apparent differences in substrate specificity influenced by structure and net charge are observed within these carriers and may be expected also among AUX1/LAX proteins (Fischer et al. 2002). AUX1/LAX proteins are proton-gradient-driven secondary transporters, where auxin influx is mediated as cotransport with proton(s). AUX1/LAX proteins are supposed to transport auxin in its anionic form. However, it has not been unequivocally confirmed whether translocation of the IAA^- anion across PM is accompanied by one or two protons (Lomax et al. 1995). Members of the AAAP family are predicted to function as integral membrane proteins whose 10–11 transmembrane helices (TMHs) repeatedly span the PM (Bennett et al. 1996). Topology mapping based on YFP-labeling have proven 11 TMDs with intracellular N-terminus in the structure of AUX1 (Swarup et al. 2004) (Fig. 1a).

Auxin-specific PIN carriers form the Auxin Efflux Carrier (AEC) protein family. On the basis of the fact that in none of the PIN sequences ATP-binding domains have been identified, PINs are considered to act as secondary transporters driven by electrochemical gradient. The transport action is probably based on auxin/H^+ antiport (or auxin/H^+ symport in case of intracellular PINs, such as PIN5) (TransportDB; http://www.membranetransport.org/index.html; Ren et al. 2007); however, a clear proof is still missing. Proteins of the PIN auxin efflux family consist of two transmembrane domains (TMDs) of about five TMHs separated by a hydrophilic loop (Fig. 1b); (Gälweiler et al. 1998; Luschnig et al. 1998; Müller et al. 1998). Based on the length/structure of the central loop, PINs are divided into two subfamilies: "long" PINs (PIN 1,2,3,4,7 and also 6 – see discussion in Křeček et al. 2009) that localize predominantly on the PM and "short" PINs with the reduced hydrophilic core (PIN5 and 8) that target to endomembranes, most frequently the endoplasmic reticulum (ER) (Mravec et al. 2009). Figure 1b shows the topology of *At*PIN1 as an example of canonical "long" PIN with ~330 amino acid residues in hydrophilic loop. This sequence is reduced to ~50 amino acids in "short" *At*PIN5. The predicted number of transmembrane spans is based on the bioinformatics analysis. Currently accepted models suggesting ten TMHs with both extracellular termini are supported by indirect proofs such as nonfunctional GFP labeling of both PIN1-termini (J. Friml, personal communication) or a high degree of identity/homology in individual TMHs (Křeček et al. 2009).

Auxin efflux is executed also by primary active transport. ABC transporters comprise an extremely diverse class of proteins that uses the energy of ATP hydrolysis to translocate substrate across membranes. Typically, an ABC transporter is composed of four parts: two transmembrane domains (each composed of six TMHs) and two ATP-binding domains (nucleotide-binding domains, NBDs; nucleotide-binding folds NBFs). Aligning of amino acid sequences groups most ABC proteins into eight major subfamilies (Verrier et al. 2008). Until now members of B and G subgroups have been identified in auxin flow (Noh et al. 2001; Růžička et al. 2010). Figure 1c shows the typical secondary structure of ABCB protein

(TMD-NBD-TMD-NBD). Two NBDs cooperatively bind and hydrolyze ATP whereas TMDs are involved in substrate recognition. Yang and Murphy (2009) computed models of ABCB19 and ABCB4 carriers based on the crystal structure of Sav1866, a bacterial multidrug resistance ABC transporter (Dawson and Locher 2006). In addition to two putative IAA binding sites identified in the auxin exporter ABCB19, a unique third, putative IAA-interacting site was found in auxin importer ABCB4 (see Sect. 2.2). This might explain the observed activation of ABCB4 export activity by its substrate resulting in changeable direction of auxin transport.

2.2 *Auxin Uptake/Influx Carriers*

Native auxin IAA and also some synthetic auxins (such as naphthalene-1-acetic acid, NAA) can penetrate through PM and so they can enter cells via passive diffusion. However, as passive diffusion is slow, in some cells, the need for quicker and massive uptake of auxin from the apoplast may arise (e.g., in the lateral root cap). Therefore, an active uptake process for auxin is needed here in parallel to its passive flow (Kramer and Bennett 2006). This active auxin uptake processes can be carried out by at least three types of carriers. The longest known is the group of AUX1/LAX permeases (reviewed by Kerr and Bennett 2007). The *aux1* loss-of-function mutant was characterized by Maher and Martindale (1980), and the *AUX1* gene was isolated and its product assigned to auxin uptake by Bennett et al. (1996). The physiological function of AUX1/LAXes was characterized in relation to many physiological processes (see part 4 below). The biochemical proof for the cellular auxin-influx function of AUX1 was provided by heterologous overexpression of AUX1 in Xenopus oocytes (Yang et al. 2006) and confirmed in *Schizosaccharomyces pombe* (Yang and Murphy 2009).

The genome of Arabidopsis encodes one AUX1 and three Like AUX1 (LAX1, LAX2, and LAX3) proteins, which share approximately 80%-similarity on the amino acid level (Parry et al. 2001). The AUX1/LAX sequences are highly similar also among plant species – at both the nucleotide and the amino acid levels. The only major differences between the predicted protein sequences are at both the amino and the carboxyl termini, and phylogenetic analysis based on multiple alignments revealed two distinct subfamilies of AUX/LAX proteins that differ predominantly in cytosolic hydrophilic loops (Parry et al. 2001; Swarup et al. 2004; Hoyerová et al. 2008). The changes on specific positions in the loops may imply alterations in biochemical properties, which are determined by the particular amino acid residues. As such, they might indicate differences between the two subfamilies of the AUX/LAX proteins, for example, resulting in interactions with various intracellular proteins and, thus, in modifications of intracellular transport machinery and/or signaling (Hoyerová et al. 2008).

The other group of auxin transporters, which may act as auxin influx carriers, are ABCB transporters, namely ABCB4/PGP4 (Terasaka et al. 2005). As shown using heterologous overexpression in *S. pombe* (Yang and Murphy 2009), this transporter exhibits a complex transport behavior. If the intracellular concentration of auxin is low, it may work as an uptake carrier; if the auxin level inside cells increases, ABCB4 might change the direction of auxin transport across the PM and work as an auxin efflux carrier. This property may be important in such tissues and physiological situations when a fine-tuned auxin balance is needed (e.g., root hair elongation; Cho et al. 2007).

Recently and perhaps unexpectedly, also a nitrate transporter of Arabidopsis, NRT1.1/Chl1 (TC #2.A.17.3.1), has been shown to facilitate the transport of auxin into cells (Krouk et al. 2010). This PM-localized secondary transporter belongs to the Proton-dependent Oligopeptide Transporter (POT) family in Major Faciliator Superfamily. Twelve TMHs are predicted in the NRT1.1 sequence. NRT1.1 has been proved indispensable for nitrate signaling related to root growth, and it has been proposed to act also as a nitrate sensor (Walch-Liu and Forde 2008; Ho et al. 2009). Nitrate is able to inhibit NRT1.1-dependent auxin uptake; however, auxin does not affect nitrate transport by NRT1.1. This finding suggests that nitrate and auxin do not act as simple competitors here. Anyway, the action of NRT1.1 may represent a new mechanism combining environmental stimuli, nutrient sensing, and hormonal signaling in roots (Krouk et al. 2010).

2.3 Auxin Efflux Carriers

The first transporters from the plant-specific PIN family were ascribed to polar auxin transport at the end of nineties (Gälweiler et al. 1998; Luschnig et al. 1998), even if the first information about respective mutants (*pin-formed 1, pin1*) was reported ca. 10 years earlier (in Goto et al. 1987) and mutants themselves were characterized in the early nineties (Okada et al. 1991). As mentioned earlier, PINs are secondary transporters but the direct experimental evidence identifying the quality of symported/antiported ion is still missing (cf. Sect. 2.1). In Arabidopsis, the PIN family comprises eight members and separates into two subfamilies (see above). Canonical "long" PINs (reviews by Křeček et al. 2009; Tanaka et al. 2006; Vieten et al. 2007; Zažímalová et al. 2007) are localized on the PM mostly in a polar manner, and namely for PIN 1, 2, 4, 6, and 7 the rate-limiting role in intercellular auxin transport was proved (Petrášek et al. 2006; Blakeslee et al. 2007). Their polar localization provides intercellular auxin flow with clear directionality (Wiśniewska et al. 2006), which is crucial for many auxin-dependent processes in plant development (see part 4 below). The localization of canonical PINs is not stationary as they are subjected to the constitutive cycling between the PM and endosomal compartments (Geldner et al. 2001; Dhonukshe et al. 2007); so they can be removed rapidly to other parts of the cell via a transcytosis-like mechanism (Kleine-Vehn et al. 2008a) – for example, as a reaction on environmental changes (see below). All PINs seem to transport auxins directly. For "long" PINs, a wide functional redundancy has been described (Vieten et al. 2005), making relatively weak phenotypes of most *pin* mutants understandable.

There are some PIN-like proteins in other than plant kingdoms, which show only very little sequence or functional similarity to plant PINs (Titapiwatanakun et al. 2009; Yang and Murphy 2009). There is about 15% sequence similarity between plant PINs and related proteins from yeast, and almost no auxin transport activity has been shown for the auxin effluxer-like protein (AEL1) from *S. pombe* (Titapiwatanakun et al. 2009). The similarity between plant PINs and bacterial PIN-like proteins is even lower.

In contrast to plant- and auxin-specific PINs, the ABC superfamily is one of the largest and most ubiquitous transporter families in all kingdoms and particular members of this protein family recognize many various molecules as their substrates (Geisler and Murphy 2006; Verrier et al. 2008). Again in contrast to PINs, there is a very high structural similarity between prokaryotic and eukaryotic members of this superfamily (Becker et al. 2009). The auxin transporters from this superfamily are phospho-glycoproteins belonging to its ABCB subgroup – namely ABCB1, 4, and 19. Their involvement in auxin transport was first suggested on the basis of studies on hypocotyl elongation for PGP1/ABCB1 (Sidler et al. 1998), later also for PGP19/MDR1/ABCB19 (Noh et al. 2001), and in the screen for binding to the inhibitor of polar auxin transport, 1-naphthylphthalamic acid (NPA; Murphy et al. 2002). The auxin efflux functions for the PGP1/ABCB1 and PGP19/MDR1/ABCB19 were shown by Geisler et al. (2005) and Bouchard et al. (2006), respectively. The phenotypes of *abcb1* and *abcb19* mutants resemble those of the *twd1* mutant, which is in concert with the interaction of the C-terminal domains of ABCB1 and ABCB19 with immunophilin-like protein TWD1/FKBP42 (Geisler et al. 2003). Since the abundance of FKBP42 in cells is very low compared with ABCB1 and ABCB19 (Bailly et al. 2008), the function to induce conformational changes in ABCB1 and ABCB19 and thus to activate the ABCB membrane complexes was proposed for TWD1 (Geisler et al. 2003; Bouchard et al. 2006; Bailly et al. 2008).

Both ABCB1/PGP1 and ABCB19/PGP19 are close orthologs to mammalian ABCB1 multidrug resistance transporters and show high substrate specificity (Titapiwatanakun et al. 2009; Titapiwatanakun and Murphy 2009; Yang and Murphy 2009). They differ from PINs also in their primarily nonpolar cellular localization, and they seem to be very stable at the PM (Bailly et al. 2008; Titapiwatanakun et al. 2009). These properties point to a control of long-distance auxin transport and a localized loading of auxin into the vectorial auxin transport system by ABCBs rather than facilitating the vectorial auxin movement itself (Titapiwatanakun and Murphy 2009; Zažímalová et al. 2010).

Even though PINs and ABCBs have many opposite properties, they seem to coact in specific cell types and/or developmental situations: ABCB19/PGP19 was shown to collaborate with PIN1, stabilizing it in PM microdomains and enhance the substrate specificity (Mravec et al. 2008; Titapiwatanakun et al. 2009; Titapiwatanakun and Murphy 2009).

In contrast to ABCBs, which transport native auxin IAA, another member of the ABC superfamily, ABCG37/PIS1/PDR9, was shown to transport synthetic IAA analogs (Ito and Gray 2006) and also the putative IAA precursor – indole-3-butyric acid (IBA) (Růžička et al. 2010) but not IAA itself.

There is now plethora of data showing that AUX1/LAXes, some PINs and ABCBs are the catalysts/effectors of intercellular auxin flow within plant body. However, the action of other anion transporters cannot be fully excluded. The action in auxin transport of the nitrate transporter NRT1.1 (see above) has already been demonstrated. Analogous mechanisms might be involved also in case of organic acid transporters, for example, the malate transporter ABCB14 (Lee et al. 2008) and/or others.

2.4 Intracellular Auxin Carriers

In addition to "long" PINs, there are two members (PIN5 and 8, see Sect. 2.1) of the Arabidopsis PIN family, in which the central cytosolic loop is dramatically reduced (Fig. 1b). Probably due to the changes in the amino acid sequence around the putative tyrosine motif NPNTY, responsible for recruitment into the clathrin-coated endocytotic vesicles (Ohno et al. 1995), the "short" PINs localize to endomembranes, particularly to the ER. Thanks to the ectopic overexpression of PIN5 onto the PM in yeast, it was proven that PIN5 has the capacity to function as an auxin carrier. Therefore, its activity on the ER leads to the intracellular auxin redistribution subjecting auxin molecules to different set of auxin-metabolizing enzymes and thus resulting in the significant change in auxin metabolic profile (Mravec et al. 2009). Therefore, these findings reveal a so far unknown function of some PIN auxin transporters in the control of intracellular auxin homeostasis and thus in regulation of availability of free auxin for signaling pathways.

In Table 1, the genes for putative auxin transporters are summarized for which the auxin transporting role was proved directly, on the single cell level.

3 Regulation of Auxin Transporters' Function: How, How Much, and Where?

3.1 Levels of Regulation

It is quite a paradox that although we still do not understand in detail the mechanism of how the auxin molecule is transported by particular auxin carriers, we do know a lot about mutually interconnected regulatory networks influencing the function of these carriers. These networks often form elegant regulatory systems that have been reported to play a role during various stages of plant development. Examples of these networks will be presented later in this review (see Sect. 4).

Table 1 Auxin transporters with function in auxin transport demonstrated at the single cell level. The experimental evidence is based on the auxin transport (accumulation) measurements after the mutation or the overproduction of the respective gene or protein, respectively. Heterologous and homologous expressions of candidate genes are listed. IAA, its synthetic analogs, as well as its putative precursor indole-3-butyric acid (IBA) have been used. For details on the particular assays, see references listed. (*At Arabidopsis thaliana; Zm Zea mays*)

Function	Gene	Organism used to test the auxin transport function	References
Auxin influx			
	AtAUX1 At2g38120	*Schizosaccharomyces pombe*, *Xenopus laevis* oocytes	Yang et al. (2006), Yang and Murphy (2009)
	AtLAX3 At1g77690	*Xenopus laevis* oocytes, *Homo sapiens* U20S cells	Swarup et al. (2008)
	AtABCB4 (*PGP4, MDR4*) At2G47000	*Schizosaccharomyces pombe (at low levels of auxin)*, *Homo sapiens* Hela cells	Terasaka et al. (2005), Yang and Murphy (2009)
	AtANT1 At3G11900	*Saccharomyces cerevisiae*	Chen et al. (2001)
	AtNRT1.1 At1G12110	*Saccharomyces cerevisiae*, *Xenopus laevis* oocytes	Krouk et al. (2010)
Auxin efflux			
	AtPIN1 At1g73590	*Arabidopsis thaliana* suspension cells, *Nicotiana tabacum* suspension cells, *Schizosaccharomyces pombe*	Petrášek et al. (2006), Mravec et al. (2008), Yang and Murphy (2009)
	AtPIN2/EIR1/ *AGR1/WAV6* At5G57090	*Saccharomyces cerevisiae*, *Schizosaccharomyces pombe*, *Homo sapiens* Hela cells	Luschnig et al. (1998), Petrášek et al. (2006), Yang and Murphy (2009)
	AtPIN3 At1g70940	*Nicotiana tabacum* suspension cells	Lee and Cho (2006)
	AtPIN4 At2G01420	*Nicotiana tabacum* suspension cells	Petrášek et al. (2006)
	AtPIN5 At5G16530	*Saccharomyces cerevisiae*	Mravec et al. (2009)
	AtPIN6 At1G77110	*Nicotiana tabacum* suspension cells	Petrášek et al. (2006)
	AtPIN7 At1G23080	*Nicotiana tabacum* suspension cells, *Saccharomyces cerevisiae*, *Homo sapiens* Hela cells	Petrášek et al. (2006)
	AtABCB1 (*PGP1*) At2G36910	*Arabidopsis thaliana* protoplasts, *Saccharomyces cerevisiae*, *Schizosaccharomyces pombe*, *Homo sapiens Hela cells*	Geisler et al. (2005), Bouchard et al. (2006), Yang and Murphy (2009)
	AtABCB4 (*PGP4, MDR4*) At2G47000	*Nicotiana tabacum* suspension cells, *Schizosaccharomyces pombe* (at higher levels of auxin)	Cho et al. (2007), Yang and Murphy (2009)
	AtABCB19 (*MDR1, MDR11, PGP19*) At3G28860	*Arabidopsis thaliana* protoplasts; *Nicotiana tabacum* suspension cells, *Saccharomyces cerevisiae*, *Schizosaccharomyces pombe*, *Homo sapiens* Hela cells	Geisler et al. (2005), Petrášek et al. (2006), Bouchard et al. (2006), Yang and Murphy (2009)
	ZmTM20	*Xenopus laevis* oocytes	Jahrmann et al. (2005)
	AtABCG37/PIS1/ *PDR9*	*Arabidopsis thaliana* protoplasts, *Saccharomyces cerevisiae*, *Homo sapiens* Hela cells	Růžička et al. (2010)

The regulation of all individual auxin-transporting systems is very complex. Importantly, auxin itself is among the most effective regulators making the whole story even more complicated. Earlier physiological observations of auxin effects on organogenesis and differentiation led to the formulation of the canalization hypothesis (Sachs 1981) supposing the positive feedback effect of auxin on its own transport and polarity of this transport. To substantiate these earlier findings, molecular biology approaches were employed and it was shown that the formation of certain auxin levels in the tissue initiates many downstream effects triggering regulatory loops with or without gene expression involved (Benjamins and Scheres 2008).

Generally speaking, three main levels of regulation are included in carrier-mediated auxin transport: (1) regulation of particular carriers' abundance (transcription, translation, and degradation), (2) their targeting to the specific position at the PM and intracellular trafficking, and (3) regulation of their transport activity (post-translational modifications, inhibitors, regulation of pH gradient across PM, composition of PM, and interactions among individual transporters).

3.2 The Abundance of Auxin Carriers: Transcription, Translation, and Degradation

As for many other proteins, differential expression as well as degradation of auxin carriers is under tight control of various signals triggered during plant development and modified by the environment. However, detailed knowledge on the molecular mechanisms affecting the gene expression of auxin carriers for developmentally-important environmental signals (light quality and quantity, temperature, water and nutrient status, etc.) is mostly missing. More information is available for control by phytohormones, other endogenous signals, and for the auxin itself in particular. Auxin has been reported to bind to its specific receptor TIR1 (TRANSPORT INHIBITOR RESPONSE 1) belonging to the group of F-box proteins, subunits of the SCF E3-ligase protein complex (Kepinski and Leyser 2005; Dharmasiri et al. 2005; Badescu and Napier 2006; Quint and Gray 2006). This binding activates nuclear proteasome-mediated degradation of auxin-inducible Aux/IAA transcriptional regulators with subsequent derepression of auxin response factors (ARFs) resulting in activated transcription of auxin-inducible genes (review by Leyser 2006). Described mechanism represents an elegant and efficient control of gene expression triggered by auxin and is common for several other phytohormones (review by Bishopp et al. 2006; Lechner et al. 2006; Santner and Estelle 2009).

The gene expression of all three basic groups of auxin transporters (PINs, ABCBs, and AUX1/LAXes) is positively stimulated by auxin (Paponov et al. 2008). The only exception is PIN5, whose expression was downregulated upon IAA treatment (Mravec et al. 2009). The transcriptional regulation of the overall abundance of auxin carrier proteins is under strict control of auxin through the above-mentioned

Aux/IAA-dependent signaling. Various timing of the transcriptional response as well as its modulation by the developmental context was reported in *Arabidopsis thaliana* (Noh et al. 2001; Geisler et al. 2005; Terasaka et al. 2005; Vanneste et al. 2005; Vieten et al. 2005; Paponov et al. 2008). While the variable expression of individual members of PIN family might reflect their functional redundancy (Blilou et al. 2005; Vieten et al. 2005), ABCBs protein abundance usually reflects their mRNA expression levels (for review see Titapiwatanakun and Murphy 2009), ABCB19 being even extremely stable (Titapiwatanakun et al. 2009).

Other plant hormones such as ethylene (Růžička et al. 2007), cytokinins (Dello Ioio et al. 2008; Pernisová et al. 2009; Růžička et al. 2009), brassinosteroids (Nemhauser et al. 2004; Goda et al. 2004; Li et al. 2005b), and gibberellins (Ogawa et al. 2003) interfere with the levels of gene expression of PINs and AUX1.

Although there is less information on the auxin carriers from other plants, developmentally and hormonally-regulated gene expressions have been experimentally proved for both AUX1/LAX and PIN homologs from rice (Wang et al. 2009), pea (Hoshino et al. 2003, 2004), hybrid aspen (Schrader et al. 2003), medicago (Schnabel and Frugoli 2004), cucumber (Kamada et al. 2003), and lupin (Oliveros-Valenzuela et al. 2007). The expression of AUX1/LAX homologs was analyzed in maize (Hochholdinger et al. 2000), medicago (de Billy et al. 2001), and casuarina (Peret et al. 2007). PIN homologs were described in pea (Chawla and DeMason 2004), maize (Carraro et al. 2006; Gallavotti et al. 2008), mustard greens (Ni et al. 2002), and rice (Xu et al. 2005; Wang et al. 2009), and ABCB homologs in maize and sorghum (Multani et al. 2003).

At least for PIN proteins, their overall abundance is further controlled at the posttranscriptional level by MODULATOR OF PIN (MOP) proteins (Malenica et al. 2007) and also by specific proteasome-mediated proteolysis as it has been reported for PIN2 (Abas et al. 2006).

3.3 *Intracellular Trafficking and Targeting of Auxin Carriers*

The transcriptional regulation of the intracellular abundance of auxin carriers, as described earlier, is not enough to provide plant cells from various tissues with the information on the auxin flow and its polarity needed for the transduction of various developmental cues. This information mainly comes from the differential distribution of auxin carriers in the PM, membrane vesicles, and endomembranes. Since all intercellular auxin transporters are integral PM proteins, upon correct folding they are present in the membrane vesicles, which are targeted to the PM where they exert their transporting action. However, for some auxin carriers, the mechanism of recycling of membrane vesicles between PM and the endosomal space of the cell is well-described. So, dynamic targeting of newly-synthesized carriers, the recycling of membrane vesicles carrying auxin carriers, as well as the stabilization of some auxin carriers in the PM are important in the fine tuning of auxin gradients.

The best characterized is the trafficking and recycling of PIN proteins. Its discovery is connected with the studies that used brefeldin A (BFA), the inhibitor of the guanine-nucleotide exchange factor (GEF) for ADP-ribosylation factors (ARF) GTPases (ARF-GEF). BFA selectively blocks the anterograde transport of membrane vesicles toward PM, and so after its application only endocytotic processes are apparent. This helped to uncover constitutive cycling of PINs between the PM and the endosomal space (Steinmann et al. 1999; Geldner et al. 2001), as suggested previously by Robinson et al. (1999). It was shown that the anterograde transport of membrane vesicles carrying PIN1 toward PM is regulated by the BFA-target GNOM, one of the ARF-GEFs (Steinmann et al. 1999; Geldner et al. 2003) and that other ARFs might contribute to the trafficking of other PINs (Xu and Scheres 2005). The endocytosis step of PIN1 and PIN2 was described to be mediated by the clathrin-coated vesicles (Dhonukshe et al. 2007) and to be influenced by both sterol composition of the PM (Willemsen et al. 2003; Men et al. 2008) and the activity of phosphoinositide signaling system, as shown by the importance of PLDζ2 activity for the endocytosis of PINs (Li and Xue 2007). Interestingly, the endocytosis-dependent cycling of proteins in plant cells is regulated by auxin efflux inhibitors (Geldner et al. 2001) as well as by auxin itself (Paciorek et al. 2005). By inhibition of endocytosis of PINs, auxin increases the abundance and thus the capacity of efflux transporters at the PM, which results in the increased efflux of auxin from the cell. Thus, this mechanism constitutes nontranscriptional feedback-type of regulation of auxin transport.

The mechanisms of PIN recycling are needed for dynamic changes of PIN protein localization by the processes of transcytosis, where individual PIN carriers might be trafficked using ARF-GEF-dependent machinery to various domains in the PM (Kleine-Vehn et al. 2008a). ARF-GEF sorting is also involved in the trafficking of PIN2 to the lytic vacuolar pathway through prevacuolar compartments. This process is regulated by light that seems to be needed for the PM localization of PIN2, while in darkness PIN2 is degraded through vacuolar pathway (Laxmi et al. 2008). PINs might be retrieved again from these compartments with the assistance of the retromer complex subunits SORTIN NEXIN1 (SNX1) and VACUOLAR PROTEIN SORTING 29 (VPS29) (Jaillais et al. 2006; Kleine-Vehn et al. 2008b).

With respect to the trafficking of PIN proteins, the involvement of the octameric complex exocyst, the effector of Ras (rat sarcoma) superfamily of monomeric GTPases, tethering, and targeting post-Golgi vesicles to the PM is highly probable. Exocyst has been characterized also in plants (Hala et al. 2008), and it possibly defines highly dynamic recycling domains in the PM needed for spatial information for polarized secretion (Žárský et al. 2009). It acts with the assistance of plant homologs of Rho GTPases (Rho of plants, ROP). Importantly, PIN1 recruitment through the secretory system has been recently shown to be assisted by the coiled coil-scaffold protein, the interactor of constitutive active ROP 1 (ICR1) (Hazak et al. 2010). ICR1 is needed for the recruitment of PIN1 to the polar domains in the PM and its localization and expression is under the control of auxin, suggesting the auxin-modulated link between polarized protein secretion and auxin-driven morphogenesis.

Possibly the best characterized regulation of PIN trafficking is their phosphorylation status. The characterization of *A. thaliana* mutants with phenotypes typical for altered auxin transport *root curling in NPA 1* (*rcn1*) and *pinoid* (*pid*) led to the identification of regulatory subunit of protein phosphatase-2A (PP2A), a heterotrimeric serine/threonine protein phosphatase (Garbers et al. 1996; Deruère et al. 1999) and serine/threonine protein kinase PID (Bennett et al. 1995; Christensen et al. 2000). It was demonstrated that the activity of PID controls the polarity of PIN targeting (Friml et al. 2004) and enhances auxin efflux (Lee and Cho 2006). Moreover, *A. thaliana* 3-phosphoinositide-dependent protein kinase 1 (PDK1) has been shown to stimulate the activity of PID kinase, providing the evidence that upstream phospholipid signaling might play a role in the regulation of auxin transport (Zegzouti et al. 2006). Most importantly, PP2A and PID act antagonistically on the phosphorylation state of the central hydrophilic loop of "long" PINs mediating their apical–basal polar targeting (Michniewicz et al. 2007). Recently, the phosphorylation at the evolutionary conserved phosphorylation site localized in the cytosolic loop of PIN1 has been shown to determine the localization of PIN1 and redirect the auxin flow in Arabidopsis roots (Zhang et al. 2010). The analysis of GNOM-dependent and phosphorylation-dependent trafficking of PINs suggested that these two pathways are independent, but collaborate in the setting the auxin transport polarity up (Kleine-Vehn et al. 2009). It should be noted that auxin itself regulates PID expression (Benjamins et al. 2001) and PIN polar localization through TIR1 signaling (Sauer et al. 2006), providing another example of auxin-dependent regulatory loop controlling the deposition of PIN proteins.

Although BFA was originally reported to interfere with the localization of AUX1 (Grebe et al. 2002), direct comparison with the trafficking of PINs in Arabidopsis root protophloem cells uncovered novel GNOM-independent, BFA-insensitive, and sterol composition-dependent endosomal pathways for the trafficking of AUX1 (Kleine-Vehn et al. 2006). Its asymmetrical distribution seems to be assisted by the ER-accessory protein AUXIN RESISTANT 4 (AXR4) playing the role probably in posttranslational modifications of AUX1 (Dharmasiri et al. 2006). Conceptually, the independent regulation of trafficking of putative auxin influx and efflux carriers might be crucial in physiological processes such as root gravitropism, where AUX1 in concert with PIN2 mediates basipetal auxin transport (Swarup et al. 2005).

In comparison with PIN and AUX1 proteins, ABCB transporters are more stable in the PM, although some evidence on their trafficking is available (summarized in Blakeslee et al. 2005). The biogenesis of ABCB19 at the ER-Golgi interface involves the activity of GNOM-LIKE (GNL1) ARF-GEF. This activity is needed for correct glycosylation of ABCB19 and subsequent targeting (Titapiwatanakun et al. 2009) at the PM, where ABCB19 is localized in the sterol-rich PM microdomains (Murphy et al. 2002; Blakeslee et al. 2007). The interaction of ABCB19 with PIN1 (Blakeslee et al. 2007) might be important for the stabilization of PIN proteins by ABCBs in these domains

(Titapiwatanakun et al. 2009) and therefore for the inhibition of PIN recycling or degradation.

3.4 Regulation of the Auxin Carrier Proteins' Activity

Regulation of auxin transporting activity of the particular auxin carrier protein at the PM depends on its direct interactions with other integral and peripheral PM or cytosolic proteins, including enzymes and regulatory proteins. However, there are quite a few examples of protein–protein interactions that directly regulate the activity of auxin carriers. The comprehensive analysis of *A. thaliana* membrane protein interactome that is being presently analyzed using split ubiquitin system (http://www.associomics.org/index.php) will hopefully yield promising set of candidates for further testing. Additionally, pH gradient, chemical inhibitors, and membrane composition might play an important role here.

As shown above for PIN auxin carriers, the phosphorylation status of their molecules is critical for their differential distribution. With respect to the regulatory interaction at the PM, it seems that PM-localized PIN proteins are also targets for phosphorylation by another group of AGC kinases, D6 protein kinases (Zourelidou et al. 2009) that regulate directly their activity at the PM.

One of the best characterized protein–protein interactions for auxin carriers at the PM is the interaction of ABCB1 and ABCB19 with the immunophilin-like FK506-binding protein (FKBP) TWISTED DWARF1 (TWD1/FKBP42; Geisler et al. 2003). This protein was originally copurified with ABCB proteins (Murphy et al. 2002) and later suggested to induce regulatory conformational changes in ABCB1 and ABCB19 (Geisler et al. 2003; Bouchard et al. 2006). Its role lies mainly in the activation of complexes with ABCBs so that they could transport auxin, and their modulation by auxin transport inhibitors, such as NPA and quercetin (Bailly et al. 2008).

In agreement with the chemiosmotic hypothesis, the gradient of pH is one of the most critical conditions in the maintenance of auxin flow across the membrane. The upregulation of Arabidopsis vacuolar H^+-pyrophosphatase (AVP1) is active in this process decreasing (in collaboration with PM H^+-ATPase) the pH in the apoplast and thus the activity of PIN proteins (Li et al. 2005a). In addition, PM-located potassium transporter TINY ROOT HAIR 1 (TRH1) might be involved in auxin transport in the root tip directly by transporting auxin or indirectly by changing the gradient of ions across membrane (Vicente-Agullo et al. 2004).

From the historical perspective, one of the most fruitful approaches in studying the regulation of auxin transport machinery is the application of inhibitors. The most widely used inhibitor of auxin efflux, that is, NPA belongs to a group of inhibitors known as phytotropins (Rubery 1990). The application of NPA to the plant tissues results typically in the increase of auxin accumulation, presumably because of the inhibition of auxin efflux activity (for review see Morris et al. 2004).

Detailed knowledge of the mechanism by which NPA and other phytotropins inhibit auxin efflux at the PM is still lacking. Older results suggested the existence of NPA-binding protein at the cytoplasmic side of the PM (NBP; Sussman and Gardner 1980), where it is anchored with the help of actin cytoskeleton (Cox and Muday 1994; Dixon et al. 1996; Butler et al. 1998). Lately, more general action of some phytotropins based on the inhibition of endocytotic processes and underlying actin dynamics has been reported (Geldner et al. 2001; Dhonukshe et al. 2008); however, the concentrations of NPA needed to affect these processes were much higher than those effective to inhibit cellular auxin efflux (Petrášek et al. 2003). It has been shown that the activity of PIN7 after its inducible overexpression in tobacco cells is inhibited significantly by NPA, while ABCB19 showed about 20% of its auxin efflux activity to be insensitive to NPA in the same system (Petrášek et al. 2006), suggesting differential sensitivity of these transport pathways to NPA. There exist high-affinity NPA-binding sites for ABCB transporters at the PM that were experimentally verified (Rojas-Pierce et al. 2007; Bailly et al. 2008) together with low-affinity ones (Murphy et al. 2002), the latter being probably responsible for the more general inhibitory effect of phytotropins on endocytotic processes and actin cytoskeleton dynamics (Geldner et al. 2001; Dhonukshe et al. 2008). As pointed out by Nick et al. (2009), the action of 2,3,5-triiodobenzoic acid (TIBA), another inhibitor of auxin transport, is known to primarily affect actomyosin system (Rahman et al. 2007). The effort to find naturally occurring substances analogous to phytotropins led to the finding that flavonoids, endogenous polyphenolic compounds, modulate auxin transport and tropic responses (Murphy et al. 2000; Peer and Murphy 2007; Santelia et al. 2008) as well as the activity of both ABCB1 and ABCB19 (Noh et al. 2001; Murphy et al. 2002; Geisler et al. 2003; Rojas-Pierce et al. 2007; Bailly et al. 2008). Flavonoids (as well as NPA) seem to inhibit the interaction of ABCB proteins with the above-mentioned TWD1/FKBP42 peripheral PM immunophilin-like protein (Bailly et al. 2008). For AUX1/LAX auxin influx carriers, synthetic inhibitor of auxin influx 1-naphthoxyacetic acid (Imhoff et al. 2000; Parry et al. 2001) has been established and possible inhibitory naturally occurring substance chromosaponin was uncovered (Rahman et al. 2001).

The composition of PM itself constitutes the environment for protein–protein interactions and thus determines how effective will be the resulting flux of auxin across membrane. The sterol composition of membranes reflecting the activity of STEROL METHYL TRANSFERASE1 (SMT1) was shown to be crucial for the positioning of PIN1 and PIN3, but not AUX1 in the PM (Willemsen et al. 2003). Also subsequent step in sterol biosynthesis catalyzed by cyclopropylsterol isomerase1 (CPI1) is needed for correct apical localization of PIN2 in root epidermis (Men et al. 2008). The composition of membrane is further important for the localization of ABCB19 as it has been shown to be present in the detergent resistant fraction from *A. thaliana* (for review see Titapiwatanakun and Murphy 2009; Titapiwatanakun et al. 2009). Such sterol- and sphingolipid-rich PM microdomains seemingly constitute important places, where ABCB19 and PIN1 might interact physically (Blakeslee et al. 2007, see above).

4 Auxin Transporters in Plant Development: When and Why?

4.1 The Formation of Auxin Gradients: Morphogenic Action of Auxin

Since auxin affects many processes at both genomic and nongenomic levels, developmentally-regulated auxin gradients are very important in the setting of the cell fate within the particular tissue. As mentioned also in the Section 2.1, based on the character of dissociated molecule of auxin (IAA^-), classical physiological experiments predicted the existence of active transporters – auxin efflux carriers, responsible for the efflux of auxin from cells. Moreover, their asymmetrical distribution between the opposite sites of the cells was suggested to explain the directionality of the auxin flow. Detailed characterization of individual auxin transporters has then contributed substantially to the understanding of their cooperation during individual phases of plant development. Although at the moment there is no reliable method to track auxin gradients inside individual cells in vivo, the spectrum of more or less indirect methods is being used to follow auxin gradients in tissues (for review see Petrášek and Friml 2009).

As mentioned in Sect. 3.3, the localization of auxin transporters at various domains in the PM is critical in the setting of the directionality of particular auxin flow. As shown in Fig. 2 for PIN proteins, in principle they might transport auxin both across PM and perhaps also into membrane vesicles containing these transporters. Interestingly, for PIN proteins with the long hydrophilic loop, their localization in vesicles would not prevent this loop from the regulatory phosphorylations (see Fig. 2). Then, the dynamics of these vesicles would be critical for the overall auxin flow across the cell, so that its rate would not depend only on the diffusion inside cells. There are two experimental proofs for this "vesicular secretion" model (Baluška et al. 2003, 2008). First, auxin was immunolocalized in these vesicles (Schlicht et al. 2006) and second, a significant portion of auxin transport measured noninvasively in the Arabidopsis root tip was mediated by PLDζ2-dependent vesicle secretion (Mancuso et al. 2007); this was also reported to be important, among others, for the trafficking of PIN2 (Li and Xue 2007). However, there are no kinetic data on the contribution of this way of auxin transport at the single cell level. Other auxin transporters that possibly act inside the cell are transporters with short hydrophilic loop from PIN family, namely PIN5 for which IAA import into the endoplasmic reticulum has been reported (Mravec et al. 2009).

In general, auxin transporters contribute substantially, through the formation of auxin gradients, to the proper setting of developmental signals in embryogenesis, organogenesis, vascular tissue formation, and tropisms. In all of these examples, auxin maxima generated by auxin biosynthesis and/or by its release from conjugates, as well as activities of auxin transporters trigger morphogenetic response in a form of unequal cell division or differential cell expansion, elongation, and/or differentiation. These processes often generate the establishment of new axis or tissue polarity.

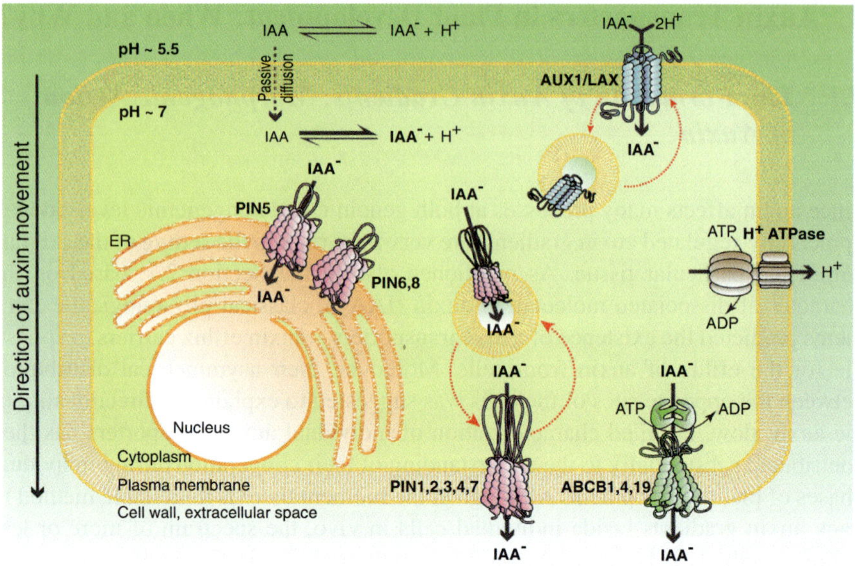

Fig. 2 Subcellular localization of auxin transporters. The scheme depicts proteins that are involved in auxin transport machinery. AUX1/LAX influx carriers are shown in *blue*, PIN efflux carriers are symbolized in *pink*, and the difference between PM- and ER-localized PINs is demonstrated by the distinct size of their cytosolic *loops*. Auxin efflux carriers of ABCB-type are shown in *green*. PM-H^+-ATPase is drawn in *pale brown*. *Full black arrows* represent the direction of carrier-mediated auxin transport (and proton translocation by PM-ATPase), ABCB4 being suggested to transport auxin in both directions. *Black dashed arrow* symbolizes passive diffusion, *curved red dashed arrows* represent constitutive (re)cycling between PM and endosomal compartments.
There is a possibility that vesicles are used also for the vesicle-mediated intracellular auxin transport (Baluška et al. 2003, see also in the text), where PIN carriers might be involved in the translocation of auxin molecules into the vesicle. Vesicle itself would then function as a "vehicle" to deliver auxin to the PM where, after fusion of the vesicle with the PM, auxin molecules would be excreted to the extracellular space. *ER* endoplasmic reticulum

Because of this activity, auxin is often called plant morphogen, although it does not fully fit into the definition of the term morphogen in animals. Although plants and animals evolved multicellularity independently, some parallelism in the morphogenic action of auxin in plants and, for example, serotonin during the early animal embryogenesis (Levin 2006) suggests fundamentally conserved modules, which could be used in various contexts.

4.2 Auxin Transporters During Embryogenesis

During embryogenesis of seed plants, all important structural features of plant body are established. On the basis of the analysis of *A. thaliana* mutants and localization studies, it has been shown that individual phases of embryogenesis are accompanied by concerted action of PIN auxin efflux carriers PIN1, PIN3, PIN4, and PIN7 (Friml et al. 2002a, 2003) maintaining the auxin maxima needed for apical–basal axis establishment. The importance of the cooperation of these carriers seems to be in buffering situations, where auxin levels are changed by locally induced biosynthesis or conjugation (Weijers et al. 2005). Moreover, they seem to be functionally redundant as single *pin* mutants can still perform embryogenesis while *pin1pin3pin4pin7* quadruple mutants are defective in the overall apical–basal polarity establishment (Benková et al. 2003; Friml et al. 2003). In contrast to multiple *pin* mutants, *abcb1abcb19* mutants do not have any defects in embryogenesis suggesting that PIN proteins play a crucial role here. But then, similar functional redundancy observed for *pin1pin3pin4pin7* quadruple mutants was revealed also in *aux1lax1-lax2lax3* quadruple mutant embryos, which had extremely disorganized radicle apex (Ugartechea-Chirino et al. 2010). Clear auxin concentration gradients are already manifested soon after the first anticlinal division of fertilized zygote (Fig. 3a). Auxin

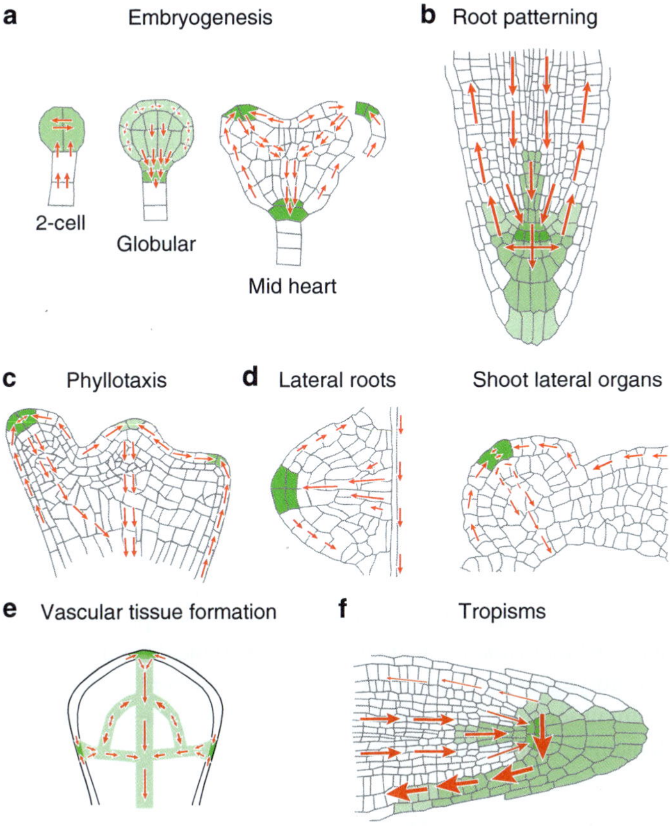

Fig. 3 Developmental role of auxin transporters in the establishment and maintenance of auxin gradients. Auxin maxima (*green*) and auxin carrier-mediated transport (*red arrows*) during embryogenesis (**a**), root patterning (**b**), shoot patterning (**c**), lateral organ (root and shoot) formation (**d**), vascular development (**e**) and tropisms (**f**). Adapted from Petrášek and Friml (2009) and Benková et al. (2003).

maximum in the apical cell is supported by the activity of apically localized PIN7 in the suspensor-forming cells. At this stage, PIN1 is responsible for the distribution of auxin in the forming proembryo being localized to the PM by endocytic recycling processes (Dhonukshe et al. 2008). As it has been shown recently (Mravec et al. 2008) both ABCB1 and ABCB19 contribute to the auxin efflux during these early phases of proembryo formation. While ABCB1 is localized in all suspensor and proembryonal cells, ABCB19 is restricted to cells forming suspensor and later at dermatogen stage also to the lower tier cells. Both ABCB1 and ABCB19 have no obvious subcellular polarity. Later, during early globular phase, PIN1 starts to be localized preferentially at the bottom PMs of cells in the embryo. Together with the supportive activity of PIN4, there is a clear redirection of auxin flow toward the hypophysis, the precursor of root meristem. Simultaneously, the polarity of PIN7 is shifted from apical to the basal localization in the uppermost suspensor cells by endocytic recycling process. At this moment, apical to basal polarity of auxin flow is clearly established. Precisely controlled concentration of auxin at this moment is critical as has been documented by heavy defects in apical–basal polarity specification in mutants of auxin binding F-box proteins TIR1 and other AFBs (Dharmasiri et al. 2005) and downstream transcriptional regulators MP/ARF5 and BDL/IAA12 (Hardtke and Berleth 1998; Hamann et al. 2002). During development of heart stage of the embryo, auxin maxima are formed in both cotyledon primordia and in the hypophysis, precursor of root meristem. Auxin transport toward hypophysis as well as toward newly formed cotyledon primordia is maintained preferentially by PIN1 activity (Benková et al. 2003) restricted in the precursors of columella and provasculature initials. The pattern of apolarly-localized ABCB19 is complementary to PIN1 being restricted to the outer layers and vascular precursors surrounding the embryo. Since the triple mutant *pin1abcb1abcb19* has, in contrast to single mutant *pin1* or double mutant *abcb1abcb19,* severe defects in the establishment of proper auxin maxima leading to fused cotyledons, it is obvious that there must be synergistic interaction between PIN1 and ABCB proteins (Mravec et al. 2008). The possible scenario of this interaction during embryogenesis would include the activity of protein phosphatase PP2A. This phosphatase, together with protein kinase PID (Furutani et al. 2004), has been shown to be responsible for the regulation of polarized targeting of PIN1 in the inner embryo (Michniewicz et al. 2007). Although mutants in PP2A (*rcn1*) do not have embryogenic phenotype (Garbers et al. 1996), triple mutant *rcn1abcb1abcb19* had strong defects typical for aberrant auxin transport during embryogenesis (Mravec et al. 2008). Therefore, ABCB/PGP-mediated transport of auxin seems to be necessary during embryogenesis, forming a functionally complementary system to PIN-dependent transport.

4.3 Auxin Transporters During Root and Shoot Development

Auxin plays an important role in the patterning of both shoot and root apices as well as in the emergence and subsequent development of lateral organs (leafs and lateral roots). Active cell-to-cell carrier-mediated auxin flow in root and shoot is actually mirroring each other. In the root apex, auxin coming through vasculature in the inner tissues is reversed in the columella initials and lateral root cap toward epidermal cells, and it is then transported upward (Fig. 3b). In the shoot apex, the situation is reverse, auxin comes through epidermal cells being distributed also into leaf primordia, and then it is redirected downward through vasculature (Fig. 3c). Auxin concentration maxima always mark the position of the organ initiation and later they mark the tip of the organ primordia (Benková et al. 2003; Fig. 3d). In agreement with that, the local application of auxin was shown to trigger the formation of leaf or flower (Reinhardt et al. 2000) or lateral roots (Dubrovsky et al. 2008; Laskowski et al. 1995). Finally, auxin maxima are needed for the vascular tissue differentiation (Fig. 3e), where they represent major positional signals as suggested by the *de novo* formation of vasculature after auxin local application (Sachs 1991). In general, all three groups of auxin transporters contribute significantly to the auxin transport during root and shoot development.

In the root tip (Fig. 3b), auxin is transported acropetally in the stele with basally-localized PIN1, PIN3, and PIN7 (Blilou et al. 2005; Friml et al. 2002a). AUX1 helps to maintain this flow by its activity in the unloading of auxin from the phloem to the protophloem cells (Swarup et al. 2001). The redirection of auxin flow from the columella cells to the lateral root cap and epidermis is maintained by PIN3 and PIN7, where apically-localized PIN2 maintains the basipetal auxin flow (Müller et al. 1998; Friml et al. 2003) with the help of AUX1 (Swarup et al. 2001) and ABCB4 (Terasaka et al. 2005; Wu et al. 2007). PIN1, PIN3, and PIN7 might recycle the portion of auxin from the epidermal layers to the vasculature (Blilou et al. 2005). This cooperative action of PINs and AUX1 restricts the expression domain for one of the main auxin-inducible transcription factor PLETHORA (PLT), the master regulator of root tip cell fate (Blilou et al. 2005).The role of ABCB1 being present in all root cells except the columella (Mravec et al. 2008) is to support the generation of auxin maximum in the root tip and ABCB19 by its action in the endodermis and pericycle helps to separate the acropetal and basipetal auxin transport (Blakeslee et al. 2007; Wu et al. 2007).

The early phases of post-germination development of shoot have been characterized with respect to the localization of auxin carriers only recently. Both AUX1/LAX proteins (Vandenbussche et al. 2010) and PIN proteins (Žádníková et al. 2010) have been reported to play a role during apical hook development and maintenance. During subsequent development of the seedling, ABCB19 seems to be important for the deetiolation and transporting auxin to the developing cotyledons (Lewis et al. 2009; Wu et al. 2010). During later phases of development, the situation in the shoot apical meristem is almost reversal to the situation in the root tip. Auxin is transported through the epidermis to the organ initials and then it is canalized into the basipetal stream of auxin in the shoot, where PIN1 is the main transporter (Gälweiler et al. 1998), and ABCB1 and ABCB19 are the transporters concentrating the auxin flow in the vasculature (Geisler et al. 2005; Blakeslee et al. 2007). It should be also noted that the cooperative action of PIN1, AUX1, LAX1, LAX2, and LAX3 (Benková et al. 2003; Heisler et al. 2005; Bainbridge et al. 2008) in the shoot apical meristem is important for the generation of auxin maxima needed for the successive initiation of lateral organs in periodic phyllotactic pattern (Fig. 3c).

The initiation of lateral organ formation and development (Fig. 3d) is another example of cooperation between individual auxin transporters. In lateral root formation, the process starts in the pericycle cells, where the redistribution of PIN1 to the outer lateral PM (Kleine-Vehn et al. 2008a) and the generation of auxin maxima accompany the switch of polarity of cell division plane from anticlinal to periclinal (Benková et al 2003). Gradual formation of auxin maximum in the tip of developing organ is further supported by the activity of PIN2 transporting auxin out of the tip through epidermal cells (Fig. 3d). The role of AUX1 in this process is mainly in the providing auxin by its unloading from the phloem (Marchant et al. 2002). LAX3 seems to play a role in the generation of auxin maxima around developing primordium that triggers the production of cell wall remodeling enzymes (Swarup et al. 2008). Both ABCB1 and ABCB19 are also required for the process as documented by defects in single *abcb* mutants and double *pin abcb* mutants (Mravec et al. 2008). In lateral shoot formation (Fig. 3d), as already described earlier for the initiation phase from the shoot apical meristem, the activity of PIN1 is the major determinant in the transport of auxin through epidermis to the lateral organ tip and further through inner tissues in the developing vasculature back to the existing vasculature of the shoot (Benková et al. 2003; Heisler et al. 2005). This is further supported by the activity of ABCB1 and ABCB19 (Noh et al. 2001).

As already mentioned earlier, for the processes of auxin transport through inner tissues of shoot, the generation of new vascular strands is tightly connected to the auxin-transporter-generated auxin gradients. During the formation of vascular veins in leaves (Fig. 3e), PIN1 activity and polar localization are responsible for the generation of auxin maxima in so-called convergence points in the epidermis, from where the formation of new vein is started (Scarpella et al. 2006). Consequently, during the formation of other veins, PIN1 proteins are always oriented toward the older veins generating the auxin gradient needed for the differentiation of vasculature. Supporting role of AUX1 lies in its role in the leaf phloem loading (Marchant et al. 2002) and ABCB19 by its presence in the vascular bundle sheet cells helps to prevent the leakage of auxin out of the stream in the vasculature (Blakeslee et al. 2007).

4.4 Auxin Transporters During Tropisms

It is obvious that responses of plants to tropic stimuli, namely phototropism and gravitropism, were among the first processes studied with respect to the potential existence of regulatory compound(s) that would transduce these stimuli. Pioneering experiments of Darwin (1880) led to the formulation of the concept of plant hormones. Subsequently, the differential distribution of auxin during positive gravitropism of root, negative gravitropism of stem, and positive phototropism of stem has been shown to activate asymmetrical growth leading to the bending of the organ (Went 1974). Positive gravitropic bending of the root tip (Fig. 3f) is characterized by the relocalization of PIN3 in columella cells in the direction of gravity stimulus (Friml et al. 2002b). This leads to the reorientation of auxin flow to the lower lateral root cap cells and epidermal cells, where PIN2 (Luschnig et al. 1998; Müller et al. 1998) and AUX1 (Bennett et al. 1996; Swarup et al. 2001, 2005) cooperate to transport auxin. Resulting excess of auxin inhibits growth at the lower side and the root bends toward the gravity vector (Swarup et al. 2005). Supporting regulatory mechanisms include specific vacuolar degradation of PIN2 at the upper side of the root (Abas et al. 2006; Kleine-Vehn et al. 2008b) and the activities of ABCB1, ABCB4, and ABCB19 (Lewis et al. 2007).

Much less information is available for the negative gravitropism of stem. On the basis of the mutant analysis (Friml et al. 2002b), PIN3 has been suggested to play a role in the endodermal cells and its relocation upon gravistimulus is expected, but needs to be experimentally proven.

At the moment, there is no detailed information about the role of individual auxin transporters in the phototropism. However, analyses of mutants suggested that all of them will be somehow involved, including PIN3 (Friml et al. 2002b), AUX1 (Stone et al. 2008), ABCB1 (Lin and Wang 2005), and ABCB19 (Lin and Wang 2005; Nagashima et al. 2008a, b; Noh et al. 2003).

Concerning other tropical stimuli, it should be mentioned that the auxin transporters from PIN family possibly play a role in transducing auxin signal during root thigmotropism (Chen et al. 2009), but seemingly they are inactive in the hydrotropical responses (Miyazawa et al. 2009).

5 Quantitative Aspects of Auxin Flow: Not Just the Form but Also the Amount Matters

As described earlier, until now there is a lot of information about auxin transport and its significance for both preprogrammed development as well as for the responses of plants to environmental stimuli. Nevertheless, these data originate from various plants and their parts, which were studied during various phases of their development and under various conditions. Even though the major amount of information comes from studies at the tissue level on one model plant species – *A. thaliana* and its mutants (Benjamins and Scheres 2008; Petrášek and Friml 2009; Koornneef and Meinke 2010), the information on the cellular level is still not sufficient. The diversity of cell types – even within the tiny organs of Arabidopsis – does not enable detailed studies of auxin movement at the cellular, subcellular, and molecular levels. Moreover, it should not be forgotten that not only the quality but namely the quantity of auxin flow is the dominant trigger determining consequent growth and development. So, the information is needed, which would be complementary to qualitative and semiquantitative studies available on tissue level.

This type of quantitative "cellular" data can be acquired using cell suspension cultures and also yeast and animal cells. Unfortunately, the obvious first candidates – i.e., Arabidopsis cell lines – suffer from low quality for quantitative measurements as they tend to form large cell clumps; therefore, other materials have been used, such as tobacco BY-2 and VBI-0 cells, as well as yeast strains, Xenopus oocytes and/or HeLa cells (e.g., Blakeslee et al. 2007; Cho et al. 2007; Paciorek et al. 2005; Petrášek et al. 2006; Yang et al. 2006; Fu et al. 2007; Cho et al. 2007; Yang and Murphy 2009; Mravec et al. 2009; Krouk et al. 2010; see also Table 1). In these systems, the particular transport function for various auxin transporters was proven and the first information about the auxin-transport capacity for various transport processes has been provided.

The quantitative data obtained on the cellular level in this way can be subjected to mathematical analysis and the hypothetical models of quantitative aspects of intra and intercellular auxin flow can be provided. The validity of such models can be then tested by another set of experimental data until there is an agreement between the data predicted by the model and experimental behavior of the system. However, the modeling of auxin flow within and through a single cell is complicated by metabolic transformations of auxin taking place in various cell compartments. Anyway, this type of approach as such represents a complement to the models created by simulations on the tissue level (e.g., Mitchison 1980, 1981 for canalization hypothesis; Jonsson et al. 2006; Smith et al. 2006; Merks et al. 2007 for phyllotaxis; Kramer 2004; Feugier et al. 2005; Fujita and Mochizuki 2006; Prusinkiewicz and Rolland-Lagan 2006 for vein formation) and has a potential to contribute significantly to the understanding of the robustness of intra and intercellular auxin flow with all developmental implications.

6 Evolutionary Aspects: Many Transporters and What Was the First?

The data available suggest an ancient origin of auxin and its signaling role in the Plantae supergroup (Johri 2008; Lau et al. 2009), and nowadays, there is convincing evidence that polar auxin transport is involved in the key developmental processes not only in vascular plants, but also in Bryophytes and even in Charophytes. The new establishment of the auxin-dependent axial and even polarized growth could have supported the evolution of some adaptation processes, which are crucial for higher plants and their development (Cook et al. 2003).

Since the ABCB transporters seem to be highly conserved throughout phyla (Peer and Murphy 2007), they probably represent the evolutionarily ancient auxin transport system. In contrast to them, the first representatives of the PIN transporter family have been found only in the genomes of land plants, beginning with the moss *Physcomitrella patens* and the club moss *Selaginella moellendorffii* (reviewed Zažímalová et al. 2007; Křeček et al. 2009). However, the complete genomic information from algae, which are more closely related to land plants (e.g., Streptophyta) as well as from the liverworts (land plants more ancient than the club mosses) is missing and so the evolutionary origin of PINs is still not clear. Nevertheless, in Arabidopsis and other higher plants, there is a relatively high degree of similarity between individual members of the PIN gene family (32–85% mutual identity) suggesting the origin from a single ancestral sequence (Paponov et al. 2005).

Almost nothing is known about the evolution of the AUX1/LAX family of auxin influx carriers, even though they are expectedly thought to develop from the ancient amino-acid-permease family facilitating the uptake of amino acids such as tryptophan (Bennett et al. 1996; Young et al. 1999).

The total number of various types of auxin transporters may be surprising at first. However, taking into account the crucial role of polar auxin transport and auxin gradients in the formation of plant body, the existence of multiple auxin-transport-catalysts (at least 16 different carriers in *A. thaliana* to date), furthermore exploiting various sources of energy for their action, is logical. So, the evolutionary pressure supported the formation of the complex auxin transport system in which individual elements (transporters) depend on various and independent sources of energy, have various cellular dynamics, and prove distinct polarity as well as stability on the PM (Zažímalová et al. 2010). While AUX1/LAXes are supposed to provide cells with enhanced auxin uptake when and where needed (Kramer 2004; Kramer and Bennett 2006), the ABCB transporters with their mostly apolar localization and high degree of stability on the PM (Blakeslee et al. 2007; Titapiwatanakun and Murphy 2009; Titapiwatanakun et al. 2009) seem to support the basal auxin flow. "Long" PINs are able to endow the auxin transport with directionality and – due to their dynamic localization on the PM – also with the adaptability necessary for control of preprogrammed developmental events as well as for growth responses to environmental stimuli (Benjamins and Scheres 2008; Vanneste and Friml 2009). Last but not least, "short" PINs seem to regulate intracellular auxin distribution and thus its availability for signaling cascades (Mravec et al. 2009).

7 Conclusions: There Is No Proper Development Without Auxin (Transporters)

Past two decades have contributed substantially to the understanding of the carrier-mediated cell-to-cell transport of plant hormone (plant growth regulatory substance) auxin. Molecular-biological approaches, mainly at *A. thaliana*, led to the characterization of corresponding mutants, to isolation of respective genes as well as to physiological characterization of the three main groups of auxin transporters. In parallel, molecular function of these carriers was tested using various simplified, homologous and heterologous expression systems. On the basis of this combined knowledge, now there is a generally accepted fact that auxin carriers are indispensable to redirect, strengthen, or simply maintain the cell-to-cell auxin flow. Together, they form highly flexible auxin transport network underlying all developmental processes in plants.

However, there is one major failure in the effort to understand the mechanism of transport of auxin molecule through these carriers: Although we know how important for plant development auxin transporters are, at the moment we are still far from complete understanding of details of carrier-mediated auxin flow. There is still lack of simplified experimental setup(s), where the auxin transport could be characterized on the biochemical level – that is, both qualitatively (e.g., specificity) and quantitatively (e.g., capacity and kinetics) and in direct relation to a particular type of carrier. Therefore, in near future, the crucial task will be to find the experimental setup for the detailed biochemical characterization of auxin transporters. These data obtained can be used for the precise determination the energetic demands for active auxin transport, for mapping the specificity of individual carriers for auxin and similar compounds, for the identification of their cofactors and interactors, etc. This knowledge will then allow better utilization and manipulation of the auxin transport system for improvement of crops, medicinal plant production, micropropagation in forestry and horticulture, and/or many other possible applications.

Acknowledgements The authors acknowledge the support for their work from the Ministry of Education, Youth and Sports of the Czech Republic, project LN06034. Figures were drawn by G. Rzewuski (http://www.bioartworks.com).

References

Abas L, Benjamins R, Malenica N, Paciorek T, Wirniewska J, Moulinier-Anzola JC, Sieberer T, Friml J, Luschnig C (2006) Intracellular trafficking and proteolysis of the Arabidopsis auxin-efflux facilitator PIN2 are involved in root gravitropism. Nat Cell Biol 8:249–256

Badescu GO, Napier RM (2006) Receptors for auxin: will it all end in TIRs? Trends Plant Sci 11:217–223

Bailly A, Sovero V, Vincenzetti V, Santelia D, Bartnik D, Koenig BW, Mancuso S, Martinoia E, Geisler M (2008) Modulation of P-glycoproteins by auxin transport inhibitors is mediated by interaction with immunophilins. J Biol Chem 283:21817–21826

Bainbridge K, Guyomarc'h S, Bayer E, Swarup R, Bennett M, Mandel T, Kuhlemeier C (2008) Auxin influx carriers stabilize phyllotactic patterning. Genes Dev 22:810–823

Baluška F, Šámaj J, Menzel D (2003) Polar transport of auxin: carrier-mediated flux across the plasma membrane or neurotransmitter-like secretion? Trends Cell Biol 13:282–285

Baluška F, Schlicht M, Volkmann D, Mancuso S (2008) Vesicular secretion of auxin: evidences and implications. Plant Signal Behav 3:254–256

Becker P, Hakenbeck R, Henrich B (2009) An ABC transporter of Streptococcus pneumoniae involved in susceptibility to vancoresmycin and bacitracin. Antimicrob Agents Chemother 53:2034–2041

Benjamins R, Scheres B (2008) Auxin: the looping star in plant development. Annu Rev Plant Biol 59:443–465

Benjamins R, Quint A, Weijers D, Hooykaas P, Offringa R (2001) The PINOID protein kinase regulates organ development in Arabidopsis by enhancing polar auxin transport. Development 128:4057–4067

Benková E, Michniewicz M, Sauer M, Teichmann T, Seifertová D, Jürgens G, Friml J (2003) Local, efflux-dependent auxin gradients as a common module for plant organ formation. Cell 115:591–602

Bennett SRM, Alvarez J, Bossinger G, Smyth DR (1995) Morphogenesis in PINOID mutants of *Arabidopsis thaliana*. Plant J 8:505–520
Bennett MJ, Marchant A, Green HG, May ST, Ward SP, Millner PA, Walker AR, Schulz B, Feldmann KA (1996) Arabidopsis AUX1 gene: a permease-like regulator of root gravitropism. Science 273:948–950
Bishopp A, Mahonen AP, Helariutta Y (2006) Signs of change: hormone receptors that regulate plant development. Development 133:1857–1869
Blakeslee J, Peer W, Murphy A (2005) MDR/PGP auxin transport proteins and andocytic cycling. In: Šamaj J, Baluška F, Menzel D (eds) Plant endocytosis. Springer, Berlin, pp 159–176
Blakeslee JJ, Bandyopadhyay A, Lee OR, Mravec J, Titapiwatanakun B, Sauer M, Makam SN, Cheng Y, Bouchard R, Adamec J, Geisler M, Nagashima A, Sakai T, Martinoia E, Friml J, Peer WA, Murphy AS (2007) Interactions among PIN-FORMED and P-glycoprotein auxin transporters in Arabidopsis. Plant Cell 19:131–147
Blilou I, Xu J, Wildwater M, Willemsen V, Paponov I, Friml J, Heidstra R, Aida M, Palme K, Scheres B (2005) The PIN auxin efflux facilitator network controls growth and patterning in Arabidopsis roots. Nature 433:39–44
Bouchard R, Bailly A, Blakeslee JJ, Oehring SC, Vincenzetti V, Lee OR, Paponov I, Palme K, Mancuso S, Murphy AS, Schulz B, Geisler M (2006) Immunophilin-like TWISTED DWARF1 modulates auxin efflux activities of Arabidopsis P-glycoproteins. J Biol Chem 281:30603–30612
Butler JH, Hu S, Brady SR, Dixon MW, Muday GK (1998) In vitro and in vivo evidence for actin association of the naphthylphthalamic acid-binding protein from zucchini hypocotyls. Plant J 13:291–301
Carraro N, Forestan C, Canova S, Traas J, Varotto S (2006) ZmPIN1a and ZmPIN1b encode two novel putative candidates for polar auxin transport and plant architecture determination of maize. Plant Physiol 142:254–264
Chawla R, DeMason D (2004) Molecular expression of PsPIN1, a putative auxin efflux carrier gene from pea (Pisum sativum L.). Plant Growth Regul 44:1–14
Chen LS, Ortiz-Lopez A, Jung A, Bush DR (2001) ANT1, an aromatic and neutral amino acid transporter in Arabidopsis. Plant Physiol 125:1813–1820
Chen Z, Noir S, Kwaaitaal M, Hartmann HA, Wu M-J, Mudgil Y, Sukumar P, Muday G, Panstruga R, Jones AM (2009) Two seven-transmembrane domain mildew resistance locus O proteins cofunction in Arabidopsis root thigmomorphogenesis. Plant Cell 21:1972–1991
Cho M, Lee S, Cho H (2007) P-glycoprotein4 displays auxin efflux transporter-like action in Arabidopsis root hair cells and tobacco cells. Plant Cell 19:3930–3943
Christensen SK, Dagenais N, Chory J, Weigel D (2000) Regulation of auxin response by the protein kinase PINOID. Cell 100:469–478
Cook T, Poli D, Cohen J (2003) Did auxin play a crucial role in the evolution of novel body plans during the late silurian–early devonian radiation of land plants? In: Hemsley A, Poole I (eds) The evolution of plant physiology. Elsevier, Oxford, pp 85–107
Cox D, Muday G (1994) NPA binding activity is peripheral to the plasma membrane and is associated with the cytoskeleton. Plant Cell 6:1941–1953
Darwin CR (1880) The power of movement in plants. John Murray, London
Dawson R, Locher K (2006) Structure of a bacterial multidrug ABC transporter. Nature 443:180–185
de Billy F, Grosjean C, May S, Bennett M, Cullimore JV (2001) Expression studies on AUX1-like genes in Medicago truncatula suggest that auxin is required at two steps in early nodule development. Mol Plant-Microbe Interact 14:267–277
Dello Ioio R, Nakamura K, Moubayidin L, Perilli S, Taniguchi M, Morita MT, Aoyama T, Costantino P, Sabatini S (2008) A genetic framework for the control of cell division and differentiation in the root meristem. Science 322:1380–1384
Deruère J, Jackson K, Garbers C, Soll D, DeLong A (1999) The RCN1-encoded A subunit of protein phosphatase 2A increases phosphatase activity in vivo. Plant J 20:389–399
Dharmasiri N, Dharmasiri S, Weijers D, Lechner E, Yamada M, Hobbie L, Ehrismann JS, Jürgens G, Estelle M (2005) Plant development is regulated by a family of auxin receptor F box proteins. Dev Cell 9:109–119

Dharmasiri S, Swarup R, Mockaitis K, Dharmasiri N, Singh SK, Kowalchyk M, Marchant A, Mills S, Sandberg G, Bennett MJ, Estelle M (2006) AXR4 is required for localization of the auxin influx facilitator AUX1. Science 312:1218–1220

Dhonukshe P, Aniento F, Hwang I, Robinson D, Mravec J, Stierhof Y, Friml J (2007) Clathrin-mediated constitutive endocytosis of PIN auxin efflux carriers in Arabidopsis. Curr Biol 17:520–527

Dhonukshe P, Grigoriev I, Fischer R, Tominaga M, Robinson DG, Hašek J, Paciorek T, Petrášek J, Seifertová D, Tejos R, Meisel LA, Zažímalová E, Gadella TWJ, Stierhof YD, Ueda T, Oiwa K, Akhmanova A, Brock R, Spang A, Friml J (2008) Auxin transport inhibitors impair vesicle motility and actin cytoskeleton dynamics in diverse eukaryotes. Proc Natl Acad Sci USA 105:4489–4494

Dixon MW, Jacobson JA, Cady CT, Muday GK (1996) Cytoplasmic orientation of the naphthylphthalamic acid-binding protein in Zucchini plasma membrane vesicles. Plant Physiol 112:421–432

Dubrovsky JG, Sauer M, Napsucialy-Mendivil S, Ivanchenko MG, Friml J, Shishkova S, Celenza J, Benková E (2008) Auxin acts as a local morphogenetic trigger to specify lateral root founder cells. Proc Natl Acad Sci USA 105:8790–8794

Feugier F, Mochizuki A, Iwasa Y (2005) Self-organization of the vascular system in plant leaves: inter-dependent dynamics of auxin flux and carrier proteins. J Theor Biol 236:366–375

Fischer W, Loo D, Koch W, Ludewig U, Boorer K, Tegeder M, Rentsch D, Wright E, Frommer W (2002) Low and high affinity amino acid H+-cotransporters for cellular import of neutral and charged amino acids. Plant J 29:717–731

Friml J, Benková E, Blilou I, Wiśniewska J, Hamann T, Ljung K, Woody S, Sandberg G, Scheres B, Jürgens G, Palme K (2002a) AtPIN4 mediates sink-driven auxin gradients and root patterning in Arabidopsis. Cell 108:661–673

Friml J, Wiśniewska J, Benková E, Mendgen K, Palme K (2002b) Lateral relocation of auxin efflux regulator PIN3 mediates tropism in Arabidopsis. Nature 415:806–809

Friml J, Vieten A, Sauer M, Weijers D, Schwarz H, Hamann T, Offringa R, Jürgens G (2003) Efflux-dependent auxin gradients establish the apical-basal axis of Arabidopsis. Nature 426:147–153

Friml J, Yang X, Michniewicz M, Weijers D, Quint A, Tietz O, Benjamins R, Ouwerkerk PBF, Ljung K, Sandberg G, Hooykaas PJJ, Palme K, Offringa R (2004) A PINOID-dependent binary switch in apical-basal PIN polar targeting directs auxin efflux. Science 306:862–865

Fu D, van Dam E, Brymora A, Duggin I, Robinson P, Roufogalis B (2007) The small GTPases Rab5 and RalA regulate intracellular traffic of P-glycoprotein. Biochim Biophys Acta Mol Cell Res 1773:1062–1072

Fujita H, Mochizuki A (2006) Pattern formation of leaf veins by the positive feedback regulation between auxin flow and auxin efflux carrier. J Theor Biol 241:541–551

Furutani M, Vernoux T, Traas J, Kato T, Tasaka M, Aida M (2004) PIN-FORMED1 and PINOID regulate boundary formation and cotyledon development in Arabidopsis embryogenesis. Development 131:5021–5030

Gallavotti A, Yang Y, Schmidt R, Jackson D (2008) The relationship between auxin transport and maize branching. Plant Physiol 147:1913–1923

Gälweiler L, Guan CH, Müller A, Wisman E, Mendgen K, Yephremov A, Palme K (1998) Regulation of polar auxin transport by AtPIN1 in Arabidopsis vascular tissue. Science 282:2226–2230

Garbers C, DeLong A, Deruère J, Bernasconi P, Soll D (1996) A mutation in protein phosphatase 2A regulatory subunit A affects auxin transport in Arabidopsis. EMBO J 15:2115–2124

Geisler M, Murphy AS (2006) The ABC of auxin transport: the role of p-glycoproteins in plant development. FEBS Lett 580:1094–1102

Geisler M, Kolukisaoglu HU, Bouchard R, Billion K, Berger J, Saal B, Frangne N, Koncz-Kalman Z, Koncz C, Dudler R, Blakeslee JJ, Murphy AS, Martinoia E, Schulz B (2003) TWISTED DWARF1, a unique plasma membrane-anchored immunophilin-like protein, interacts with

Arabidopsis multidrug resistance-like transporters AtPGP1 and AtPGP19. Mol Biol Cell 14:4238–4249

Geisler M, Blakeslee J, Bouchard R, Lee O, Vincenzetti V, Bandyopadhyay A, Titapiwatanakun B, Peer W, Bailly A, Richards E, Ejendal K, Smith A, Baroux C, Grossniklaus U, Müller A, Hrycyna C, Dudler R, Murphy A, Martinoia E (2005) Cellular efflux of auxin catalyzed by the Arabidopsis MDR/PGP transporter AtPGP1. Plant J 44:179–194

Geldner N, Friml J, Stierhof YD, Jürgens G, Palme K (2001) Auxin transport inhibitors block PIN1 cycling and vesicle traficking. Nature 413:425–428

Geldner N, Anders N, Wolters H, Keicher J, Kornberger W, Müller P, Delbarre A, Ueda T, Nakano A, Jürgens G (2003) The Arabidopsis GNOM ARF-GEF mediates endosomal recycling, auxin transport, and auxin-dependent plant growth. Cell 112:219–230

Goda H, Sawa S, Asami T, Fujioka S, Shimada Y, Yoshida S (2004) Comprehensive comparison of auxin-regulated and brassinosteroid-regulated genes in Arabidopsis. Plant Physiol 134:1555–1573

Goldsmith MHM (1977) Polar transport of auxin. Annu Rev Plant Physiol Plant Mol Biol 28:439–478

Goto N, Starke M, Kranz A (1987) Effect of gibberellins on flower development of the pin-formed mutant of Arabidopsis thaliana. Arabidopsis Inf Serv 23:66–71

Grebe M, Friml J, Swarup R, Ljung K, Sandberg G, Terlou M, Palme K, Bennett MJ, Scheres B (2002) Cell polarity signaling in Arabidopsis involves a BFA-sensitive auxin influx pathway. Curr Biol 12:329–334

Guilfoyle TJ, Hagen G (2007) Auxin response factors. Curr Opin Plant Biol 10:453–460

Hála M, Cole R, Synek L, Drdová E, Pečenková T, Nordheim A, Lamkemeyer T, Madlung J, Hochholdinger F, Fowler J, Žárský V (2008) An exocyst complex functions in plant cell growth in Arabidopsis and tobacco. Plant Cell 20:1330–1345

Hamann T, Benková E, Baurle I, Kientz M, Jürgens G (2002) The Arabidopsis BODENLOS gene encodes an auxin response protein inhibiting MONOPTEROS-mediated embryo patterning. Genes Dev 16:1610–1615

Hardtke CS, Berleth T (1998) The Arabidopsis gene MONOPTEROS encodes a transcription factor mediating embryo axis formation and vascular development. EMBO J 17:1405–1411

Hazak O, Bloch D, Poraty L, Sternberg H, Zhang J, Friml J, Yalovsky S (2010) A rho scaffold integrates the secretory system with feedback mechanisms in regulation of auxin distribution. PLoS Biol 8(1):e1000282

Heisler MG, Ohno C, Das P, Sieber P, Reddy GV, Long JA, Meyerowitz EM (2005) Patterns of auxin transport and gene expression during primordium development revealed by live imaging of the Arabidopsis inflorescence meristem. Curr Biol 15:1899–1911

Ho CH, Lin SH, Hu HC, Tsay YF (2009) CHL1 functions as a nitrate sensor in plants. Cell 138:1184–1194

Hochholdinger F, Wulff D, Reuter K, Park WJ, Feix G (2000) Tissue-specific expression of AUX1 in maize roots. J Plant Physiol 157:315–319

Hoshino T, Hitotsubashi R, Miyamoto K, Tanimoto E, Ueda J (2003) Expression of PIN and AUX1 genes encoding putative carrier proteins for auxin polar transport in etiolated pea epicotyls [correction of epicotyles] under simulated microgravity conditions on a three-dimensional clinostat. Biol Sci Space 17:175–176

Hoshino T, Miyamoto K, Ueda J (2004) Automorphosis and auxin polar transport of etiolated pea seedlings under microgravity conditions. Biol Sci Space 18:94–95

Hoyerová K, Perry L, Hand P, Laňková M, Kocábek T, May S, Kottová J, Pačes J, Napier R, Zažímalová E (2008) Functional characterization of PaLAX1, a putative auxin permease, in heterologous plant systems. Plant Physiol 146:1128–1141

Imhoff V, Müller P, Guern J, Delbarre A (2000) Inhibitors of the carrier-mediated influx of auxin in suspension-cultured tobacco cells. Planta 210:580–588

Ito H, Gray WM (2006) A gain-of-function mutation in the Arabidopsis pleiotropic drug resistance transporter PDR9 confers resistance to auxinic herbicides. Plant Physiol 142:63–74

Jahrmann T, Bastida M, Pineda M, Gasol E, Ludevid M, Palacin M, Puigdomenech P (2005) Studies on the function of TM20, a transmembrane protein present in cereal embryos. Planta 222:80–90

Jaillais Y, Fobis-Loisy I, Miege C, Rollin C, Gaude T (2006) AtSNX1 defines an endosome for auxin-carrier trafficking in Arabidopsis. Nature 443:106–109

Johri M (2008) Hormonal regulation in green plant lineage families. Physiol Mol Biol Plants 14:23–38

Jonsson H, Heisler M, Shapiro B, Meyerowitz E, Mjolsness E (2006) An auxin-driven polarized transport model for phyllotaxis. Proc Natl Acad Sci USA 103:1633–1638

Kamada M, Yamasaki S, Fujii N, Higashitani A, Takahashi H (2003) Gravity-induced modification of auxin transport and distribution for peg formation in cucumber seedlings: possible roles for CS-AUX1 and CS-PIN1. Planta 218:15–26

Kepinski S, Leyser O (2005) The Arabidopsis F-box protein TIR1 is an auxin receptor. Nature 435:446–451

Kerr ID, Bennett MJ (2007) New insight into the biochemical mechanisms regulating auxin transport in plants. Biochem J 401:613–622

Kleine-Vehn J, Dhonukshe P, Swarup R, Bennett M, Friml J (2006) Subcellular trafficking of the Arabidopsis auxin influx carrier AUX1 uses a novel pathway distinct from PIN1. Plant Cell 18:3171–3181

Kleine-Vehn J, Dhonukshe P, Sauer M, Brewer PB, Wiśniewska J, Paciorek T, Benková E, Friml J (2008a) ARF GEF-dependent transcytosis and polar delivery of PIN auxin carriers in Arabidopsis. Curr Biol 18:526–531

Kleine-Vehn J, Leitner J, Zwiewka M, Sauer M, Abas L, Luschnig C, Friml J (2008b) Differential degradation of PIN2 auxin efflux carrier by retromer-dependent vacuolar targeting. Proc Natl Acad Sci USA 105:17812–17817

Kleine-Vehn J, Huang F, Naramoto S, Zhang J, Michniewicz M, Offringa R, Friml J (2009) PIN auxin efflux carrier polarity is regulated by PINOID kinase-mediated recruitment into GNOM-independent trafficking in Arabidopsis. Plant Cell 21:3839–3849

Koornneef M, Meinke D (2010) The development of Arabidopsis as a model plant. Plant J 61:909–921

Kramer EM (2004) PIN and AUX/LAX proteins: their role in auxin accumulation. Trends Plant Sci 9:578–582

Kramer E, Bennett M (2006) Auxin transport: a field in flux. Trends Plant Sci 11:382–386

Křeček P, Skůpa P, Libus J, Naramoto S, Tejos R, Friml J, Zažímalová E (2009) The PIN-FORMED (PIN) protein family of auxin transporters. Genome Biol 10:249

Krouk G, Lacombe B, Bielach A, Perrine-Walker F, Malinská K, Mounier E, Hoyerová K, Tillard P, Leon S, Ljung K, Zažímalová E, Benková E, Nacry P, Gojon A (2010) Nitrate-regulated auxin transport by NRT1.1 defines a mechanism for nutrient sensing in plants. Dev Cell 18(6):927–937

Laskowski MJ, Williams ME, Nusbaum HC, Sussex IM (1995) Formation of lateral root-meristems is a 2-stage process. Development 121:3303–3310

Lau S, Shao N, Bock R, Jürgens G, De Smet I (2009) Auxin signaling in algal lineages: fact or myth? Trends Plant Sci 14:182–188

Laxmi A, Pan J, Morsy M, Chen R (2008) Light plays an essential role in intracellular distribution of auxin efflux carrier PIN2 in Arabidopsis thaliana. PLoS ONE 3:e1510

Lechner E, Achard P, Vansiri A, Potuschak T, Genschik P (2006) F-box proteins everywhere. Curr Opin Plant Biol 9:631–638

Lee SH, Cho HT (2006) PINOID positively regulates auxin efflux in Arabidopsis root hair cells and tobacco cells. Plant Cell 18:1604–1616

Lee M, Choi Y, Burla B, Kim Y, Jeon B, Maeshima M, Yoo J, Martinoia E, Lee Y (2008) The ABC transporter AtABCB14 is a malate importer and modulates stomatal response to CO2. Nat Cell Biol 10:1217–1223

Levin M (2006) Is the early left-right axis like a plant, a kidney, or a neuron? The integration of physiological signals in embryonic asymmetry. Birth Defects Res C Embryo Today 78:191–223

Lewis DR, Miller ND, Splitt BL, Wu GS, Spalding EP (2007) Separating the roles of acropetal and basipetal auxin transport on gravitropism with mutations in two Arabidopsis multidrug resistance-like ABC transporter genes. Plant Cell 19:1838–1850

Lewis DR, Wu G, Ljung K, Spalding EP (2009) Auxin transport into cotyledons and cotyledon growth depend similarly on the ABCB19 Multidrug Resistance-like transporter. Plant J 60: 91–101

Leyser O (2006) Dynamic integration of auxin transport and signalling. Curr Biol 16:R424–R433

Li G, Xue HW (2007) Arabidopsis PLDzeta2 regulates vesicle trafficking and is required for auxin response. Plant Cell 19:281–295

Li JS, Yang HB, Peer WA, Richter G, Blakeslee J, Bandyopadhyay A, Titapiwantakun B, Undurraga S, Khodakovskaya M, Richards EL, Krizek B, Murphy AS, Gilroy S, Gaxiola R (2005a) Arabidopsis H+ -PPase AVP1 regulates auxin-mediated organ development. Science 310:121–125

Li L, Xu J, Xu ZH, Xue HW (2005b) Brassinosteroids stimulate plant tropisms through modulation of polar auxin transport in Brassica and Arabidopsis. Plant Cell 17:2738–2753

Lin R, Wang H (2005) Two homologous ATP-binding cassette transporter proteins, AtMDR1 and AtPGP1, regulate Arabidopsis photomorphogenesis and root development by mediating polar auxin transport. Plant Physiol 138:949–964

Lomax T, Muday G, Rubery P (1995) Auxin transport. In: Davis PJ (ed) Plant hormones. Kluwer, Dordrecht, pp 509–530

Luschnig C, Gaxiola RA, Grisafi P, Fink GR (1998) EIR1, a root-specific protein involved in auxin transport, is required for gravitropism in Arabidopsis thaliana. Genes Dev 12:2175–2187

Maher E, Martindale S (1980) Mutants of Arabidopsis thaliana with altered responses to auxins and gravity. Biochem Genet 18:1041–1053

Malenica N, Abas L, Benjamins R, Kitakura S, Sigmund HF, Jun KS, Hauser MT, Friml J, Luschnig C (2007) MODULATOR OF PIN genes control steady-state levels of Arabidopsis PIN proteins. Plant J 51:537–550

Mancuso S, Marras A, Mugnai S, Schlicht M, Žárský V, Li G, Song L, Xue H, Baluška F (2007) Phospholipase dzeta2 drives vesicular secretion of auxin for its polar cell-cell transport in the transition zone of the root apex. Plant Signal Behav 2:240–244

Marchant A, Bhalerao R, Casimiro I, Eklof J, Casero PJ, Bennett M, Sandberg G (2002) AUX1 promotes lateral root formation by facilitating indole-3-acetic acid distribution between sink and source tissues in the Arabidopsis seedling. Plant Cell 14:589–597

Men SZ, Boutte Y, Ikeda Y, Li XG, Palme K, Stierhof YD, Hartmann MA, Moritz T, Grebe M (2008) Sterol-dependent endocytosis mediates post-cytokinetic acquisition of PIN2 auxin efflux carrier polarity. Nat Cell Biol 10:237–244

Merks RMH, Van de Peer Y, Inze D, Beemster GTS (2007) Canalization without flux sensors: a traveling-wave hypothesis. Trends Plant Sci 12:384–390

Michniewicz M, Zago MK, Abas L, Weijers D, Schweighofer A, Meskiene I, Heisler MG, Ohno C, Zhang J, Huang F, Schwab R, Weigel D, Meyerowitz EM, Luschnig C, Offringa R, Friml J (2007) Antagonistic regulation of PIN phosphorylation by PP2A and PINOID directs auxin flux. Cell 130:1044–1056

Mitchison G (1980) Model for vein formation in higher-plants. Proc R Soc Lond B 207:79–109

Mitchison G (1981) The polar transport of auxin and vein patterns in plants. Philos Trans R Soc Lond B Biol Sci 295:461–471

Miyazawa Y, Takahashi A, Kobayashi A, Kaneyasu T, Fujii N, Takahashi H (2009) GNOM-mediated vesicular trafficking plays an essential role in hydrotropism of Arabidopsis roots. Plant Physiol 149:835–840

Morris DA, Friml J, Zažímalová E (2004) The transport of auxins. In: Davies PJ (ed) Plant hormones: biosynthesis, signal transduction, action! Kluwer, Dordrecht, pp 437–470

Mravec J, Kubeš M, Bielach A, Gaykova V, Petrášek J, Skůpa P, Chand S, Benková E, Zažímalová E, Friml J (2008) Interaction of PIN and PGP transport mechanisms in auxin distribution-dependent development. Development 135:3345–3354

Mravec J, Skůpa P, Bailly A, Hoyerová K, Křeček P, Bielach A, Petrášek J, Zhang J, Gaykova V, Stierhof YD, Dobrev PI, Schwarzerová K, Rolčik J, Seifertová D, Luschnig C, Benková E, Zažímalová E, Geisler M, Friml J (2009) Subcellular homeostasis of phytohormone auxin is mediated by the ER-localized PIN5 transporter. Nature 459:1136–1140

Müller A, Guan CH, Gälweiler L, Tanzler P, Huijser P, Marchant A, Parry G, Bennett M, Wisman E, Palme K (1998) AtPIN2 defines a locus of Arabidopsis for root gravitropism control. EMBO J 17:6903–6911

Multani DS, Briggs SP, Chamberlin MA, Blakeslee JJ, Murphy AS, Johal GS (2003) Loss of an MDR transporter in compact stalks of maize br2 and sorghum dw3 mutants. Science 302:81–84

Murphy A, Peer WA, Taiz L (2000) Regulation of auxin transport by aminopeptidases and endogenous flavonoids. Planta 211:315–324

Murphy AS, Hoogner KR, Peer WA, Taiz L (2002) Identification, purification, and molecular cloning of N-1-naphthylphthalmic acid-binding plasma membrane-associated aminopeptidases from Arabidopsis. Plant Physiol 128:935–950

Nagashima A, Suzuki G, Uehara Y, Saji K, Furukawa T, Koshiba T, Sekimoto M, Fujioka S, Kuroha T, Kojima M, Sakakibara H, Fujisawa N, Okada K, Sakai T (2008a) Phytochromes and cryptochromes regulate the differential growth of Arabidopsis hypocotyls in both a PGP19-dependent and a PGP19-independent manner. Plant J 53:516–529

Nagashima A, Uehara Y, Sakai T (2008b) The ABC subfamily B auxin transporter AtABCB19 is involved in the inhibitory effects of N-1-Naphthyphthalamic acid on the phototropic and gravitropic responses of Arabidopsis hypocotyls. Plant Cell Physiol 49:1250–1255

Nemhauser JL, Mockler TC, Chory J (2004) Interdependency of brassinosteroid and auxin signaling in Arabidopsis. PLoS Biol 2:E258

Ni WM, Chen XY, Xu ZH, Xue HW (2002) Isolation and functional analysis of a Brassica juncea gene encoding a component of auxin efflux carrier. Cell Res 12:235–245

Nick P, Han M, An G (2009) Auxin stimulates its own transport by shaping actin filaments. Plant Physiol 151:155–167

Noh B, Murphy A, Spalding E (2001) Multidrug resistance-like genes of Arabidopsis required for auxin transport and auxin-mediated development. Plant Cell 13:2441–2454

Noh B, Bandyopadhyay A, Peer WA, Spalding EP, Murphy AS (2003) Enhanced gravi- and phototropism in plant mdr mutants mislocalizing the auxin efflux protein PIN1. Nature 423:999–1002

Ogawa M, Hanada A, Yamauchi Y, Kuwahara A, Kamiya Y, Yamaguchi S (2003) Gibberellin biosynthesis and response during Arabidopsis seed germination. Plant Cell 15:1591–1604

Ohno H, Stewart J, Fournier M, Bosshart H, Rhee I, Miyatake S, Saito T, Gallusser A, Kirchhausen T, Bonifacino J (1995) Interaction of tyrosine-based sorting signals with clathrin-associated proteins. Science 269:1872–1875

Okada K, Ueda J, Komaki MK, Bell CJ, Shimura Y (1991) Requirement of the auxin polar transport-system in early stages of Arabidopsis floral bud formation. Plant Cell 3:677–684

Oliveros-Valenzuela MR, Reyes D, Sanchez-Bravo J, Acosta M, Nicolas C (2007) The expression of genes coding for auxin carriers in different tissues and along the organ can explain variations in auxin transport and the growth pattern in etiolated lupin hypocotyls. Planta 227:133–142

Paciorek T, Zažímalová E, Ruthardt N, Petrášek J, Stierhof YD, Kleine-Vehn J, Morris DA, Emans N, Juergens G, Geldner N, Friml J (2005) Auxin inhibits endocytosis and promotes its own efflux from cells. Nature 435:1251–1256

Paponov IA, Teale WD, Trebar M, Blilou K, Palme K (2005) The PIN auxin efflux facilitators: evolutionary and functional perspectives. Trends Plant Sci 10:170–177

Paponov IA, Paponov M, Teale W, Menges M, Chakrabortee S, Murray JA, Palme K (2008) Comprehensive transcriptome analysis of auxin responses in Arabidopsis. Mol Plant 1:321–337

Parry G, Delbarre A, Marchant A, Swarup R, Napier R, Perrot-Rechenmann C, Bennett MJ (2001) Novel auxin transport inhibitors phenocopy the auxin influx carrier mutation aux1. Plant J 25:399–406

Pasquier C, Promponas V, Palaios G, Hamodrakas J, Hamodrakas S (1999) A novel method for predicting transmembrane segments in proteins based on a statistical analysis of the SwissProt database: the PRED-TMR algorithm. Protein Eng 12:381–385

Peer WA, Murphy AS (2007) Flavonoids and auxin transport: modulators or regulators? Trends Plant Sci 12:556–563

Petrášek J, Černá A, Schwarzerová K, Elčkner M, Morris DA, Zažímalová E (2003) Do Phytotropins Inhibit Auxin Efflux by Impairing Vesicle Traffic? Plant Physiol 131:254–263

Peret B, Swarup R, Jansen L, Devos G, Auguy F, Collin M, Santi C, Hocher V, Franche C, Bogusz D, Bennett M, Laplaze L (2007) Auxin influx activity is associated with Frankia infection during actinorhizal nodule formation in Casuarina glauca. Plant Physiol 144:1852–1862

Petrášek J, Friml J (2009) Auxin transport routes in plant development. Development 136:2675–2688

Pernisová M, Klíma P, Horák J, Válková M, Malbeck J, Souček P, Reichman P, Hoyerová K, Dubová J, Friml J, Zažímalová E, Hejátko J (2009) Cytokinins modulate auxin-induced organogenesis in plants via regulation of the auxin efflux. Proc Natl Acad Sci USA 106:609–3614

Petrášek J, Mravec J, Bouchard R, Blakeslee JJ, Abas M, Seifertová D, Wiśniewska J, Tadele Z, Kubeš M, Čovanová M, Dhonukshe P, Skůpa P, Benková E, Perry L, Křeček P, Lee OR, Fink GR, Geisler M, Murphy AS, Luschnig C, Zažímalová E, Friml J (2006) PIN proteins perform a rate-limiting function in cellular auxin efflux. Science 312:914–918

Prusinkiewicz P, Rolland-Lagan A (2006) Modeling plant morphogenesis. Curr Opin Plant Biol 9:83–88

Quint M, Gray WM (2006) Auxin signaling. Curr Opin Plant Biol 9:448–453

Rahman A, Ahamed A, Amakawa T, Goto N, Tsurumi S (2001) Chromosaponin I specifically interacts with AUX1 protein in regulating the gravitropic response of arabidopsis roots. Plant Physiol 125:990–1000

Rahman A, Bannigan A, Sulaman W, Pechter P, Blancaflor E, Bascin T (2007) Auxin, actin and growth of the Arabidopsis thalian. Plant J 50:514–528

Raven JA (1975) Transport of indoleacetic acid in plant-cells in relation to pH and electrical potential gradients, and its significance for polar IAA transport. New Phytol 74:163–172

Reinhardt D, Mandel T, Kuhlemeier C (2000) Auxin regulates the initiation and radial position of plant lateral organs. Plant Cell 12:507–518

Ren Q, Chen K, Paulsen I (2007) TransportDB: a comprehensive database resource for cytoplasmic membrane transport systems and outer membrane channels. Nucleic Acids Res 35: D274–D279

Robinson JS, Albert AC, Morris DA (1999) Differential effects of brefeldin A and cycloheximide on the activity of auxin efflux carriers in Cucurbita pepo L. J Plant Physiol 155:678–684

Rojas-Pierce M, Titapiwatanakun B, Sohn EJ, Fang F, Larive CK, Blakeslee J, Cheng Y, Cuttler S, Peer WA, Murphy AS, Raikhel NV (2007) Arabidopsis P-glycoprotein19 participates in the inhibition of gravitropism by gravacin. Chem Biol 14:1366–1376

Rubery PH (1990) Phytotropins-receptors and endogenous ligands. Hormone perception and signal transduction in animals and plants. Symp Soc Exp Biol 44:119–146

Rubery PH, Sheldrake AR (1974) Carrier-mediated auxin auxin transport. Planta 118:101–121

Růžička K, Ljung K, Vanneste S, Podhorská R, Beeckman T, Friml J, Benková E (2007) Ethylene regulates root growth through effects on auxin biosynthesis and transport-dependent auxin distribution. Plant Cell 19:2197–2212

Růžička K, Šimášková M, Duclercq J, Petrášek J, Zažímalová E, Simon S, Friml J, Van Montagu MCE, Benková E (2009) Cytokinin regulates root meristem activity via modulation of the polar auxin transport. Proc Natl Acad Sci USA 106:4284–4289

Růžička K, Strader L, Bailly A, Blakeslee J, Nejedlá E, Langowski L, Yang H, Fujika H, Ito H, Syono K, Gray W, Hejátko J, Martinoia E, Geisler M, Murphy A, Bartel B, Friml J (2010)

Arabidopsis PIS1 encodes the ABCG37 transporter of the auxin precursor indole-3-butyric acid. Proc Natl Acad Sci USA 107:10749–10753

Sachs T (1981) The control of the patterned differentiation of vascular tissues. Adv Bot Res Inc Adv Plant Pathol 9:151–262

Sachs T (1991) Pattern formation in plant tissues. Cambridge University Press, New York

Santelia D, Vincenzetti V, Azzarello E, Bovet L, Fukao Y, Düchtig P, Mancuso S, Martinoia E, Geisler M (2005) MDR-like ABC transporter AtPGP4 is involved in auxin-mediated lateral root and root hair development. FEBS Lett 579:5399–5406

Santelia D, Henrichs S, Vincenzetti V, Sauer M, Bigler L, Klein M, Bailly A, Lee Y, Friml J, Geisler M, Martinoia E (2008) Flavonoids redirect PIN-mediated polar auxin fluxes during root gravitropic responses. J Biol Chem 283:31218–31226

Santner A, Estelle M (2009) Recent advances and emerging trends in plant hormone signalling. Nature 459:1071–1078

Sauer M, Balla J, Luschnig C, Wiśniewska J, Reinöhl V, Friml J, Benková E (2006) Canalization of auxin flow by Aux/IAA-ARF-dependent feedback regulation of PIN polarity. Genes Dev 20:2902–2911

Scarpella E, Marcos D, Friml J, Berleth T (2006) Control of leaf vascular patterning by polar auxin transport. Genes Dev 20:1015–1027

Schlicht M, Strnad M, Scanlon MJ, Mancuso S, Hochholdinger F, Palme K, Volkmann D, Menzel D, Baluška F (2006) Auxin immunolocalization implicates vesicular neurotransmitter-like mode of polar auxin transport in root apices. Plant Signal Behav 1:122–133

Schnabel EL, Frugoli JF (2004) The PIN and LAX families of auxin transport genes in Medicago truncatula. Mol Genet Genomics 272:420–432

Schrader J, Baba K, May ST, Palme K, Bennett M, Bhalerao RP, Sandberg G (2003) Polar auxin transport in the wood-forming tissues of hybrid aspen is under simultaneous control of developmental and environmental signals. Proc Natl Acad Sci USA 100:10096–10101

Sidler M, Hassa P, Hasan S, Ringli C, Dudler R (1998) Involvement of an ABC transporter in a developmental pathway regulating hypocotyl cell elongation in the light. Plant Cell 10:1623–1636

Smith R, Guyomarc'h S, Mandel T, Reinhardt D, Kuhlemeier C, Prusinkiewicz P (2006) A plausible model of phyllotaxis. Proc Natl Acad Sci USA 103:1301–1306

Steinmann T, Geldner N, Grebe M, Mangold S, Jackson C, Paris S, Gälweiler L, Palme K, Jürgens G (1999) Coordinated polar localization of auxin efflux carrier PIN1 by GNOM ARF GEF. Science 286:316–318

Stone BB, Stowe-Evans EL, Harper RM, Celaya RB, Ljung K, Sandberg R, Liscum E (2008) Disruptions in AUX1-dependent auxin influx alter hypocotyl phototropism in Arabidopsis. Mol Plant 1:129–144

Sussman MR, Gardner G (1980) Solubilization of the receptor for N-1-naphthylphthalamic acid. Plant Physiol 66:1074–1078

Swarup R, Friml J, Marchant A, Ljung K, Sandberg G, Palme K, Bennett M (2001) Localization of the auxin permease AUX1 suggests two functionally distinct hormone transport pathways operate in the Arabidopsis root apex. Genes Dev 15:2648–2653

Swarup R, Kargul J, Marchant A, Zadik D, Rahman A, Mills R, Yemm A, May S, Williams L, Millner P, Tsurumi S, Moore I, Napier R, Kerr ID, Bennett MJ (2004) Structure-function analysis of the presumptive Arabidopsis auxin permease AUX1. Plant Cell 16:3069–3083

Swarup R, Kramer EM, Perry P, Knox K, Leyser HMO, Haseloff J, Beemster GTS, Bhalerao R, Bennett MJ (2005) Root gravitropism requires lateral root cap and epidermal cells for transport and response to a mobile auxin signal. Nat Cell Biol 7:1057–1065

Swarup K, Benková E, Swarup R, Casimiro I, Peret B, Yang Y, Parry G, Nielsen E, De Smet I, Vanneste S, Levesque MP, Carrier D, James N, Calvo V, Ljung K, Kramer E, Roberts R, Graham N, Marillonnet S, Patel K, Jones JDG, Taylor CG, Schachtman DP, May S, Sandberg G, Benfey P, Friml J, Kerr I, Beeckman T, Laplaze L, Bennett MJ (2008) The auxin influx carrier LAX3 promotes lateral root emergence. Nat Cell Biol 10:946–954

Tanaka H, Dhonukshe P, Brewer PB, Friml J (2006) Spatiotemporal asymmetric auxin distribution: a means to coordinate plant development. Cell Mol Life Sci 63:2738–2754

Terasaka K, Blakeslee JJ, Titapiwatanakun B, Peer WA, Bandyopadhyay A, Makam SN, Lee OR, Richards EL, Murphy AS, Sato F, Yazaki K (2005) PGP4, an ATP binding cassette P-glycoprotein, catalyzes auxin transport in Arabidopsis thaliana roots. Plant Cell 17:2922–2939

Titapiwatanakun B, Murphy AS (2009) Post-transcriptional regulation of auxin transport proteins: cellular trafficking, protein phosphorylation, protein maturation, ubiquitination, and membrane composition. J Exp Bot 60:1093–1107

Titapiwatanakun B, Blakeslee JJ, Bandyopadhyay A, Yang H, Mravec J, Sauer M, Cheng Y, Adamec J, Nagashima A, Geisler M, Sakai T, Friml J, Peer WA, Murphy AS (2009) ABCB19/PGP19 stabilises PIN1 in membrane microdomains in Arabidopsis. Plant J 57:27–44

Tusnady G, Simon I (1998) Principles governing amino acid composition of integral membrane proteins: Application to topology prediction. J Mol Biol 283:489–506

Ugartechea-Chirino Y, Swarup R, Swarup K, Peret B, Whitworth M, Bennett M, Bougourd S (2010) The AUX1 LAX family of auxin influx carriers is required for the establishment of embryonic root cell organization in Arabidopsis thaliana. Ann Bot 105:277–289

Vandenbussche F, Petrášek J, Žádníková P, Hoyerová K, Pešek B, Raz V, Swarup R, Bennett M, Zažímalová E, Benková E, Van Der Straeten D (2010) The auxin influx carriers AUX1 and LAX3 are involved in auxin-ethylene interactions during apical hook development in Arabidopsis thaliana seedlings. Development 137:597–606

Vanneste S, Friml J (2009) Auxin: a trigger for change in plant development. Cell 136:1005–1016

Vanneste S, Maes L, De Smet I, Himanen K, Naudts M, Inze D, Beeckman T (2005) Auxin regulation of cell cycle and its role during lateral root initiation. Physiol Plant 123:139–146

Verrier PJ, Bird D, Buria B, Dassa E, Forestier C, Geisler M, Klein M, Kolukisaoglu U, Lee Y, Martinoia E, Murphy A, Rea PA, Samuels L, Schulz B, Spalding EP, Yazaki K, Theodoulou FL (2008) Plant ABC proteins – a unified nomenclature and updated inventory. Trends Plant Sci 13:151–159

Vicente-Agullo F, Rigas S, Desbrosses G, Dolan L, Hatzopoulos P, Grabov A (2004) Potassium carrier TRH1 is required for auxin transport in Arabidopsis roots. Plant J 40:523–535

Vieten A, Vanneste S, Wiśniewska J, Benková E, Benjamins R, Beeckman T, Luschnig C, Friml J (2005) Functional redundancy of PIN proteins is accompanied by auxindependent cross-regulation of PIN expression. Development 132:4521–4531

Vieten A, Sauer M, Brewer PB, Friml J (2007) Molecular and cellular aspects of auxin-transport-mediated development. Trends Plant Sci 12:160–168

Walch-Liu P, Forde BG (2008) Nitrate signalling mediated by the NRT1.1 nitrate transporter antagonises L-glutamate-induced changes in root architecture. Plant J 54:820–828

Wang J, Hu H, Wang G, Li J, Chen J, Wu P (2009) Expression of PIN genes in rice (Oryza sativa L.): tissue specificity and regulation by hormones. Mol Plant 2:823–831

Weijers D, Sauer M, Meurette O, Friml J, Ljung K, Sandberg G, Hooykaas P, Offringa R (2005) Maintenance of embryonic auxin distribution for apical-basal patterning by PIN-FORMED-dependent auxin transport in Arabidopsis. Plant Cell 17:2517–2526

Went FW (1974) Reflections and speculations. Annu Rev Plant Physiol Plant Mol Biol 25:1–26

Willemsen V, Friml J, Grebe M, van den Toorn A, Palme K, Scheres B (2003) Cell polarity and PIN protein positioning in Arabidopsis require STEROL METHYLTRANSFERASE1 function. Plant Cell 15:612–625

Wiśniewska J, Xu J, Seifertová D, Brewer P, Růžička K, Blilou I, Rouquie D, Scheres B, Friml J (2006) Polar PIN localization directs auxin flow in plants. Science 312:883

Wu GS, Lewis DR, Spalding EP (2007) Mutations in Arabidopsis multidrug resistance-like ABC transporters separate the roles of acropetal and basipetal auxin transport in lateral root development. Plant Cell 19:1826–1837

Wu G, Cameron JN, Ljung K, Spalding EP (2010) A role for ABCB19-mediated polar auxin transport in seedling photomorphogenesis mediated by cryptochrome 1 and phytochrome B. Plant J 62(2):179–191

Xu J, Scheres B (2005) Dissection of Arabidopsis ADP-RIBOSYLATION FACTOR 1 function in epidermal cell polarity. Plant Cell 17:525–536

Xu M, Zhu L, Shou HX, Wu P (2005) A PIN1 family gene, OsPIN1, involved in auxin-dependent adventitious root emergence and tillering in rice. Plant Cell Physiol 46:1674–1681

Yang HB, Murphy AS (2009) Functional expression and characterization of Arabidopsis ABCB, AUX 1 and PIN auxin transporters in Schizosaccharomyces pombe. Plant J 59:179–191

Yang Y, Hammes U, Taylor C, Schachtman D, Nielsen E (2006) High-affinity auxin transport by the AUX1 influx carrier protein. Curr Biol 16:1123–1127

Young G, Jack D, Smith D, Saier M (1999) The amino acid/auxin: proton symport permease family. Biochim Biophys Acta Biomembr 1415:306–322

Žádníková P, Petrášek J, Marhavý P, Raz V, Vandenbussche F, Ding Z, Schwarzerová K, Morita M, Tasaka M, Hejátko J, Van Der Straeten D, Friml J, Benková E (2010) Role of PIN-mediated auxin efflux in apical hook development of Arabidopsis thaliana. Development 137:607–617

Žárský V, Cvrčková F, Potocký M, Hála M (2009) Exocytosis and cell polarity in plants – exocyst and recycling domains. New Phytol 183:255–272

Zažímalová E, Křeček P, Skůpa P, Hoyerová K, Petrášek J (2007) Polar transport of the plant hormone auxin – the role of PIN-FORMED (PIN) proteins. Cell Mol Life Sci 64:1621–1637

Zažímalová E, Murphy A, Yang H, Hoyerová K, Hošek P (2010) Auxin transporters-why so many? Cold Spring Harb Perspect Biol 2:a001552

Zegzouti H, Anthony RG, Jahchan N, Bogre L, Christensen SK (2006) Phosphorylation and activation of PINOID by the phospholipid signaling kinase 3-phosphoinositidedependent protein kinase 1 (PDK1) in Arabidopsis. Proc Natl Acad Sci USA 103:6404–6409

Zhang J, Nodzynski T, Pěnčik A, Rolčik J, Friml J (2010) PIN phosphorylation is sufficient to mediate PIN polarity and direct auxin transport. Proc Natl Acad Sci USA 107:918–922

Zourelidou M, Müller I, Willige BC, Nill C, Jikumaru Y, Li HB, Schwechheimer C (2009) The polarly localized D6 PROTEIN KINASE is required for efficient auxin transport in Arabidopsis thaliana. Development 136:627–636

Part V
Membrane Structures and Development, Trafficking and Lipid-Transporter Interactions

V-ATPases: Rotary Engines for Transport and Traffic

Karin Schumacher and Melanie Krebs

Abstract Vacuolar-type H^+-ATPases (V-ATPase), complex molecular machines that use rotary motion produced by ATP hydrolysis to pump protons across membranes, are essential for what can be defined as cellular logistics. The flow of goods between compartments of the eukaryotic cell is achieved either by protein-mediated membrane transport or via vesicular trafficking. Over the past years, it has become increasingly clear that V-ATPases do not only energize secondary active transport across a wide variety of membranes but are also important regulators of vesicle trafficking and protein targeting. In this chapter, we thus pay particular attention to the dual function of the V-ATPase in transport and trafficking.

1 Introduction

Almost 30 years ago, research on the plant vacuolar-type H^+-ATPase (V-ATPase) was initiated by seminal reports of anion-sensitive H^+-ATPase activity associated with microsomal membranes (Hager and Helmle 1981; Walker and Leigh 1981; Churchill and Sze 1983). Since then, it has been firmly established that these complex molecular machines, which use rotary motion produced by ATP hydrolysis to pump protons across membranes, are essential for what can be defined as cellular logistics. In the business world, logistics means the management of the flow of goods between its points of origin and of consumption in order to meet the requirements of the system. In the world of a eukaryotic cell, the flow of goods between compartments is achieved either by protein-mediated membrane transport or via vesicular trafficking. Over the past years, it has become increasingly clear that V-ATPases do not only energize secondary active transport across a wide

K. Schumacher (✉) and M. Krebs
Heidelberg Institute for Plant Sciences (HIP), Universität Heidelberg, Im Neuenheimer Feld 230, 69120 Heidelberg, Germany
e-mail: karin.schumacher@hip.uni-heidelberg.de

M. Geisler and K. Venema (eds.), *Transporters and Pumps in Plant Signaling*, Signaling and Communication in Plants 7,
DOI 10.1007/978-3-642-14369-4_10,

variety of membranes but are also important regulators of vesicle trafficking and protein targeting. After reviewing what is currently known about structure, mechanism, and regulation of the plant V-ATPase, we pay particular attention to its dual function in transport and trafficking.

2 The V-ATPase Structure and Mechanism of a Complex Proton Pump

2.1 Subunit Composition

V-ATPases are found throughout all kingdoms of life and share a common ancestor with the F-ATPases and the archaebacterial A-ATPases. Like their congeners, V-ATPases are nanometer-scale molecular machines composed of two subcomplexes: The peripheral V_1 complex responsible for ATP hydrolysis, and the membrane-integral V_0 part responsible for proton translocation (see Fig. 1).

The complete inventory of V-ATPase subunits was greatly facilitated by the analysis of yeast mutants that display the conditional-lethal Vma$^-$ phenotype characterized by failure to grow at neutral pH and high calcium concentrations (Anraku et al. 1989; Kane et al. 1992). Yeast genes encoding eight V_1 subunits (A–H) and five V_0 subunits (a, c, c′, c″, d) were identified. With the exception of

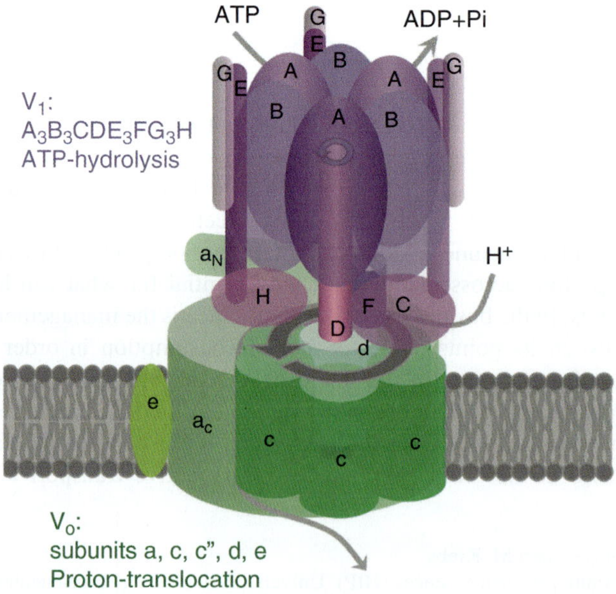

Fig. 1 Schematic model of the V-ATPase modified after Krebs et al. (2010)

subunit c′, related sequences are readily found in all eukaryotic genomes including the members of the viridiplantae sequenced so far (see Table 1). An additional protein first identified in insects and mammals was later also found in yeast where it was shown to be a genuine subunit of the vacuolar V_0 complex assigned as subunite (Ludwig et al. 1998; Merzendorfer et al. 1999; Sambade and Kane 2004).

Table 1 Comparative analysis of VHA-genes in green plants

	VHA-A	VHA-B	VHA-C	VHA-D	VHA-E	VHA-F	VHA-G	VHA-H
M. esculenta	2	4	2	2	2	1	4	2
R. communis	1	2	1	2	1	1	3	1
P. trichocarpa	2	2	2	1	3	1	5	2
M. truncatula	1	1	1	1	1	1	1	1
G. max	2	4	4	2	7	2	5	2
C. sativus	1	1	1	1	1	1	2	1
A. thaliana	1	3	1	1	3	1	3	1
A. lyrata	1	3	1	1	3	1	2	1
C. papaya	1	2	1	1	2	1	1	1
V. vinifera	2	2	2	1	2	1	4	2
M. guttatus	2	2	1	1	3	1	2	2
S. bicolor	2	2	2	2	3	2	2	2
Z. mays	2	2	3	3	4	2	1	1
O. sativa	2	2	1	1	3	2	2	1
B. distachyon	2	5	1	1	3	2	2	1
S. moellendorffii	1	1	1	1	1	1	1	1
P. patens	3	3	2	3	4	1	2	2
C. reinhartdii	1	1	1	1	1	1	1	1

	VHA-a	VHA-c	VHA-c″	VHA-d	VHA-e
M. esculenta	4	7	2	3	2
R. communis	3	4	1	1	1
P. trichocarpa	4	6	2	2	2
M. truncatula	4	5	1	2	1
G. max	8	10	2	2	4
C. sativus	3	3	2	1	2
A. thaliana	3	5	2	2	2
A. lyrata	3	5	2	2	2
C. papaya	4	3	2	1	1
V. vinifera	3	3	2	1	2
M. guttatus	3	3	1	2	2
S. bicolor	2	3	1	1	1
Z. mays	2	4	2	1	2
O. sativa	3	4	1	1	1
B. distachyon	3	4	1	1	1
S. moellendorffii	2	2	1	1	1
P. patens	5	5	4	1	2
C. reinhartdii	3	1	1	1	1

Represented are numbers of orthologs present in the genomes of *Manihot esculenta, Ricinus communis, Populus trichocarpa, Medicago truncatula, Glycine max, Cucumis sativus* (Eurosid I); *Arabidopsis thaliana, Arabidopsis lyrata, Carica papaya* (Eurosid II); *Vitis vinifera, Mimulus guttatus* (Eudicot); *Sorghum bicolor, Zea mays, Oryza sativa, Brachypodium distachyon* (Monocots); *Selaginella moellendorffii* (Lycophyte), *Physcomitrella patens* (Bryophyte) and *Chlamydomonas reinhartdii* (Chlorophyte)

The presence of genes coding proteins with similarity to subunits identified in other organisms does of course not proof that they are integral subunits of the plant V-ATPase. Indeed, after purification of the tobacco leaf cell tonoplast enzyme only subunits A, B, C, D, E, F, G, c, and d could be identified by matrix-assisted laser-desorption ionization mass spectrometry (MALDI-MS) and amino acid sequencing (Drobny et al. 2002). More recently, several analyses of the tonoplast proteome of Arabidopsis have been published and with the exception of c″ and e, all other subunits were identified (Carter et al. 2004; Konishi et al. 2005; Jaquinod et al. 2007). Final proof that the small hydrophobic proteins c″ and e are part of the plant V-ATPase complex is thus still missing. Based on transient expression and FRET analysis of fluorescent fusion-proteins, it has been reported that VHA-c″ and VHA-e2 are ER-localized and thus should not be considered as subunits of the final V-ATPase complex. Similarly, VHA-e1 was reported to be TGN-localized (Seidel et al. 2008). In yeast, GFP-fusions of subunits c, c′, and c″ have also been shown to be ER-localized, but as the native proteins are found in the vacuolar holocomplex, the ER-localization could also be an artifact of the GFP-fusion (Huh et al. 2003).

Many subunits of V-ATPase have sequence similarity to their counterparts in the F-ATPase; however, subunits C, H, and e are recent additions to the eukaryotic V-ATPase which were also absent in the more primitive bacterial V-ATPases. Compared to the F-ATPase or ATP-Synthase that has in large parts been crystallized, the structure of the V-ATPase holocomplex is still less well defined. However, low-resolution maps of V-ATPases from all eukaryotic kingdoms including plants obtained by electron microscopy combined with atomic structures of individual subunits determined by X-ray crystallography already indicated that although the general structure of V- and F-ATPases is similar, V-ATPases are much more complex (Muench et al. 2009). Most recently, cryo-electron microscopy followed by 3D reconstruction of the intact pumps of *Manduca sexta* (Diepholz et al. 2008) as well as yeast revealed the presence of three peripheral stators and a central stalk connecting the V_1 catalytic domain and the V_0 membrane domain (Muench et al. 2009).

The overall similar general structure of F- and V-ATPase evokes that they also share a common mechanical design (Diepholz et al. 2008). Indeed, the elegant experiments that visualized rotation of the bacterial F-ATPase by attachment of labeled actin-filaments have been successfully repeated for eukaryotic V-ATPases (Sambongi et al. 1999; Drory and Nelson 2006), but the rotation kinetics of V_1 and F_1 turned out to be different (Hirata et al. 2003). Rotational catalysis requires functional coupling of the hydrolysis motor V_1 and the proton turbine V_0, that are connected by a central shaft (D, F, d), and the peripheral stator elements (C, E, G, H, and a). ATP-hydrolysis at the catalytic nucleotide binding sites found at the interfaces of the hexameric arrangement of V_1 subunits A and B induces a rotation of the central shaft (subunits d, D, F), which is conveyed to the proteolipid ring (subunits c, c′, c″). Rotation of the c-ring against subunit a, which provides the second part of the proton channel, drives protons across the membrane. Each proteolipid subunit contains an essential glutamic acid residue, located in TM4 on c and c′ and TM3 on c″, that interacts with subunit a and undergoes reversible protonation during proton transport (Imamura et al. 2005). Co-rotation of the two domains is prevented by a static connection between V_1 and V_0 that is provided by subunits a, E, G, C, and H (Qi et al. 2007; Grüber and Marshansky 2008; Saroussi and Nelson 2009; Nakanishi-Matsui et al. 2010).

The complex stator is proposed to be part of a "gearbox" mechanism that controls physical coupling between the rotor axle and the proton-translocating proteolipid-ring. Such a transmission would provide the structural basis for the variable ATP/H^+-coupling ratio typical of V-ATPases. Using patch clamp analysis, the coupling ratio of the red bead V-ATPase has been determined thermodynamically and was found to range between 1.75 and 3.28 H^+ per ATP and was strictly dependent on cytosolic and vacuolar pH (Davies et al. 1994).

The task of pumping protons across a membrane is also achieved by much simpler and less costly enzymes like the P-type H^+-ATPase and the H^+-PPase. It is therefore tempting to speculate that the V-ATPase has been kept in business by features like its variable coupling rate and the presence of multiple subunit isoforms. The regulatory potential of such a complex machine may provide the flexible framework that has allowed adapting this ancient rotary engine to the requirements of a modern eukaryotic cell. In the following sections, we will thus review what is currently known about different levels of V-ATPase regulation.

3 Levels of V-ATPase Regulation

3.1 V-ATPase Isoforms: The Even More Complex Pump of Higher Plants

The awesome power of yeast genetics has clearly sped up identification of plant V-ATPase subunit composition. However, the fact that in plants, like in all higher eukaryotes, most of the V-ATPase subunits are encoded by gene families, adds an additional layer of complexity that cannot be addressed in the yeast model. The first full complement of V-ATPase genes was established for Arabidopsis, where the 13 subunits are encoded by a total of 27 VHA-genes (Sze et al. 2002). Since then, genome sequences from most major branches of the viridiplantae kingdom became available (http://www.phytozome.net) providing the opportunity for comparative genomics of V-ATPase subunit isoforms (see Table 1). Total numbers range from 14 VHA-genes in Chlamydomonas up to 54 in Glycine max (soybean). Both Arabidopsis and soybean have undergone at least two rounds of genome duplication (Sze et al. 2002; Seoighe and Gehring 2004; Schmutz et al. 2010). In contrast to Arabidopsis, in which many duplicated VHA-genes have vanished (Sze et al. 2002), soybean seems to have kept most of its VHA-genes as is also true for most other duplicated genes. Remarkably, with the exception of VHA-a and VHA-c, a single copy of all other subunits seems to be sufficient in one or the other higher plant species indicating that isoform specialization will in many cases be restricted to a certain subgroup or even species-specific.

VHA-A and VHA-c belong to the most highly conserved eukaryotic proteins and it is therefore not surprising that the isoforms found within a species show very little, if any, divergence at the protein level. In the case of the two Arabidopsis genes VHA-c1 and VHA-c3, identical proteins are encoded but whereas VHA-c1 is expressed ubiquitously, VHA-c3 expression is limited to root caps (Padmanaban et al. 2004). Similarly, VHA-A1 of tomato is found ubiquitously whereas VHA-A2 is restricted to roots and fruits (Bageshwar et al. 2005). In such cases, it seems more likely that the underlying gene duplications were followed by changes in regulatory sequences that provided the opportunity to evolve more complex and adaptive expression patterns. However, although the two tomato VHA-A isoforms are 95% identical, in the tissues in which they are coexpressed, they do not occur in the same V-ATPase complexes, indicating differences in protein function (Bageshwar et al. 2005). Moreover, the presence of organelle-specific VHA-A isoforms was reported in carrot (Gogarten et al. 1992), underlining that functional specialization found in one species cannot be generalized unless the particular isoform can be traced to a common ancestor. Similarly, functional differentiation has also been reported for Arabidopsis VHA-B1 and VHA-B3 (Cho et al. 2006) that date back to a recent duplication in the Arabidopsis genome and are 98% identical (Cho et al. 2006; Sze et al. 2002).

In the case of the VHA-E isoforms a distinct clade comprising sequences from both mono- and dicotyledonous plants was identified that contained Arabidopsis *VHA-E2*. This isoform is exclusively expressed during male gametophyte development (Dettmer et al. 2005, 2010). In silico, analysis revealed that, like *AtVHA-E2*, the two monocot genes represented in this clade, *OsVHA-E2* and *SbVHA-E2*, are expressed only during pollen development. It thus seems likely that the whole clade represents a specialized VHA-E isoform dedicated to male gametophyte development. However, *VHA-E2*-like sequences are missing in several genomes and a null-allele of *VHA-E2* does not affect pollen development indicating that this isoform is not essential for pollen development (Dettmer et al. 2010). The duplication that gave rise to the two other Arabidopsis isogenes *VHA-E1* and *VHA-E3* is restricted to the Brassicaceae. Whereas *VHA-E1* is expressed ubiquitously, *VHA-E3* expression is restricted to the epidermal cell layer with the exception of guard cells and is highly regulated by the phytohormone methyl-jasmonate (Dettmer et al. 2010). If the different expression patterns are linked to divergent properties of the proteins remains to be determined. Interestingly, V-ATPases isolated from pea epicotyls contained two isoforms of VHA-E and showed differences in the kinetics of ATP hydrolysis (Dettmer et al. 2010).

Although it has been pointed out early on that the V-ATPase is found in many compartments of the plant endomembrane system (Herman et al. 1994), it is still frequently perceived as a purely "vacuolar" enzyme. This can in part be explained by the fact that, in the absence of functional evidence, V-ATPases in other compartments could simply be in transit to their final destination at the tonoplast. In yeast, the two isoforms of subunit a, Vph1p (Vacuolar acidification-defective 1) and Stv1p (Similar To VPH) are differentially localized. Whereas Vph1p resides at the tonoplast, Stv1p cycles continuously between a late Golgi compartment and prevacuolar endosomes (Manolson et al. 1992; Kawamura et al. 2000). Moreover, the two yeast isoforms confer different stability and kinetic properties to the holocomplex (Manolson et al. 1994). A systematic analysis of the subcellular localization of the three Arabidopsis VHA-a isoforms using GFP-fusion proteins expressed under the endogenous promoters revealed that they are responsible for differential localization. Indeed, the distribution in Arabidopsis is very similar to the one found in yeast. VHA-a2 and VHA-a3 are found at the tonoplast, whereas VHA-a1 is localized in the trans-Golgi network (TGN, see Fig. 2). As in yeast Stv1p, the targeting information resides in the N-terminal cytosolic domain (Dettmer et al. 2006). Differential localization of subunit a isoforms is also found in mammals (Toyomura et al. 2003) and in Paramecium where 17 (!) isoforms of subunit a have been found in seven different cellular locations (Wassmer et al. 2006).

As indicated by phylogenetic analysis, differential localization of V-ATPase complexes via the subunit a seems to have arisen independently in yeast and plants. As shown in Fig. 3, the VHA-a1 clade includes mono and dicot sequences but has no members from mosses or algae although they also have multiple VHA-a isoforms. Based on the fact that an antibody against one of the VHA-a isoforms of Mesembryanthemum labeled the ER but failed to detect tonoplast V-ATPase complexes in maize root cells, an ER-specific VHA-a isoform has been proposed in maize (Kluge et al. 2004). It thus remains to be demonstrated where VHA-a1 from monocots is localized. However, given the important function of the V-ATPase in the TGN of Arabidopsis (see below) it seems likely that a dedicated TGN-isoform is also present in monocots.

Differences in V-ATPase activity between plant organs and tissues or after changes in the nutritional status have been reported and could be correlated either with changes in the overall structure of the complex or with the presence of individual protein isoforms (Dietz et al. 2001). Determining how these isoforms affect the kinetic and regulatory properties of the holocomplex will remain challenging, as long as it relies exclusively on measuring enzyme activity in individual tissues or at different time-points. The analysis of mutants lacking individual isoforms could greatly facilitate this task.

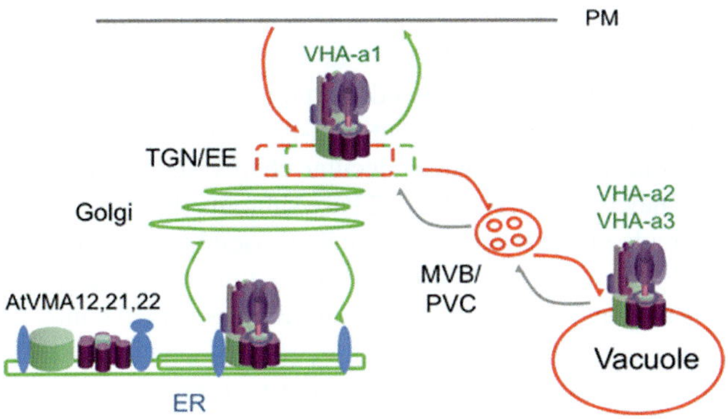

Fig. 2 Cartoon depicting assembly and localization of the Arabidopsis V-ATPase

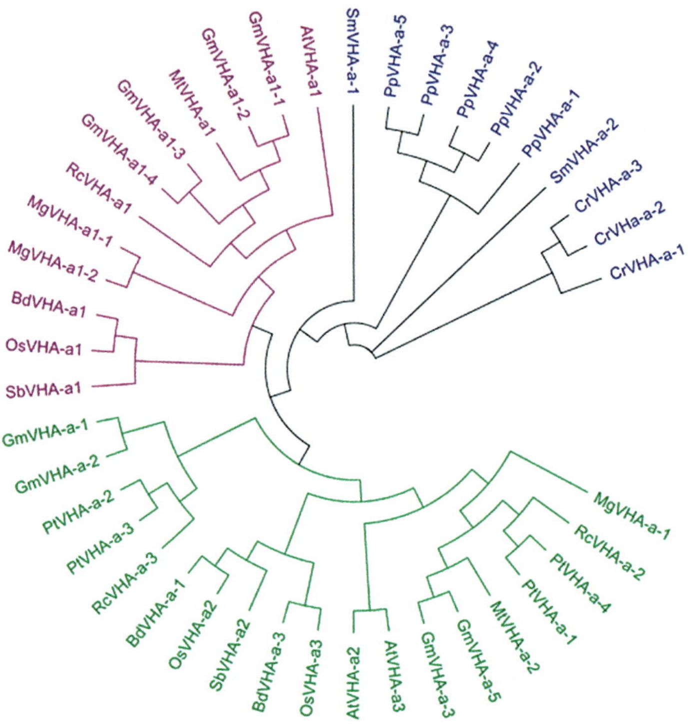

Fig. 3 Phylogenetic tree of VHA-a proteins from gree plants. Neighbor joining tree for VHA-a sequences of *M. esculenta* (Me), *R. comunis* (Rc) , *P. trichocarpa* (Pt), *M. truncatula* (Mt), *G. max* (Gm), *C. sativus* (Cs), *A. thaliana* (At), *A. lyrata* (Al), *C. papaya* (Cp), *V. vinifera* (Vv), *M. guttatus* (Mg), *S. bicolor* (Sb), *Z. mays* (Zm), *O. sativa* (Os), *B. distachyon* (Bd), *S. moellendorffii* (Sm), *P. patens* (Pp), *C. reinhartdii* (Cr)

3.2 Assembly of the V-ATPase: Putting the Pieces Together

The potential to form V-ATPases with different subunit composition raises the question how these complexes are formed. In plants and other higher eukaryotes the assembly process is largely unaddressed. A monoclonal antibody against VHA-A form oat was shown to precipitate the entire V-ATPase complex as well as calnexin and BiP, suggesting that these two molecular chaperones are involved in folding and assembly of newly synthesized V-ATPase subunits (Li et al. 1998). In addition, these results provide rather strong evidence that assembly of the two subcomplexes V_1 and V_0 takes place at the ER. The pH in the ER lumen is generally assumed to be neutral and it is thus an open question if such fully assembled complexes at the ER are active.

In contrast to V_1 that is generally assumed to form via self-assembly in the cytosol (Kane 1999), the membrane-integral V_0 complex of yeast is put together in the ER-membrane only when a set of dedicated chaperones is present. The four

V-ATPase assembly factors (Vma12p, Vma21p, Vma22p, and Pkr1p) are all localized to the ER membrane, and are not part of the final V_0V_1 holocomplex (Kane 1999; Graham et al. 2003; Davis-Kaplan et al. 2006). Vma12p is a 25 kDa protein with two transmembrane domains that recruits Vma22p, a 21 kDa hydrophilic protein to the ER-membrane (Hirata et al. 1993; Jackson and Stevens 1997; Davis-Kaplan et al. 2006). Vma12p and Vma22p form a complex and transiently interact with subunit a of the V_0 subcomplex during assembly in the ER (Hill and Stevens 1995). Vma21p encodes an 8.5 kDa protein with two predicted transmembrane domains that interacts with the "c ring" of the V_0 subcomplex. After V_0 assembly is complete, Vma21p acts as an export chaperone moving along with the V_0 subcomplex to the Golgi complex but is retrieved back to the ER via a di-lysine motif at the C terminus for more rounds of V_0 subcomplex assembly (Malkus et al. 2004).

Despite the high degree of sequence conservation for the V-ATPase subunits, orthologs of the yeast assembly factors are not readily identified in the genomes of higher eukaryotes. The identification of functional VMA21p-orthologs, first from Arabidopsis (Neubert et al. 2008) and later form humans (Ramachandran et al. 2009) provided evidence that the V_0-assembly machinery is conserved although the individual proteins have substantially diverged. Potential orthologs of VMA22p and VMA12p have also been identified in Arabidopsis (our own unpublished results) and it now remains to be determined if this machinery is involved in assembling V_0 complexes with specific subunit composition.

3.3 *Reversible Dissociation: Pulling the Complex Apart*

Reversible dissociation of the V-ATPase complex represents a fast and efficient mechanism of regulating acidification of intracellular compartments in vivo. In yeast, dissociation occurs rapidly in response to glucose depletion and might thus help to prevent ATP-depletion. Dissociation of V_1 and V_0 is not dependent on conventional glucose-signaling pathways and requires an intact microtubular network (Parra and Kane 1998; Xu and Forgac 2001). Reassembly is microtubule- and protein-biosynthesis independent but involves the so-called RAVE (regulator of the H^+-ATPase of vacuolar and endosomal membrane)- complex, which is also involved in the normal assembly pathway of the V-ATPase (Seol et al. 2001; Smardon et al. 2002). The yet unknown dissociation signal is proposed to be mediated via subunit C (Smardon et al. 2002; Smardon and Kane 2007; Voss et al. 2007). whereas subunit H is implicated in the silencing of ATPase activity in the released V_1 (Parra et al. 2000).

Reversible assembly occurs not only in yeast but also in insects and mammalian cells (Parra et al. 2000; Beyenbach and Wieczorek 2006; Kane 2006). Although the mechanism seems to be conserved among higher eukaryotes, no direct evidence for its existence in plants has been reported. However, the facts that RAVE-related proteins are present in plant genomes (Kane and Smardon 2003) and significant amounts of free V_1 complexes are found in the cytosol of Arabidopsis (Schumacher et al. 1999) might be taken as leads worth following. Moreover, a recent study reported that the two glycolytic enzymes enolase and aldolase are associated with the tonoplast V-ATPase of *Mesembryanthemum crystallinum* (Barkla et al. 2009). In yeast, the physical interaction between aldolase and V-ATPase is essential for assembly and activity of the proton pump indicating that a direct physical link to glycolytic enzymes is also involved in the glucose-regulated reversible assembly mechanism (Lu et al. 2004, 2007).

3.4 Redox Regulation: An On- and Off-Switch?

Based on in vitro studies, V-ATPases from mammals, yeast, and plants are subject to oxidative inactivation and activity can be recovered by reducing agents (Feng and Forgac 1992a, b, 1994). Three cysteine residues are conserved in subunit A of all eukaryotes, and formation of an intramolecular disulfide bond between Cysteine 254, which resides within the nucleotide binding P-loop motif, and Cysteine 532 of subunit A was suggested to be responsible for oxidative inhibition whereas disulfide formation between Cys279 and Cys535 is hypothesized to recover the activity and to represent the active state of the V-ATPase (Feng and Forgac 1992a, b, 1994; Dietz et al. 2001). In plants, redox-dependent activity changes have been reported to involve disulfide bridge formation not only in VHA-A but also in the stalk subunit VHA-E (Tavakoli et al. 2001). To elucidate the in vivo relevance of redox regulation of the plant V-ATPase, the conserved cysteines in VHA-A were replaced by site-directed mutagenesis and the resulting constructs were assayed for complementation of a null allele of Arabidopsis *VHA-A* (Dettmer et al. 2005), see below. Treatments with oxidizing agents revealed a complete absence of oxidative inhibition in the transgenic line Cys256Ser whereas oxidative inhibition was unaffected in Cys279Ser and Cys535Ser lines indicating that intramolecular disulfide bridges in VHA-A are not involved in oxidative inhibition (our own unpublished results).

3.5 Fine-Tuning by Protein Phosphorylation?

Given that between 1/10 and ½ of the protein repertoire of a cell is phosphorylated at some point during its lifetime, it is not surprising that studies of the phosphoproteome of tonoplast proteins identified in vivo phosphorylation sites in several V-ATPase subunits [A, B, C, D, H, and a; (Endler et al. 2009)]. In vitro phosphorylation of a plant V-ATPase has been demonstrated in two cases. Phosphorylation of VHA-A by an unknown tonoplast-associated protein kinase has been implicated in binding of 14-3-3 proteins during blue light activation of the V-ATPase from etiolated barley seedlings (Klychnikov et al. 2007). The respective phosphorylation site has not been identified and in vivo phosphorylation needs to be confirmed.

WNK8, a protein kinase identified by its specific interaction with VHA-C, phosphorylates this subunit at multiple sites as well as other V_1-subunits (Hong-Hermesdorf et al. 2006) but the in vivo relevance of this interaction remains to be addressed. Conversely, binding of SOS2/CIPK24, a member of the Calcineurin B-like interacting protein kinase family, to the V-ATPase has been demonstrated in vivo. SOS2/CIPK24 is an important regulator of salt tolerance, and its interaction with the V-ATPase is enhanced under salt stress and proton transport activity is reduced in *sos2* mutant tonoplast vesicles (Batelli et al. 2007). Although these data strongly suggest regulation of the V-ATPase, direct evidence for SOS2/CIPK24-mediated phosphorylation was not provided in this study.

4 Genetic Studies: The Plant V-ATPase Is Essential: But Where?

V-ATPase deficient mutants, identified in a number of multicellular eukaryotes ranging from *Caenorhabditis elegans* to mouse, have shown that V-ATPase activity is essential for embryonic or larval development in animals (Allan et al. 2005; Hinton et al. 2009). However, unlike animals, plants possess a second endomembrane proton pump, the vacuolar H^+-PPase (V-PPase). Reverse genetics in Arabidopsis has provided unequivocal evidence that, despite the presence of the H^+-PPase, the V-ATPase is essential. A T-DNA insertion allele of VHA-A, the single copy gene encoding the catalytic subunit of the Arabidopsis V-ATPase, causes complete male and partial female gametophytic lethality. Electron microscopy of mutant pollen grains revealed changes in the morphology of Golgi stacks and Golgi-derived vesicles around whereas vacuoles were unaffected in *vha-A* pollen (Dettmer et al. 2005). Embryos lacking the constitutively expressed isoform VHA-E1 do not develop past the globular stage and display a range of defects including multinucleate cells, cell wall stubs, and abnormal division planes characteristic of cytokinesis-defective mutants. Again, Golgi-stacks in *vha-E1(tuff)*-mutants displayed an abnormal organization whereas vacuolar morphology was not affected (Strompen et al. 2005).

The *det3* mutant, originally identified as a potential negative regulator of photomorphogenesis because of the short hypocotyls of dark-grown seedlings, is caused by a weak allele of *VHA-C* (Schumacher et al. 1999). Although *det3* mutants still have about 40% of total V-ATPase activity found in the wild type, cell expansion is severely reduced leading to severe dwarfism in adult plants. Similarly, RNAi-mediated inhibition of VHA-c1 in Arabidopsis or antisense-mediated inhibition of VHA-A in carrot caused reduced cell expansion (Gogarten et al. 1992; Padmanaban et al. 2004). Given the dual localization of the Arabidopsis V-ATPase in the TGN and the tonoplast, reduced cell expansion could be caused by either reduced turgor pressure due to a lack of osmolyte transport into the vacuole or reduced cell wall synthesis brought about by a defect in the secretory system.

Indirect evidence in favor of the latter came from the analysis of plants lacking activity of vacuolar H^{+}/Ca^{2+}-antiporters. In *cax1* mutants and *cax1/cax3* double mutants, V-ATPase activity at the tonoplast is, for yet unknown reasons, reduced to similar levels as in *det3* (Cheng et al. 2003, 2005). However, growth of *det3* is much more severely reduced, indicating that tonoplast V-ATPase activity is not the limiting factor. Indeed, neither in *vha-a3* single nor in *vha-a2 vha-a3* double mutants that completely lack tonoplast V-ATPase is cell expansion as severely affected as in *det3*. In contrast, inducible RNAi of *VHA-a1* expression inhibits hypocotyl elongation and causes the characteristic changes in Golgi morphology previously described for cells treated with the V-ATPase inhibitor Concanamycin A (Dettmer et al. 2005, 2006), demonstrating clearly that it is the V-ATPase activity in the TGN that limits cell expansion.

Interestingly, a null-allele of *ClC-d*, a gene for a TGN-localized anion transporter, renders Arabidopsis cell expansion hypersensitive to ConcA (Fecht-Bartenbach et al. 2007). This result suggests that plant ClC-transporters might provide the electric shunt for the efficient acidification of intracellular compartments by the V-ATPase as it has been previously demonstrated for the mammalian ClC-family (Jentsch 2007).

Finally, null-alleles of *VHA-a1* are lethal (our own unpublished results) and it can thus be concluded that the V-ATPase is strictly required to establish and maintain a functional secretory system.

4.1 The V-ATPase in the TGN/EE: Essential for Endocytic and Secretory Traffic

The use of VHA-a1-GFP as a marker protein has revealed that the TGN not only is the central sorting station of the secretory pathway but also functions as an early endosome (Dettmer et al. 2006). The TGN/EE thus represents the main hub for protein trafficking in plant cells and it is of great interest to understand the function of the V-ATPase in this highly dynamic and independent compartment (Viotti et al. 2010). Inhibition of the V-ATPase by ConcA interferes with secretory and endocytic trafficking of various cargo molecules and leads to a loss of TGN-identity at the ultrastructural level but the molecular basis of these effects is not understood (Matsuoka et al. 1997; Dettmer et al. 2006).

Luminal acidification has long been known to be important for receptor–ligand dissociation or activation of hydrolytic enzymes. However, more direct connections between the V-ATPase and the vesicle trafficking machinery are becoming apparent and provide leads for further studies of the V-ATPase in the plant TGN/EE. Membrane-recruitment of components of ARF and ARF-GEF proteins can depend on luminal acidification (Aniento et al. 1996; Maranda et al. 2001) but it was long unknown, how this pH information is transmitted across the membrane. Recently, it was shown that subunits of the mammalian V-ATPase interact directly and in a luminal pH-dependent manner with proteins regulating endocytic trafficking (Hurtado-Lorenzo et al. 2006). The V-ATPase can thus function as a pH-sensor that communicates luminal pH to cytosolic proteins. Remarkably, both Wnt and Notch signaling involve specific adaptor proteins that mediate regulation of endosomal V-ATPase by the receptors. In both cases, V-ATPase-mediated acidification was shown to be essential for signaling (Yan et al. 2009; Cruciat et al. 2010). Last but not least, a V-ATPase independent function for V_0 subunits in membrane fusion that was first demonstrated for homotypic vacuole fusion in yeast has also been identified during synaptic vesicle fusion in Drosophila and apical secretion of exosomes in *C. elegans*, indicating that it could be a common feature of membrane fusion events (Hiesinger et al. 2005; Liegeois et al. 2006; Bayer et al. 2003).

4.2 The Vacuolar V-ATPase: Important but Not Essential

A unique feature of plant cells is the presence of a large and multifunctional acidic vacuole. All functions associated with this unique compartment require massive fluxes of ions and metabolites that are channeled by a battery of vacuolar transport proteins (Martinoia et al. 2007). Transport across the delimiting membrane, the tonoplast, is energized by two proton pumps, the V-ATPase and the vacuolar H^+-pyrophosphatase (V-PPase) (Gaxiola et al. 2007). The combined activity of the two pumps creates the proton gradient and the membrane potential that is used to transport compounds against their concentration or electrochemical gradients (Hedrich et al. 1989; Martinoia et al. 2007). Both enzymes are among the most abundant tonoplast proteins (Carter et al. 2004; Jaquinod et al. 2007) indicating that the amount of energy invested into vacuolar transport is substantial. PP_i is a by-product of several biosynthetic processes and it has therefore been argued that the V-PPase is the predominant proton pump in the vacuoles of young, growing cells (Nakanishi and Maeshima 1998). The V-PPase has also been discussed as a back-up system for the V-ATPase under ATP-limiting conditions like anoxia and cold stress (Maeshima 2000). Due to their different energy sources it is generally assumed that the combined action of the two enzymes enables plants to maintain transport into the vacuole even under stressful conditions.

The *vha-a2 vha-a3* double mutant, which lacks the two tonoplast-localized isoforms of the membrane-integral V-ATPase subunit VHA-a, provides the unique opportunity to dissect the relative contributions of V-ATPase and V-PPase during development and under different growth conditions. The primary phenotype of the *vha-a2 vha-a3* mutant is a more alkaline vacuolar pH. Compared to vacuoles in wild-type root cells which have a pH of 5.9, *vha-a2 vha-a3* vacuoles were found shifted to pH 6.4 (Krebs et al. 2010). Activity of the V-PPase was not increased indicating that a lack of tonoplast V-ATPase activity does not lead to increased V-PPase activity. Interestingly, vacuolar pH in the *avp1* knock-out mutant was found to be increased by only 0.2–0.3 units (Li et al. 2005). Assuming a cytosolic pH of 7.4, the proton concentration in *vha-a2 vha-a3* mutant vacuoles is still tenfold higher than in the cytosol. It remains to be determined if the remaining proton gradient is solely due to the activity of the V-PPase or if TGN/EE derived vesicles acidified by the activity of VHA-a1 contribute to vacuolar acidification. In any case, activity of the V-PPase is obviously sufficient during gametophyte, embryo, and early seedling development and becomes limiting only during vegetative and reproductive growth.

A striking feature of the *vha-a2 vha-a3* mutant is its day-length-dependent growth retardation caused by a severe shift in nitrogen-metabolism (Krebs et al. 2010). Nitrate, a major plant nutrient, is accumulated and stored in the vacuole from where it can be retrieved according to metabolic demands (Miller et al. 2007). Uptake into the vacuole is achieved by proton-coupled anion transport (Schumaker and Sze 1987), which has been shown to be mediated by AtCLCa, a member of the ClC-family of anion channels that acts as a NO_3^-/H^+ exchanger (De Angeli et al. 2006). In the *vha-a2 vha-a3* mutant nitrate content is strongly reduced whereas nitrate reductase activity and glutamine levels are increased in the *vha-a2 vha-a3* mutant. Nitrate assimilation is a light-dependent and highly energy-consuming process and thus photosynthesis-derived reduction equivalents seem to be the growth-limiting factor when the mutant is grown under restricted light periods (Krebs et al. 2010). Other ions that are transported into the vacuole via proton-coupled transport include calcium, zinc, and sodium (Martinoia et al. 2007). In accordance with the reduced proton gradient, the *vha-a2 vha-a3* mutant has reduced calcium content and is less zinc tolerant than the wild type. Surprisingly, it does not show enhanced salt sensitivity implying that the vacuolar V-ATPase may not be limiting for vacuolar sodium accumulation (Krebs et al. 2010).

4.3 The V-ATPase and Its Role in Salt Tolerance

The vacuolar Na^+/H^+-antiport system is assumed to be an important component of the system that removes sodium from the cytosol. Sodium regulates the expression of different subunits of the V-ATPase from both halophytes and glycophytes at the transcript and protein levels and transport activity has been shown to increase under salt stress (Dietz et al. 2001; Kluge et al. 2003). Moreover, plants with reduced V-ATPase activity, like *det3* and RNAi-lines for *vha-c3*, are more salt sensitive (Batelli et al. 2007; Padmanaban et al. 2004). How is it possible that under the same conditions the *vha-a2 vha-a3* mutant is not affected? The enhanced salt sensitivity in RNAi-lines for *VHA-a1* resolves this apparent contradiction and points to an important function of the endosomal V-ATPase in salt tolerance (Krebs et al. 2010). The important question that now needs to be addressed is if the endosomal system simply is a target of sodium toxicity as it has been shown in yeast (Hernandez et al. 2009) or if endosomal Na^+/H^+-antiporters might contribute to removal of sodium from the cytosol.

5 Conclusion

pH is one of the most fundamental cellular parameters that affects most biological processes. Moreover, the chemiosmotic circuits in plants depend on proton gradients, thus forming an inseparable connection between secondary active transport and pH control. Due to this dual function, proton pumps such as the V-ATPase are of pivotal importance, and substantial progress in understanding their structure, function, and regulation has been achieved. We are beginning to see how the V-ATPase is integrated in the diverse cellular and metabolic networks and many tools have been established that will allow to dissect the underlying molecular mechanisms. What is now urgently needed are tools to monitor pH and V-ATPase activity in vivo, in different compartments and in a noninvasive manner.

References

Allan AK, Du J, Davies SA, Dow JA (2005) Genome-wide survey of V-ATPase genes in Drosophila reveals a conserved renal phenotype for lethal alleles. Physiol Genomics 22:128–138

Aniento F, Gu F, Parton RG, Gruenberg J (1996) An endosomal beta COP is involved in the pH-dependent formation of transport vesicles destined for late endosomes. J Cell Biol 133:29–41

Anraku Y, Umemoto N, Hirata R, Wada Y (1989) Structure and function of the yeast vacuolar membrane proton ATPase. J Bioenerg Biomembr 21:589–603

Bageshwar UK, Taneja-Bageshwar S, Moharram HM, Binzel ML (2005) Two isoforms of the A subunit of the vacuolar H(+)-ATPase in Lycopersicon esculentum: highly similar proteins but divergent patterns of tissue localization. Planta 220:632–643

Barkla BJ, Vera-Estrella R, Hernández-Coronado M, Pantoja O (2009) Quantitative proteomics of the tonoplast reveals a role for glycolytic enzymes in salt tolerance. Plant Cell 21:4044–4058

Batelli G, Verslues PE, Agius F, Qiu Q, Fujii H, Pan S, Schumaker KS, Grillo S, Zhu J-K (2007) SOS2 promotes salt tolerance in part by interacting with the vacuolar H+-ATPase and upregulating its transport activity. Mol Cell Biol 27:7781–7790

Bayer MJ, Reese C, Buhler S, Peters C, Mayer A (2003) Vacuole membrane fusion: V0 functions after trans-SNARE pairing and is coupled to the Ca2+-releasing channel. J Cell Biol 162:211–222

Beyenbach KW, Wieczorek H (2006) The V-type H+ ATPase: molecular structure and function, physiological roles and regulation. J Exp Biol 209:577–589

Carter C, Pan S, Zouhar J, Avila EL, Girke T, Raikhel NV (2004) The vegetative vacuole proteome of Arabidopsis thaliana reveals predicted and unexpected proteins. Plant Cell 16:3285–3303

Cheng NH, Pittman JK, Barkla BJ, Shigaki T, Hirschi KD (2003) The Arabidopsis cax1 mutant exhibits impaired ion homeostasis, development, and hormonal responses and reveals interplay among vacuolar transporters. Plant Cell 15:347–364

Cheng NH, Pittman JK, Shigaki T, Lachmansingh J, LeClere S, Lahner B, Salt DE, Hirschi KD (2005) Functional association of Arabidopsis CAX1 and CAX3 is required for normal growth and ion homeostasis. Plant Physiol 138:2048–2060

Cho YH, Yoo SD, Sheen J (2006) Regulatory functions of nuclear hexokinase1 complex in glucose signaling. Cell 127:579–589

Churchill KA, Sze H (1983) Anion-sensitive, H-pumping ATPase in membrane vesicles from oat roots. Plant Physiol 71:610–617

Cruciat C-M, Ohkawara B, Acebron SP, Karaulanov E, Reinhard C, Ingelfinger D, Boutros M, Niehrs C (2010) Requirement of prorenin receptor and vacuolar H+-ATPase-mediated acidification for Wnt signaling. Science 327:459–463

Davies JM, Hunt I, Sanders D (1994) Vacuolar H(+)-pumping ATPase variable transport coupling ratio controlled by pH. Proc Natl Acad Sci USA 91:8547–8551

Davis-Kaplan SR, Compton MA, Flannery AR, Ward DM, Kaplan J, Stevens TH, Graham LA (2006) PKR1 encodes an assembly factor for the yeast V-type ATPase. J Biol Chem 281:32025–32035

De Angeli A, Monachello D, Ephritikhine G, Frachisse JM, Thomine S, Gambale F, Barbier-Brygoo H (2006) The nitrate/proton antiporter AtCLCa mediates nitrate accumulation in plant vacuoles. Nature 442:939–942

Dettmer J, Schubert D, Calvo-Weimar O, Stierhof Y-D, Schmidt R, Schumacher K (2005) Essential role of the V-ATPase in male gametophyte development. Plant J 41:117–124

Dettmer J, Hong-Hermesdorf A, Stierhof YD, Schumacher K (2006) Vacuolar H+-ATPase activity is required for endocytic and secretory trafficking in Arabidopsis. Plant Cell 18:715–730

Dettmer J, Liu T-Y, Schumacher K (2010) Functional analysis of Arabidopsis V-ATPase subunit VHA-E isoforms. Eur J Cell Biol 89:152–156

Diepholz M, Venzke D, Prinz S, Batisse C, Flörchinger B, Rössle M, Svergun DI, Böttcher B, Féthière J (2008) A different conformation for EGC stator subcomplex in solution and in the assembled yeast V-ATPase: possible implications for regulatory disassembly. Structure 16:1789–1798

Dietz KJ, Tavakoli N, Kluge C, Mimura T, Sharma SS, Harris GC, Chardonnens AN, Golldack D (2001) Significance of the V-type ATPase for the adaptation to stressful growth conditions and its regulation on the molecular and biochemical level. J Exp Bot 52:1969–1980

Drobny M, Schnölzer M, Fiedler S, Lüttge U, Fischer-Schliebs E, Christian A-L, Ratajczak R (2002) Phenotypic subunit composition of the tobacco (Nicotiana tabacum L.) vacuolar-type H (+)-translocating ATPase. Biochim Biophys Acta 1564:243–255

Drory O, Nelson N (2006) The emerging structure of vacuolar ATPases. Physiology (Bethesda) 21:317–325

Endler A, Reiland S, Gerrits B, Schmidt UG, Baginsky S, Martinoia E (2009) In vivo phosphorylation sites of barley tonoplast proteins identified by a phosphoproteomic approach. Proteomics 9:310–321

Fecht-Bartenbach JV, Bogner M, Krebs M, Stierhof YD, Schumacher K, Ludewig U (2007) Function of the anion transporter AtCLC-d in the trans-Golgi network. Plant J 50(3):466–474

Feng Y, Forgac M (1992a) Cysteine 254 of the 73-kDa A subunit is responsible for inhibition of the coated vesicle (H+)-ATPase upon modification by sulfhydryl reagents. J Biol Chem 267:5817–5822

Feng Y, Forgac M (1992b) A novel mechanism for regulation of vacuolar acidification. J Biol Chem 267:19769–19772

Feng Y, Forgac M (1994) Inhibition of vacuolar H(+)-ATPase by disulfide bond formation between cysteine 254 and cysteine 532 in subunit A. J Biol Chem 269:13224–13230

Forgac M (1999) The vacuolar H+-ATPase of clathrin-coated vesicles is reversibly inhibited by S-nitrosoglutathione. J Biol Chem 274:1301–1305

Gaxiola RA, Palmgren MG, Schumacher K (2007) Plant proton pumps. FEBS Lett 581:2204–2214

Gogarten JP, Fichmann J, Braun Y, Morgan L, Styles P, Taiz SL, DeLapp K, Taiz L (1992) The use of antisense mRNA to inhibit the tonoplast H+ ATPase in carrot. Plant Cell 4:851–864

Graham LA, Hill KJ, Stevens TH (1998) Assembly of the yeast vacuolar H+-ATPase occurs in the endoplasmic reticulum and requires a Vma12p/Vma22p assembly complex. J Cell Biol 142:39–49

Graham LA, Flannery AR, Stevens TH (2003) Structure and assembly of the yeast V-ATPase. J Bioenerg Biomembr 35:301–312

Grüber G, Marshansky V (2008) New insights into structure-function relationships between archeal ATP synthase (A1A0) and vacuolar type ATPase (V1V0). Bioessays 30:1096–1109

Hager A, Helmle M (1981) Properties of an ATP-fueled, Cl- dependent proton pump localized in membranes of microsomal vesicles from maize coleoptiles. Z Naturforsch 36c:997–1008

Hedrich R, Kurkdjian A, Guern J, Flugge UI (1989) Comparative studies on the electrical properties of the H+ translocating ATPase and pyrophosphatase of the vacuolar-lysosomal compartment. EMBO J 8:2835–2841

Herman E, Li X, Su R, Larsen P, Hsu H, Sze H (1994) Vacuolar-Type H+ -ATPases are associated with the endoplasmic reticulum and provacuoles of root tip cells. Plant Physiol 106:1313–1324

Hernandez A, Jiang X, Cubero B, Nieto PM, Bressan RA, Hasegawa PM, Pardo JM (2009) Mutants of the Arabidopsis thaliana cation/H+ antiporter AtNHX1 conferring increased salt tolerance in yeast: the endosome/prevacuolar compartment is a target for salt toxicity. J Biol Chem 284:14276–14285

Hiesinger PR, Fayyazuddin A, Mehta SQ, Rosenmund T, Schulze KL, Zhai RG, Verstreken P, Cao Y, Zhou Y, Kunz J, Bellen HJ (2005) The v-ATPase V0 subunit a1 is required for a late step in synaptic vesicle exocytosis in Drosophila. Cell 121:607–620

Hill KJ, Stevens TH (1995) Vma22p is a novel endoplasmic reticulum-associated protein required for assembly of the yeast vacuolar H(+)-ATPase complex. J Biol Chem 270:22329–22336

Hinton A, Bond S, Forgac M (2009) V-ATPase functions in normal and disease processes. Pflugers Arch 457:589–598

Hirata R, Umemoto N, Ho MN, Ohya Y, Stevens TH, Anraku Y (1993) VMA12 is essential for assembly of the vacuolar H(+)-ATPase subunits onto the vacuolar membrane in Saccharomyces cerevisiae. J Biol Chem 268:961–967

Hirata R, Graham LA, Takatsuki A, Stevens TH, Anraku Y (1997) VMA11 and VMA16 encode second and third proteolipid subunits of the Saccharomyces cerevisiae vacuolar membrane H+-ATPase. J Biol Chem 272:4795–4803

Hirata T, Iwamoto-Kihara A, Sun-Wada G-H, Okajima T, Wada Y, Futai M (2003) Subunit rotation of vacuolar-type proton pumping ATPase: relative rotation of the G and C subunits. J Biol Chem 278:23714–23719

Hong-Hermesdorf A, Brüx A, Grüber A, Grüber G, Schumacher K (2006) A WNK kinase binds and phosphorylates V-ATPase subunit C. FEBS Lett 580:932–939

Huh WK, Falvo JV, Gerke LC, Carroll AS, Howson RW, Weissman JS, O'Shea EK (2003) Global analysis of protein localization in budding yeast. Nature 425:686–691

Hurtado-Lorenzo A, Skinner M, El Annan J, Futai M, Sun-Wada G-H, Bourgoin S, Casanova J, Wildeman A, Bechoua S, Ausiello DA, Marshansky V (2006) V-ATPase interacts with ARNO and Arf6 in early endosomes and regulates the protein degradative pathway. Nat Cell Biol 8:124–136

Imamura H, Takeda M, Funamoto S, Shimabukuro K, Yoshida M, Yokoyama K (2005) Rotation scheme of V1-motor is different from that of F1-motor. Proc Natl Acad Sci USA 102:17929–17933

Jackson DD, Stevens TH (1997) VMA12 encodes a yeast endoplasmic reticulum protein required for vacuolar H+-ATPase assembly. J Biol Chem 272:25928–25934

Jaquinod M, Villiers F, Kieffer-Jaquinod S, Hugouvieux V, Bruley C, Garin J, Bourguignon J (2007) A proteomics dissection of Arabidopsis thaliana vacuoles isolated from cell culture. Mol Cell Proteomics 6:394–412

Jentsch TJ (2007) Chloride and the endosomal-lysosomal pathway: emerging roles of CLC chloride transporters. J Physiol 578:633–640

Kane PM (1999) Biosynthesis and regulation of the yeast vacuolar H+-ATPase. J Bioenerg Biomembr 31:49–56

Kane PM (2006) The where, when, and how of organelle acidification by the yeast vacuolar H+-ATPase. Microbiol Mol Biol Rev 70:177–191

Kane PM, Smardon AM (2003) Assembly and regulation of the yeast vacuolar H+-ATPase. J Bioenerg Biomembr 35:313–321

Kane PM, Kuehn MC, Howald-Stevenson I, Stevens TH (1992) Assembly and targeting of peripheral and integral membrane subunits of the yeast vacuolar H(+)-ATPase. J Biol Chem 267:447–454

Kawamura Y, Arakawa K, Maeshima M, Yoshida S (2000) Tissue specificity of E subunit isoforms of plant vacuolar H(+)-ATPase and existence of isotype enzymes. J Biol Chem 275:6515–6522

Kawasaki-Nishi S, Bowers K, Nishi T, Forgac M, Stevens TH (2001) The amino-terminal domain of the vacuolar proton-translocating ATPase a subunit controls targeting and in vivo dissociation, and the carboxyl-terminal domain affects coupling of proton transport and ATP hydrolysis. J Biol Chem 276:47411–47420

Kluge C, Lahr J, Hanitzsch M, Bolte S, Golldack D, Dietz K-J (2003) New insight into the structure and regulation of the plant vacuolar H+-ATPase. J Bioenerg Biomembr 35:377–388

Kluge C, Seidel T, Bolte S, Sharma SS, Hanitzsch M, Satiat-Jeunemaitre B, Ross J, Sauer M, Golldack D, Dietz K-J (2004) Subcellular distribution of the V-ATPase complex in plant cells, and in vivo localisation of the 100 kDa subunit VHA-a within the complex. BMC Cell Biol 5:29

Klychnikov OI, Li KW, Lill H, de Boer AH (2007) The V-ATPase from etiolated barley (Hordeum vulgare L.) shoots is activated by blue light and interacts with 14-3-3 proteins. J Exp Bot 58:1013–1023

Konishi H, Maeshima M, Komatsu S (2005) Characterization of vacuolar membrane proteins changed in rice root treated with gibberellin. J Proteome Res 4:1775–1780

Krebs M, Beyhl D, Görlich E, Al-Rasheid KAS, Marten I, Stierhof Y-D, Hedrich R, Schumacher K (2010) Arabidopsis V-ATPase activity at the tonoplast is required for efficient nutrient storage but not for sodium accumulation. Proc Natl Acad Sci USA 107:3251–3256

Li X, Su RT, Hsu HT, Sze H (1998) The molecular chaperone calnexin associates with the vacuolar H(+)-ATPase from oat seedlings. Plant Cell 10:119–130

Li J, Yang H, Peer WA, Richter G, Blakeslee J, Bandyopadhyay A, Titapiwantakun B, Undurraga S, Khodakovskaya M, Richards EL, Krizek B, Murphy AS, Gilroy S, Gaxiola R (2005) Arabidopsis H+-PPase AVP1 regulates auxin-mediated organ development. Science 310:121–125

Liegeois S, Benedetto A, Garnier JM, Schwab Y, Labouesse M (2006) The V0-ATPase mediates apical secretion of exosomes containing Hedgehog-related proteins in Caenorhabditis elegans. J Cell Biol 173:949–961

Lu M, Sautin YY, Holliday LS, Gluck SL (2004) The glycolytic enzyme aldolase mediates assembly, expression, and activity of vacuolar H+-ATPase. J Biol Chem 279:8732–8739

Lu M, Ammar D, Ives H, Albrecht F, Gluck SL (2007) Physical interaction between aldolase and vacuolar H+-ATPase is essential for the assembly and activity of the proton pump. J Biol Chem 282:24495–24503

Ludwig J, Kerscher S, Brandt U, Pfeiffer K, Getlawi F, Apps DK, Schagger H (1998) Identification and characterization of a novel 9.2-kDa membrane sector-associated protein of vacuolar proton-ATPase from chromaffin granules. J Biol Chem 273:10939–10947

Maeshima M (2000) Vacuolar H(+)-pyrophosphatase. Biochim Biophys Acta 1465:37–51

Malkus P, Graham LA, Stevens TH, Schekman R (2004) Role of Vma21p in assembly and transport of the yeast vacuolar ATPase. Mol Biol Cell 15:5075–5091

Manolson MF, Proteau D, Preston RA, Stenbit A, Roberts BT, Hoyt MA, Preuss D, Mulholland J, Botstein D, Jones EW (1992) The VPH1 gene encodes a 95-kDa integral membrane polypeptide required for in vivo assembly and activity of the yeast vacuolar H(+)-ATPase. J Biol Chem 267:14294–14303

Manolson MF, Wu B, Proteau D, Taillon BE, Roberts BT, Hoyt MA, Jones EW (1994) STV1 gene encodes functional homologue of 95-kDa yeast vacuolar H(+)-ATPase subunit Vph1p. J Biol Chem 269:14064–14074

Maranda B, Brown D, Bourgoin S, Casanova JE, Vinay P, Ausiello DA, Marshansky V (2001) Intra-endosomal pH-sensitive recruitment of the Arf-nucleotide exchange factor ARNO and Arf6 from cytoplasm to proximal tubule endosomes. J Biol Chem 276:18540–18550

Martinoia E, Maeshima M, Neuhaus HE (2007) Vacuolar transporters and their essential role in plant metabolism. J Exp Bot 58:83–102

Matsuoka K, Higuchi T, Maeshima M, Nakamura K (1997) A vacuolar-type H+-ATPase in a nonvacuolar organelle is required for the sorting of soluble vacuolar protein precursors in tobacco cells. Plant Cell 9:533–546

Merzendorfer H, Huss M, Schmid R, Harvey WR, Wieczorek H (1999) A novel insect V-ATPase subunit M9.7 is glycosylated extensively. J Biol Chem 274:17372–17378

Miller AJ, Fan X, Orsel M, Smith SJ, Wells DM (2007) Nitrate transport and signalling. J Exp Bot 58:2297–2306

Muench SP, Huss M, Song CF, Phillips C, Wieczorek H, Trinick J, Harrison MA (2009) Cryo-electron microscopy of the vacuolar ATPase motor reveals its mechanical and regulatory complexity. J Mol Biol 386:989–999

Nakanishi Y, Maeshima M (1998) Molecular cloning of vacuolar H(+)-pyrophosphatase and its developmental expression in growing hypocotyl of mung bean. Plant Physiol 116:589–597

Nakanishi-Matsui M, Sekiya M, Nakamoto RK, Futai M (2010) The Mechanism of Rotating Proton Pumping ATPases. Biochim Biophys Acta 1797:1343-1352

Neubert C, Graham LA, Black-Maier EW, Coonrod EM, Liu T-Y, Stierhof Y-D, Seidel T, Stevens TH, Schumacher K (2008) Arabidopsis has two functional orthologs of the yeast V-ATPase assembly factor Vma21p. Traffic 9:1618–1628

Padmanaban S, Lin X, Perera I, Kawamura Y, Sze H (2004) Differential expression of vacuolar H+-ATPase subunit c genes in tissues active in membrane trafficking and their roles in plant growth as revealed by RNAi. Plant Physiol 134:1514–1526

Parra KJ, Kane PM (1998) Reversible association between the V1 and V0 domains of yeast vacuolar H+-ATPase is an unconventional glucose-induced effect. Mol Cell Biol 18:7064–7074

Parra KJ, Keenan KL, Kane PM (2000) The H subunit (Vma13p) of the yeast V-ATPase inhibits the ATPase activity of cytosolic V1 complexes. J Biol Chem 275:21761–21767

Qi J, Wang Y, Forgac M (2007) The vacuolar (H+)-ATPase: subunit arrangement and in vivo regulation. J Bioenerg Biomembr 39:423–426

Ramachandran N, Munteanu I, Wang P, Aubourg P, Rilstone JJ, Israelian N, Naranian T, Paroutis P, Guo R, Ren Z-P, Nishino I, Chabrol B, Pellissier J-F, Minetti C, Udd B, Fardeau M, Tailor CS, Mahuran DJ, Kissel JT, Kalimo H, Levy N, Manolson MF, Ackerley CA, Minassian BA (2009) VMA21 deficiency causes an autophagic myopathy by compromising V-ATPase activity and lysosomal acidification. Cell 137:235–246

Sambade M, Kane PM (2004) The yeast vacuolar proton-translocating ATPase contains a subunit homologous to the Manduca sexta and bovine e subunits that is essential for function. J Biol Chem 279:17361–17365

Sambongi Y, Iko Y, Tanabe M, Omote H, Iwamoto-Kihara A, Ueda I, Yanagida T, Wada Y, Futai M (1999) Mechanical rotation of the c subunit oligomer in ATP synthase (F0F1): direct observation. Science 286:1722–1724

Saroussi S, Nelson N (2009) The little we know on the structure and machinery of V-ATPase. J Exp Biol 212:1604–1610

Schmutz J, Cannon SB, Schlueter J, Ma J, Mitros T, Nelson W, Hyten DL, Song Q, Thelen JJ, Cheng J, Xu D, Hellsten U, May GD, Yu Y, Sakurai T, Umezawa T, Bhattacharyya MK, Sandhu D, Valliyodan B, Lindquist E, Peto M, Grant D, Shu S, Goodstein D, Barry K, Futrell-Griggs M, Abernathy B, Du J, Tian Z, Zhu L, Gill N, Joshi T, Libault M, Sethuraman A, Zhang XC, Shinozaki K, Nguyen HT, Wing RA, Cregan P, Specht J, Grimwood J, Rokhsar D, Stacey G, Shoemaker RC, Jackson SA (2010) Genome sequence of the palaeopolyploid soybean. Nature 463:178–183

Schumacher K, Vafeados D, McCarthy M, Sze H, Wilkins T, Chory J (1999) The Arabidopsis det3 mutant reveals a central role for the vacuolar H(+)-ATPase in plant growth and development. Genes Dev 13:3259–3270

Schumaker KS, Sze H (1987) Decrease of pH gradients in tonoplast vesicles by NO(3) and Cl: evidence for H-coupled anion transport. Plant Physiol 83:490–496

Seidel T, Schnitzer D, Golldack D, Sauer M, Dietz K-J (2008) Organelle-specific isoenzymes of plant V-ATPase as revealed by in vivo-FRET analysis. BMC Cell Biol 9:28

Seoighe C, Gehring C (2004) Genome duplication led to highly selective expansion of the Arabidopsis thaliana proteome. Trends Genet 20:461–464

Seol JH, Shevchenko A, Shevchenko A, Deshaies RJ (2001) Skp1 forms multiple protein complexes, including RAVE, a regulator of V-ATPase assembly. Nat Cell Biol 3:384–391

Smardon AM, Kane PM (2007) RAVE is essential for the efficient assembly of the C subunit with the vacuolar H(+)-ATPase. J Biol Chem 282:26185–26194

Smardon AM, Tarsio M, Kane PM (2002) The RAVE complex is essential for stable assembly of the yeast V-ATPase. J Biol Chem 277:13831–13839

Sze H, Schumacher K, Müller ML, Padmanaban S, Taiz L (2002) A simple nomenclature for a complex proton pump: VHA genes encode the vacuolar H(+)-ATPase. Trends Plant Sci 7:157–161

Tavakoli N, Kluge C, Golldack D, Mimura T, Dietz KJ (2001) Reversible redox control of plant vacuolar H+-ATPase activity is related to disulfide bridge formation in subunit E as well as subunit A. Plant J 28:51–59

Toyomura T, Murata Y, Yamamoto A, Oka T, Sun-Wada GH, Wada Y, Futai M (2003) From lysosomes to the plasma membrane: localization of vacuolar-type H+ -ATPase with the a3 isoform during osteoclast differentiation. J Biol Chem 278:22023–22030

Viotti C, Bubeck J, Stierhof YD, Krebs M, Langhans M, van den Berg W, van Dongen W, Richter S, Geldner N, Takano J, de Vries S, Juergens G, Schumacher K (2010) Endocytic and Secretory Traffic in Arabidopsis Merge in the Trans-Golgi Network/Early Endosome, an Independent and Highly Dynamic Organelle. Plant Cell 22:1344–1357

Voss M, Vitavska O, Walz B, Wieczorek H, Baumann O (2007) Stimulus-induced phosphorylation of vacuolar H(+)-ATPase by protein kinase A. J Biol Chem 282:33735–33742

Walker R, Leigh R (1981) Characterization of a salt-stimulated ATPase activity associated with vacuoles isolated from storage roots of red beet (Beta vulgaris L.). Planta 153:140–149

Wassmer T, Kissmehl R, Cohen J, Plattner H (2006) Seventeen a-subunit isoforms of paramecium V-ATPase provide high specialization in localization and function. Mol Biol Cell 17:917–930

Xu T, Forgac M (2001) Microtubules are involved in glucose-dependent dissociation of the yeast vacuolar [H+]-ATPase in vivo. J Biol Chem 276:24855–24861

Yan Y, Denef N, Schüpbach T (2009) The vacuolar proton pump, V-ATPase, is required for notch signaling and endosomal trafficking in Drosophila. Dev Cell 17:387–402

Type IV (P4) and V (P5) P-ATPases in Lipid Translocation and Membrane Trafficking

Rosa L. López-Marqués, Danny M. Sørensen, and Michael G. Palmgren

Abstract The least investigated members of the P-type superfamily of primary active pumps belong to the type IV and V subfamilies, which are unique to eukaryotic organisms. Although accumulating data indicates a central role for these transporters in the secretory pathway, very little is known about their biochemical properties or regulation. In fact, even the transported substrate is a matter of intense debate. In this chapter, present knowledge concerning the role of P4- and P5-type pumps in lipid transport and membrane trafficking will be discussed.

1 Introduction

The P-type ATPase superfamily of proteins comprises a large number of primary cation transporters with the shared feature of forming a phosphorylated intermediate during catalysis (hence P-type) (Axelsen and Palmgren 1998). This superfamily is divided phylogenetically into five subfamilies characterised by their substrate specificity. Thus, the P1 group comprises K^+-ATPases from bacteria (Ballal et al. 2007) and heavy metal pumps (Argüello et al. 2007). P2-ATPases include, among others, the Na^+/K^+-ATPase responsible for establishing membrane potential across the membranes in mammalian cells (Glynn 2002), and Ca^{2+}-ATPases maintaining intracellular calcium homeostasis (Sze et al. 2000; Boursiac and Harper 2007). Members of the P3 subfamily are plasma membrane H^+-ATPases involved in the establishment of transplasma membrane electrochemical gradients of plant and fungi (see Chapter "Plant Proton Pumps: Regulatory Circuits Involving H^+ PATPase and H^+-PPase").

R.L. López-Marqués (✉), D.M. Sørensen, and M.G. Palmgren
Center for Membranes in Cell and Disease-PUMPKIN, Laboratory of Transport Biology, Department of Plant Biology and Biotechnology, Faculty of Life Sciences, University of Copenhagen, Thorvaldsensvej 40, 1871 Frederiksberg C, Denmark
e-mail: rlo@life.ku.dk, danny@life.ku.dk, palmgren@life.ku.dk

M. Geisler and K. Venema (eds.), *Transporters and Pumps in Plant Signaling*, Signaling and Communication in Plants 7,
DOI 10.1007/978-3-642-14369-4_11,

P4- and P5-ATPases are the least characterised subgroups of the P-type superfamily. Both P4- and P5A-ATPases seem to play a role in membrane trafficking, as mutants in different organisms show defects related to correct functioning of the secretory and/or endocytic machinery. However, the mechanism by which these proteins affect vesicular transport is still unclear. Although increasing evidence suggests P4 pumps to be phospholipid translocators, the nature of the transported substrate for P5 pumps remains a mystery. P5-ATPases can be phylogenetically divided into two distinct subgroups (P5A and P5B-ATPases) having different subcellular localisations and physiological roles (Møller et al. 2007). Whereas P5A-ATPases are found in every single eukaryotic genome analysed so far, P5B-ATPases appear to have been lost in some lineages including land plants (Møller et al. 2007), indicating that P5B-type function is not required in all multicellular eukaryotes. In this chapter, we will review the present knowledge concerning the physiological role of the intriguing subfamilies of P4- and P5A-ATPases in plants.

2 P4-ATPases

2.1 General Features

P4-ATPases (type IV subfamily) is typically the largest subgroup of P-type ATPases in plants. These include the model plant *Arabidopsis thaliana*, which genome contains 12 P4 family members out of a total of 46 P-type pumps, and *Oryza sativa*, which contains at least nine P4-ATPases out of 41 identified P-type pumps (Gomés et al. 2000; Axelsen and Palmgren 2001; Thever and Saier 2009). These transporters are suggested to be phospholipid translocators and, besides plants, are only present in other eukaryotes (Axelsen and Palmgren 1998). They are predicted to have ten transmembrane domains and share most of the amino acid motifs that characterise other P-type ATPase families, including the aspartyl phosphorylation motif (DKTGTLT). However, they present modifications of several other conserved regions, like the SPDEx(A/S)(F/L)(V/L) and GxT(A/G)(I/V)ED(KlR)LQ motifs, which seem to be specific for this family (Axelsen and Palmgren 1998; Gomes et al. 2000). They also lack the anionic residues involved in substrate binding in cation transporting P-type ATPases, such as the Ca^{2+}-ATPases (Catty et al. 1997; Halleck et al. 1998; Muthusamy et al. 2009). In P4-ATPases, these anionic residues are substituted by hydrophobic and non-charged amino acids, suggesting a different nature of the transported substrate or a different transport mechanism.

2.2 P4-ATPases in Plants and Their Subcellular Localisation

In the model plant *A. thaliana*, the subfamily of P4-ATPases comprises 12 members, named ALA1–ALA12 (for Aminophospholipid ATPase 1–12). Only ALA2 and ALA3 have been characterised with respect to their subcellular localisation to date. ALA3 is present in Golgi membranes (Poulsen et al. 2008a), while ALA2 is targeted to the prevacuolar compartment (López-Marqués et al. 2010). However, the putative localisation of some other family members has been suggested on the basis of large-scale studies. Several ALAs seem to be localised at the plasma membrane, including ALA1, ALA10, ALA11 and ALA12 (Benschop et al. 2007; Nühse et al. 2003; Nühse et al. 2004; Dunkley et al. 2006; Heazlewood et al. 2007). In addition, ALA1 has also been found in tonoplast membranes (Whiteman et al. 2008) and ALA2 in chloroplasts (Froehlich et al. 2003).

In *Oryza sativa*, 12 putative P4-ATPases have been annotated (Ouyang et al. 2007) and at least two have been found in the plasma membrane during a proteomic analysis (Natera et al. 2008), but no experimental confirmation of these data is available to date.

2.3 Tissue-specific Expression of Arabidopsis *P4-ATPases*

As for the subcellular localisation, most of the information about tissue-specific expression of P4-ATPases in *Arabidopsis* comes from microarrays and other large-scale datasets. These results have been confirmed by promoter expression analysis for two members of the family, ALA1 and ALA3. ALA1 was found to be expressed in the vascular tissue, the root of seedling cotyledons and the cotyledonary node, while it seems to be absent in hypocotyls, emerging lateral roots and root hairs. In flowers, it is expressed in the vascular bundle of sepals, but not in petals (Gomés et al. 2000). ALA3, on its side, is also expressed in a wide variety of tissues, including flowers, siliques, the vascular tissue of young leafs and roots, stomatal guard cells and, very specifically, at the columella root cap of both primary and emerging lateral roots, but not in the rest of the root tip (Poulsen et al. 2008a). Microarray data show that ALA6, ALA7 and ALA12 are almost completely restricted to pollen, while the remaining ALAs can be found almost all over the plant with some individual peculiarities (Genevestigator 2008, https://www.genevestigator.com/gv/index.jsp). ALA10 and ALA11 seem to be absent from inflorescences and seeds and show a high expression in the root endodermis. ALA2 is specially represented in senescent leaves. ALA8, ALA4 and ALA9 are highly expressed in seeds, with ALA4 and ALA9 concentration in seed coats. ALA9 is also expressed in pollen and ALA5, on its side, is present in seedling radicles and the root hair zone. There also seems to be a differential expression of ALAs along the different developmental stages of the plant. ALA1, ALA3, ALA10, ALA11 and ALA12 have a peak of expression during early flowering stages with a subsequent decay, while ALA9 is heavily expressed at all stages during flowering. ALA2, ALA3 and ALA5, as well as the pollen-specific ALA6, ALA7 and ALA12, increase expression at very late development stages, while ALA3, ALA4, ALA5, ALA9 and ALA10 seem to be highly represented during development of the first leaves.

2.4 The Cdc50p Homologue Family As β-subunits for P4-ATPases

2.4.1 General Features

P4-ATPases seem to require the presence of an accessory protein for their catalytic activity. On one hand, yeast mutants lacking members of the Cdc50p family of proteins show phenotypes that mimic those derived from lack of P4-ATPase expression (Saito et al. 2004; Pérez-Victoria et al. 2006). On the other hand, functional complementation of yeast mutants lacking one or more endogenous P4-ATPases by some of their plant counterparts requires the presence of a plant protein homologous to yeast Cdc50p (Gomés et al. 2000; Poulsen et al. 2008a; López-Marqués et al. 2010). No other plant P-type ATPase subfamily described so far seems to require the presence of a β-subunit for functionality. The Cdc50 family of proteins comprises five members in *Arabidopsis*, named ALIS1–ALIS5 (for ALA Interacting Subunit 1–5) (Poulsen et al. 2008a). As their homologues in other organisms, they are characterised by the presence of two transmembrane domains, separated by a large exoplasmic lumenal loop, which is heavily *N*-glycosylated and stabilised by formation of disulfide bridges (Puts and Holthius 2009). This structure resembles a fusion of the β- and γ-subunits of the mammalian Na^+/K^+-ATPase (Fig. 1), which might indicate similar physiological roles for the accessory proteins (Poulsen et al. 2008b).

2.4.2 Putative Role of P4-ATPase β-subunits

As for the β- and γ-subunits of Na^+/K^+ pumps, a role for Cdc50 proteins in folding and trafficking of the P-type pump has been suggested. In fact, exit of P4-ATPases from the ER requires the presence of a β-subunit in yeast, mammals, parasites and plants (Saito et al. 2004; Paulusma et al. 2008; Pérez-Victoria et al. 2006; López-Marqués et al. 2010). Interestingly, once out of the ER, the final localisation of the P4-ATPase in a concrete organelle membrane seems to be determined by the nature of the pump itself and not by the interacting β-subunit (López-Marqués et al. 2010).

Indirect evidence in yeast indicates that Cdc50 proteins might also be involved in determining the substrate specificity of the P4-ATPase (Puts and Holthius 2009). Thus, plasma membrane-localised Dnf1p and Dnf2p interact with the same Cdc50p homologue (LEM3p) and share substrate specificities (Pomorski et al. 2003), while trans-Golgi Drs2p and Dnf3p interact, respectively, with Cdc50p and Crf1p and transport different phospholipids (Alder-Baerens et al. 2006; Saito et al. 2004; Furuta et al. 2007). In contrast, a recent study has demonstrated that *Arabidopsis* ALA2 and ALA3 can make use of any of three different ALIS to promote transport of phospholipids, but each ALA protein shows a characteristic substrate preference that is independent of the nature of the β-subunits (see also Sect. 2.5;

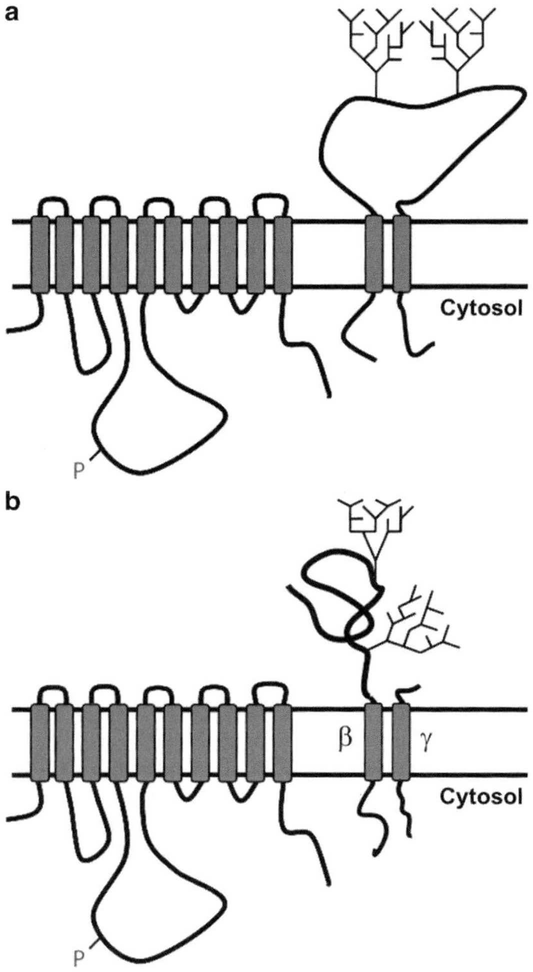

Fig. 1 Structural comparison between P4- and Na^+/K^+-ATPases and their accessory proteins. (**a**) Schematic representation of a P4-ATPase and its β-subunit. The β-subunit has two transmembrane domains flanking a long luminal loop known to be glycosylated. (**b**) Schematic representation of the Na^+/K^+-ATPase of mammalian cells and its accessory β- and γ-subunits. The β-subunit has one transmembrane domain and a large glycosylated exodomain, while the γ-subunit consists of a single transmembrane domain. Both P-type pumps share a common secondary structure with ten transmembrane domains and the big cytosolic loop between transmembrane domains 4 and 5 contains the aspartyl residue that is characteristically phosphorylated during the catalytic cycle (represented with a P). Their corresponding accessory proteins are also structurally related, with the β-subunit for P4-ATPases resembling a fusion between the two Na^+/K^+-ATPase accessory proteins

López-Marqués et al. 2010). These results suggest that not only the subcellular localisation but also the substrate specificity is controlled by the P4-ATPase, at least in plants (López-Marqués et al. 2010). Even when substrate specificity is determined by the α-catalytic subunit, it cannot be ruled out that the β-subunit is necessary for modulating the kinetics or other biochemical properties of the protein complex. Indeed, yeast Cdc50p is required for Drs2p to form the characteristic phosphorylated intermediate during the ATPase catalytic cycle, and the affinity of both interaction partners for each other fluctuates along this cycle (Lenoir et al. 2009). Interestingly, some P4-ATPases, like *Arabidopsis* ALA1, do not seem to require the presence of a β-subunit (Gomés et al. 2000). However, characterisation of ALA1 activity was carried out by heterologous expression in yeast, and it cannot be ruled out that this protein is capable of interacting with an endogenous Cdc50.

2.5 Role of Plant P4-ATPases in Phospholipid Translocation

Although the nature of the transported substrate for P4-ATPases is still a matter of debate, all investigated P4-ATPases from yeast (not including the endosomal Neo1p), mammals, parasites, worms and plants have been shown to be capable of promoting transport (flipping) of phospholipids or phospholipid analogs (Pomorski et al. 2003; Chen et al. 2006; Paulusma et al. 2008; Poulsen et al. 2008a; López-Marqués et al. 2010). For Arabidosis, heterologous expression of ALA1, ALA2 or ALA3 in yeast mutant strains lacking endogenous P4-ATPases results in increased phospholipid translocation across membranes. Thus, reconstituted yeast microsomal membranes containing ALA1 internalise fluorescent analogs of phosphatidylserine (PS) and phosphatidylethanolamine (PE) more effectively than a control sample (Gomés et al. 2000). Translocation of fluorescent PS across the plasma membrane of living yeast cells is increased by expression of ALA2 and ALA3. In addition, cells expressing ALA3 also show improved uptake of PE and phosphatidylcholine (PC) analogs across the plasma membrane (López-Marqués et al. 2010). This phospholipid translocation activity is not restricted to fluorescent lipids. By using cytotoxic peptides that bind to specific phospholipids in the outer leaflet of the plasma membrane, it has been possible to prove internalisation of natural PS in yeast cells expressing ALA2 and natural PS and PE in cells expressing ALA3 (López-Marqués et al. 2010). Nevertheless, as most of the evidence for the phospholipid flipping activity of P4-ATPases comes from measurements of intact living cells or fractionated membranes, it cannot be ruled out that transport occurs through a simporter, whose activity is dependent on the presence of a P4-ATPase (Poulsen et al. 2008b). However, two very recent independent publications have reported the successful reconstitution of highly purified P4-ATPases in artificial vesicles with a controlled lipid composition, which argues for a direct role of these proteins in phospholipid translocation (Coleman et al. 2009; Zhou and Graham 2009).

2.6 Role of Plant P4-ATPases in Membrane Trafficking

Yeast mutants lacking one or more P4-ATPases present phenotypes related to defects in vesicle budding. For instance, the absence of plasma membrane Dnf1p and Dnf2p causes a cold-sensitive defect in the biogenesis of endocytic vesicles (Pomorski et al. 2003). Absence of Neo1p affects vesicle budding from endosomes (Hua and Graham 2003; Wicky et al. 2004) and an exocytic pathway responsible for sorting cargo from the trans-Golgi network into a late endosome pathway (Hua et al. 2002; Singer-Kruger et al. 2008). Lack of Golgi-localised Drs2p reduces clathrin-coated vesicle budding on the trans-Golgi and affects endocytosis (Chen et al. 1999; Hua et al. 2002; Natarajan et al. 2004; Saito et al. 2004). Phenotypes of mutant plants lacking or having a reduced expression of P4-ATPases resemble the yeast phenotypes. For example, *Arabidopsis ala3* knock-out mutant plants are unable to produce slime vesicles from the trans-Golgi network in peripheral columella cells at the root tip (Poulsen et al. 2008a). This causes a defect in root development, prevents the shedding of border-like cells and affects general plant health, which results in impaired shoot and root growth. These ala3 mutants also present aberrant trichome expansion and a reduced in vitro growth rate for pollen tubes (Zhang and Oppenheimer 2009). Both these processes depend on rapid cell expansion, in which membrane trafficking plays a central role. Thus, it has been proposed that a similar vesicle-production defect like the one found in peripheral columella root cap cells is responsible for these pollen and trichome phenotypes. ALA1 antisense plants are sensitive to chilling and grow smaller than a wild type plant at temperatures around 15°C. The mechanism of this chilling sensitivity has not been further investigated, but interestingly, it phenocopies the cold sensitivity observed for Δ*drs2* yeast strains (Tang et al. 1996; Saito et al. 2004). No other P4-ATPase mutants have been characterised to date in plants. Some interaction partners within the elements of the vesiculation machinery have been identified for the yeast proteins (Chantallat et al. 2004; Wicky et al. 2004; Singer-Krueger et al. 2008; Liu et al. 2008; Natarajan et al. 2009), but no report is available in the literature on possible interaction partners for plant P4-ATPases.

3 P5-ATPases

3.1 General Features

P5-ATPases (type V subfamily) is the least investigated group of P-type ATPases. As for the P4 subfamily, their members are found in every single eukaryotic genome analysed so far but do not seem to have homologues in bacterial genomes (Møller et al. 2007). P5 ATPases can be recognised by a PPxxPxx motif located in transmembrane helix M4. In P-type Ca^{2+}- and Na^{+}/K^{+}-ATPases, whose structures are well known, this motif is conserved as Pxxx(P/L)xx. These conserved prolines have been suggested to be involved in unwinding of transmembrane segment M4 to facilitate exposure of backbone carbonyl oxygen atoms that are involved in coordination of the transported cation substrate (Toyoshima et al. 2000; Morth et al. 2007; Pedersen et al. 2007; Shinoda et al. 2009). The additional proline present in the P5-specific motif suggests a markedly different conformation of M4 in these pumps. Differences inside the motif have further been linked to the divergence of the P5 group into two distinct subgroups: P5A and P5B (Møller et al. 2007). P5A sequences contain a PP(E/D)xPx(D/E) motif with two conserved negatively charged residues, whereas P5B sequences are characterised by a PP(A/V)xP(A/V)x motif with two conserved hydrophobic residues. The presence and absence of two negative charges at this position is likely to have a strong impact on the electrical field around the unwound helix region, which can be expected to differentially influence the substrate specificity of the two subgroups (Møller et al. 2007). Although this idea is attractive, experimental confirmation is still lacking, as no substrate has been identified to date for any member of the P5-ATPase family (see Sect. 3.2). Interestingly, in organisms like yeast that contain proteins representative of both P5-ATPase subgroups, these present marked differences in subcellular localisation and do not seem to share any major physiological functions (Vashist et al. 2002; Suzuki 2001; Cronin et al. 2002; Gitler et al. 2009; Schmidt et al. 2009), suggesting that P5A and P5B pumps might play very different roles in the eukaryotic cell. Moreover, P5B-ATPases are completely absent in some lineages, including land plants (Møller et al. 2007). This indicates that P5B-type function is not required in all multicellular eukaryotes and that plants have evolved in such a way that either other mechanisms readily take over for the lost function of the P5B subgroup or gene loss is associated with an advanced trait found only in these lineages.

3.2 *Physiological Role of P5A-ATPases*

Very little is known about the physiological role of plant P5A-ATPases. The single member of this family identified in *A. thaliana*, MIA/PDR2, is localised to the ER membrane (Axelsen and Palmgren 2001; Jakobsen et al. 2005; Dunkley et al. 2006). Loss of gene function results in decreased fertility due to collapsed pollen grains (Jakobsen et al. 2005) and inability to maintain root stem cells during limiting phosphate conditions (Ticconi et al. 2009). *mia* knock-out mutants also display an altered expression pattern of genes involved in protein secretion, protein folding and solute transport. Finally, cation concentration in these mutant lines differs from that of a wild type *Arabidopsis*, indicating that ER homeostasis is likely to be perturbed (Jakobsen et al. 2005). These distinct effects suggest that P5A-ATPases either play different physiological roles in various plant tissues or have a unique and central role in cell metabolism, so that loss of P5A-ATPase function results in broad and diverse phenotypes. Unfortunately, no follow-up reports concerning

P5A-ATPases in plants are available, but the results described for *mia* mutants match those found in yeast.

As *Arabidopsis*, *Saccharomyces cerevisiae* contains only one P5A-ATPase, Spf1p, which resides in the ER (Vashist et al. 2002; Cronin et al. 2002) and *cis*-Golgi membranes (Suzuki 2001). Several studies have demonstrated the requirement of Spf1p for correct functioning of ER-related processes and its involvement in the unfolded-protein response (Travers et al. 2000; Suzuki 2001; Vashist et al. 2002; Ng et al. 2000). Moreover, *spf1* mutant cells present an abnormal cell wall composition (Suzuki and Shimma 1999; Suzuki 2001) and seem to have a defective glycosylation machinery (Suzuki and Shimma 1999), which suggests an inability of the mutated yeast to fully glycosylate cell wall components. In line with these findings, a large-scale genetic network analysis has identified genetic interactions between SPF1 and genes related to cell wall biosynthesis, including N-linked glycosylation (Tong et al. 2001; Tong et al. 2004). In this analysis, SPF1 was also shown to genetically interact with genes related to cytoskeletal reorganisation. Likewise, in the fission yeast *Schizosaccharomyces pombe*, the Spf1p homologue Cta4p is also involved in cell shape control and microtubule dynamics (Façanha et al. 2002). These data indicate that Spf1p is probably involved in other processes outside of the ER compartment. In further support of this notion, SPF1 also interacts genetically with GET1, a subunit of the GET complex involved in retrieval of HDEL signal-containing proteins from the Golgi (Ando and Suzuki 2005), and PMR1, a Ca^{2+}/Mn^{2+} P-type ATPase found in the Golgi membrane (Suzuki and Shimma 1999; Vashist et al. 2002; Cronin et al. 2002). The latter interaction is also conserved between the orthologous genes in fission yeast (Furune et al. 2008), indicating a general role for P5A-ATPases in the ER to Golgi transport system.

Taken all together, the diverse mutant phenotypes found in yeast and plants suggest a role of P5A-ATPases in the early part of the secretory pathway. Loss of gene function leads to broad unspecific phenotypes related to an inability to control ER homeostasis. The downstream effects include loss of protein processing integrity, loss of cell shape dynamics and probably also a defective ER to Golgi transport system.

3.3 What Is the Substrate of P5A-ATPases?

Although P5A-ATPases (and also P5B-ATPases) represent a biochemically uncharacterised subgroup of P-type pumps, the results from phenotypic analyses can provide clues about the nature of the transported substrate(s). In *Saccharomyces* cells, several studies have suggested a connection between SPF1 and calcium homeostasis (Cronin et al. 2000; Cronin et al. 2002). In line with these results, *Schizosaccharomyces* Cta4p, localised to the ER and the nuclear envelope, also seems to control nuclear calcium levels (Façanha et al. 2002; Furune et al. 2008). These findings might be explained at the molecular level if P5A-ATPases act as Ca^{2+} transporters. However, no biochemical evidence exists to support this conclusion, and several factors indicate that this might not be the case: (a) the phenotypes of *spf1* null mutants cannot be functionally complemented by expression of P-type Ca^{2+}-ATPases from yeast, mammalians or plants (Cronin et al. 2002); (b) the ATPase activity of purified Spf1p seems to be independent of calcium and several other tested substrate ions (Cronin et al. 2002); and (c) if Spf1p indeed transports Ca^{2+}, a novel binding site must be present in the P5 pumps, as only a single among several residues important for coordination of Ca^{2+} in the well-known sarcoplasmic reticulum Ca^{2+}-ATPase (SERCA) is conserved in P5 ATPases (Toyoshima et al. 2000). The possibility therefore remains that the impact of Spf1p on Ca^{2+} homeostasis is of a secondary nature. Indeed, defects in homeostasis of just a single ion sequestered by the secretory pathway are known to be able to cause broad changes affecting many downstream targets, including a general alteration of cellular ion homeostasis (Yadav et al. 2007). Thorough biochemical characterisation of P5A-ATPases is therefore required to assign a substrate specificity to these pumps.

4 Conclusions

Although P4- and P5A-ATPases seem to have functions that are vital for the correct functioning of plant cells, very little is known about their biochemical properties, and their physiological role is still unclear. Accumulating evidence suggests that P4-ATPases act as phospholipid translocators across membranes, and that this translocation is key for vesicle production from different organelle membranes. However, the mechanism by which these proteins translocate phospholipids is a mystery, and there is no clear evidence of how lipid flipping is linked to vesicle production. The structural characterisation of these proteins in complex with their β-subunits is becoming the major challenge in order to understand these processes. The nature of the transported substrate for P5A-ATPases is completely unknown and the matter of intense debate. Although a defined physiological role is still to be determined, evidence obtained from different organisms seems to point to a role in regulation of ER homeostasis. Due to the pleiotropic nature of P5A-ATPase mutant phenotypes, future research can be anticipated to be focusing on the biochemical characterisation of these proteins in order to shed some light on their primary role inside the cell.

References

Alder-Baerens N, Lisman Q, Luong L, Pomorski T, Holthuis JC (2006) Loss of P4 ATPases Drs2p and Dnf3p disrupts aminophospholipid transport and asymmetry in yeast post-Golgi secretory vesicles. Mol Biol Cell 17:1632–1642

Ando A, Suzuki C (2005) Cooperative function of the CHD5-like protein Mdm39p with a P-type ATPase Spf1p in the maintenance of ER homeostasis in *Saccharomyces cerevisiae*. Mol Genet Genomics 273:497–506

Argüello JM, Eren E, González-Guerrero M (2007) The structure and function of heavy metal transport P1B-ATPases. Biometals 20:233–248

Axelsen KB, Palmgren MG (1998) Evolution of substrate specificities in the P-type ATPase superfamily. J Mol Evol 46:84–101

Axelsen KB, Palmgren MG (2001) Inventory of the superfamily of P-type ion pumps in Arabidopsis. Plant Physiol 126:696–706

Ballal A, Basu B, Apte SK (2007) The Kdp-ATPase system and its regulation. J Biosci 32:559–568

Benschop JJ, Mohammed S, O'Flaherty M, Heck AJ, Slijper M, Menke FL (2007) Quantitative phosphoproteomics of early elicitor signaling in Arabidopsis. Mol Cell Proteomics 6:1198–1214

Boursiac Y, Harper JF (2007) The origin and function of calmodulin regulated Ca^{2+} pumps in plants. J Bioenerg Biomembr 39:409–414

Catty P, de Kerchove d'Exaerde A, Goffeau A (1997) The complete inventory of the yeast *Saccharomyces cerevisiae* P-type transport ATPases. FEBS Lett 409:325–332

Chantalat S, Park SK, Hua Z, Liu K, Gobin R, Peyroche A, Rambourg A, Graham TR, Jackson CL (2004) The Arf activator Gea2p and the P-type ATPase Drs2p interact at the Golgi in *Saccharomyces cerevisiae*. J Cell Sci 117:711–722

Chen C-Y, Ingram MF, Rosal PH, Graham TR (1999) Role for Drs2p, a P-type ATPase and potential aminophospholipid translocase, in yeast late Golgi function. J Cell Biol 147:1223–1236

Chen S, Wang J, Muthusamy BP, Liu K, Zare S, Andersen RJ, Graham TR (2006) Roles for the Drs2p–Cdc50p complex in protein transport and phosphatidylserine asymmetry of the yeast plasma membrane. Traffic 7:1503–1517

Coleman JA, Kwok MCM, Molday RS (2009) Localization, purification, and functional reconstitution of the P4-ATPase Atp8a2, a phosphatidylserine flippase in photoreceptor disc membranes. J Biol Chem 28:32670–32679

Cronin SR, Khoury A, Ferry DK, Hampton RY (2000) Regulation of HMG-CoA reductase degradation requires the P-type ATPase Cod1p/Spf1p. J Cell Biol 148:915–923

Cronin SR, Rao R, Hampton RY (2002) Cod1p/Spf1p is a P-type ATPase involved in ER function and Ca2+ homeostasis. J Cell Biol 157:1017–1028

Dunkley TP, Hester S, Shadforth IP, Runions J, Weimar T, Hanton SL, Griffin JL, Bessant C, Brandizzi F, Hawes C, Watson RB, Dupree P, Lilley KS (2006) Mapping the Arabidopsis organelle proteome. Proc Natl Acad Sci USA 103:6518–6523

Façanha AL, Appelgren H, Tabish M, Orokov L, Ekwall E (2002) The endoplasmic reticulum cation P-type ATPase Cta4p is required for control of cell shape and microtubule dynamics. J Cell Biol 157:1029–1039

Froehlich JE, Wilkerson CG, Ray WK, McAndrew RS, Osteryoung KW, Gage DA, Phinney BS (2003) Proteomic study of the *Arabidopsis thaliana* chloroplastic envelope membrane utilizing alternatives to traditional two-dimensional electrophoresis. J Proteome Res 2:413–425

Furune T, Hashimoto K, Ishiguro J (2008) Characterization of a fission yeast P-5-type ATPase homologue that is essential for Ca2+/Mn2+ homeostasis in the absence of P-2-type ATPases. Genes Genet Syst 83:373–381

Furuta N, Fujimura-Kamada K, Saito K, Yamamoto T, Tanaka K (2007) Endocytic recycling in yeast is regulated by putative phospholipid translocases and the Ypt31p/32p-Rcy1p pathway. Mol Biol Cell 18:295–312

Genevestigator (2008) NEBION/ETH Zurich. https://www.genevestigator.com/gv/index.jsp. Cited 23 March 2010

Gitler AD, Chesi A, Geddie ML, Strathearn KE, Hamamichi S, Caldwell GA et al (2009) Alpha-synuclein is part of a diverse and highly conserved interaction network that includes PARK9 and manganese toxicity. Nat Genet 41:308–315

Glynn IM (2002) A hundred years of sodium pumping. Annu Rev Physiol 64:1–18

Gomes E, Jakobsen MK, Axelsen KB, Geisler M, Palmgren MG (2000) Chilling tolerance in Arabidopsis involves ALA1, a member of a new family of putative aminophospholipid translocases. Plant Cell 12:2441–2454

Halleck MS, Pradhan D, Blackman C, Berkes C, Williamson P, Schlegel RA (1998) Multiple members of a third subfamily of P-type ATPases identified by genomic sequences and ESTs. Genome Res 8:354–361

Heazlewood JL, Verboom RE, Tonti-Filippini J, Small I, Millar AH (2007) SUBA: the Arabidopsis subcellular database. Nucleic Acids Res 35:D213–D218

Hua Z, Graham TR (2003) Requirement for Neo1p in retrograde transport from the Golgi complex to the endoplasmic reticulum. Mol Biol Cell 13:3162–3177

Hua Z, Fatheddin P, Graham TR (2002) An essential subfamily of Drs2p-related P-type ATPases is required for protein trafficking between Golgi complex and endosomal/vacuolar system. Mol Biol Cell 13:3162–3177

Jakobsen MK, Poulsen LR, Schulz A, Fleurat-Lessard P, Møller A, Husted S, Schiøtt S, Amtmann A, Palmgren MG (2005) Pollen development and fertilization in Arabidopsis is dependent on the male gametogenesis impaired anthers gene encoding a Type V P-type ATPase. Genes Dev 19:2757–2769

Lenoir G, Williamson P, Puts CF, Holthuis JCM (2009) Cdc50p plays a vital role in the ATPase reaction cycle of the putative aminophospholipid transporter Drs2p. J Biol Chem 284: 17956–17967

Liu K, Surendhran K, Nothwehr SF, Graham TR (2008) P4-ATPase requirement for AP-1/clathrin function in protein transport from the trans-Golgi network and early endosomes. Mol Biol Cell 19:3526–3535

Lopez-Marques RL, Poulsen LR, Hanisch S, Meffert K, Buch-Pedersen MJ, Jakobsen MK, Pomorski TG, Palmgren MG (2010) Intracellular targeting signals and lipid specificity determinants of the ALA/ALIS P4-ATPase complex reside in the catalytic ALA α-subunit. Mol Biol Cell 21:791–801

Møller AB, Asp T, Holm PB, Palmgren MG (2007) Phylogenetic analysis of P5 P-type ATPases, a eukaryotic lineage of secretory pathway pumps. Mol Phylogenet Evol 46:619–634

Morth JP, Pedersen BP, Toustrup-Jensen MS, Sørensen TL-M, Petersen J, Andersen J, Vilsen B, Nissen P (2007) Crystal structure of the sodium-potassium pump. Nature 450:1043–1049

Muthusamy BP, Natarajan P, Zhou X, Graham TR (2009) Linking phospholipid flippases to vesicle-mediated protein transport. Biochim Biophys Acta 1791(7):612–619

Natarajan P, Wang J, Hua Z, Graham TR (2004) Drs2p-coupled aminophospholipid translocase activity in yeast Golgi membranes and relationship to in vivo function. Proc Natl Acad Sci USA 101:10614–10619

Natarajan P, Liu K, Patil DV, Sciorra VA, Jackson CL, Graham TR (2009) Regulation of a Golgi flippase by phosphoinositides and an ArfGEF. Nat Cell Biol 11:1421–1426

Natera SHA, Ford KL, Cassin AM, Patterson JH, Newbigin EJ, Bacic A (2008) Analysis of the Oryza sativa plasma membrane proteome using combined protein and peptide fractionation approaches in conjunction with mass spectrometry. J Proteome Res 7:1159–1187

Ng TW, Spear ED, Walter P (2000) The unfolded protein response regulates multiple aspects of secretory and membrane protein biogenesis and endoplasmic reticulum quality control. J Cell Biol 150:77–88

Nühse TS, Stensballe A, Jensen ON, Peck SC (2003) Large-scale analysis of in vivo phosphorylated membrane proteins by immobilized metal ion affinity chromatography and mass spectrometry. Mol Cell Proteomics 2:1234–1243

Nühse TS, Stensballe A, Jensen ON, Peck SC (2004) Phosphoproteomics of the Arabidopsis plasma membrane and a new phosphorylation site database. Plant Cell 16:2394–2405

Ouyang S, Zhu W, Hamilton J, Lin H, Campbell M, Childs K, Thibaud-Nissen F, Malek RL, Lee Y, Zheng L, Orvis J, Haas B, Wortman J, Buell CR (2007) The TIGR rice genome annotation resource: improvements and new features. Nucleic Acids Res 35(Database issue):D883–D887

Paulusma CC, Folmer DE, Ho-Mok KS, de Waart DR, Hilarius PM, Verhoeven AJ, Oude Elferink RP (2008) ATP8B1 requires an accessory protein for endoplasmic reticulum exit and plasma membrane lipid flippase activity. Hepatology 47:268–278

Pedersen BP, Buch-Pedersen MJ, Morth P, Palmgren MG, Nissen P (2007) Crystal structure of the plasma membrane proton pump. Nature 450:1111–1115

Pérez-Victoria FJ, Sánchez-Cañete MP, Castanys S, Gamarro F (2006) Phospholipid translocation and miltefosine potency require both L. donovani miltefosine transporter and the new protein LdRos3 in Leishmania parasites. J Biol Chem 281:23766–23775

Pomorski T, Lombardi R, Riezman H, Devaux PF, van Meer G, Holthuis JCM (2003) Drs2p-related P-type ATPases Dnf1p and Dnf2p are required for phospholipid translocation across the yeast plasma membrane and serve a role in endocytosis. Mol Biol Cell 14:1240–1254

Poulsen LR, López-Marqués RL, McDowell SC, Okkeri J, Licht D, Schulz A, Pomorski T, Harper JF, Palmgren MG (2008a) The Arabidopsis P_4-ATPase ALA3 localizes to the golgi and requires a β-subunit to function in lipid translocation and secretory vesicle formation. Plant Cell 20: 658–676

Poulsen LR, López-Marqués RL, Palmgren MG (2008b) Flippases: still more questions than answers. Cell Mol Life Sci 65:3119–3125

Puts CF, Holthius JCM (2009) Mechanism and significance of P_4 ATPase-catalyzed lipid transport: Lessons from a Na^+/K^+-pump. Biochim Biophys Acta 1791:603–611

Saito K, Fujimura-Kamada K, Furuta N, Kato U, Umeda M, Tanaka K (2004) Cdc50p, a protein required for polarized growth, associates with the Drs2p P-Type ATPase implicated in phospholipid translocation in *Saccharomyces cerevisiae*. Mol Biol Cell 15:3418–3432

Schmidt K, Wolfe DM, Stiller B, Pearce DA (2009) Cd^{2+}, Mn^{2+}, Ni^{2+} and Se^{2+} toxicity to *Saccharomyces cereviosiae* lacking YPK9p the orthologue of human ATP13A2. Biochem Biophys Res Commun 383:198–202

Shinoda T, Ogawa H, Cornelius F, Toyoshima C (2009) Crystal structure of the sodium–potassium pump at 2.4 Å resolution. Nature 459(7245):446–450

Singer-Kruger B, Lasic M, Burger AM, Hausser A, Pipkorn R, Wang Y (2008) Yeast and human Ysl2p/hMon2 interact with Gga adaptors and mediate their subcellular distribution. EMBO J 27:1423–1435

Suzuki C (2001) Immunochemical and mutational analysis of P-type ATPase spf1p involved in the yeast secretory pathway. Biosci Biotechnol Biochem 11:2405–2411

Suzuki C, Shimma YI (1999) P-type ATPase spf1 mutants show a novel resistance mechanism for the killer toxin SMKT. Mol Microbiol 32(4):813–823

Sze H, Liang F, Hwang I, Curran AC, Harper JF (2000) Diversity and regulation of plant Ca^{2+} pumps: insights from expression in yeast. Annu Rev Plant Physiol Plant Mol Biol 51: 433–462

Tang X, Halleck MS, Schlegel RA, Williamson P (1996) A subfamily of P-type ATPases with aminophospholipid transporting activity. Science 272:1495–1497

Thever MD, Saier MH (2009) Bioinformatic Characterization of P-type ATPases encoded within the fully sequenced genomes of 26 eukaryotes. J Membr Biol 229:115–130

Ticconi CA, Lucero RD, Sakhonwasee S, Adamson AW, Creff A, Nussaume L, Desnos T, Abel S (2009) ER-resident proteins PDR2 and LPR1 mediate the developmental response of root meristems to phosphate availability. Proc Natl Acad Sci USA 106:14174–14179

Tong AH, Evangelista M, Parsons AB, Xu H, Bader GD, Page N, Robinson M, Raghibizadeh S, Hogue CW, Bussey H, Andrews B, Tyers M, Boone C (2001) Systematic genetic analysis with ordered arrays of yeast deletion mutants. Science 294:2364–2368

Tong AH, Lesage G, Bader GD, Ding H et al (2004) Global mapping of the yeast genetic interaction network. Science 303:808–813

Toyoshima C, Nakasako M, Nomura H, Ogawa H (2000) Crystal structure of the calcium pump of sarcoplasmic reticulum at 2.6Å resolution. Nature 405:647–655

Travers KJ, Patil CK, Wodicka L, Lockhart DJ, Weissman JS, Walter P (2000) Functional and genomic analysis reveal an essential coordination between the unfolded protein response and ER-associated degradation. Cell 101:249–258

Vashist S, Frank CG, Jakob CA, Ng DTW (2002) Two distinctly localized P-Type ATPases collaborate to maintain organelle homeostasis required for glycoprotein processing and quality control. Mol Biol Cell 21:3955–3966

Whiteman SA, Serazetdinova L, Jones AM, Sanders D, Rathjen J, Peck SC, Maathuis FJ (2008) Identification of novel proteins and phosphorylation sites in a tonoplast enriched membrane fraction of *Arabidopsis thaliana*. Proteomics 8:3536–3547

Wicky S, Schwarz H, Singer-Krüger B (2004) Molecular interactions of yeast Neo1p, an essential member of the Drs2 family of aminophospholipid translocases, and its role in membrane trafficking within the endomembrane system. Mol Cell Biol 24:7402–7418

Yadav J, Muend S, Zhang Y, Rao R (2007) A phenomics approach in yeast links proton and calcium pump function in the Golgi. Mol Biol Cell 18:1480–1489

Zhang X, Oppenheimer DG (2009) Irregular trichome branch 2 (ITB2) encodes a putative aminophospholipid translocase that regulates trichome branch elongation in Arabidopsis. Plant J 60:195–206

Zhou X, Graham TR (2009) Reconstitution of phospholipid translocase activity with purified Drs2p, a type-IV P-type ATPase from budding yeast. Proc Nat Acad Sci USA 106:16586–16591

Peroxisomal Transport Systems: Roles in Signaling and Metabolism

Frederica L. Theodoulou, Xuebin Zhang, Carine De Marcos Lousa, Yvonne Nyathi, and Alison Baker

Abstract Peroxisomes perform a range of different functions, including β-oxidation of fatty acids and synthesis and degradation of bioactive molecules. A notable feature of peroxisomes is their role in metabolic pathways which are shared between several subcellular compartments, including mitochondria, chloroplasts and cytosol. Transport across the peroxisomal membrane is therefore central to the co-ordination of metabolism. Although transport processes are required for import of substrates and cofactors, export of intermediates and products and the operation of redox shuttles, relatively few peroxisomal transporters have been identified to date. This chapter reviews the current evidence for and against different peroxisomal transport processes.

1 Introduction

Peroxisomes are near-ubiquitous organelles, which are characterised by an essentially oxidative metabolism and bound by a single membrane derived from the ER. Peroxisomes have no DNA, and their constituent matrix proteins and most of their membrane proteins are imported post-translationally by a dedicated import machinery (Lanyon-Hogg et al. 2010). Since their discovery, a wide range of biological functions has been ascribed to these organelles, including fatty acid breakdown, the glyoxylate cycle, photorespiration, and metabolism of hormones and reactive oxygen species (Table 1; Kaur et al. 2009). A key feature of plant peroxisomes is their plasticity, with enzymatic content and prevailing functions depending on

F.L. Theodoulou (✉) and X. Zhang
Biological Chemistry Department, Rothamsted Research, Harpenden AL5 2JQ, UK
e-mail: freddie.theodoulou@bbsrc.ac.uk

C. De Marcos Lousa, Y. Nyathi, and A. Baker
Centre for Plant Sciences, University of Leeds, Leeds LS2 9JT, UK

M. Geisler and K. Venema (eds.), *Transporters and Pumps in Plant Signaling*, Signaling and Communication in Plants 7,
DOI 10.1007/978-3-642-14369-4_12, © Springer-Verlag Berlin Heidelberg 2011

Table 1 Peroxisome functions in plants

Pathway/function	Associated enzymes/transporters	References
β-oxidation of fatty acids	Transporters: ABC transporter (PXA1/CTS/PED3/ACN2), adenine nucleotide translocator (PNC1, PNC2) Acyl-activating enzymes (AAE; LACS6, LACS7) Core β-oxidation: acyl-CoA oxidases (ACX1-6), multifunctional proteins (MFP, AIM1), 3-ketoacyl-CoA thiolases (KAT1, KAT2/PED1, KAT5). Auxiliary enzymes: enoyl-CoA isomerase (ECI); 2,4-dienoyl-CoA reductase (DECR)/short-chain dehydrogenase/reductase (SDRb); enoyl-CoA hydratase (ECH)	Graham and Eastmond (2002), Baker et al. (2006), Goepfert and Poirier (2007), Graham (2008), Kaur et al. (2009)
Glyoxylate cycle (and acetate metabolism)	Aconitase; citrate synthase, isocitrate lyase, malate synthase; metabolite shuttles. AAE7/ACN1	Kunze et al. (2006), Turner et al. (2005), Pracharoenwattana et al. (2005, 2007), Graham (2008)
Photorespiration	Glycolate oxidase (GOX); catalase (CAT); ser:glyoxylate transaminase (SGT); glu: glyoxylate transaminase (GGT); hydroxypyruvate reductase (HPR); malate dehydrogenase (PMDH);	Reumann and Weber (2006), Foyer et al. (2009), Pracharoenwattana et al. (2007, 2010)
Jasmonate biosynthesis	CTS; OPR3; OPCL1; other AAE; ACX1 ACX5; AIM1 (MFP); KAT2; thioesterase?	Schaller and Stintzi (2009)
Indole-3-butyric acid metabolism	PXA1/CTS/PED3; AAE; IBR3 (putative acyl-CoA dehydrogenase); ACX3; IBR10/ECI2 (hydratase); IBR1 (short-chain dehydrogenase/reductase); AIM 1; thiolase (KAT1,2,5); thioesterase?	Zolman et al. (2000, 2001a, 2007, 2008)
2,4-DB metabolism	AAE18; IBR1; ACX3, ACX4; AIM1; KAT2	Wiszniewski et al. (2009), Kaur et al. (2009)
ROS scavenging and detoxification	Catalase (CAT1-3); ascorbate peroxidase; monodehydroascorbate reductase (MDAR); dehydroascorbate reductase (DHAR); glutathione reductase (GR); G-6-P DH; 6-phosphogluconate dehydrogenase; NADP-isocitrate DH; 6-phosphogluconolactonase; GST (GSTT1-3); superoxide dismutase (SOD)	del Río et al. (2006), Nyathi and Baker (2006), Kaur et al. (2009)
ROS generation		

(*continued*)

Table 1 (continued)

Pathway/function	Associated enzymes/transporters	References
	Acyl-CoA oxidase; glycolate oxidase; sulphite oxidase; sarcosine oxidase; Cu-Zn SOD; Mn SOD; MDAR; PMP18; PMP29	Byrne et al. (2009), Nyathi and Baker (2006)
RON generation	Not known	Prado et al. (2004), del Río et al. (2006), Nyathi and Baker (2006), Corpas et al. (2009)
Pathogen response	SGT; PEN2 glycosyl hydrolase (myrosinase?); see also ROS generation and SA biosynthesis	Taler et al. (2004), Lipka et al. (2005), Westphal et al. (2008), Clay et al. (2009), Bednarek et al. (2009)
Polyamine catabolism	Polyamine oxidase, (PAO); copper-containing amine oxidase (CuAO); betaine aldehyde dehydrogenase (BADH)	Eubel et al. (2008), Kamada-Nobusada et al. (2008), Moschou et al. (2008), Reumann et al. (2007, 2009)
Sulphite oxidation	Sulphite oxidase (SO); catalase	Nakamura et al. (2002), Nowak et al. (2004), Hänsch and Mendel (2005), Hänsch et al. (2006), Lang et al. (2007)
Branched chain amino acid metabolism	β-hydroxyisobutyryl-CoA hydrolase (CHY1); sarcosine oxidase	Graham and Eastmond (2002), Zolman et al. (2001b), Lange et al. (2004), Goyer et al. (2004)
Ureide degradation	Uricase; 2-oxo-4-hydroxy-4-carboxy-5-ureidoimidazoline; (OHCU) decarboxylase; 5-hydroxyisourate (HIU) hydrolase (legumes only?)	Hennebry et al. (2006), Reumann et al. (2007, 2009), Eubel et al. (2008)
Salicylic acid biosynthesis (speculative)	Core β-oxidation; AAE isoforms	Reumann et al. (2004), Kienow et al. (2008)
Isopropanoid mevalonic acid pathway (speculative)	Acetoacyl-CoA thiolase; possibly other enzymes	Carrie et al. (2008), Reumann et al. (2007), Kaur et al. (2009), Sapir-Mir et al. (2008)
Phylloquinone biosynthesis (speculative)	AAE14 (dual targeted to peroxisome and chloroplast); naphthoate synthase;	Babujee et al. (2010)

Arabidopsis protein names are given in upper case.

cell type and developmental stage (Hayashi and Nishimura 2006). Whilst some metabolic pathways, such as β-oxidation, are confined to the peroxisome in plants, more commonly, metabolic pathways are shared between peroxisomes and other cellular compartments. Thus peroxisomes have been described as "organelles at the crossroads" (Erdmann et al. 1997). Transport across the peroxisomal membrane is therefore paramount in the co-ordination of metabolism between different compartments and the efficient functioning of metabolic pathways.

2 The Peroxisomal Membrane as a Boundary: Permeability and Porins vs. Selective Transporters

Despite the obvious importance of peroxisomal transport processes, the role of the peroxisomal membrane as a permeability barrier to metabolites has been a matter of considerable controversy. Two, apparently contradictory, models have been suggested (Antonenkov and Hiltunen 2006; Rottensteiner and Theodoulou 2006; Visser et al. 2007). Firstly, it has been proposed that peroxisomal membranes contain non-selective channels and are freely permeable to solutes, as is the case for the outer mitochondrial membrane. In contrast, a second school of thought proposes that the peroxisomal contains a complement of selective transporters, in common with the inner mitochondrial membrane. However, these models are not mutually exclusive: indeed, recent evidence supports the existence of both types of transporter in the peroxisomal membrane, which has been termed the “two channel” concept of peroxisomal membrane permeability (Antonenkov and Hiltunen 2006).

2.1 Evidence for Peroxisomal Porins

Early research with isolated peroxisomes and detergent-permeabilised cells suggested that peroxisomal enzymes lack structure-linked latency in vitro, indicating that they must be freely accessible to substrates (Verleur and Wanders 1993, and refs therein). These studies, together with a subsequent investigation of peroxisomal permeability using radiolabelled solutes, led to the concept of peroxisomal porins, proteins which form relatively non-specific channels in the peroxisomal membrane (van Veldhoven et al. 1987; Reumann 2000). The porin concept has met with some criticism, since it has been asserted that non-selective pores are incompatible with the control required for the efficient operation of metabolic pathways. However, it has been proposed that that compartmentation of peroxisomal metabolism is in fact not dependent on the function of the boundary membrane but rather to the organisation of peroxisomal enzymes in complexes, since peroxisomes with osmotically-shocked membranes could sustain rates of photorespiratory metabolism comparable to those required in vivo (Heupel and Heldt 1994; Reumann 2000). Although both these notions challenge the classical view of organelle membranes as semi-permeable barriers to the movement of solutes, the existence of porins is supported by an increasing body of experimental evidence: channel-forming activities have been demonstrated in preparations of peroxisomes from plants, animals and yeast (Sulter et al. 1993; Reumann et al. 1995, 1997, 1998; Antonenkov et al. 2005, 2009; Grunau et al. 2009). The most comprehensive evidence to date is for the mouse 22 kDa integral peroxisomal membrane protein, Pxmp2 (Rokka et al. 2009). Peroxisomes of *Pxmp2* knockout mice had reduced permeability to solutes in vitro and in vivo, as evidenced by altered osmotic behaviour and increased latency of oxidase enzymes. Both recombinant and native Pxmp2 exhibited channel-forming activities consistent with a peroxisomal channel, which permits diffusion of ions and solutes with molecular masses up to 300 Da (Rokka et al. 2009).

The molecular identification of plant porins remains elusive: none has yet been identified by forward or reverse genetics and the hydrophobic character of peroxisomal membrane proteins combined with their low abundance biases against identification by proteomic techniques. However, a voltage-dependent anion selective channel homologue was identified in a proteomic study of soybean peroxisomes (Arai et al. 2008a) and a candidate porin, Arabidopsis PMP22 has been localised to the peroxisomal membrane (Tugal et al. 1999), though neither has been characterised functionally.

2.2 *Evidence for Specific Transporters*

Studies with intact yeast cells demonstrated that the peroxisomal membrane is not freely permeable to certain solutes (van Roermund et al. 1995). Accordingly, careful studies with isolated peroxisomes have provided convincing evidence that, for mammalian peroxisomes at least, the membrane is freely permeable to solutes with molecular masses less than 300 Da (e.g. urate, glycolate, other organic acids, etc.), but has restricted permeability to larger compounds such as cofactors and substrates of beta-oxidation (ATP, NAD/H, NADP/H, CoA and acetyl-CoA species) (Antonenkov et al. 2004a; Rokka et al. 2009). Antonenkov, Hiltunen and co-workers also demonstrated that lysis of peroxisomes following or during isolation is due to the permeability of peroxisomes to low molecular weight osmotica such as sucrose and can be partially prevented using higher molecular weight osmoprotectants such as polyethylene glycol (Antonenkov et al. 2004b). A comparable set of experiments has not yet been published for plant peroxisomes. However, genetic and biochemical evidence for peroxisomal transporters has emerged in recent years and is summarised below.

3 Import of Substrates, Cofactors and Co-Substrates for β-Oxidation

β-oxidation comprises a series of reactions which result in the repeated cleavage of acetate units from the thiol end of fatty acyl-CoA molecules: for each turn of the β-oxidation spiral, the fatty acyl chain is shortened by two carbon units and a molecule of acetyl CoA is generated (Baker et al. 2006; Fig. 1). Although β-oxidation was originally discovered as the pathway for breakdown of fatty acids (Knoop 1904), it has subsequently been shown to have a wider range of roles in plants, including the metabolism of signaling molecules (Baker et al. 2006; Poirier et al. 2006; Goepfert and Poirier 2007). This metabolic flexibility is possible due to the presence either of isoforms of "core" β-oxidation enzymes with differing substrate specificity or enzymes with broad substrate specificity and also to the existence of ancillary enzymes, such as reductases, dehydrogenases, isomerases and acyl-activating

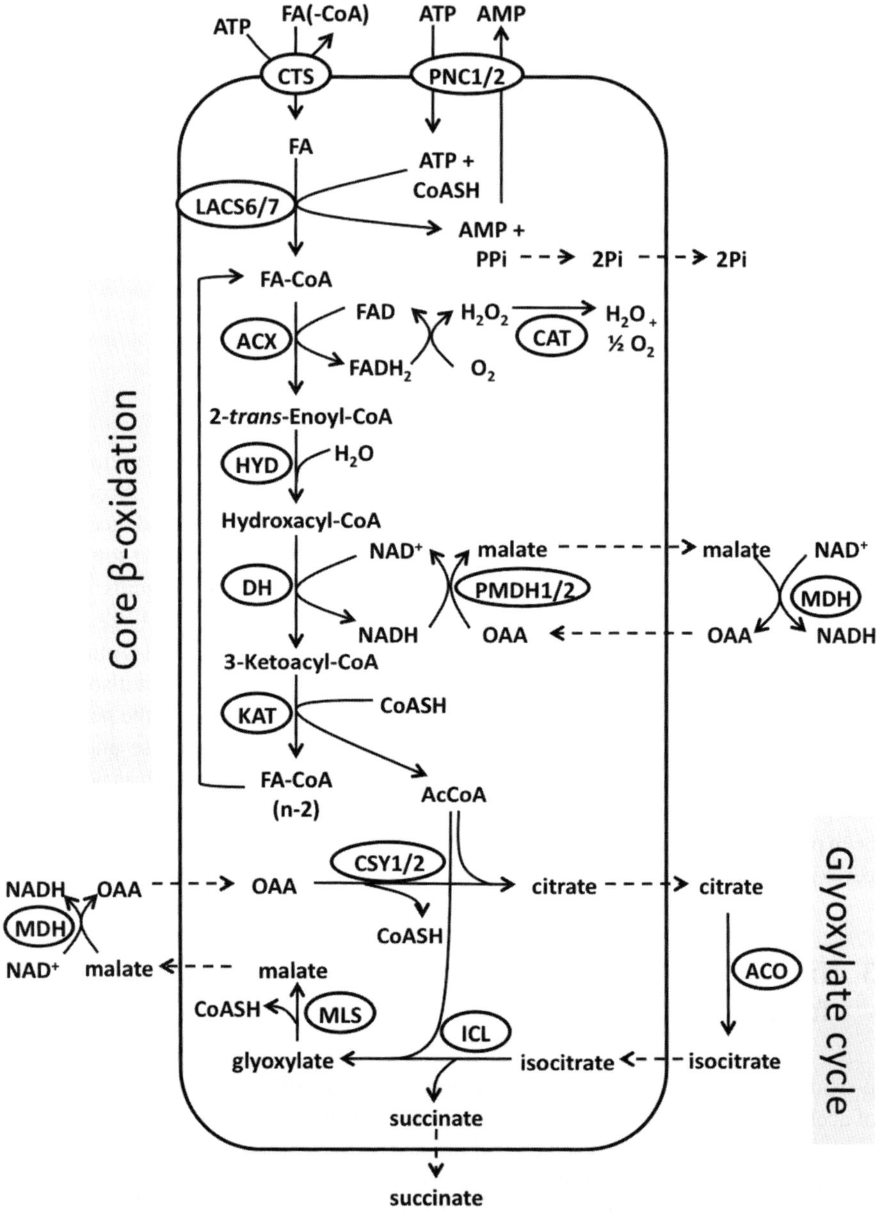

Fig. 1 β-oxidation, glyoxylate cycle and associated transport processes. Fatty acids (FA) or fatty acyl-CoAs (FA-CoA) are imported by the ABC transporter, COMATOSE (CTS). In the case of CoA esters, it is possible, though unproven, that the CoA moiety is cleaved off by peroxisomal thioesterases, or even by CTS (not shown). Long chain acyl-CoA synthetases 6 and 7 (LACS6/7) catalyse ATP-dependent formation of FA-CoA. ATP is imported by peroxisomal nucleotide carriers 1 and 2 (PNC1/2), in counter-exchange for AMP. Pyrophosphate generated by acyl-CoA synthetases probably decomposes into two molecules of orthophosphate which may be

enzymes, which allow "non-standard" substrates to pass though the β-oxidation spiral (Graham and Eastmond 2002; Reumann et al. 2004). As such, β-oxidation plays key roles in growth and development, throughout the plant life cycle (Table 2) (Footitt et al 2006, 2007a; Graham 2008; Kunz et al. 2009; Pinfield-Wells et al. 2005).

3.1 ABC Transporters

Import of substrates for β-oxidation requires the ATP Binding Cassette (ABC) transporter, COMATOSE (CTS; also known as PED3, AtPXA1, ACN2, AtABCD1). CTS was identified in several independent forward genetic screens, selecting for mutants impaired in germination potential, for sugar-dependent mutants, and for mutants resistant to indole butyric acid (IBA), 2,4-dichlorophenoxy-butyric acid (2,4-DB) and fluoroacetate (Eastmond 2006; Footitt et al. 2002; Hayashi et al. 2002; Hooks et al. 2007; Russell et al. 2000; Zolman et al. 2001a). Thus, analysis of *cts* null mutants has provided a great deal of insight into the physiological and biochemical functions of CTS and, by extension, β-oxidation. *cts* mutants do not germinate in the absence of classical dormancy-breaking treatments and remain in a physiological state that is intermediate between that of dormant and non-dormant wild-type seeds (Footitt et al. 2006). Accordingly, transcriptome analysis revealed that CTS is required for the expression of a subset of genes late in phase II of germination (Carrera et al. 2007). *cts* seeds can be made to germinate by mechanically disrupting the testae and plating on media containing sugar. Null mutants fail to complete seedling establishment in the absence of an exogenous energy source such as sucrose, since they are unable to break down storage triacylglycerol (TAG) to provide energy and carbon skeletons before the photosynthetic apparatus is functional. Interestingly, the inability to rescue the germination phenotype by sucrose alone implies a role for CTS which is distinct from TAG mobilisation

Fig. 1 (Continued) exported from the peroxisome by an as-yet uncharacterised phosphate transporter. "Core" β-oxidation is initiated by acyl CoA oxidase (ACX), a FAD-requiring enzyme which yields a 2-*trans*-enoyl CoA. The regeneration of FAD produces H_2O_2 which is degraded by catalase (CAT). The subsequent 2-*trans*-enoyl CoA hydratase (HYD) and hydroxyacyl-CoA dehydrogenase (DH) reactions are catalysed by multifunctional proteins in plants. The DH reaction produces NADH, which is reoxidised by peroxisomal malate dehydrogenases (PMDH1/2). Malate is thought to be exported from the peroxisome for conversion to oxaloacetate (OAA) by cytosolic malate dehydrogenase (MDH) at the expense of mitochondrial reducing power. Peroxisomal hydroxypyruvate reductase also contributes to NAD^+ re-oxidation in Arabidopsis seedlings (not shown). The final step of β-oxidation is catalysed by 3-ketoacyl-CoA thiolase (KAT), which generates acetyl CoA (AcCoA) plus FA-CoA shortened by 2 carbons. Acetyl CoA enters the glyoxylate cycle, which yields 4-carbon compounds via the sequential action of peroxisomal citrate synthase (CSY1/2), cytosolic aconitase (ACO), the glyoxylate cycle enzymes, isocitrate lyase (ICL) and malate synthase (MLS), followed by cytosolic malate dehydrogenase (MDH). This requires import and export of organic acids which then participate in the TCA cycle or gluconeogenesis. Transport steps for which the transporter has not yet been identified are indicated by dashed arrows, but are probably mediated by porin-like proteins

Table 2 Physiological and biochemical roles of β-oxidation in plants

Physiological role	Biochemical function	References
Embryo development	Unknown	Rylott et al. (2003)
Seedling establishment	Mobilisation of seed TAG for energy and carbon skeletons	Graham (2008)
Germination	Mobilisation of seed TAG; metabolism of unknown signaling compounds? Required for gene expression, including ABI5.	Baker et al. (2006), Footitt et al. (2006), Pinfield-Wells et al. (2005), Pracharoenwattana et al. (2005), Carrera et al. (2007), Kanai et al. (2010)
Fertility	JA biosynthesis, mobilisation of pollen oil reserves; IBA metabolism in filaments; inflorescence development	Footitt et al. (2007a, b), Richmond and Bleeker (1999), Theodoulou et al. (2005)
Wound response	JA biosynthesis	Theodoulou et al. (2005)
Root and cotyledon growth	IBA metabolism	Strader et al. 2010, Zolman et al. (2000; 2001a)
Pathogen response	SA metabolism?	Reumann et al. (2004)
	Acetate metabolism	Hooks et al. (2007)
Senescence and carbon starvation	Mobilisation of membrane lipids; branched chain amino acid degradation	Kunz et al. (2009), Slocombe et al. (2009), Lucas et al. (2007), Zolman et al. (2001b), Lange et al. (2004)

(Baker et al. 2006; Footitt et al. 2006; Pinfield-Wells et al. 2005). In agreement with this, it has recently been shown that CTS promotes seed germination by suppressing expression of polygalacturonase inhibitor proteins in a pathway that involves the transcription factor, ABI5 (Kanai et al. 2010). Thus CTS may promote radicle protrusion from the seed coat in WT seeds. Following establishment, *cts* plants are able to complete a full life cycle but have altered root morphology, are smaller than wild type plants and are impaired in fertilisation (Zolman et al. 2001a; Footitt et al. 2007a). CTS also plays a role in dark-induced senescence (Kunz et al. 2009; Slocombe et al. 2009). These phenotypes are attributable to different functions of β-oxidation during the life cycle of Arabidopsis.

The identification of CTS alleles in screens for IBA- and 2,4-DB- resistant mutants indicated a potential role for CTS in auxin metabolism (Hayashi et al. 1998, 2002; Zolman et al. 2000, 2001a). IBA and 2,4-DB are metabolised by one round of β-oxidation to produce indole acetic acid (IAA) and 2,4-dichlorophenoxy acetic acid (2,4-D), respectively (Fig. 2). These compounds cause stunting of roots, which is a readily-scorable phenotype. β-oxidation of IBA is not the sole biosynthetic route to IAA in Arabidopsis (Ljun et al. 2002), but this branch of the pathway appears to be important at several distinct developmental stages. *cts* mutants make fewer lateral roots than WT, unless supplied with exogenous IAA, suggesting a role in promoting lateral root formation (Zolman et al. 2001a) and the IBA to IAA conversion also contributes to stamen elongation, since the short filament phenotype of *cts* alleles can be rescued by application of exogenous NAA (Footitt et al. 2007a). Recently, analysis of IBA response mutants has revealed a role for IBA-derived IAA in driving root hair and cotyledon cell expansion (Strader et al. 2010).

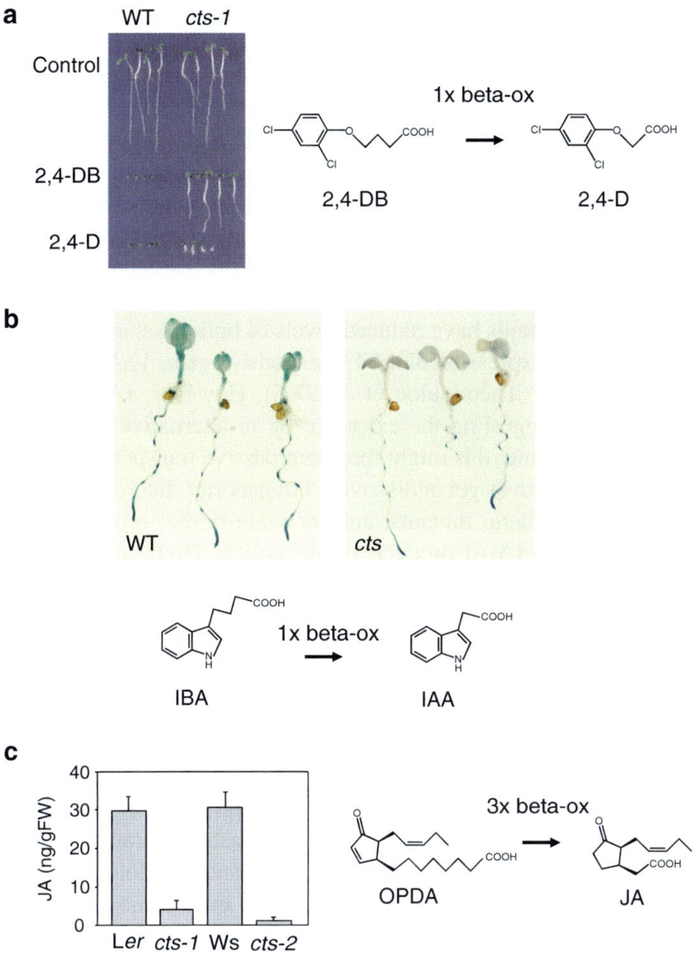

Fig. 2 Role of CTS in β-oxidation of ring-containing compounds (**a**) In wild type seedlings, 2,4-dichlorophenoxybutyric acid (2,4-DB) undergoes one round of β-oxidation to produce the auxin-like herbicide, 2,4-dichlorophenoxyacetic acid (2,4-D), which stunts roots. *cts* mutants are resistant to 2,4-DB because they cannot convert it to the bioactive 2,4-D. The same principle applies to metabolism of indole butyric acid (IBA). (**b**) Histochemical staining of WT and *cts* seedlings expressing the auxin reporter, *DR5::GUS* indicate that auxin levels are reduced in plants which cannot import IBA into the peroxisome and convert it to indole acetic acid (IAA) by β-oxidation. (**c**) Leaves of *cts* plants have reduced basal JA levels. JA synthesis is completed in the peroxisome, via the reduction of 12-oxo-phytodienoic acid (OPDA) to 3-oxo-2(2′[Z]-pentenyl)-cyclopentane-1-octanoic acid (OPC:8) followed by activation and three cycles of β-oxidation. Acknowledgements: the photograph in panel (**a**) is reproduced with permission from Footitt et al. (2002). The graph in panel (**c**) is reproduced with permission from Theodoulou et al. (2005)

Histochemical staining of *cts* and WT plants expressing the auxin reporter, *DR5::GUS* is consistent with this (Fig. 2; Theodoulou and Zhang, unpublished data).

The fact that CTS apparently handles ring-containing molecules with a short acyl chain suggested that it might be a broad-specificity transporter, which would accept other substrates. This led to the hypothesis that CTS may play a role in jasmonic acid (JA) biosynthesis. JA synthesis is initiated in the chloroplast, where 12-oxophytodienoic acid (OPDA) is produced from linolenic acid (18:3) in several steps (Schaller and Stintzi 2009). OPDA is then transferred from the plastid to the peroxisome, where it undergoes reduction to 3-oxo-2(2′[Z]-pentenyl)-cyclopentane-1-octanoic acid (OPC:8), followed by activation and three rounds of β-oxidation to yield JA (Fig. 2). *cts* mutants have reduced levels of both basal and wound-inducible JA and exhibit reduced expression of the JA-responsive gene, *VSP2*, consistent with a role in JA biosynthesis (Theodoulou et al 2005). However, JA is not completely absent in *cts* tissues, suggesting the existence of an alternative route for import of OPDA into the peroxisome: this might represent passive transport by anion trapping, but could also be due to an as yet undiscovered transporter. Interestingly, *cts* mutants, unlike other JA biosynthetic mutants, are not male-sterile, probably because they have sufficient residual JA to produce fertile pollen. However, in common with other β-oxidation alleles, they do exhibit defects in fertilisation unrelated to JA-dependent phenomena (Footitt et al. 2007a, b). Transmission of *cts* through the male gametophyte is considerably reduced and pollen tubes of *cts* mutants grown in vitro are shorter than WT unless supplied with sucrose. This probably reflects their inability to mobilise pollen lipids, but measurements of pollen tube growth in vivo suggest that β-oxidation also plays a role in the female sporophytic tissues (Footitt et al. 2007a). The senescence phenotypes of *cts* alleles are also attributable to impaired lipid metabolism: in dark-grown leaves, β-oxidation provides metabolic energy via respiration of free fatty acids and chloroplast membrane lipids (Kunz et al. 2009; Slocombe et al. 2009). Further biochemical and physiological functions for CTS have been proposed, including a possible role in phytanoyl CoA degradation (Ishizaki et al. 2005; Baker et al. 2006) and a potential role in salicylic acid metabolism (Reumann et al. 2004). However, these await experimental confirmation.

Taking all the available evidence together, it is likely that CTS is a transporter with broad substrate specificity, which mediates import of diverse substrates (known and unknown) for β-oxidation, with differing physiological outputs. Multi-specificity is a classical feature of certain ABC transporters but is by no means the only possibility: for example, CTS could be a regulator of other transport processes. The mammalian ABC transporter superfamily contains atypical proteins, which have intrinsic channel activity (cystic fibrosis transmembrane conductance regulator; CFTR) or channel regulatory functions (sulfonylurea receptor; SUR and P-glycoprotein) (Dean et al. 2001). However, one piece of evidence in favour of a pump with multiple substrates arises from in vivo studies of WT seedlings: in addition to its inhibitory effect on root growth, IBA reduces hypocotyl extension in the dark, an effect which is potentiated by sucrose (Dietrich et al. 2009). Since lipid breakdown is retarded markedly by the presence of sucrose in the growth medium (Martin et al. 2002; Fulda et al. 2004), this is consistent with a scenario where IBA and fatty acids compete for transport by CTS and reduced flux of fatty acids through β-oxidation in the presence of sucrose enables increased conversion of IBA to IAA, resulting in hypocotyl shortening. It could also be argued that the different functions of CTS can be separated to some extent by mutagenesis may be consistent with a role as a primary pump rather than a regulator (Dietrich et al. 2009).

Assuming that CTS acts as a broad specificity pump, the question of the biochemical identity of its substrates arises: free acids or acyl-CoA esters? Activation of substrates such as fatty acids by esterification to CoA is a prerequisite for β-oxidation but potentially, this could occur inside or outside the peroxisome. Plants contain a large family of acyl activating enzymes with differing substrate specificities, which are distributed in different intracellular compartments, including plastids, microsomes, cytosol and peroxisomes (Shockey et al. 2003; Reumann et al. 2004). In baker's yeast, long chain fatty acids (LCFA) are activated outside the peroxisome and their CoA esters imported by the heterodimeric ABC transporter, Pxa1p/Pxa2p, which is homologous to CTS (Shani et al. 1995; Hettema et al. 1996; Shani and Valle 1996; Swartzman et al. 1996). In contrast, short and medium chain FA cross the peroxisome by an unknown mechanism (possibly passive transport, although a requirement for the peroxin, Pex11 has been suggested; van Roermund et al. 2000) and are activated by the peroxisomal acyl CoA synthetase, Faa2p (Hettema et al. 1996). Although transport data have not been published to date, experiments employing selective solubilisation of the plasma membrane suggest that long chain fatty acyl-CoAs (LCFA-CoA) and not free acids are the substrates of Pxa1p/Pxa2p (Verleur et al. 1997a). By analogy with yeast, it seems likely that CTS is also a transporter of FA-CoA, and the observation that FA-CoA are accumulated in *cts* cotyledons supports this notion, though it is also possible that the CoA pool simply represents a sink for fatty acids which cannot be esterified into membrane lipids (Footitt et al. 2002). However, genetic experiments are more consistent with free FA as substrates. Arabidopsis has two, redundant peroxisomal acyl CoA synthetases, LACS6 and 7, which handle fatty acids with a range of different chain lengths (Fulda et al. 2002). The seedling establishment phenotype of the *lacs 6 lacs7* double mutant is identical to that of *cts*, suggesting that CTS and LACS operate in the same, rather than parallel pathways (Fulda et al. 2004). Similarly, the identification of CTS and a peroxisomal acetyl CoA synthetase in a screen for fluoroacetate resistant mutants is consistent with transport of a free acid (acetate) followed by peroxisomal activation (Turner et al. 2005; Hooks et al. 2007). Knockdown of the peroxisomal adenine nucleotide translocators also supports this hypothesis (see below). To rationalise these apparently contradictory possibilities, it has been suggested that CTS may cleave the CoA moiety during the transport cycle (Fulda et al. 2004); an alternative hypothesis is that acyl-CoAs are cleaved upon import into the peroxisome by thioesterases and therefore require re-activation before entering the β-oxidation spiral (Hunt and Alexson 2008).

Expression of CTS in baker's yeast is a first step to resolving this debate. CTS has recently been shown to complement the yeast *Δpxa1 Δpxa2* double mutant for growth on oleate and β-oxidation of a range of fatty acids. Moreover, peroxisomes expressing recombinant CTS exhibit ATPase activity, which could be stimulated by addition of FA-CoA, but not free FA (Nyathi et al. 2010). So-called "substrate stimulation" of basal ATPase activity is a hallmark feature of many ABC transporters but does not constitute unequivocal proof that FA-CoA are transport substrates. Indeed, it has been suggested that CoA esters may play regulatory roles in plant cells, as has been shown in mammalian systems (Graham et al. 2002). Ultimately, transport studies with reconstituted protein will be required to determine precisely the molecular species, which is transported across the peroxisomal membrane.

3.2 Adenine Nucleotide Translocator

A peroxisomal pool of ATP is required for the activation of substrates prior to β-oxidation. Proteomic studies and homology searches have revealed two peroxisomal adenine nucleotide carriers in Arabidopsis, named PNC1 and 2, which belong to the mitochondrial carrier family (MCF) of solute transporters (Arai et al. 2008b; Linka et al. 2008). PNC1 and 2 both complement a yeast mutant deficient in peroxisomal ATP uptake and studies employing recombinant protein demonstrated ATP transport in strict counter-exchange with ATP, ADP or AMP (Linka et al. 2008). Under physiological conditions, it is likely that PNC1 and 2 facilitate ATP/AMP exchange to support the activity of acyl CoA synthetases (Fig. 1). Accordingly, plants in which expression of both genes is reduced by RNAi exhibit phenotypes similar to those of severe β-oxidation mutants, with defects in storage oil mobilisation, seedling growth and auxin metabolism (Arai et al. 2008b; Linka et al. 2008). This indicates that there is no other ATP-generating system in plant peroxisomes. Additionally, the RNAi plants have a growth phenotype, which is not rescued by exogenous sucrose, indicating functions for the peroxisomal ATP pool beyond β-oxidation, consistent with the identification of kinases and other ATP-utilising enzymes in the plant peroxisomal proteome (Reumann et al. 2007).

A third member of the MCF family (encoded by At2g39970) is also present in the Arabidopsis peroxisomal membrane, although this does not appear to play a role in ATP import, as judged by lack of yeast complementation and analysis of the recombinant protein (Linka et al. 2008). The function of this protein remains to be determined.

3.3 The Peroxisomal CoA Budget

In addition to ATP, peroxisomal acyl CoA synthetases also require free CoA and CoA is a cofactor for 3-ketoacyl-CoA thiolase in the final step of β-oxidation. CoA is released during the glyoxylate cycle as a product of the citrate synthase and malate reactions and many texts discuss "acetyl CoA export" from peroxisomes but it should be noted that citrate and/or succinate and not acetyl CoA are the exported species (Fig. 1, and see below). Precise details regarding the establishment and maintenance of the peroxisomal CoA pool remain to be determined: according to one school of thought, the peroxisomal membrane is impermeable to free CoA (Antonenkov et al. 2004a, b; van Roermund et al. 1995) and it has been argued that the peroxisome has a discrete CoA pool which is established upon organelle biogenesis and which is not supported by net import. However, an Arabidopsis mutant defective in CoA biosynthesis requires sucrose for seedling establishment and exhibits other hallmarks of impairment in β-oxidation, such as retention of FA and FA-CoA (Rubio et al. 2006). This indicates a need for continued CoA biosynthesis during the process of seedling establishment and argues against a scenario in which a "biogenesis pool" of CoA is sufficient to maintain high rates of FA β-oxidation. To date, no peroxisomal transporter for free CoA has been identified but in yeast at least, CoA is imported into peroxisomes in the form of long chain acyl-CoAs via Pxa1p/Pxa2p (Hettema et al. 1996; Verleur et al. 1997a). Medium and short-chain FA however, are reliant on the peroxisomal CoA pool to enter β-oxidation, which has implications for the control of flux through this pathway and suggests that CoA supply could be a limiting factor. Although it is energetically costly, removal of the CoA moiety might therefore play a role in the regulation of β-oxidation; indeed, characterisation of peroxisomal thioesterases in mammalian systems supports this hypothesis (Hunt and Alexson 2008). In plants, the identification of multiple acyl CoA synthetases with specificity for different intermediates of JA biosynthesis and the detection of de-esterified JA intermediates in plant extracts both argue that CoA is repeatedly cleaved from intermediates and re-esterified during β-oxidation (Koo and Howe 2007; Kienow et al. 2008; Schaller and Stintzi 2009). Mammalian peroxisomes also contain a small family of Nudix hydrolases, enzymes able to degrade acyl-CoAs and free CoA. These may contribute to the regulation of the CoA pool by determining availability of free CoA and possibly by removing slowly-metabolised CoA species which inhibit flux though β-oxidation (Antonenkov and Hiltunen 2006; Hunt and Alexson 2008).

3.4 Phosphate Transport

Activation of fatty acids and other substrates by esterification to CoA generates pyrophosphate, which is thought to decompose into two molecules of inorganic phosphate. However, the yeast adenine nucleotide translocator does not exchange ATP for phosphate (Palmieri et al. 2001), implying the existence of an alternative export route for this by-product of β-oxidation. Studies with proteoliposomes isolated from bovine kidney peroxisomes demonstrated a phosphate transport activity in the peroxisomal membrane, but the corresponding gene has not yet been cloned (Visser et al. 2005). Thus, it is plausible but as yet unproven, that plant peroxisomes also contain a phosphate transporter.

3.5 Import of Other Cofactors

Core β-oxidation requires FAD and NAD^+ as cofactors and the auxiliary enzyme, Δ^2-Δ^4-dienoyl CoA reductase (required for β-oxidation of unsaturated fatty acids with double bond at even-numbered positions) uses $NADP^+$. As for CoA, there is at present no experimental evidence to support the existence of peroxisomal transporters for these cofactors (Antonenkov et al. 2004a, b; van Roermund et al. 1995). Whilst FAD has been shown to be imported with the folded acyl-CoA oxidase protein (Titorenko et al. 2002) and also with pre-assembled oligomeric alcohol oxidase (Ozimek et al. 2003), the source of the peroxisomal pools of nicotinamide cofactors is unknown. These could however, be co-imported with folded proteins, by analogy with peroxisomal FAD-containing enzymes. The appropriate redox state of these cofactors is maintained by a series of metabolites shuttles (see below).

4 Glyoxylate Cycle and Fatty Acid Respiration

Acetyl CoA produced by β-oxidation is converted to 4-carbon compounds by the glyoxylate cycle (Fig. 1). Following export from the peroxisome, these intermediates can enter the mitochondrial TCA cycle to provide metabolic energy or are used in gluconeogenesis. In plants lacking functional isocitrate lyase or malate synthase, acetyl units from β-oxidation can be respired and the glyoxylate produced in the *mls* mutant is transferred to the photorespiratory pathway. Consequently, *icl* and *mls* mutants do not exhibit the strong phenotypes characteristic of fatty acid β-oxidation mutants, which are dependent on sucrose for seedling establishment (Eastmond et al. 2000; Cornah et al. 2004). In contrast, the strong phenotype of the peroxisomal citrate synthase double mutant, *csy2 csy3* provides strong evidence that the "export" of acetyl units from the peroxisome is absolutely dependent on their conversion to citrate (Pracharoenwattana et al. 2005). This is in contrast to baker's yeast, in which a carnitine shuttle operates in addition to citrate export (van Roermund et al. 1999).

4.1 Metabolite and Redox Shuttles; Transport Requirements of the Glyoxylate Cycle

Examination of Fig. 1 reveals that the operation of the glyoxylate cycle in plants requires several hypothetical membrane transport steps: export of malate, citrate and succinate, and import of oxaloacetate and isocitrate (reviewed in: Kunze et al. 2006). It is also been proposed that oxaloacetate is not imported during the glyoxylate cycle, but is generated from aspartate and 2-oxoglutarate, with the generation of glutamate in a transamination reaction catalysed by aspartate amino transferase (Mettler and Beevers 1980). However, the operation of a malate/2-oxoglutarate shuttle has been disputed, based on subsequent evidence (Schmitt and Edwards 1983; Verleur et al. 1997b). Peroxisomal malate dehydrogenase, which catalyses re-oxidation of NADH generated by the dehydrogenase reaction of β-oxidation, also requires the export and import of malate and oxaloacetate, respectively (Pracharoenwattana et al. 2007). A role for hydroxypyruvate reductase as an alternative route for NADH re-oxidation has recently been demonstrated (Pracharoenwattana et al. 2010) which implies the presence of activities permitting the export of glycerate and glycine and the import of serine. These transport steps are also important in photorespiration (see below).

Currently, the molecular identity of the glyoxylate cycle metabolite transporters is unknown, but it is thought that anion-selective porins mediate the traffic of these metabolic intermediates (Reumann 2000). Electrophysiological studies of isolated peroxisomes from spinach and castor bean revealed the presence of a pore-forming channel with specificity for organic anions including malate, oxaloacetate, succinate, glycolate, glycerate, glutamate and 2-oxoglutarate (Reumann et al. 1995, 1997, 1998).

5 Photorespiration

Photorespiration is initiated when Rubisco accepts oxygen, rather than CO_2 as a substrate, resulting in the formation of phosphoglycolate from ribulose-1,5-bisphosphate (Reumann and Weber 2006; Foyer et al. 2009). Following a chloroplastic dephosphorylation step, the resulting glycolate is transferred to the peroxisome, where it is oxidised to glyoxylate, with the concomitant generation of H_2O_2. Glyoxylate undergoes transamination by two peroxisomal aminotransferases, glutamine:glyoxylate amino transferase (GGT) and serine:glyoxylate amino transferase (SGT) to yield glycine and hydroxypyruvate (Fig. 3). The glycine produced is converted to serine in mitochondria, which enters the peroxisome and is used in the SGT reaction. Hydroxypyruvate is reduced at the expense of NADH to glycerate, which is then returned to the chloroplast to enter the Calvin cycle (Reumann and Weber 2006). Thus, photorespiration requires several peroxisomal transport steps to transfer metabolites between the chloroplasts, mitochondria and peroxisomes. The peroxisomal transport steps are probably accomplished by porins, as judged by the permeability of leaf peroxisome channels to photorespiratory intermediates (Reumann et al. 1998). It was originally thought that peroxisomal malate dehydrogenase was responsible for NADH regeneration, but the two isoforms of this enzyme only play a relatively minor role, since the *pmdh1 pmdh2* double mutant is not markedly impaired in photorespiration (Cousins et al. 2008). Additional mechanisms for supply of reductant must therefore exist or alternatively, the peroxisomal HPR reaction is circumvented by a cytosolic step (Timm et al. 2008).

6 Peroxisomal pH

An important question in peroxisomal transport is the existence of a pH gradient across the peroxisomal membrane, since this is a critical factor in determining the rate of potential passive transport of solutes. However, peroxisomal pH has not

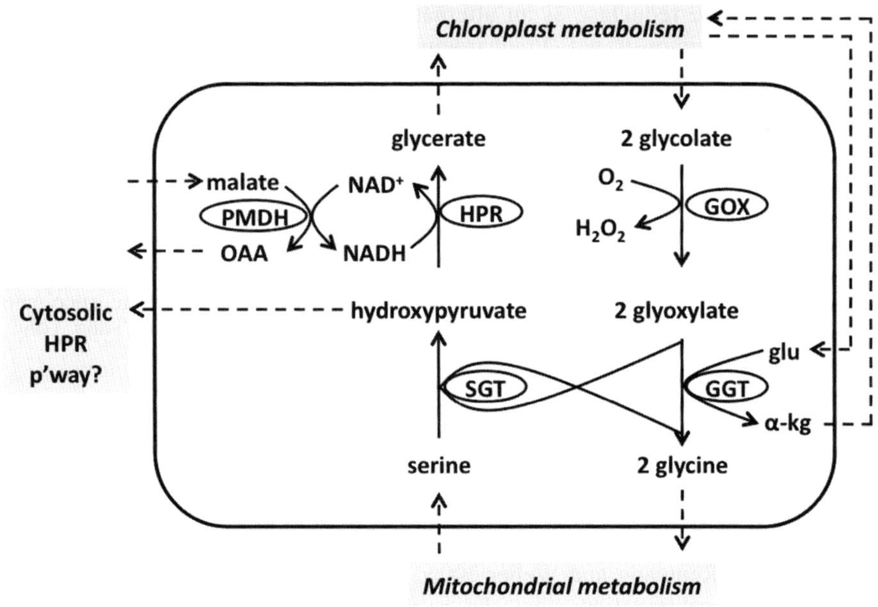

Fig. 3 Role of the peroxisome in photorespiration and associated transport processes. Photorespiration is initiated in the chloroplast by the conversion of ribulose-1,5-bisphosphate to phosphoglycolate by the oxygenating activity of Rubisco. Phosphoglycolate is then dephosphorylated and glycolate is transferred from the chloroplast to the peroxisome, where it is converted to glycerate in several enzymatic steps. Abbreviations: GOX glycolate oxidase, GGT glutamine:glyoxylate amino transferase, α-kg α-ketoglutarate; SGT serine:glyoxylate amino transferase, HPR hydroxypyruvate reductase, PMDR peroxisomal malate dehydrogenase, OAA oxaloacetate. Although H_2O_2 can be detoxified by peroxisomal catalase, it may also interact non-enzymatically with glyoxylate and hydroxypyruvate to yield formate and glycolate, respectively (not shown). Re-drawn from Cousins et al. (2008)

been measured in plant cells and reports of peroxisomal pH measurements in mammals and yeasts have been extremely contradictory (reviewed in Rottensteiner and Theodoulou 2006). Basic pH values have been reported both for human fibroblasts and baker's yeast (Dansen et al. 2000; van Roermund et al. 2004), another report concluded that the peroxisomal pH of fibroblasts and Chinese hamster ovary cells adapts to that of the cytosol (Jankowski et al. 2001) and a further study reported that the peroxisome of baker's yeast is acidic (Lasorsa et al. 2004). These markedly different conclusions may reflect the fact that all these studies employed different experimental approaches to pH measurement. It should also be noted that peroxisomal pH may vary in response to prevailing metabolic conditions and may also differ between organisms, for example, the methylotropic yeast, *Hansenula polymorpha*, has been reported to have an acidic peroxisome lumen, which is required for the enzymology of methanol utilisation (van der Klei et al. 1991). The question of peroxisomal pH and its maintenance therefore remains to be resolved. Despite extensive proteomic investigations, there is no evidence for a proton pump in plant peroxisomal membranes, so the manner by which a pH gradient could be generated is called into question. Two studies conclude that the adenine nucleotide translocator is instrumental in establishment of the pH gradient across the yeast peroxisomal membrane, but invoke different mechanisms (Lasorsa et al. 2004; van Roermund et al. 2004). It has also been proposed that the pH gradient between the cytosol and the peroxisome lumen is created via a Donnan equilibrium (Antonenkov and Hiltunen 2006; Rokka et al. 2009). In this scenario, electroneutrality is maintained by the equilibration of ions across the membrane to balance the charge on impermeable macromolecules, including lumen proteins and bulky solutes (such as cofactors) (Price et al. 2001). This requires the free permeation of ions (including H^+ and OH^-) across the peroxisome membrane and the gradient across the membrane depends on the differences in overall charges of molecules such as proteins, which are unable to cross the membrane. In this context, it is interesting to note that the basic pI of many peroxisome proteins has been invoked as evidence for a basic pH lumen (Dansen et al. 2000): the Donnan equilibrium hypothesis predicts that positively-charged matrix proteins would attract negatively-charged solutes, thus forming an inside-basic pH gradient.

7 Peroxisomes as a Source of Signaling Molecules

Peroxisomes generate signaling molecules that fall into four broad classes: bioactive molecules derived from β-oxidation, reactive oxygen species (ROS), reactive nitrogen species (RON) and changes in the peroxisomal redox state (Nyathi and Baker 2006). Both ROS and RON can diffuse freely across membranes and do not require transporters, however, various transport steps are implicated in metabolism associated with ROS and RON generation and scavenging pathways.

RON generation plant peroxisomes is not well understood, but it is well established that the enzymatic complement of peroxisomes has a significant capacity to generate reactive oxygen species, including superoxide and H_2O_2 (Table 1; reviewed in Nyathi and Baker 2006; del Río et al. 2006). Although it has been suggested that peroxisomes are the major site of H_2O_2 production in C3 plants during photorespiration (Foyer and Noctor 2003), and that this signal impacts on transcription (Vandenabeele et al. 2004), the wider physiological relevance of peroxisomal ROS signaling is not yet fully understood. However, peroxisomes possess an efficient ROS scavenging system, to minimise potentially deleterious effects of oxidative damage. In addition to superoxide dismutase and catalase, current evidence indicates that peroxisomes are also equipped with an ascorbate-glutathione cycle, comprising ascorbate peroxidase, monodehydroascorbate reductase, dehydroascorbate reductase and glutathione reductase (Nyathi and Baker 2006; del Río et al. 2006; Kaur et al. 2009). Additionally, peroxisomes contain three theta-class glutathione transferases, which exhibit glutathione peroxidase activity (Reumann et al. 2007; Dixon et al. 2009). The presence of an ascorbate-glutathione cycle requires intraperoxisomal pools of glutathione and ascorbate, but it is not clear how these are generated as transporters for these antioxidants have not been identified in the peroxisomal membrane. It should be noted that the ascorbate-glutathione cycle does not result in net consumption of antioxidants (and could in theory, at least, be sustained by a "biogenesis pool"), but does result in oxidation of cofactors. NADH is thought to be regenerated by a glucose-6-phosphate dehydrogenase-dependent mechanism and NADPH via an isocitrate/2-oxoglutarate shuttle, which require transport of carboxylates across the peroxisomal membrane (Nyathi and Baker 2006; Rottensteiner and Theodoulou 2006; Kaur et al. 2009). The operation of these regenerating systems is supported by enzymatic measurements (Corpas et al. 1998, 1999) and in silico predictions of enzyme location (Reumann et al. 2004, 2007, 2009; Eubel et al. 2008), but they have yet to be tested, for example, by specific knock-down of different components.

The ROS generating and scavenging activities of peroxisome offer considerable scope for alterations in the peroxisomal redox state, as determined by ratios of NAD(P):NAD(P)H; GSH:GSSG; ascorbate:dehydroascorbate. In other compartments, these redox couples have potent signaling activities (Noctor 2006), but the significance with respect to peroxisomal metabolism remains to be determined.

8 Conclusions

Although a great deal has been learnt about plant peroxisomal functions in recent years and the molecular details of transport processes associated with peroxisomal metabolism are beginning to be uncovered, much remains to be discovered. Open questions include: the nature and regulation of intraperoxisomal pH, the identity of peroxisomal transporters responsible for exporting metabolites and products which are generated in this organelle and the regulation of diverse transport processes. A key question is how the balance between different functions is maintained: in several cases, enzymes and transporters handle metabolites belonging to different pathways and very little is known of how the operation of these pathways is managed. We are still some way from understanding how transport processes are regulated to co-ordinate peroxisomal metabolism with that of other cellular compartments, but with many more experimental tools at our disposal, this promises to be an intriguing area for future investigation.

References

Antonenkov VD, Hiltunen JK (2006) Peroxisomal membrane permeability and solute transfer. Biochim Biophys Acta 1763:1697–1706

Antonenkov VD, Sormunen RT, Hiltunen JK (2004a) The rat liver peroxisomal membrane forms a permeability barrier for cofactors but not for small metabolites in vitro. J Cell Sci 117:5633–5642

Antonenkov VD, Sormunen RT, Hiltunen JK (2004b) The behavior of peroxisomes in vitro: mammalian peroxisomes are osmotically sensitive particles. Am J Physiol Cell Physiol 287: C1623–C1635

Antonenkov VD, Rokka A, Sormunen RT, Benz R, Hiltunen JK (2005) Solute traffic across mammalian peroxisomal membrane – single channel conductance monitoring reveals pore-forming activities in peroxisomes. Cell Mol Life Sci 62:2886–2895

Antonenkov VD, Mindthoff S, Grunau S, Erdmann R, Hiltunen JK (2009) An involvement of yeast peroxisomal channels in transmembrane transfer of glyoxylate cycle intermediates. Int J Biochem Cell Biol 41:2546–2554

Arai Y, Hayashi M, Nishimura M (2008a) Proteomic analysis of highly purified peroxisomes from etiolated soybean cotyledons. Plant Cell Physiol 49:526–539

Arai Y, Hayashi M, Nishimura M (2008b) Proteomic identification and characterization of a novel peroxisomal adenine nucleotide transporter supplying ATP for fatty acid beta-oxidation in soybean and Arabidopsis. Plant Cell 20:3227–3240

Babujee L, Wurtz V, Ma C, Lueder F, Soni P, van Dorsselaer A, Reumann S (2010) The proteome map of spinach leaf peroxisomes indicates partial compartmentalization of phylloquinone (vitamin K1) biosynthesis in plant peroxisomes. J Exp Bot 61:1441–1453

Baker A, Graham IA, Holdsworth M, Smith SM, Theodoulou FL (2006) Chewing the fat: beta-oxidation in signalling and development. Trends Plant Sci 11:124–132

Bednarek P, Pislewska-Bednarek M, Svatos A, Schneider B, Doubsky J, Mansurova M, Humphry M, Consonni C, Panstruga R, Sanchez-Vallet A, Molina A, Schulze-Lefert P (2009) A glucosinolate metabolism pathway in living plant cells mediates broad-spectrum antifungal defense. Science 323:101–106

Byrne RS, Hänsch R, Mendel RR, Hille R (2009) Oxidative half-reaction of *Arabidopsis thaliana* sulfite oxidase: generation of superoxide by a peroxisomal enzyme. J Biol Chem 284:35479–35484

Carrera E, Holman T, Medhurst A, Peer W, Schmuths H, Footitt S, Theodoulou FL, Holdsworth MJ (2007) Gene expression profiling reveals defined functions of the ATP-binding cassette transporter COMATOSE late in phase II of germination. Plant Physiol 143:1669–1679

Carrie C, Murcha MW, Kuehn K, Duncan O, Barthet M, Smith PM, Eubel H, Meyer E, Day DA, Millar AH, Whelan J (2008) Type II NAD(P)H dehydrogenases are targeted to mitochondria and chloroplasts or peroxisomes in *Arabidopsis thaliana*.. FEBS Lett 582:3073–3079

Clay NK, Adio AM, Denoux C, Jander G, Ausubel FM (2009) Glucosinolate metabolites required for an Arabidopsis innate immune response. Science 323:95–101

Cornah JE, Germain V, Ward JL, Beale MH, Smith SM (2004) Lipid utilization, gluconeogenesis, and seedling growth in Arabidopsis mutants lacking the glyoxylate cycle enzyme malate synthase. J Biol Chem 279:42916–42923

Corpas FJ, Barroso JB, Sandalio LM, Distefano S, Palma JM, Lupiáñez JA, Del Río LA (1998) A dehydrogenase-mediated recycling system of NADPH in plant peroxisomes. Biochem J 330:777–784

Corpas FJ, Barroso JB, Sandalio LM, Palma JM, Lupiáñez JA, del Río LA (1999) Peroxisomal NADP-dependent isocitrate dehydrogenase. Characterization and activity regulation during natural senescence. Plant Physiol 121:921–928

Corpas FJ, Hayashi M, Mano S, Nishimura M, Barroso JB (2009) Peroxisomes are required for in vivo nitric oxide accumulation in the cytosol following salinity stress of Arabidopsis plants. Plant Physiol 151:2083–2094

Cousins AB, Pracharoenwattana I, Zhou W, Smith SM, Badger MR (2008) Peroxisomal malate dehydrogenase is not essential for photorespiration in Arabidopsis but its absence causes an increase in the stoichiometry of photorespiratory CO_2 release. Plant Physiol 148:786–795

Dansen TB, Wirtz KW, Wanders RJ, Pap EH (2000) Peroxisomes in human fibroblasts have a basic pH. Nat Cell Biol 2:51–53

Dean M, Rzhetsky A, Allikmets R (2001) The human ATP-binding cassette (ABC) transporter superfamily. Genome Res 11:1156–1166

del Río LA, Sandalio LM, Corpas FJ, Palma JM, Barroso JB (2006) Reactive oxygen species and reactive nitrogen species in peroxisomes. Production, scavenging, and role in cell signalling. Plant Physiol 141:330–335

Dietrich D, Schmuths H, Lousa Cde M, Baldwin JM, Baldwin SA, Baker A, Theodoulou FL, Holdsworth MJ (2009) Mutations in the Arabidopsis peroxisomal ABC transporter COMATOSE allow differentiation between multiple functions in planta: insights from an allelic series. Mol Biol Cell 20:530–543

Dixon DP, Hawkins T, Hussey PJ, Edwards R (2009) Enzyme activities and subcellular localization of members of the Arabidopsis glutathione transferase superfamily. J Exp Bot 60:1207–1218

Eastmond PJ (2006) SUGAR-DEPENDENT1 encodes a patatin domain triacylglycerol lipase that initiates storage oil breakdown in germinating Arabidopsis seeds. Plant Cell 18:665–675

Eastmond PJ, Germain V, Lange PR, Bryce JH, Smith SM, Graham IA (2000) Postgerminative growth and lipid catabolism in oilseeds lacking the glyoxylate cycle. Proc Natl Acad Sci USA 97:5669–5674

Erdmann R, Veenhuis M, Kunau WH (1997) Peroxisomes: organelles at the crossroads. Trends Cell Biol 7:400–407

Eubel H, Meyer EH, Taylor NL, Bussell JD, O'Toole N, Heazlewood JL, Castleden I, Small ID, Smith SM, Millar AH (2008) Novel proteins, putative membrane transporters, and an integrated metabolic network are revealed by quantitative proteomic analysis of Arabidopsis cell culture peroxisomes. Plant Physiol 148:1809–1829

Footitt S, Slocombe SP, Larner V, Kurup S, Wu Y, Larson T, Graham I, Baker A, Holdsworth M (2002) Control of germination and lipid mobilization by COMATOSE, the Arabidopsis homologue of human ALDP. EMBO J 21:2912–2922

Footitt S, Marquez J, Schmuths H, Baker A, Theodoulou FL, Holdsworth M (2006) Analysis of the role of COMATOSE and peroxisomal beta-oxidation in the determination of germination potential in Arabidopsis. J Exp Bot 57:2805–2814

Footitt S, Dietrich D, Fait A, Fernie AR, Holdsworth MJ, Baker A, Theodoulou FL (2007a) The COMATOSE ATP-binding cassette transporter is required for full fertility in Arabidopsis. Plant Physiol 144:1467–1480

Footitt S, Cornah JE, Pracharoenwattana I, Bryce JH, Smith SM (2007b) The Arabidopsis 3-ketoacyl-CoA thiolase-2 (*kat2-1*) mutant exhibits increased flowering but reduced reproductive success. J Exp Bot 58:2959–2968

Foyer CH, Noctor G (2003) Redox sensing and signalling associated with reactive oxygen in chloroplasts, peroxisomes and mitochondria. Physiol Plant 119:355–364

Foyer CH, Bloom AJ, Queval G, Noctor G (2009) Photorespiratory metabolism: genes, mutants, energetics, and redox signalling. Annu Rev Plant Biol 60:455–484

Fulda M, Shockey J, Werber M, Wolter FP, Heinz E (2002) Two long-chain acyl-CoA synthetases from *Arabidopsis thaliana* involved in peroxisomal fatty acid beta-oxidation. Plant J 32:93–103

Fulda M, Schnurr J, Abbadi A, Heinz E, Browse J (2004) Peroxisomal Acyl-CoA synthetase activity is essential for seedling development in *Arabidopsis thaliana*. Plant Cell 16:394–405

Goepfert S, Poirier Y (2007) Beta-oxidation in fatty acid degradation and beyond. Curr Opin Plant Biol 10:245–251

Goyer A, Johnson TL, Olsen LJ, Collakova E, Shachar-Hill Y, Rhodes D, Hanson AD (2004) Characterization and metabolic function of a peroxisomal sarcosine and pipecolate oxidase from Arabidopsis. J Biol Chem 279:16947–16953

Graham IA (2008) Seed storage oil mobilization. Annu Rev Plant Biol 59:115–142

Graham IA, Eastmond PJ (2002) Pathways of straight and branched chain fatty acid catabolism in higher plants. Prog Lipid Res 41:156–181

Graham IA, Li Y, Larson TR (2002) Acyl-CoA measurements in plants suggest a role in regulating various cellular processes. Biochem Soc Trans 30:1095–1099

Grunau S, Mindthoff S, Rottensteiner H, Sormunen RT, Hiltunen JK, Erdmann R, Antonenkov VD (2009) Channel-forming activities of peroxisomal membrane proteins from the yeast *Saccharomyces cerevisiae*.. FEBS J 276:1698–1708

Hänsch R, Mendel RR (2005) Sulfite oxidation in plant peroxisomes. Photosynth Res 86:337–343

Hänsch R, Lang C, Riebeseel E, Lindigkeit R, Gessler A, Rennenberg H, Mendel RR (2006) Plant sulfite oxidase as novel producer of H_2O_2: combination of enzyme catalysis with a subsequent non-enzymatic reaction step. J Biol Chem 281:6884–6888

Hayashi M, Toriyama K, Kondo M, Nishimura M (1998) 2, 4-Dichlorophenoxybutyric acid-resistant mutants of Arabidopsis have defects in glyoxysomal fatty acid beta-oxidation. Plant Cell 10:183–195

Hayashi M, Nito K, Takei-Hoshi R, Yagi M, Kondo M, Suenaga A, Yamaya T, Nishimura M (2002) Ped3p is a peroxisomal ATP-binding cassette transporter that might supply substrates for fatty acid beta-oxidation. Plant Cell Physiol 43:1–11

Hayashi M, Nishimura M (2006) Arabidopsis thaliana–a model organism to study plant peroxisomes. Biochim Biophys Acta 1763:1382–1391

Hennebry SC, Wright HM, Likic VA, Richardson SJ (2006) Structural and functional evolution of transthyretin and transthyretin-like proteins. Proteins 64:1024–1045

Hettema EH, van Roermund CW, Distel B, van den Berg M, Vilela C, Rodrigues-Pousada C, Wanders RJ, Tabak HF (1996) The ABC transporter proteins Pat1 and Pat2 are required for import of long-chain fatty acids into peroxisomes of *Saccharomyces cerevisiae*.. EMBO J 15:3813–3822

Heupel R, Heldt HW (1994) Protein organization in the matrix of leaf peroxisomes. A multi-enzyme complex involved in photorespiratory metabolism. Eur J Biochem 220:165–172

Hooks MA, Turner JE, Murphy EC, Johnston KA, Burr S, Jarosławski S (2007) The Arabidopsis ALDP protein homologue COMATOSE is instrumental in peroxisomal acetate metabolism. Biochem J 406:399–406

Hunt MC, Alexson SE (2008) Novel functions of acyl-CoA thioesterases and acyltransferases as auxiliary enzymes in peroxisomal lipid metabolism. Prog Lipid Res 7:405–421

Ishizaki K, Larson TR, Schauer N, Fernie AR, Graham IA, Leaver CJ (2005) The critical role of Arabidopsis electron-transfer flavoprotein:ubiquinone oxidoreductase during dark-induced starvation. Plant Cell 17:2587–2600

Jankowski A, Kim JH, Collins RF, Daneman R, Walton P, Grinstein S (2001) In situ measurements of the pH of mammalian peroxisomes using the fluorescent protein pHluorin. J Biol Chem 276:48748–48753

Kamada-Nobusada T, Hayashi M, Fukazawa M, Sakakibara H, Nishimura M (2008) A putative peroxisomal polyamine oxidase, AtPAO4, is involved in polyamine catabolism in *Arabidopsis thaliana*.. Plant Cell Physiol 49:1272–1282

Kanai M, Nishimura M, Hayashi M (2010) A peroxisomal ABC transporter promotes seed germination by inducing pectin degradation under the control of ABI5. Plant J 62(6):936–947

Kaur N, Reumann S, Hu J (2009) Peroxisome biogenesis and function. In The Arabidopsis book, American Society of Plant Biologists, Rockville, MD. doi: 10.1199/tab.0123, http://www.aspb.org/publications/arabidopsis/. Accessed 11 Sep 2009

Kienow L, Schneider K, Bartsch M, Stuible HP, Weng H, Miersch O, Wasternack C, Kombrink E (2008) Jasmonates meet fatty acids: functional analysis of a new acyl-coenzyme A synthetase family from *Arabidopsis thaliana*.. J Exp Bot 59:403–419

Knoop F (1904) Der Abbau aromatischer Fettsöuren im Tierkörper. Beitr Chem Physiol Pathol 6:150–162

Koo AJ, Howe GA (2007) Role of peroxisomal beta-oxidation in the production of plant signaling compounds. Plant Signal Behav 2:20–22

Kunz HH, Scharnewski M, Feussner K, Feussner I, Flügge UI, Fulda M, Gierth M (2009) The ABC transporter PXA1 and peroxisomal beta-oxidation are vital for metabolism in mature leaves of Arabidopsis during extended darkness. Plant Cell 21:2733–2749

Kunze M, Pracharoenwattana I, Smith SM, Hartig A (2006) A central role for the peroxisomal membrane in glyoxylate cycle function. Biochim Biophys Acta 1763:1441–1452

Lang C, Popko J, Wirtz M, Hell R, Herschbach C, Kreuzwieser J, Rennenberg H, Mendel RR, Hänsch R (2007) Sulphite oxidase as key enzyme for protecting plants against sulphur dioxide. Plant Cell Environ 30:447–455

Lange PR, Eastmond PJ, Madagan K, Graham IA (2004) An Arabidopsis mutant disrupted in valine catabolism is also compromised in peroxisomal fatty acid beta-oxidation. FEBS Lett 571:147–153

Lanyon-Hogg T, Warriner SL, Baker A (2010) Getting a camel through the eye of a needle: the import of folded proteins into peroxisomes. Biol Cell 102:245–263

Lasorsa FM, Scarcia P, Erdmann R, Palmieri F, Rottensteiner H, Palmieri L (2004) The yeast peroxisomal adenine nucleotide transporter: characterization of two transport modes and involvement in Delta pH formation across peroxisomal membranes. Biochem J 381:581–585

Lipka V, Dittgen J, Bednarek P, Bhat R, Wiermer M, Stein M, Landtag J, Brandt W, Rosahl S, Scheel D, Llorente F, Molina A, Parker J, Somerville S, Schulze-Lefert P (2005) Pre- and postinvasion defenses both contribute to nonhost resistance in Arabidopsis. Science 310:1180–1183

Linka N, Theodoulou FL, Haslam RP, Linka M, Napier JA, Neuhaus HE, Weber AP (2008) Peroxisomal ATP import is essential for seedling development in *Arabidopsis thaliana*.. Plant Cell 20:3241–3257

Ljun K, Hul AK, Kowalczyk M, Marchant A, Celenza J, Cohen JD, Sandberg G (2002) Biosynthesis, conjugation, catabolism and homeostasis of indole-3-acetic acid in *Arabidopsis thaliana*. Plant Mol Biol 50:309–332

Lucas KA, Filley JR, Erb JM, Graybill ER, Hawes JW (2007) Peroxisomal metabolism of propionic acid and isobutyric acid in plants. J Biol Chem 282:24980–24989

Martin T, Oswald O, Graham IA (2002) Arabidopsis seedling growth, storage lipid mobilization, and photosynthetic gene expression are regulated by carbon: nitrogen availability. Plant Physiol 128:472–481

Mettler IJ, Beevers H (1980) Oxidation of NADH in glyoxysomes by a malate–aspartate shuttle. Plant Physiol 66:555–560

Moschou PN, Sanmartin M, Andriopoulou AH, Rojo E, Sanchez-Serrano JJ, Roubelakis-Angelakis KA (2008) Bridging the gap between plant and mammalian polyamine catabolism: a novel peroxisomal polyamine oxidase responsible for a full back-conversion pathway in Arabidopsis. Plant Physiol 147:1845–1857

Nakamura T, Meyer C, Sano H (2002) Molecular cloning and characterization of plant genes encoding novel peroxisomal molybdoenzymes of the sulphite oxidase family. J Exp Bot 53:1833–1836

Noctor G (2006) Metabolic signalling in defence and stress: the central roles of soluble redox couples. Plant Cell Environ 29:409–425

Nowak K, Luniak N, Witt C, Wüstefeld Y, Wachter A, Mendel RR, Hänsch R (2004) Peroxisomal localization of sulfite oxidase separates it from chloroplast-based sulfur assimilation. Plant Cell Physiol 45:1889–1894

Nyathi Y, Baker A (2006) Plant peroxisomes as a source of signalling molecules. Biochim Biophys Acta 1763:1478–1495

Nyathi Y, De Marcos Lousa C, van Roermund CW, Wanders RJ, Johnson B, Baldwin SA, Theodoulou FL, Baker A (2010) The Arabidopsis peroxisomal ABC transporter, comatose, complements the *Saccharomyces cerevisiae pxa1 pxa2*Δ mutant for metabolism of long chain fatty acids and exhibits fatty acyl-CoA stimulated ATPase activity. J Biol Chem (in press)

Ozimek P, van Dijk R, Latchev K, Gancedo C, Wang DY, van der Klei IJ, Veenhuis M (2003) Pyruvate carboxylase is an essential protein in the assembly of yeast peroxisomal oligomeric alcohol oxidase. Mol Biol Cell 14:786–797

Palmieri L, Rottensteiner H, Girzalsky W, Scarcia P, Palmieri F, Erdmann R (2001) Identification and functional reconstitution of the yeast peroxisomal adenine nucleotide transporter. EMBO J 20:5049–5059

Pinfield-Wells H, Rylott EL, Gilday AD, Graham S, Job K, Larson TR, Graham IA (2005) Sucrose rescues seedling establishment but not germination of Arabidopsis mutants disrupted in peroxisomal fatty acid catabolism. Plant J 43:861–872

Poirier Y, Antonenkov VD, Glumoff T, Hiltunen JK (2006) Peroxisomal beta-oxidation–a metabolic pathway with multiple functions. Biochim Biophys Acta 1763:1413–1426
Pracharoenwattana I, Cornah JE, Smith SM (2005) Arabidopsis peroxisomal citrate synthase is required for fatty acid respiration and seed germination. Plant Cell 17:2037–2048
Pracharoenwattana I, Cornah JE, Smith SM (2007) Arabidopsis peroxisomal malate dehydrogenase functions in beta-oxidation but not in the glyoxylate cycle. Plant J 50:381–390
Pracharoenwattana I, Zhou W, Smith SM (2010) Fatty acid beta-oxidation in germinating Arabidopsis seeds is supported by peroxisomal hydroxypyruvate reductase when malate dehydrogenase is absent. Plant Mol Biol 72:101–109
Prado AM, Porterfield DM, Feijó JA (2004) Nitric oxide is involved in growth regulation and reorientation of pollen tubes. Development 131:2707–2714
Price NC, Dwek RA, Ratcliffe RG, Wormald MR (2001) Principles and problems in physical chemistry for biochemists. Oxford University Press, Oxford, p401
Reumann S (2000) The structural properties of plant peroxisomes and their metabolic significance. Biol Chem 381:639–648
Reumann S, Weber AP (2006) Plant peroxisomes respire in the light: some gaps of the photorespiratory C2 cycle have become filled–others remain. Biochim Biophys Acta 1763:1496–1510
Reumann S, Maier E, Benz R, Heldt HW (1995) The membrane of leaf peroxisomes contains a porin-like channel. J Biol Chem 270:17559–17565
Reumann S, Bettermann M, Benz R, Heldt HW (1997) Evidence for the presence of a porin in the membrane of glyoxysomes of castor bean. Plant Physiol 115:891–899
Reumann S, Maier E, Heldt HW, Benz R (1998) Permeability properties of the porin of spinach leaf peroxisomes. Eur J Biochem 251:359–366
Reumann S, Ma C, Lemke S, Babujee L (2004) AraPerox. A database of putative Arabidopsis proteins from plant peroxisomes. Plant Physiol 136:2587–2608
Reumann S, Babujee L, Ma C, Wienkoop S, Siemsen T, Antonicelli GE, Rasche N, Lüder F, Weckwerth W, Jahn O (2007) Proteome analysis of Arabidopsis leaf peroxisomes reveals novel targeting peptides, metabolic pathways, and defense mechanisms. Plant Cell 19:3170–3193
Reumann S, Quan S, Aung K, Yang P, Manandhar-Shrestha K, Holbrook D, Linka N, Switzenberg R, Wilkerson CG, Weber AP, Olsen LJ, Hu J (2009) In-depth proteome analysis of Arabidopsis leaf peroxisomes combined with in vivo subcellular targeting verification indicates novel metabolic and regulatory functions of peroxisomes. Plant Physiol 150:125–143
Richmond TA, Bleecker AB (1999) A defect in beta-oxidation causes abnormal inflorescence development in Arabidopsis. Plant Cell 11:1911–1924
Rottensteiner H, Theodoulou FL (2006) The ins and outs of peroxisomes: co-ordination of membrane transport and peroxisomal metabolism. Biochim Biophys Acta 1763:1527–1540
Rokka A, Antonenkov VD, Soininen R, Immonen HL, Pirilä PL, Bergmann U, Sormunen RT, Weckström M, Benz R, Hiltunen JK (2009) Pxmp2 is a channel-forming protein in mammalian peroxisomal membrane. PLoS ONE 4:e5090
Rubio S, Larson TR, Gonzalez-Guzman M, Alejandro S, Graham IA, Serrano R, Rodriguez PL (2006) An Arabidopsis mutant impaired in coenzyme A biosynthesis is sugar dependent for seedling establishment. Plant Physiol 140:830–843
Russell L, Larner V, Kurup S, Bougourd S, Holdsworth M (2000) The Arabidopsis COMATOSE locus regulates germination potential. Development 127:3759–3767
Rylott EL, Rogers CA, Gilday AD, Edgell T, Larson TR, Graham IA (2003) Arabidopsis mutants in short- and medium-chain acyl-CoA oxidase activities accumulate acyl-CoAs and reveal that fatty acid beta-oxidation is essential for embryo development. J Biol Chem 278:21370–21377
Sapir-Mir M, Mett A, Belausov E, Tal-Meshulam S, Frydman A, Gidoni D, Eyal Y (2008) Peroxisomal localization of Arabidopsis isopentenyl diphosphate isomerases suggests that

part of the plant isoprenoid mevalonic acid pathway is compartmentalized to peroxisomes. Plant Physiol 148:1219–1228
Schaller A, Stintzi A (2009) Enzymes in jasmonate biosynthesis – structure, function, regulation. Phytochemistry 70:1532–1538
Schmitt MR, Edwards GE (1983) Provisions of reductant for the hydroxypyruvate to glycerate conversion in leaf peroxisomes: a critical evaluation of the proposed malate/aspartate shuttle. Plant Physiol 72:728–734
Shani N, Watkins PA, Valle D (1995) PXA1, a possible *Saccharomyces cerevisiae* ortholog of the human adrenoleukodystrophy gene. Proc Natl Acad Sci USA 92:6012–6016
Shani N, Valle D (1996) A *Saccharomyces cerevisiae* homolog of the human adrenoleukodystrophy transporter is a heterodimer of two half ATP-binding cassette transporters. Proc Natl Acad Sci USA 93:11901–11906
Shockey JM, Fulda MS, Browse J (2003) Arabidopsis contains a large superfamily of acyl-activating enzymes. Phylogenetic and biochemical analysis reveals a new class of acyl-coenzyme a synthetases. Plant Physiol 132:1065–1076
Slocombe SP, Cornah J, Pinfield-Wells H, Soady K, Zhang Q, Gilday A, Dyer JM, Graham IA (2009) Oil accumulation in leaves directed by modification of fatty acid breakdown and lipid synthesis pathways. Plant Biotechnol J 7:694–703
Strader LC, Culler AH, Cohen JD, Bartel B (2010) Conversion of endogenous indole-3-butyric acid to indole-3-acetic acid drives cell expansion in Arabidopsis seedlings. Plant Physiol 153:1577–1586
Sulter GJ, Verheyden K, Mannaerts G, Harder W, Veenhuis M (1993) The in vitro permeability of yeast peroxisomal membranes is caused by a 31 kDa integral membrane protein. Yeast 9:733–742
Swartzman EE, Viswanathan MN, Thorner J (1996) The PAL1 gene product is a peroxisomal ATP-binding cassette transporter in the yeast *Saccharomyces cerevisiae*.. J Cell Biol 132:549–563
Taler D, Galperin M, Benjamin I, Cohen Y, Kenigsbuch D (2004) Plant eR genes that encode photorespiratory enzymes confer resistance against disease. Plant Cell 16:172–184
Theodoulou FL, Job K, Slocombe SP, Footitt S, Holdsworth M, Baker A, Larson TR, Graham IA (2005) Jasmonic acid levels are reduced in COMATOSE ATP-binding cassette transporter mutants. Implications for transport of jasmonate precursors into peroxisomes. Plant Physiol 137:835–840
Timm S, Nunes-Nesi A, Pärnik T, Morgenthal K, Wienkoop S, Keerberg O, Weckwerth W, Kleczkowski LA, Fernie AR, Bauwe H (2008) A cytosolic pathway for the conversion of hydroxypyruvate to glycerate during photorespiration in Arabidopsis. Plant Cell 20:2848–2859
Titorenko VI, Nicaud JM, Wang H, Chan H, Rachubinski RA (2002) Acyl-CoA oxidase is imported as a heteropentameric, cofactor-containing complex into peroxisomes of *Yarrowia lipolytica*.. J Cell Biol 156:481–494
Tugal HB, Pool M, Baker A (1999) Arabidopsis 22-kilodalton peroxisomal membrane protein. Nucleotide sequence analysis and biochemical characterization. Plant Physiol 120:309–320
Turner JE, Greville K, Murphy EC, Hooks MA (2005) Characterization of Arabidopsis fluoroacetate-resistant mutants reveals the principal mechanism of acetate activation for entry into the glyoxylate cyclel. J Biol Chem 280:2780–2787
Vandenabeele S, Vanderauwera S, Vuylsteke M, Rombauts S, Langebartels C, Seidlitz HK, Zabeau M, Van Montagu M, Inzé D, Van Breusegem F (2004) Catalase deficiency drastically affects gene expression induced by high light in *Arabidopsis thalianal*. Plant J 39:45–58
van der Klei IJ, Harder W, Veenhuis M (1991) Methanol metabolism in a peroxisome-deficient mutant of *Hansenula polymorpha*: a physiological study. Arch Microbiol 156:15–23
van Roermund CW, Elgersma Y, Singh N, Wanders RJ, Tabak HF (1995) The membrane of peroxisomes in *Saccharomyces cerevisiae* is impermeable to NAD(H) and acetyl-CoA under in vivo conditionsl. EMBO J 14:3480–3486

van Roermund CW, Hettema EH, Kal AJ, van den Berg M, Tabak HF, Wanders RJ (1998) Peroxisomal beta-oxidation of polyunsaturated fatty acids in *Saccharomyces cerevisiae*: isocitrate dehydrogenase provides NADPH for reduction of double bonds at even positionsl. EMBO J 17:677–687

van Roermund CW, Hettema EH, van den Berg M, Tabak HF, Wanders RJ (1999) Molecular characterization of carnitine-dependent transport of acetyl-CoA from peroxisomes to mitochondria in *Saccharomyces cerevisiae* and identification of a plasma membrane carnitine transporter, Agp2pl. EMBO J 18:5843–5852

van Roermund CW, Tabak HF, van Den Berg M, Wanders RJ, Hettema EH (2000) Pex11p plays a primary role in medium-chain fatty acid oxidation, a process that affects peroxisome number and size in *Saccharomyces cerevisiael*. J Cell Biol 150:489–498

van Roermund CW, de Jong M, Ijlst L, van Marle J, Dansen TB, Wanders RJ, Waterham HR (2004) The peroxisomal lumen in *Saccharomyces cerevisiae* is alkalinel. J Cell Sci 117:4231–4237

Van Veldhoven PP, Just WW, Mannaerts GP (1987) Permeability of the peroxisomal membrane to cofactors of beta-oxidation. Evidence for the presence of a pore-forming proteinl. J Biol Chem 262:4310–4318

Verleur N, Wanders RJ (1993) Permeability properties of peroxisomes in digitonin-permeabilized rat hepatocytes. Evidence for free permeability towards a variety of substratesl. Eur J Biochem 218:75–82

Verleur N, Hettema EH, van Roermund CW, Tabak HF, Wanders RJ (1997a) Transport of activated fatty acids by the peroxisomal ATP-binding-cassette transporter Pxa2 in a semi-intact yeast cell systeml. Eur J Biochem 249:657–661

Verleur N, Elgersma Y, Van Roermund CW, Tabak HF, Wanders RJ (1997b) Cytosolic aspartate aminotransferase encoded by the AAT2 gene is targeted to the peroxisomes in oleate-grown *Saccharomyces cerevisiael*. Eur J Biochem 247:972–980

Visser WF, Van Roermund CW, Ijlst L, Hellingwerf KJ, Wanders RJ, Waterham HR (2005) Demonstration and characterization of phosphate transport in mammalian peroxisomesl. Biochem J 389:717–722

Visser WF, van Roermund CW, Ijlst L, Waterham HR, Wanders RJ (2007) Metabolite transport across the peroxisomal membranel. Biochem J 401:365–375

Westphal L, Scheel D, Rosahl S (2008) The *coi1-16* mutant harbors a second site mutation rendering PEN2 nonfunctional. Plant Cell 20:824–826

Wiszniewski AA, Zhou W, Smith SM, Bussell JD (2009) Identification of two Arabidopsis genes encoding a peroxisomal oxidoreductase-like protein and an acyl-CoA synthetase-like protein that are required for responses to pro-auxinsl. Plant Mol Biol 69:503–515

Zolman BK, Yoder A, Bartel B (2000) Genetic analysis of indole-3-butyric acid responses in *Arabidopsis thaliana* reveals four mutant classesl. Genetics 156:1323–1337

Zolman BK, Silva ID, Bartel B (2001a) The Arabidopsis *pxa1* mutant is defective in an ATP-binding cassette transporter-like protein required for peroxisomal fatty acid beta-oxidationl. Plant Physiol 127:1266–1278

Zolman BK, Monroe-Augustus M, Thompson B, Hawes JW, Krukenberg KA, Matsuda SP, Bartel B (2001b) *chy1*, an Arabidopsis mutant with impaired beta-oxidation, is defective in a peroxisomal beta-hydroxyisobutyryl-CoA hydrolasel. J Biol Chem 276:31037–31046

Zolman BK, Nyberg M, Bartel B (2007) IBR3, a novel peroxisomal acyl-CoA dehydrogenase-like protein required for indole-3-butyric acid responsel. Plant Mol Biol 64:59–72

Zolman BK, Martinez N, Millius A, Adham AR, Bartel B (2008) Identification and characterization of Arabidopsis indole-3-butyric acid response mutants defective in novel peroxisomal enzymesl. Genetics 180:237–251

Regulation of Plant Transporters by Lipids and Microdomains

F. Simon-Plas, S. Mongrand, and D. Wipf

Abstract Transporters in the broad sense, that is, carriers, pumps, and channels, are proteins inserted in a lipid bilayer separating two cellular compartments. This lipid bilayer is not only the physical support of such proteins, but also a powerful way to regulate their activity. This chapter will first summarize the different means by which lipids can regulate the activity of transmembrane proteins (including the physical properties of the bilayer, its dynamic lateral compartmentalization, and the presence of particular lipid species acting as cofactors). It will then illustrate these general rules with examples of such regulations found in plant literature and, as a reference, in animal studies.

1 General Principles of the Regulation of Transporters and Channels Through Lipid Annulus, Cofactors and Membrane Raft Microdomains

Integral membrane proteins operate in an environment made up, in part, of the surrounding lipid bilayer. The composition of the lipid bilayer must therefore be close to optimal for functioning of the proteins in the membrane. Effects of lipid structure on membrane protein function can be described in terms of molecular interactions between the lipid and protein molecules such as hydrophobic effects, hydrogen bonding, or charge interactions or in terms of physical properties of the lipid bilayer such as lipid fluidity, membrane tension, and so on (Lee 2004).

F. Simon-Plas (✉) and D. Wipf
Laboratoire Plante-Microbe-Environnement, UMR INRA 1088/CNRS 5184/Université de Bourgogne, 17 rue Sully, 21065 Dijon, France
e-mail: simon@dijon.inra.fr

S. Mongrand
Laboratoire de Biogenèse Membranaire, UMR 5200 CNRS-Université de Bordeaux, 146 rue Léo Saignat, 33076 Bordeaux, France

M. Geisler and K. Venema (eds.), *Transporters and Pumps in Plant Signaling*, Signaling and Communication in Plants 7,
DOI 10.1007/978-3-642-14369-4_13, © Springer-Verlag Berlin Heidelberg 2011

The lipid molecules in a membrane not in contact with a protein at all are usually referred to as bulk lipids. While the bulk of the lipid molecules in contact with an intrinsic membrane protein act as a solvent for the protein, interacting with the protein relatively nonspecifically, some proteins also interact with much greater specificity with a small number of lipid molecules, these lipid molecules often being essential for protein activity and acting like a traditional cofactor. The solvent lipids have been referred to as boundary lipids, or as annular lipids to denote the fact that they form an annular shell of lipid around the protein (Lee 2003). Cofactor lipids, often bound between transmembrane α-helices either within a protein or at protein–protein interfaces in multisubunit proteins, have then been referred to as nonannular lipids (Simmonds et al. 1982). Most of the lipid molecules resolved in high-resolution crystal structures of membrane proteins are likely to be nonannular lipids, their strong binding to the protein leading to immobilization of at least part of the lipid molecule so that they appear in the high-resolution structure (Lee 2003). For example, Fig. 1a shows a high-resolution structure of the aquaporin tetramer AQP0, which reveals close contacts between membrane lipid and the border of the proteins, the lipid fatty acyl chains likely covering the whole hydrophobic surface of AQP0 tetramer (Gonen et al. 2005).

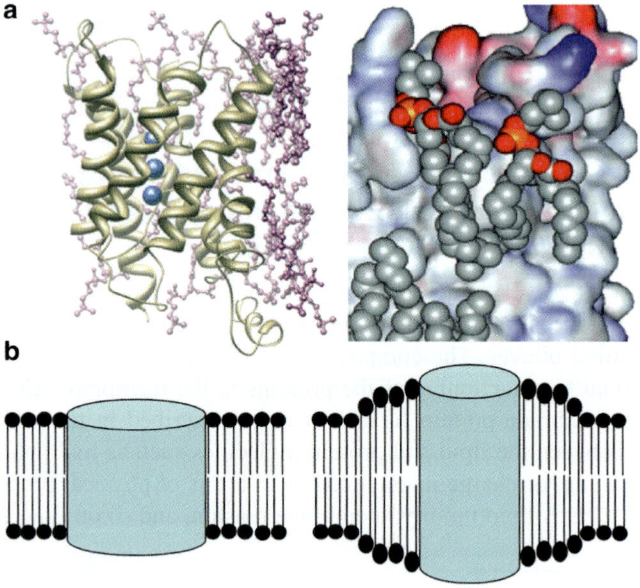

Fig. 1 Annular lipids circling proteins and change of thickness of the lipid bilayer around integral membrane protein. (**a**) R*ight*, Structure of the aquaporine tetramer (AQP0) showing the lipid molecules (in *violet*) covering the faces of the proteins. *Left*, close up of the interaction between the membrane lipids and the hydrophobic face of AQP0; From Protein DataBank (PDB) code 2B6O. (**b**) Schema showing that the hydrophobic thickness of the lipid bilayer match well the hydrophobic thickness of any protein embedded in the bilayer; this process may regulate channel activity

Moreover, at the whole cell scale, the lipid composition of the bulk is not homogeneous, since it has been demonstrated for years that the plasma membrane of eukaryotes was indeed not only a "fluid mosaïc" as proposed by Singer and Nicolson (1972) but also a dynamic mosaïc of domains of various sizes, lipid composition, and biophysical properties. The possibility for a given protein to be associated or not to such domains, according to physiological conditions, is also a potent modulator of its activity.

In the first part of this chapter, we explain in more detail some elements of the following three ways in which lipids can regulate membrane transporters in a broad sense (channels, pumps, and transporters): physical properties of the lipid bilayer, presence of particular lipids, and association with particular domains of the membrane.

1.1 Lipids as Solvent: Annular Lipids

1.1.1 Importance of the Polar Head

The lipid–water interface is not a sharp barrier between a slab of hydrocarbon chain and a polar region made up of the lipid head groups and water. Rather, the properties of the interface change gradually over a distance of some 15Å, the head group region having sufficient thickness to accommodate, for example, an α-helix lying parallel to the bilayer surface (White and Wimley 1999). Packing in the lipid headgroup region will depend on the lipid fatty acyl chain structure as well as the structure of the lipid headgroup itself (see examples of lipid structures in Fig. 2). For example, the area occupied in the bilayer surface by a molecule of dipalmitoylphosphatidylcholine [di(C16:0)PC] in the liquid crystalline phase at 50°C is 64Å^2, whereas the area occupied by a molecule of dioleoylphosphatidylcholine [di (C18:1)PC] is 72.5Å^2 (Nagle and Tristam-Naggle 2000). Differences in the areas occupied by different lipid molecules in the bilayer surface will lead to different patterns of hydrogen bonding and hydration in the headgroup region of the bilayer. Thus, the extent of hydration is very different for bilayers of phosphatidylcholine and phosphatidylethanolamine; at full hydration, a bilayer of di(C16:0)PC takes up about 23 molecules of water per molecule of lipid (Nagle and Wiener 1988), whereas a bilayer of di(C12:0)PE takes up only about ten molecules of water per molecule of lipid (McIntosh and Simon 1986). The structure of the lipid headgroup region could thus affect the structure of a protein penetrating into this region of the bilayer, because of the requirements of the polypeptide backbone and of any polar residues for hydrogen bonding, tending to drive the formation of secondary structures such as α-helix and β-sheet. Changes in the lipid headgroup region could then lead to changes in structure at the ends of the transmembrane α-helices and to consequent changes in the packing of the helices. Finally, the lipid headgroup region could affect the activity of a membrane protein by changing the concentrations of charged molecules or ions close to the surface of the membrane.

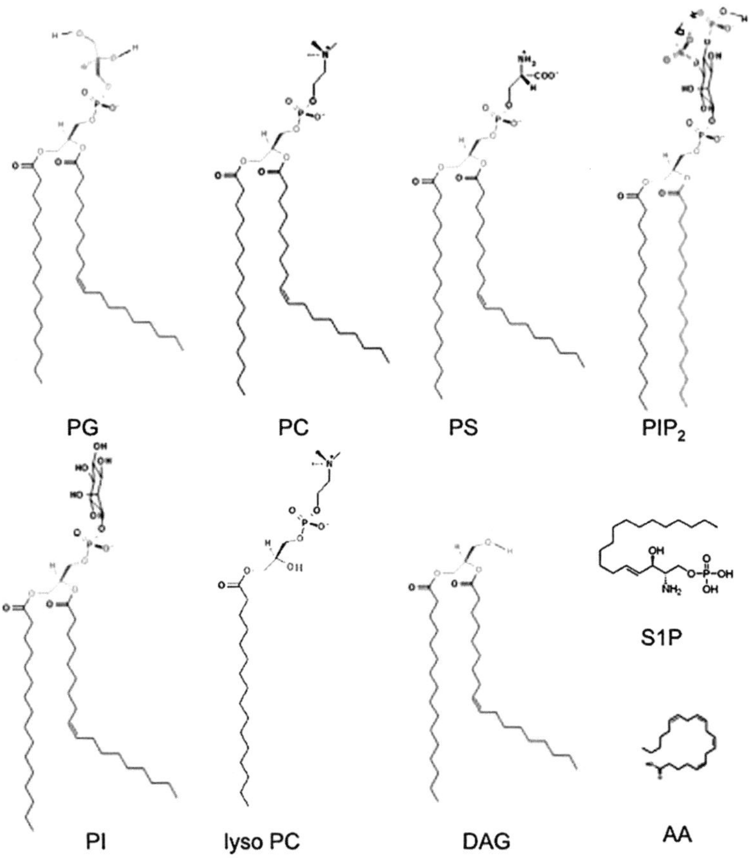

Fig. 2 Chemical structure of some lipids involved in the regulation of channel and transporters. Abbreviations are: *PG* phosphatidylglycerol; *PS* phosphatidylserine; *PIP2* phosphatidylinositol (4, 5) bisphosphate; *LPC* lyso phosphatidylcholine; *DAG* Diacylglycerol; *S1P* Sphingosine 1-Phosphate; *AA* Arachidonic acid

Incorporation of a negatively charged lipid into a membrane will increase the negative charge on the surface of the membrane and thus, increase the concentration of positively charged molecules or ions close to the surface of the membrane and, correspondingly, decrease the concentration of negatively charged molecules or ions close to the surface.

1.1.2 Importance of the Hydrophobic Thickness

An obvious and important property of a lipid bilayer is the thickness of the hydrophobic core of the bilayer, generally taken, for glycerophospholipids, to correspond to the separation between the glycerol backbone regions on the two sides of the bilayer. The hydrophobic thickness of the lipid bilayer is expected to match well the hydrophobic thickness of any protein embedded in the bilayer, because of the high cost of exposing either fatty acyl chains or hydrophobic amino acids to water. Any mismatch between the hydrophobic thickness of the lipid bilayer and the protein would be expected to lead to distortion of the lipid bilayer, or the protein, or both, to minimize the mismatch. The strongest evidence that membrane proteins do distort significantly in response to hydrophobic mismatch (as illustrated in Fig. 1b) comes from the observed changes in activity for a variety of membrane proteins in the presence of lipids with changes in fatty acyl chain length (Starling et al. 1993; Pilot et al. 2001; Weiss et al. 2003). Many of the observed profiles of activity against membrane thickness are similar, with highest activity in phospholipids with a chain length of about C18, matching the average chain length of most biological membranes, while lipids with shorter or longer chains support lower activities. Clearly, there will be no universal relationship between hydrophobic mismatch and effects on enzyme activity; the effect of mismatch on activity will be unique for each particular membrane protein, depending on the change in structure resulting from the mismatch and on the effect that that particular change in structure has on activity. If a membrane protein can adopt more than one conformation, each with a different hydrophobic thickness, then a change in the thickness of the surrounding lipid bilayer would be expected to shift the equilibrium between the various conformations towards the one that best matches the hydrophobic thickness of the bilayer.

1.1.3 Effect of Membrane Viscosity

Proteins, in carrying out their functions, undergo changes in shape. Any molecule or part of a molecule moving in a liquid environment experiences some frictional drag, opposing the movement. The resistance to motion through the liquid is expressed in terms of the viscosity g or its inverse, fluidity. In a lipid bilayer, the resistance to motion will come predominantly from the lipid fatty acyl chains. Molecular dynamics simulation suggests an effective viscosity of about 1 cP (1 P = 0.1 Pa s) for chain motion in a liquid crystalline bilayer, a value comparable to that for a simple alkane (Pastor et al. 1991; Venable et al. 1993). Experimental studies of the effects of membrane viscosity on the activities of membrane proteins are difficult because membrane viscosity can only be changed at a constant temperature by changing the composition of the membrane, and these changes in composition could directly affect the properties of the proteins in the membrane through changes in lipid–protein interaction. Any direct, static effects of lipid composition need to be sorted out before any possible dynamic influence of lipid on enzymatic rates can be clearly revealed. While it is undoubtedly true that many organisms change the lipids in their membranes in response to a change in temperature in such a way as to reduce the effects of the temperature change on the viscosity of their membranes, it would not be true, in general, to say that this actually maintains a constant viscosity in the membrane (Cossins 1994).

1.1.4 Effect of Interfacial Curvature and Elastic Strain

Most, if not all, biological membranes contain lipids, such as phosphatidylethanolamine, that, in isolation, prefer to adopt a curved, hexagonal HII phase rather than the normal, planar, bilayer phase (Cullis and de Kruijff 1979) and it has been suggested that bacteria control the lipid compositions of their membranes to maintain a constant proportion of lipids favoring the hexagonal HII phase (for a review, see Cronan (2003)). Mixtures of two lipids, one preferring the bilayer phase and one the hexagonal HII phase, adopt a bilayer phase if the mixture contains more than about 20 mol% of the bilayer-preferring lipid (Boni and Hui 1983). The presence of an intrinsic membrane protein also has a strong tendency to force a bilayer phase onto phospholipids preferring the hexagonal HII phase. It is therefore assumed that the lipid molecules surrounding an intrinsic membrane protein in a membrane will all be in the bilayer phase, even when the lipid molecules would, in isolation, adopt a hexagonal HII phase. The fact that phospholipids preferring to adopt a curved structure are forced to adopt a planar structure has been said to result in a state of "curvature frustration" for these lipids, and it has been suggested that this could be important for the proper function of the membrane (Gruner 1985). A mechanism by which stored curvature elastic energy could be linked to a mismatch between the hydrophobic thicknesses of a membrane protein and the surrounding lipid bilayer has been suggested (Botelho et al. 2002). One possible response of a system when the hydrophobic thickness of a protein is greater than that of the surrounding lipid bilayer is for the fatty acyl chains of the lipids around the protein to stretch to provide hydrophobic matching. The lipids around the protein will show negative curvature. For phospholipids that favor a planar bilayer structure, this will be unfavorable, but formation of a membrane with negative curvature will be favorable for phospholipids, such as phosphatidylethanolamine. Thus, if a membrane protein can adopt two conformational states, in the first of which its hydrophobic thickness matches that of the planar bilayer and in the second its hydrophobic thickness is greater than that of the planar bilayer, then the presence of phosphatidylethanolamine will favor the thicker form.

1.1.5 Effect of Membrane Tension: Mechanosensitive Ion Channels and Osmoregulated Transporters

One of the most fundamental homeostatic processes in prokaryotic organisms is regulation of the balance between the internal and external osmotic forces across the cytoplasmic membrane. Exposure of bacterial cells to a medium of low osmotic pressure leads to an influx of water into the cell, to cell expansion, and to an increase in tension within the cell membrane that would result in cell lysis unless the osmotic gradient is reduced in some way. For this reason bacteria contain mechanosensitive channels that, when activated by stretching of the membrane, allow a rapid and nonselective flux of solute out of the cell, thus preventing lysis of the cell. The most studied of the mechanosensitive channels is MscL, the mechanosensitive channel of large conductance (Hamill and Martinac 2001). The trigger for opening MscL is a change in the physical properties of the surrounding lipid bilayer, since an increase in membrane tension applied by negative pressure in a patch clamp pipette opens MscL in simple reconstituted membrane systems containing MscL as the only protein (Hamill and Martinac 2001); the lipid bilayer is an essential part of the gating mechanism of the channel. The opening of other ion channels, such as the two-pore domain K+ channels TREK and TRAAK is also induced by increased tension in the membrane (Patel et al. 2001).

1.1.6 Role of Asymetrical Distribution of Lipids Across the Plasma Membrane in the Permeability to Solutes

In reconstituted biological membranes, it has been recognized that the asymmetrical distribution of PM lipids between the cytosolic and apoplastic leaflets plays important roles for maintaining a low permeability to protons, solutes, and gasses (Hill et al. 1999; Hill and Zeidel 2000). The biophysical properties of the barrier resides in the exofacial leaflet, the sphingolipids playing a role in reducing membrane permeability with an absolute requirement for cholesterol to mediate this effect. Only recently, researchers investigated the asymmetrical distribution of PM lipids in plants (Tjellstrom et al. 2008). A clear asymmetry in PM phospholipids, free and conjugated sterols and the glucosylceramide exist between the cytosolic and apoplastic leaflets with a molar ratio of 65:35, 30:70, and 30:70, respectively (Tjellstrom et al. 2010). Even if this does not constitute a direct effect on transporter activity, the maintenance of different ion concentrations on both sides of the membrane, is likely to regulate, via the transmembrane potential thus established, the activity of many transport proteins.

1.2 Lipids as Cofactors: Nonannular Lipids

Some membrane proteins only show activity when in the presence of particular classes of lipids, these special lipids often copurifying with the protein and being resolved in high-resolution structures of the protein. Lipids of this type have been referred to as nonannular lipids, to distinguish them from the annular lipids that interact with the bulk hydrophobic surface of a membrane protein. Most nonannular lipids are bound between transmembrane α-helices, often between subunits in multi-subunit proteins.

However, tightly bound lipids are not only found between transmembrane α-helices. For example, a resolved phosphatidylglycerol molecule is seen bound on the surface of the transmembrane domain of the heterotrimeric nitrate reductase A (Bertero et al. 2003). Although the effect of phosphatidylglycerol on the function of nitrate reductase A has not been defined, it is likely that it plays a major structural role, holding the heterotrimeric structure together. A classical example of an essential nonannular lipid interaction is provided by the requirement for cardiolipin by a number of proteins involved in bioenergetics, including NADH dehydrogenase, cytochrome bc1, ATP synthase, cytochrome oxidase, and the ADP/ATP carrier (Jiang et al. 2000; Heimpel et al. 2001). Bovine heart cytochrome oxidase copurifies with a small number of tightly bound cardiolipin molecules whose removal leads to loss of activity (Robinson 1982). The tightly bound cardiolipin molecule found in the purple bacterial reaction center also appears to play a role in protein stability (Fyfe et al. 2004). Nonannular lipid molecules located at protein–protein interfaces could help to ensure good packing at the interface. For example, the tightly bound molecule of phosphatidylglycerol in K^+ channel from *Streptomyces lividans* (KcsA) is located at the monomer–monomer interface in the homotetrameric structure, between two Arg residues, one coming from each of the monomers at the interface. When KcsA is reconstituted into lipid vesicles, it shows a requirement for anionic phospholipids for function, with no specificity for which species of anionic lipid, which could, for example, be phosphatidylglycerol, phosphatidylserine, or cardiolipin (Heginbotham et al. 1998). It is possible that the role of the anionic phospholipid molecule is to reduce electrostatic repulsion between two Arg residues. Nonannular lipid molecules might also have functional effects on membrane proteins by affecting the movement of transmembrane α-helices in the protein.

Not all members of a particular class of membrane protein necessarily show the same requirements for nonannular phospholipids. For example, phosphatidylglycerol is enriched in the photosynthetic membranes of purple bacteria such as *Rhodobacter sphaeroides* and *Rhodospirillum rubrum* where it has been suggested to interact preferentially with the light-harvesting complex 2 (LH2) antenna pigment–protein complex, but it is not present in the LH2 complex of *Rhodopseudomonas acidophilia*, showing that it cannot play an essential role in the photosynthetic membrane (Russell et al. 2002). This is what would be expected if phosphatidylglycerol was playing a role in protein packing; differences in the amino acid sequences of various members of a family would lead to different requirements for optimal packing.

An emerging theme in lipid signaling is the involvement of PtdIns(4,5)P2 in the regulation of integral membrane proteins at the plasma membrane (for a review, see Gamper and Shapiro (2007)). These include a variety of inward rectifier and voltage-gated potassium channels, calcium channels and pumps, transient receptor potential channels, epithelial sodium channels, ion exchangers and probably, many other classes of proteins, such as receptors for chemical ligands. For some proteins, presence of binding sites for PtdIns(4,5)P2 may represent a mechanism for optimal function in their appropriate cellular context, that is, the plasma membrane. In other cases, regulation by PtdIns(4,5)P2 may allow these proteins to function as effectors of receptors coupled to PtdIns(4,5)P2 synthesis or degradation. Regulatory actions of PtdIns(4,5)P2 are well documented for several transient receptor potential channels, many of which function in sensory transduction. Phosphoinositides achieve direct signaling effects through the binding of their head groups to cytosolic proteins or cytosolic domains of membrane proteins. Thus, they can regulate the function of integral membrane proteins, or recruit to the membrane cytoskeletal and signaling components. This action is reminiscent of that of phosphotyrosine residues of membrane proteins, which mediate the recruitment of phosphotyrosine-binding modules (for example, SH2 domains). Typically, binding of proteins to phosphoinositides involves electrostatic interactions with the negative charges of the phosphate(s) on the inositol ring. In some cases, adjacent hydrophobic amino acids strengthen the interaction through a partial penetration into the bilayer. Protein surfaces that interact with phosphoinositides can consist either of clusters of basic residues within unstructured regions, such as those found in many actin regulatory proteins (for example, profilin), or of folded modules, such as the pleckstrin homology domain.

1.3 Compartmentalization of Membrane Proteins Within Particular Domains of the Membrane

1.3.1 Definition of Membrane Rafts

Over the last 10 years, another aspect of membrane biology raised from biological, biochemical and biophysical studies: the existence of membrane domains, also called "lipid rafts" (Simons and Ikonen 1997; Brown and London 1998). According to this concept, another dimension in the study of signaling events must be taken into account since biological membranes should no longer be considered as homogeneous bilayers composed of lipids and proteins. Indeed, "discrete islands" with specific location and composition exist in cellular membranes, especially in the PM. Such domains allow the segregation of active components in membranes and are involved in many cellular processes (for review see Rajendran and Simons (2005)). Characterization of membrane domains mainly rely on their properties of insolubility to detergent at cold temperature, therefore these domains are defined as Detergent-Resistant-Membranes (DRMs). They have been extensively studied because they can help to explain numerous biochemical processes occurring in cell membranes in term of specificity, selectivity and rapidity, processes that cannot be readily explained by other models. A consensus definition of lipid rafts emerging from the Keystone Symposium of Lipid Rafts is: "Membrane rafts are small (10–200 nm), heterogeneous, highly dynamic, sterol- and sphingolipid-enriched domains that compartmentalize cellular processes. Small rafts can sometimes be stabilized to form larger platforms through protein–protein and protein–lipid interactions" (Pike 2006). They have been associated with diverse functions, such as intracellular sorting of proteins, membrane trafficking, and signal transduction (Mishra and Joshi 2007).

In general, the function of rafts is thought to be brought about in any of the following three ways: (1) an increased concentration of proteins in rafts could facilitate homomeric and heteromeric interactions, (2) the lateral compartmentation of the plasma membrane might separate signaling components until a signal allows them to come together, or (3) the different lipid environment as such may affect the functions of membrane proteins. Activities of many membrane proteins are affected specifically by phospholipids and sterols. In comparison with the intense interest in rafts of mammalian and yeast cells, surprisingly little attention has been paid to potential lipid rafts of plant membranes.

1.3.2 Characterization of Plant Membrane Rafts

Plant DRM were characterized at the level of protein and lipid content in various organisms including tobacco, *Medicago*, and *Arabidopsis*. Phytosterols together with sphingolipids were found to be the main lipid components of plant DRM (Mongrand et al. 2004; Borner et al. 2005; Lefebvre et al. 2007). Structural phospholipids were largely excluded from DRM, with the notable exception of polyphosphoinositides strongly enriched in DRM (Furt et al. 2010). Interestingly, polyphosphoinositides are known to be involved in various signal transduction events and control the activity of ion transporters and channels during biosynthesis or vesicle trafficking (Liu et al. 2005; Monteiro et al. 2005), suggesting a putative role of plant membrane rafts as signaling membrane domains or membrane-docking platforms (Fig. 3).

Free phytosterols enriched in DRM, have been demonstrated to constitute the "glue" maintaining the cohesion of protein and lipid raft components. Indeed, treatment with methyl-beta cyclodextrin, which chelates free phytosterols, disrupts protein raft domains (Raffaele et al. 2009) and increases lipid acyl chain disorder and reduced the overall liquid-phase heterogeneity in correlation with the disruption of phytosterol-rich domains (Roche et al. 2008). Besides free sterols, conjugated phytosterols (Steryl glucoside, SG and Acylated steryl glucoside, ASG), typical for plant membranes, are also strongly enriched in DRMs (Lefebvre et al.

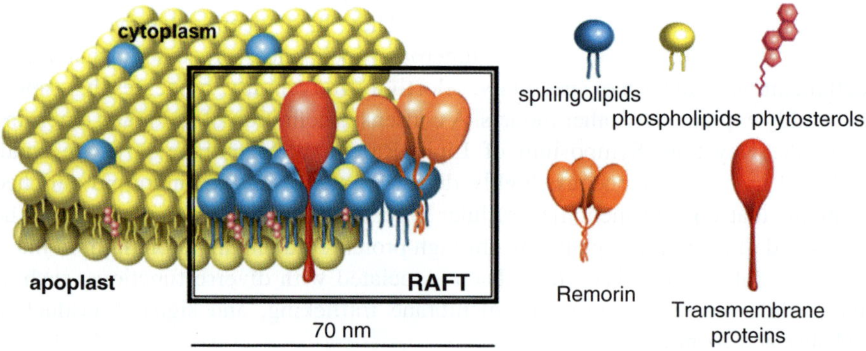

Fig. 3 Putative structure of inner leaflet plant membrane rafts. Sphingolipids and sterols strongly interact largely excluding phospholipids. The plant protein REMORIN is the first plant raft marker which locates in the inner leaflet of plasma membrane microdomains of ca. 70 nm (Raffaele et al. 2009)

2007; Furt et al. 2010). The presence of saturated fatty acyl chains (C16:0 and C18:0) in ASG (Lefebvre et al. 2007) is consistent with the hypothesis that lipid domains are closely packed with sterols and saturated lipids. SG molecules are likely oriented in membranes with the sterol moiety embedded in the hydrophobic phase of the bilayer and the sugar in the plane of the polar head groups of the phospholipids. The sterol and the fatty acyl chain moieties of ASGs are probably inside the bilayer with the sugar group appressing the hydrophilic surface.

Indirect measurement of sphingolipids present in DRM, by analysis of hydrolysed LCB, showed that sphingolipids are also main components of plant DRM. Because of the unique chemistry due to their unique long-chain base (LCB) and highly glycosylated nature of complex sphingolipids, the exact nature of the molecular species still remains elusive. Four classes of sphingolipids exist in plant membranes, namely LCB, Ceramide, glucosylcerebrosides and Glucosyl Inositol PhosphoCeramide (GIPC). Recent evidence indicates that plant sphingolipids are able to regulate ion channels and pumps, in particular, sphingosine-1-phosphate (S1P, see Fig. 2), one of the more soluble sphingolipids, generated from ceramide by the consecutive actions of ceramidase and sphingosine kinase (SphK). In animals, S1P has the ability to function intracellularly as a second messenger or in an autocrine and/or paracrine fashion to activate signaling through a family of specific G protein coupled receptors present on the cell surface. Interestingly, the significance of other LCBs and LCB phosphates (LCBPs), such as sphingosine (the direct precursor of S1P), phytosphingosine, delta8-desaturated LCBs, and phytosphingosine-1-phosphate (phyto-S1P), in modulating physiological processes in plants is also emerging.

2 Regulation of Plant Transporters by Lipids

2.1 Examples of Regulation by Association to Microdomains

2.1.1 Channels

Association with lipid rafts is an important mechanism for compartmentalization of critical signaling proteins and ion channels, especially in neurons (Ledesma et al. 1998; Martens et al. 2004). By bringing together membrane receptors and ion channels with other signaling proteins, lipid rafts are thought to facilitate the assembly of intracellular signaling cascades. Moreover, rafts are involved in numerous other cellular events such as membrane traffic, cell surface polarity, endocytic pathways, and viral infection (Simons and Ikonen 1997; Simons and Toomre 2000). Several transient receptor potential (TRP) channels have been shown to segregate into lipid rafts, including TRPC1, TRPC3, TRPC4, and TRPC5 (Ambudkar et al. 2004; Brownlow and Sage 2005). Furthermore, the activity of some TRP channels (e.g., TRPV1 and TRPC3) is sensitive to membrane cholesterol content, suggesting that raft association is pivotal for the physiological role of these channels (Graziani et al. 2006; Liu et al. 2006). The cold-activated TRPM8 channel was, for instance, reported to locate in lipid raft microdomains, both in vivo and in heterologous expression systems. It has also been demonstrated that glycosylation at the Asn934 residue facilitates this specific compartmentalization. Furthermore, lipid raft disruption increased both menthol- and cold-mediated activation of TRPM8 channels, suggesting that channel activity is higher outside of lipid raft microdomains.

In plants, evidences related to the presence of channels within microdomains are quite scarce. A cyclic nucleotide gated channel (CNGC 6), a cyclic nucleotide and calmodulin regulated ion channel and the potassium channel TORK1 have been identified in DRMs extracted from tobacco cells (Morel et al. 2006; Stanislas et al. 2009) and a voltage dependent anion channel was found associated to DRMs extracted from *Medicago truncatula* roots (Lefebvre et al. 2007). The potassium channel from *Arabidopsis thaliana* KAT1, transiently expressed in tobacco leaves, has been observed positioned in a stable manner within domains of the plasma membrane of 0.5 μm (Sutter et al. 2006). This particular targeting has been proven to be regulated by SNARE proteins. However, no clear indication was given of a regulation of any of these channels related to their compartmentalized distribution within the membrane.

2.1.2 Pumps

Several isoforms of the most representative integral protein of the plant PM, the P-type H-ATPase (PMA) have been shown to be present in DRM. All AHAs except AHA3, AHA7 and AHA have been detected associated to this fraction in *Arabidopsis* (Shahollari et al. 2004; Borner et al. 2005; Minami et al. 2009) and isoforms PMA 1, 2, 3, 4, 5, 6 and 9 have been identified in tobacco DRMs by proteomic approaches (Mongrand et al. 2004; Morel et al. 2006; Stanislas et al. 2009). Such an association of PM H^+-ATPase with membrane microdomains is quite consistent with previous immunodetection studies showing a nonhomogeneous distribution of PMA in minor veins of *Vicia faba* (Bouche-Pillon et al. 1994). It is also quite relevant regarding the presence in DRMs of many proteins potentially involved in the regulation of this enzyme such as kinases, or 14-3-3 proteins (Morel et al. 2006). Interestingly the yeast counterpart of this protein, Pma1p has been demonstrated to be enriched in DRMs in an ergosterol-dependent manner, and visualized in fusion with fluorescent proteins in distinct areas of the plasma membrane of about 300 μM diameter (Grossmann et al. 2006). However, it seems that at least some isoforms of the plant PMAs can be present both in the detergent-soluble and -insoluble fractions of the plasma membrane, preventing to consider PMAs in general as a biochemical protein marker of plant DRM (Raffaele et al. 2009).

Beyond its subcompartmentalization in distinct areas of the membrane in association to regulatory proteins, the association of the PM-H^+ATPase to microdomains could also correspond to a lipidic environment favorable to its activity. Indeed, this enzyme, and particularly its proton transport activity, has been proven to be regulated by sterols (Grandmougin-Ferjani et al. 1997). The occurrence of SG and ASG in plant plasma membrane has been suggested to modulate the activity of raft-associated H^+-ATPases (Mongrand et al. 2004; Bhat and Panstruga 2005) as they do in the case of the tonoplast ATPase (Yamagushi and Kasamo 2002).

Different proteins identified as subunits of a vacuolar H^+-ATPase are also present in DRMs as well as several members of the aquaporin family (plasma membrane intrinsic proteins (PIPs) (Borner et al. 2005; Morel et al. 2006; Kierszniowska et al. 2009; Minami et al. 2009; Stanislas et al. 2009). However, no indication about their possible regulation by such a localization is currently available.

2.1.3 Transporters

Whereas numerous studies report the regulation of membrane transporters from animal cells through their association to lipid rafts, very few evidences are presently available for plants. In the following section, we therefore describe examples obtained from animal literature and discuss to what extent this could be relevant for plants.

ABC Transporters

In a recent review, Klappe and collaborators (2009) discussed the modulation of the localization and function of the animal ABC transporters ABCA1, Pgp/ABCB1 and MRP1/ABCC1 by two relevant lipid classes, i.e., sphingolipids and cholesterol, in the context of membrane rafts.

Initially, Orlowski and collaborators (2007) suggested that ABCA1 is not localized in lipid rafts as previously defined by isolation with Triton X-100 (Lavie et al. 1998). However, when Lubrol WX was used as detergent, ABCA1 was partially localized in DRMs (Bared et al. 2004). Atshaves et al. (2007) developed a detergent-free lipid raft isolation procedure. Rafts and also non-raft membranes were isolated by means of affinity chromatography. ABCA1 was detected in lipid rafts and found to be enriched in these domains by a factor 6.6. Pgp/ABCB1 is a full ABC transporter best known for its overexpression in multidrug resistant cancer cells. Cholesterol depletion by means of methylcyclodextrin results in loss of function and a shift of Pgp/ABCB1 out of lipid raft fractions.

MRP1/ABCC1 is a full ABC transporter with three transmembrane domains and two ATP-binding domains, being one of the transporters capable of causing multidrug resistance in tumor cells (Bakos and Homolya 2007). MRP1/ABCC1 is not associated with Triton-based DRMs but rather with Lubrol-based DRMs and is also located in detergent-free lipid rafts (Hinrichs et al. 2004). A role for glycolipids in MRP1/ABCC1 localization in DRMs and its functional activity was suggested by overexpression of MRP1/ABCC1 by culturing the cells with gradually increasing colchicine concentrations in the HT29 human colon tumor cell line (Kok et al. 2000).

Two models have been proposed to explain the observed differences in ABC transporter localisation between the Triton- and the Lubrol-based DRMs (Klappe et al. 2009). Model I is based on the previously proposed layered raft model (Hinrichs et al. 2004; Pike 2004) and argues that membrane proteins abundant in the outer concentric layers of a given single type heterogeneous lipid raft are largely solubilized when only the lipid raft core is left intact using, e.g., Triton X-100 to isolate lipid rafts. Thus, a lower yield of membrane proteins is obtained with Triton X-100, compared to, e.g., Lubrol WX. Model 2 is based on the work performed by Delaunay and collaborators (Delaunay et al. 2008) who suggest that strong detergents like Triton X-100 are more effective than Lubrol WX at solubilizing lipid raft phosphatidylethanolamine, located in the inner leaflet. Solubilization of the inner leaflet results in destabilization of the transmembrane proteins. According to this model, Lubrol WX preserves the membrane inner leaflet and this may explain the high recovery of transmembrane proteins and lipids of the cytoplasmic leaflet in Lubrol-based DRMs. It is worthwhile to note that ABC transporters, are not only modulated by lipids and lipid rafts, but that they can also modulate their lipid environment by acting as transporters/flippases for lipids.

Several proteins belonging to the ABC transporter family have been identified in DRMs extracted from different plant species (Morel et al. 2006; Lefebvre et al. 2007; Kierszniowska et al. 2009; Stanislas et al. 2009). Moreover, the association of such proteins to DRMs has been proven to be sterol-dependent (Kierszniowska et al. 2009).

A striking example of the regulation of a plant ABC transporter by its association to lipid rafts concerns auxin transport. Auxin transport in plants is mediated by three independent sets of transporters and channels: (1) the PINFORMED1 (PIN) auxin efflux carrier, (2) ABC-transporter/p-glycoprotein (PGP)-type auxin efflux carriers (ABCB/PGP) and (3) AUX/LAX transport proteins. A recent report addressed the question of the role of a lipid-driven microsegregation at the level of the membrane which may lead to the formation of macrodomains at one pole of the cell, and which may regulate auxin transport (Titapiwatanakun and Murphy 2009). The presence of *Arabidopsis* auxin transport proteins in DRM fractions brings experimental support to this hypothesis, and further challenges the direct and/or indirect regulation of auxin transport by membrane lipids. First, it was shown that several ABCB/PGP, including ABCB19/PGP19, were clearly enriched in DRM, whereas PIN1 was not strikingly enriched in DRM fractions (Titapiwatanakun et al. 2009). Moreover, in abcb19 mutants, polar PIN1 localization was merely affected, but appeared more diffuse in Triton X-100-extracted roots, than in wild type roots. This suggests that ABCB19 stably located in PM microdomains, recruits PIN1 to membrane domains. Direct evidence for the stabilization and regulation of the activity of PIN1 in membrane microdomains by ABCB19/PGP19, however, is still lacking. Second, in terms of lipids, experimental evidences came from sterol mutant analyses such as sterol methyltransferase1 (smt1) and cyclopropylsterol isomerase1-1 (cpi1-1) mutants, where the polar localization of some PIN proteins is altered. For example, in cpi1-1, PIN2 can be found at both apical and basal membranes, whereas it localizes strictly at the apical domain in the wild type, likely involving clathrin-

and sterol-dependent endocytosis. Besides, it turned out that methyl-β-cylcodextrin treatment releases ABCB19 from *Arabidopsis* membranes, also suggesting a crucial role of phytosterol. It has been suggested that ABCB19 stably located in PM microdomains with PIN1, enhances PIN1 auxin transport activity. Demonstration of PIN1-mediated auxin transport in *Saccharomyces pombe*, where plant-like sterol-enriched microdomains are present, confirms this conclusion and indicates that *Arabidopsis* ABCB19 may activate PIN1 in heterologous systems (Blakeslee et al. 2007).

Amino-Acid Transporters

In the last years, numerous studies focused on the role of membrane lipids in neurotransmission. In neurons membrane rafts have been demonstrated to harbor receptors like acetylcholine receptors, glutamate receptors, and GABA receptors, as well as transporters like EAAT2, SERT, and NHE3 (Allen et al. 2007). Excitatory amino acid transporter 2 (EAAT2) is the major glutamate transporter in the central nervous system. A large portion of total EAAT2 protein is associated with cholesterol-rich lipid raft microdomains of the plasma membrane. This association with lipid rafts is important for EAAT2 trafficking and function (Butchbach et al. 2004). Tian et al. (2010) showed that depletion of membrane cholesterol by methyl-β-cyclodextrin resulted in a significant reduction of EAAT2 glutamate uptake activity. Amino acids are essential elements in animals and plants. In a genome wide analysis in yeasts, plants and animals five AA transporter superfamilies could be identified (Wipf et al. 2002; Lalonde et al. 2004). Transporters of one superfamily (ATF1, SLC38) which includes animal and plant members correspond to proteins involved in neurotransmitter transport in animals. Their close relation with plant genes suggests that not only a closely related transport mechanism could be involved in AA transport in plants, but also that the numerous results obtained for lipid regulation of neurotransmitter transporters will greatly help to unreveal the current black box on this topic in plants. Up to now, a lysine and histidine–specific transporter (Morel et al. 2006; Stanislas et al. 2009) and two oligopeptide transporters (Morel et al. 2006; Stanislas et al. 2009) have been identified in plant DRMs.

Sucrose Transporters

The proton–monosaccharide cotransporter MST1 (Morel et al. 2006; Lefebvre et al. 2007) and hexose and sucrose transporters (Minami et al. 2009; Stanislas et al. 2009) have been so far identified in plant DRMs. The hexose–proton symporter HUP1 shows a spotty distribution in the plasma membrane of the green alga *Chlorella kessleri*. As, *C. kessleri* is not transformable, Grossmann and collaborators (2006) expressed a HUP1-GFP fusion in the yeast *Saccharomyces cerevisiae*. In yeast, the protein clearly exhibited a patchy distribution in the plasma membrane which resembled the 300-nm patches of Can1-GFP, Fur4-GFP, and Sur7-GFP observed in *S. cerevisiae*. In fact, HUP1 was localized in a particular raft domain, the subdomain RMC C, which also houses Can1p, Fur4p, and Sur7p. HUP1-GFP was also present to a minor extent in the non-raft-cluster area, which possibly indicates that the heterologous protein does not carry the optimal sorting information required for yeast raft or raft cluster association. Similarly, when expressed in yeast, the plant sucrose transporter SUT1 from *Solanum tuberosum* was targeted to the plasma membrane where the protein concentrates in 200–300-nm raft-like microdomains (Krügel et al. 2008). SUT1 revealed a dramatic redox-dependent increase in sucrose transport activity when heterologously expressed in *S. cerevisiae*. Localization studies of green fluorescent protein fusion constructs showed that an oxidative environment increased the targeting of SUT1 to the raft-like microdomains. In plant membranes, *StSUT1* could be detected in the detergent-resistant membrane fraction. The concentration of the Sl SUT1-GFP fusion construct in lipid raft-like structures is disturbed in the yeast ergosterol *erg6* mutant, whereas the increase in plasma membrane targeting in the presence of oxidizing agents can still be observed in the *erg6* mutant, arguing for two distinct phenomena. Krügel and collaborators (2008) suggested that raft association of St SUT1 might be related to oligomerization and/or plasma membrane recycling and endocytosis of the protein.

Heterologous expression in yeast, which has been successful for the isolation of numerous plant transporters, may be a generally applicable and convenient method to learn about lipid requirements and potential raft properties of plant membrane proteins.

2.2 Examples of Regulation by Lipids as Cofactors

2.2.1 Channels

In animal cells, numerous examples of ion channels regulated by lipidic cofactors exist. For instance, an inhibitory effect of diacylglycerol (DAG, see Fig. 2) on cyclic nucleotide-gated channels has been observed, and could result from its insertion either into a putative hydrophobic crevice among the transmembrane domains of each subunit, or at the hydrophobic interface between the channel and the bilayer (Crary et al. 2000). On the opposite, different members of the transient receptor potential (TRP) ion channel family are activated by DAG, although the precise molecular mechanisms underlying this effect remains unknown (Hardie 2007). A complex regulation of voltage-gated calcium channels (VGCC) by the polyunsaturated free fatty acid arachidonic acid (AA) has also been evidenced: micromolar concentrations of AA inhibit the activity by stabilizing the channel in a closed state, and this requires a chain length of at least 18 carbons and multiple bonds located near the fatty acid's carboxy terminus, whereas, acting at a second site, AA increases the rate of VGCC activation kinetics (Roberts-Crowley et al. 2009). A striking point is the ability of phosphoinositides to regulate a great diversity of animal channels. Several members of the TRP family were reported to be regulated by phosphatidylinositol 4,5-biphosphate, (Fig. 2) and the hypothesis has been raised that this lipid could be a common regulator of mammalian TRP channels (Rohacs 2007, 2009). The putative mechanism underlying such a regulation has been studied in detail for a particular member of the family: TRPM4 : PIP_2 could bind directly to the channel (a PIP_2-binding pocket seems to comprise a part of the TRP domain and especially pleckstrin homology domains in the C terminus), induce sensitization to activation by Ca^{2+}, and shift the voltage dependence towards negative potentials. The complex regulation of epithelium sodium channels (Pochynyuk et al. 2008), neural KCNQ channels (Hernandez et al. 2008) and high voltage-activated Ca^{2+} channels (Michailidis et al. 2007), has also been reported.

In plants, phosphoinositides have also been identified as regulators of ion channels. The K^+-efflux channel of tobacco, NtORK (outward-rectifying K channel) has its activity inversely related to the plasma membrane PIP_2 level controlled genetically or enzymatically (Ma et al. 2009). Biophysical analysis of NtORK whole-cell outward K^+ currents, provided an insight into a possible mechanism: NtORK activity was diminished mainly through decreased maximum available conductance via the channels, and, by altering the voltage-dependent channel activation, i.e., making the channels more difficult to open. In a similar way the anion channel activity present on the plasma membrane of *V. faba* guard cells, is inhibited by exogenously administrated PIP_2 (Lee et al. 2007). As it was observed that upon light irradiation the PIP_2 level increased at the plasma membrane, the consecutive inhibition of anion channels could participate to stomatal opening. On the opposite, exogenously added PIP_2 was able to restore the activity of plant shaker-type channels of three subtypes, SKOR, KAT1 and LKT1, consecutively to their inactivation after patch excision (Liu et al. 2005). A detailed study indicated that such a channel activation required both PIP_2 and a specific range of membrane voltage, and tests with different forms of PIPs revealed that one phosphate on the inositol ring is required and sufficient for the channel activation, the position of such a phosphorylation on the inositol ring being not decisive. Finally, a non specific charge-effect could be ruled out, indicating that the effect of PIP_2 on channel activities requires both negatively charged inositol phosphate head and the lipophilic fatty acid tail. In a cellular context, effects of various inhibitors, strongly suggest that PI kinase activity plays an important role in activation of plant shaker-type channels (Liu et al. 2005).

Channels with high PIP_2 affinity have low sensitivity to PIP_2 depletion and might not respond to moderate changes in PIP_2 level induced by signaling cascades, whereas channels with low affinity will be affected by slight modulations of PIP_2 levels. The increasing number of transport proteins that have been shown to be sensitive to phosphoinositides raises the question of the mechanism conferring a specificity to such a regulation. Among the proposed hypothesis, a membrane compartmentation either of the signaling proteins or of phosphoinositides themselves might restrict diffusion so that PIP_2 synthesis and depletion are local. It is clear here that the distinction between regulation by lipid cofactors and by membrane domains becomes then quite artificial. Another possibility could be that a competition exists between all these proteins for the available PIP_2, and that those with the highest affinity could undergo the most efficient regulation (Gamper and Shapiro 2007).

Another example of regulation of a plant channel by a lipid cofactor is the observed inhibition of the activity of the inward K^+-channel of *V. faba* guard cell protoplasts by phosphatidic acid, either artificially induced or produced via an increase of phospholipase D activity triggered by ABA (Jacob et al. 1999). The precise molecular determinants of such a regulation have not yet been elucidated.

Sphingosine 1-phosphate (S1P) has also been identified as a key player in plant cell signaling since it was proven to regulate guard cell behavior via Ca^{2+} mobilization, inhibition of PM inward rectifying K^+ channels and stimulation of slow anion channels. The action of S1P on ion channels is impaired in guard cells of *Arabidopsis* plants harboring T-DNA null mutations in the sole prototypical G protein gene, GPA1, suggesting that heterotrimeric G proteins are downstream targets for S1P in plants as in mammals (Coursol et al. 2003).

2.2.2 Pumps

The plant plasma membrane H^+-ATPase plays a major role in plant physiology, through the control of nutrient uptake, intracellular pH and membrane potential. The mechanisms able to regulate its activity in various physiological conditions have thus been extensively studied (Duby and Boutry 2009). Among these mechanisms, several indications of a regulation by lipids have been given (Kasamo 2003), with some difficulties to determine without any ambiguity whether the modulations observed were due to the bulk lipids or to a specific effect of a particular lipid.

The specific requirement of the hydrolytic activity of the purified H^+-ATPase from corn roots plasma membrane, for negative phospholipids has been demonstrated. A semi-purified fraction of the enzyme was delipidated, leading to its complete inactivation, and the restoration of its activity was assayed using different combinations of phospholipids (Simon-Plas et al. 1991). While phosphatidylcholine (PC) alone, or in combination with phosphatidylethanolamine (PE) or sterol, was unable to promote hydrolysis activity, the addition of a low amount of negative phospholipids such as phosphatidylinositol (PI) of phosphatidylglycerol (PG) to PC (PC :PI 6 :1, or PC :PG 6 :1), restored an hydrolysis activity equivalent to the one observed with a complete mixture of soybean lipids (see Fig. 2 for chemical structure of lipids). In these conditions, the enzyme was only relipidated and not inserted in a lipid bilayer, making impossible the measurement of its vectorial activity.

The effect of various lipids on both activities of the enzyme (Mg-ATP hydrolysis and H^+-transport) have also been studied after reconstitution of a purified enzyme in proteoliposomes containing different combinations of lipids (Kasamo and Yamanishi 1991). A strong effect of both the polar head group and the fatty acyl chain of phospholipids used to prepare liposomes on the enzyme activities was observed, with, in particular, a remarkably high rate of H^+-pumping with proteoliposomes prepared with 1-saturated, 2-unsaturated fatty acids. However, a precise interpretation of the results obtained is quite difficult since the effects observed might result at least from three types of regulations : an effect of the lipid mixture on the efficiency of the insertion of the enzyme within the lipid bilayer, an effect on the proton leakage of the bilayer which might then affect the coupling efficiency between Mg-ATP hydrolysis and H^+-transport, and finally a direct effect on the enzyme activity.

Lysophosphatidylcholine (LPC) has also been shown to activate the plasma membrane H^+-ATPase (Pedchenko et al. 1990). The deletion of the C-terminal part of the enzyme prevented such a stimulation suggesting that LPC could activate the H^+-ATPase by displacing the autoinhibitory C-terminal domain from its interaction site (Regenberg et al. 1995). Further studies revealed that LPC binds to a site located outside the C-terminal region, in competition experiments with the fungal toxin beticolin-1, which inhibits the wild type and truncated H^+-ATPases in a similar way (Gomès et al. 1996).

2.2.3 Transporters

Changes in tertiary structure are becoming recognized as a mechanism potentially allowing proteins to achieve diversity of function (Jeffery 1999) (see Sect. 1.2). The most striking example is the prion protein, which changes from a mixed α-helical and β-sheet conformation to the β-sheet–rich, disease-causing isoform Prpsc (Pan et al. 1993; Safar et al. 1993; Pergami et al. 1996). It has been postulated that this shift requires tissue specific cofactors that stabilize Prpsc, In animals, many reports exist about lipids as cofactors that may enable protein folding variants to attain new functions. For plant transporters, less is known. Grossmann and collaborators (2006) when expressing the Chlorella hexose–proton symporter HUP1 in yeast, could observe a decreased transport activity of HUP1 in ergosterol-depleted cells. Their results indicated that sterols are important for the activity of the HUP1 protein. Whether the clustering of the transporter within the membrane or the presence of a critical amount of a specific sterol is decisive for the effect observed is still an open question. The possibility that ergosterol is a more potent cofactor for the activity of the HUP1 protein than zymosterol cannot be excluded.

3 Conclusion

During the past few years, our vision of the dynamic and regulatory processes occurring at the plasma membrane have considerably evolved. The lipid bilayer has been considered for long essentially as a structural support for membrane proteins responsible for transport and signaling processes, activity of which was supposed to be regulated mainly by post translational modifications. The concept of individual lipids and of the dynamic structure and compartmentation of the bilayer as essential regulatory elements of the membrane protein activity is now well established.

However, it is clear that precise examples illustrating such regulation come, at the moment, mainly from studies performed on bacterial or animal cells, and that similar demonstrations remain scarce in the field of plant membrane biology. There is no doubt that the recent characterization of plant membrane rafts and of proteic and lipidic markers of such domains will lead to a considerable development concerning the regulation of plant transporters through their dynamic association to particular domains of the membrane. For that purpose, biochemistry should be combined with currently developing highly resolutive imaging approaches. Considering the regulation of plant transporters by lipid cofactors, there is a crucial need for molecular evidences, which will require a high input leading to the elucidation of crystallographic structures of membrane proteins, able to reveal the tight association of a particular lipid with special parts of the protein.

References

Allen JA, Halverson-Tamboli RA, Rasenick MM (2007) Lipid raft microdomains and neurotransmitter signaling. Nat Rev Neurosci 8:128–140

Ambudkar IS, Brazer SC, Liu X, Lockwich T, Singh B (2004) Plasma membrane localization of TRPC channels: role of caveolar lipid rafts. Novartis Found Symp 258:63–70

Atshaves BP, McIntosh AL, Payne HR, Gallegos AM, Landrock K, Maeda N, Kier AB, Schroeder F (2007) SCP-2/SCP-x gene ablation alters lipid raft domains in primary cultured mouse . hepatocytes. J Lipid Res 48:2193–2211

Bakos E, Homolya L (2007) portrait of multifaceted transporter, the multidrug resistance-associated protein 1 (MRP1/ABCC1). Pflugers Arch 453:621–641

Bared SM, Buechler C, Boettcher A, Dayoub R, Sigruener A, Grandl M, Rudolph C, Dada A, Schmitz G (2004) Association of ABCA1 with syntain 13 and flotillin-1 and enhanced phagocytosis in tangier cells. Mol Biol Cell 15:5399–5407

Bertero MG, Rothery RA, Palak M, Hou C, Lim D, Blasco F, Weiner JH, Strynadka NCJ (2003) Insights into the respiratory electron transfer pathway from the structure of the nitrate reductase. Nat Struct Biol 10:681–687

Bhat RA, Panstruga R (2005) Lipid rafts in plants. Planta 1:5–19

Blakeslee JJ, Bandyopadhyay A, Lee OR, Mravec J, Titapiwatanakun B, Sauer M, Makam SN, Cheng Y, Bouchard R, Adamec J, Geisler M, Nagashima A, Sakai T, Martinoïa E, Friml J, Peer WA, Murphy AS (2007) Interactions among PIN-FORMED and P-glycoprotein auxin transporters in *Arabidopsis*.. Plant Cell 19:131–147

Boni LT, Hui SW (1983) Polymophic phase behaviour of dilinoleoyl-phosphatidylethanolamine and palmytoeloyl-phosphatidylcholine mixtures. Structural changes between hexagonal, cubic and bilayer phases. Biochim Biophys Acta 731:177–185

Borner GH, Sherrier DJ, Weimar T, Michaelson LV, Hawkins ND, Macaskill A, Napier JA, Beale MH, Lilley KS, Dupree P (2005) Analysis of detergent-resistant membranes in *Arabidopsis*. Evidence for plasma membrane lipid rafts. Plant Physiol 137:104–116

Botelho AV, Gibson NJ, Thurmond RL, Wand Y, Brown MF (2002) Conformational energetics of rhodopsin modulated by nonlamellar-forming lipids. Biochemistry 41:6354–6368

Bouche-Pillon S, Fleurat-Lessard P, Fromont JC, Serrano R, Bonnemain JL (1994) Immunolocalization of the plasma membrane H^+-ATPase in minot veins of Vicia faba in relation to phloem loading. Plant Physiol 105:691–697

Brown DA, London E (1998) Structure and origin of ordered lipid domains in biological membranes. J Membr Biol 164:103–114

Brownlow SL, Sage SO (2005) Transient receptor potential protein subunit assembly and membrane distribution in human platelets. Thromb Haemost 94:839–845

Butchbach ME, Tian G, Guo H, Lin CL (2004) Association of excitatory amino acid transporters, especially EAAT2 with cholesterol-rich lipid raft microdomains: importance for excitatory amino acid transporter localization and function. J Biol Chem 279:34388–34396

Cossins AR (1994) Temperature adaptation of biological membranes. Portland Press, London

Coursol S, Fan LM, Le Stunff H, Spiegel S, Gilroy S, Assmann SM (2003) Sphingolipid signaling in *Arabidopsis* guard cells involves heterotrimeric G proteins. Nature 423:651–654

Crary JI, Dean DM, Nguitragool W, Kurshan PT, Zimmermann AL (2000) Mechanism of inhibition of cyclic nucleotide-gated ion channels by diacylglycerol. J Gen Physiol 116: 755–768

Cronan JE (2003) Bacterial membrane lipids: where do we stand? Annu Rev Microbiol 57: 203–224

Cullis PR, de Kruijff B (1979) Lipid polymorphism and the functional roles of lipids in biological membranes. Biochim Biophys Acta 559:399–420

Delaunay JL, Breton M, Trugnan G, Maurice M (2008) Diferential solubilization of inner palsma membrane leaflet components by Lubrol WX and triton X-100. Biochim Biophys Acta 1778:105–112

Duby G, Boutry M (2009) The plant plasma membrane proton pump ATPase: a highly regulated P-ATPase with multiple physiological roles. Pflugers Arch 457:645–655

Furt F, König S, Bessoule JJ, Sargueil F, Zallot R, Stanislas T, Noirot E, Lherminier J, Simon-Plas F, Heilmann I, Mongrand S (2010) Polyphosphoinositides are enriched in plant membrane rafts and form microdomains in the plasma membrane. Plant Physiol 152(4):2173–2187

Fyfe PK, Isaacs NW, Cogdell RJ, Jones MR (2004) Disruption of a specific molecular interaction with a bound lipid affects the thermal stability of the purple bacterial reaction center. Biochim Biophys Acta 1608:11–22

Gamper N, Shapiro M (2007) Regulation of ion transport proteins by membrane phosphoinositides. Nat Rev Neurosci 8:921–934

Gomès E, Venema K, Simon-Plas F, Milat ML, Palmgren MG, Blein J-P (1996) Activation of the plant plasma membrane H^+-ATPase. Is there a direct interaction between lysophosphatidylcholine and the C-terminal part of the enzyme? FEBS Lett 398(1):48–52

Gonen T, Cheng Y, Sliz P, Hiroaki Y, Fujiyoshi Y, Harrison SC, Waltz T (2005) Lipid-protein interactions in double-layered two-dimensional AQP0 crystals. Nature 438(7068):633–638

Grandmougin-Ferjani A, Schuler-Muller I, Hartmann MA (1997) Sterol modulation of the plasma membrane H^+-ATPase activity from corn roots reconstituted into soybean lipids. Plant Physiol 113:163–174

Graziani A, Rosker C, Kohlwein SD, Zhu MX, Romanin C, Sattler W, Groschner K, Poteser M (2006) Cellular cholesterol controls TRPC3 function: evidence from a novel dominant-negative knockdown strategy. Biochem J 396:147–155

Grossmann G, Operakova M, Novakova L, Stolz J, Tanner W (2006) Lipid raft-based membrane compartmentation of a plant transport protein expressed in Saccharomyces cerevisiae. Eukaryot Cell 5:945–953

Gruner SM (1985) Intrinsic curvature hypothesis for biomembrane lipid composition: a role for non-bilayer lipids. Proc Natl Acad Sci USA 82:3665–3669

Hamill OP, Martinac B (2001) Molecular basis of mechanotransduction in living cells. Physiol Rev 81:685–740

Hardie RC (2007) TRP channels and lipids: from Drosophila to mammalian physiology. J Physiol 578:9–24

Heginbotham L, Kolmakova-Partensky L, Miller C (1998) Functional reconstitution of a prokaryotic K^+ channel. J Gen Physiol 111:741–749

Heimpel S, Basset G, Odoy S, Klingenberg M (2001) Expression of the mitochondrial ADP/ATP carrier in Escherichi coli-renaturation, reconstitution, and the effect of mutations of 10 positive residues. J Biol Chem 276:11499–11506

Hernandez CC, Zaika O, Tolstykh GP, Shapiro M (2008) Regulation of neural KCNQ channels: signaling pathways, structural motifs and functional implications. J Physiol 586:1811–1821
Hill WG, Rivers RL, Zeidel ML (1999) Role of leaflet asymmetry in the permeability of model biological membranes to protons, solutes and gases. J Gen Physiol 114:405–414
Hill WG, Zeidel ML (2000) Reconstituting the barrier properties of a water-tight epithelial membrane by design of leaflet-specific liposomes. J Biol Chem 275:30176–30185
Hinrichs JW, Klappe K, Hummel I, Kok JW (2004) ATP-binding cassette transporters are enriched in non-caveolar detergent-insoluble glycosphingolipid-enriched membrane domains (DIGs) in human multidrug-reistant cancer cells. J Biol Chem 279:5734–5738
Jacob T, Ritchie S, Assmann SM, Gilroy S (1999) Abscisic acid signal transduction in guard cells is mediated by phospholipase D activity. Proc Natl Acad Sci USA 96:12192–12197
Jeffery CJ (1999) Moolighting proteins. Trends Biochem Sci 24:8–11
Jiang F, Ryan MT, Schlame M, Zhao M, Gu Z, Klingenberg M, Pfanner N, Greenberg ML (2000) Absence of cardiolipin in the crd1 null mutant results in decreased mitochonrial membrane potential and reduced mitochondrial function. J Biol Chem 275:22387–22394
Kasamo K (2003) Regulation of plasma membrane H^{+}-ATPase activity by the membrane environment. J Plant Res 116:517–523
Kasamo K, Yamanishi H (1991) Functional reconsitution of plasma membrane H+-ATPase from mung bean (Vigna radiata L.) hypocotyls in liposomes prepared with various molecular species of phospholipids. Plant Cell Physiol 32:1219–1225
Kierszniowska S, Seiwert B, Schulze WX (2009) Definition of Arabidopsis sterol-rich membrane microdomains by differential treatment with methyl-b-cyclodextrin and quantitative proteomics. Mol Cell Proteomics 8(4):612–623
Klappe K, Hummel I, Hoekstra D, Kok JW (2009) Lipid dependence of ABC transporter localization and function. Chem Phys Lipids 161:57–64
Kok JW, Veldman RJ, Klappe K, Koning H, Filipieanu CM, Müller M (2000) Differential expression of sphingolipids in MRP1 overexpressing HT29 cells. Int J Cancer 87:172–178
Krügel U, Veenhoff LM, Langbein J, Wiederhold E, Liesche J, Friedrich T, Grimm B, Martinoïa E, Poolman B, Kühn C (2008) Transport and sorting of the Solanum tuberosum sucrose transporter SUT1 is affected by post-translational modification. Plant Cell 20:2497–2513
Lalonde S, Wipf D, Frommer WB (2004) Transport mechanisms for organic forms of carbon and nitrogen between source and sink. Annu Rev Plant Biol 55:341–372
Lavie Y, Fuicci G, Liscovitch M (1998) Up-regulation of caveolae and caveolar constituents in multidrug-resistant cancer cells. J Biol Chem 273:32380–32383
Ledesma M, Simons K, Dotti CG (1998) Neuronal polarity: essential role of protein-lipid complexes in axonal sorting. Proc Natl Acad Sci USA 95:3966–3971
Lee AG (2003) Lipid-protein interactions in biological membranes: a structural perspective. Biochim Biophys Acta 1612:1–40
Lee AG (2004) How lipids affect the activities of integral membrane proteins. Biochim Biophys Acta 1666:62–87
Lee Y, Kim YW, Jeaon BW, Park KY, Suh SJ, Seo J, Kwak JM, Martinoïa E, Hwang I, Lee Y (2007) Phosphatidylinositol 4, 5-bisphosphate is important for stomatal opening. Plant J 52:803–816
Lefebvre B, Furt F, Hartmann MA, Michaelson LV, Carde JP, Sargueil-Boiron F, Rossignol M, Napier JA, Cullimore J, Bessoule JJ, Mongrand S (2007) Characterization of lipid rafts from *Medicago truncatula* root plasma membranes: a proteomic study reveals the presence of a raft-associated redox system. Plant Physiol 144:402–418
Liu K, Li L, Luan S (2005) An essential function of phosphatidylinositol phosphates in activation of plant shaker-type K^{+} channels. Plant J 42:433–443
Liu M, Huang W, Wu D, Priestley JV (2006) TRPV1, but not P2X requires cholesterol for its function and membrane expression in rat nociceptors. Eur J Neurosci 24:1–6

Ma X, Shor O, Diminshtein S, Yu L, Im YJ, Perera I, Lomax A, Boss WF, Moran N (2009) Phosphatidylinositol(4, 5)bisphosphate inhibits K+-efflux channel activity in NT1 tobacco cultured cells. Plant Physiol 149:1127–1140

Martens JR, O'Connell K, Tamkun M (2004) Targeting of ion channels to membrane microdomains: localization of KV channels to lipid rafts. Trends Pharmacol Sci 25:16–21

McIntosh TJ, Simon SA (1986) Area per molecule and distribution of water in fully hydrated dilauroylphosphatidylethanolamine bilayers. Biochemistry 25:4948–4952

Michailidis IE, Zhang Y, Yang J (2007) The lipid connection-regulation of voltage-gated $Ca^{(2+)}$ channels by phosphoinositides. Pflugers Arch 455:147–155

Minami A, Fujiwara M, Furuto A, Fukao Y, Yamashita T, Kamo M, Kawamura Y, Uemura M (2009) Alterations in detergent-resistant plasma membrane microdomains in *Arabidopsis thaliana* during cold acclimatation. Plant Cell Physiol 50(2):341–359

Mishra S, Joshi PG (2007) Lipid raft heterogeneity: ana enigma. J Neurochem 103:135–142

Mongrand S, Morel J, Laroche J, Claverol S, Carde JP, Hartmann MA, Bonneu M, Simon-Plas F, Lessire R, Bessoule JJ (2004) Lipid rafts in higher plant cells: purification and characterization of Triton X-100-insoluble microdomains from tobacco plasma membrane. J Biol Chem 279:36277–36286

Monteiro D, Liu Q, Lisboa S, Scherer GE, Quader H, Malho R (2005) Phosphoinositides and phosphatidic acid regulate pollen tube growth and reorientation through modulation of Ca2+ and membrane secretion. J Exp Bot 56:1665–1674

Morel J, Claverol S, Mongrand S, Furt F, Fromentin J, Bessoule JJ, Blein J-P, Simon-Plas F (2006) Proteomics of plant detergent-resistant membranes. Mol Cell Proteomics 5:1396–1411

Nagle JF, Tristam-Naggle S (2000) Structure of lipid bilayers. Biochim Biophys Acta 1469:159–195

Nagle JF, Wiener MC (1988) Structure of fully hydrated lipid dispersions. Biochim Biophys Acta 942:1–10

Orlowski S, Comerà C, Tercé F, Collet X (2007) Lipid rafts: dream or reality for cholesterol transporters? Eur Biophys J 36:869–885

Pan KM, Baldwin M, Nguyen J, Gasset M, Serban A, Groth D, Mehlhorn I, Huang Z, Fletterick RJ, Cohen FE, Prusiner SB (1993) Conversion of alpha-helices into beta-sheets features in the formation of the scrapie prion proteins. Proc Natl Acad Sci USA 90:10962–10966

Pastor RW, Venable RM, Karplus M (1991) Model for the structure of the lipid bilayer. Proc Natl Acad Sci USA 88:892–896

Patel AJ, Lazdunski M, Honore E (2001) Lipid and mechano-gated 2P domain K^+ channels. Curr Opin Cell Biol 13:422–427

Pedchenko VK, Nasirova GF, Palladina TA (1990) Lysophosphatidylcholine specifically stimulates plasma membrane H^+-ATPase from corn roots. FEBS Lett 275:205–208

Pergami P, Jaffe H, Safar J (1996) Semipreparative chormatographic method to purify the normal cellular isoforms of the prion protein in nondenaturated form. Anal Biochem 236:63–73

Pike LJ (2004) Lipid rafts: heterogeneity on the high seas. Biochem J 378:281–292

Pike LJ (2006) Rafts defined: a report on the Keystone Symposium on lipid rafts and cell function. J Lipid Res 47:1597–1598

Pilot JD, East JM, Lee AG (2001) Effects of bilayer thickness on the activity of diacylglycerol kinase. Biochemistry 40:8188–8195

Pochynyuk O, Bugaj V, Stockand JD (2008) Physiologic regulation of the epithelial sodium channel by phosphatidylinositides. Curr Opin Nephrol Hypertens 17:533–540

Raffaele S, Bayer E, Lafarge D, Cluzet S, German-Retana S, Boubekeur T, Leborgne-Castel N, Carde JP, Lherminier J, Noirot E, Satiat-Jeunemaitre B, Laroche-Traineau J, Moreau P, Ott T, Maule A, Reymond P, Simon-Plas F, Farmer EE, Bessoule JJ, Mongrand S (2009) Remorin, a solanacae protein resident in membrane rafts and plasmodesmata, impairs potato virus X movement. Plant Cell 21:1541–1555

Rajendran L, Simons K (2005) Lipid rafts and membrane dynamics. J Cell Sci 118:1099–1102

Regenberg B, Villalba JM, Lanfermeier FC, Palmgre MG (1995) C-terminal deletion analysis of palnt plasma membrane H^+-ATPase: yeast as a model system for solute transport across the plant plasma membrane. Plant Cell 7:1655–1666

Roberts-Crowley ML, Mitra-Ganguli T, Liu L, Rittenhouse AR (2009) Regulation of voltage-gated Ca^{2+} channels by lipids. Cell Calcium 45:589–601

Robinson NC (1982) Specificity and binding affinity to phospholipids to the high-affinity cardiolipin sites of beef heart cytochrome c oxidase. Biochemistry 21:184–188

Roche Y, Gerbeaau-Pissot P, Buhot B, Thomas D, Bonneau L, Gresti J, Mongrand S, Perrier-Cornet JM, Simon-Plas F (2008) Depletion of phytosterols from the plant plasma membrane provides evidence for disruption of lipid rafts. FASEB J 22:3980–3991

Rohacs T (2007) Regulation of TRP channels by PIP(2). Pflugers Arch 453:753–762

Rohacs T (2009) Phosphoinositides regulation of non-canonical transient receptor channels. Cell Calcium 45:554–565

Russell NJ, Coleman JK, Howard TD, Johnston E, Cogdell RJ (2002) Rhodopseudomonas acidophilia strain 10050 contains photosynthetic LH2 antenna complexes that are not enriched with phosphatidylglycerol, and the phospholipids have a fatty acyl composition that is unusual fir purple non-sulfur bacteria. Biochim Biophys Acta 1556:247–253

Safar J, Roller P, Gajdusek D, Gibbs CJ (1993) Thermal-stability and conformational transitions of scrapie amyloid (prion) protein correlate with infectivity. Protein Sci 2:2206–2216

Shahollari B, Peskan-Berghöfer T, Oelmuller R (2004) Receptor kinases with leucine-rich repeats are enriched in Triton X-100 insoluble plasma membrane microdomains from plants. Physiol Plant 122:397–403

Simmonds AC, East JM, Jones OT, Rooney EK, Mc Whirter J, Lee AG (1982) Annular and non annular binding sites on the $(Ca^{2+}+Mg^{2+})$-ATPase. Biochim Biophys Acta 693(2):398–406

Simon-Plas F, Venema K, Grouzis JP, Gibrat R, Rigaud J, Grignon C (1991) Spontaneous insertion of plant plasma membrane (H^+)ATPase into a preformed bilayer. J Membr Biol 120:51–58

Simons K, Ikonen E (1997) Functional rafts in cell membranes. Nature 387:569–572

Simons K, Toomre D (2000) Lipid rafts and signal transduction. Nat Rev Mol Cell Biol 1:31–39

Singer SJ, Nicolson GL (1972) The fluid mosaic model of the structure of cell membranes. Science 175:720–731

Stanislas T, Bouyssié D, Rossignol M, Vesa S, Fromentin J, Morel J, Pichereaux C, Monsarrat B, Simon-Plas F (2009) Quantitative proteomics reveals a dynamic association of proteins to detergent-resistant memrbanes upon elictor signaling in tobacco. Mol Cell Proteomics 8:2186–2198

Starling AP, East JM, Lee AG (1993) Effects of phosphatidylcholine fatty acyl chain length on calcium binding and other functions of the $(Ca^{2+}-Mg^{2+})$ATPase. Biochemistry 32:1593–1600

Sutter JU, Campanoni P, Tyrrell M, Blatt MR (2006) Selective mobility and sensitivity to SNAREs is exhibited by the *Arabidopsis* KA1 K+ channel at the plasma membrane. Plant Cell 18:935–954

Tian G, Kong Q, Lai L, Ray-Chaudhury A, Lin CL (2010) Increased expression of cholesterol 24S-hydroxylase resluts in disruption of glial glutamate transporter EAAT2 association with lipid rafts: a potential role in Alzheimer's disease. J Neurochem 113(4):978–989

Titapiwatanakun B, Murphy AS (2009) Post-transcriptional regulation of auxin transport proteins: cellular trafficking, protein phhosphorylation, protein maturation, ubiquitination and membrane composition. J Exp Bot 60:1093–1107

Titapiwatanakun B, Blakeslee JJ, Bandyopadhyay A, Yang H, Mravec J, Sauer M, Cheng Y, Adamec J, Nagashima A, Geisler M, Sakai T, Friml J, Peer WA, Murphy AS (2009) ABCB19/PGP19 stabilises PIN1 in membrane microdomains in *Arabidopsis*.. Plant J 57:27–44

Tjellstrom H, Andersson MX, Larsson KE, Sandelius AS (2008) Membrane phosphlipids as a phophate reserve: the dynamic nature of phospholipid-to digalactosyl diacylglycerol exchange in higher plants. Plant Cell Environ 31:1388–1398

Tjellstrom H, Hellgren LI, Wieslander A, Sandelius AS (2010) Lipid asymmetry in plant plasma membranes: phosphate deficiency-induced phospholipid replacement is restricted to the cytosolic leaflet. FASEB J 24:1128–1138
Venable RM, Zhang Y, Hardy BJ, Pastor RW (1993) Molecular dynamics simulations of a lipid bilayer and of hexadecane: an investigation of membrane fluidity. Science 262:223–226
Weiss TM, Van der Wel PCA, Killian JA, Koeppe RE, Huang HW (2003) Hydrophobic mismatch between helices and lipid bilayers. Biophys J 84:379–385
White SH, Wimley WC (1999) Membrane protein folding and stability: physical principles. Annu Rev Biophys Biomol Struct 28:319–365
Wipf D, Benjdia M, Tegeder M, Frommer WB (2002) Characterization of a general amino acid permease from Hebeloma cylindrosporum. FEBS Lett 528:119–124
Yamagushi M, Kasamo K (2002) Modulation of proton pumping activity across proteoliposome membranes reconsituted with tonoplast H+-ATPase from cultured rice (Oryza sativa L. var Botro) cells by acyl steryl glucoside and steryl glucoside. Plant Cell Physiol 43:816–822

Index